REDUCING FUEL CONSUMPTION AND GREENHOUSE GAS EMISSIONS OF MEDIUM- AND HEAVY-DUTY VEHICLES, PHASE TWO

Final Report

Committee on Assessment of Technologies and Approaches for Reducing the Fuel Consumption of Medium- and Heavy-Duty Vehicles, Phase Two

Board on Energy and Environmental Systems

Division on Engineering and Physical Sciences

Transportation Research Board

A Consensus Study Report of
The National Academies of
SCIENCES · ENGINEERING · MEDICINE

THE NATIONAL ACADEMIES PRESS
Washington, DC
www.nap.edu

THE NATIONAL ACADEMIES PRESS **500 Fifth Street, NW** **Washington, DC 20001**

This activity was supported by cooperative agreement DTNH22-12-H-00389 from the U.S. Department of Transportation National Highway Traffic Safety Administration and the Presidents' Committee of the National Academies of Sciences, Engineering, and Medicine. Any opinions, findings, conclusions, or recommendations expressed in this publication do not necessarily reflect the views of any organization or agency that provided support for the project.

International Standard Book Number-13: 978-0-309-49635-3
International Standard Book Number-10: 0-309-49635-7
Digital Object Identifier: https://doi.org/10.17226/25542

Additional copies of this publication are available from the National Academies Press, 500 Fifth Street, NW, Keck 360, Washington, DC 20001; (800) 624-6242 or (202) 334-3313; http://www.nap.edu.

Printed in the United States of America

Suggested citation: National Academies of Sciences, Engineering, and Medicine. 2020. *Reducing the Fuel Consumption and Greenhouse Gas Emissions of Medium- and Heavy-Duty Vehicles, Phase Two: Final Report*. Washington, DC: The National Academies Press. https://doi.org/10.17226/25542.

The National Academies of

SCIENCES · ENGINEERING · MEDICINE

The **National Academy of Sciences** was established in 1863 by an Act of Congress, signed by President Lincoln, as a private, nongovernmental institution to advise the nation on issues related to science and technology. Members are elected by their peers for outstanding contributions to research. Dr. Marcia McNutt is president.

The **National Academy of Engineering** was established in 1964 under the charter of the National Academy of Sciences to bring the practices of engineering to advising the nation. Members are elected by their peers for extraordinary contributions to engineering. Dr. John L. Anderson is president.

The **National Academy of Medicine** (formerly the Institute of Medicine) was established in 1970 under the charter of the National Academy of Sciences to advise the nation on medical and health issues. Members are elected by their peers for distinguished contributions to medicine and health. Dr. Victor J. Dzau is president.

The three Academies work together as the **National Academies of Sciences, Engineering, and Medicine** to provide independent, objective analysis and advice to the nation and conduct other activities to solve complex problems and inform public policy decisions. The National Academies also encourage education and research, recognize outstanding contributions to knowledge, and increase public understanding in matters of science, engineering, and medicine.

Learn more about the National Academies of Sciences, Engineering, and Medicine at **www.nationalacademies.org.**

The National Academies of

SCIENCES · ENGINEERING · MEDICINE

Consensus Study Reports published by the National Academies of Sciences, Engineering, and Medicine document the evidence-based consensus on the study's statement of task by an authoring committee of experts. Reports typically include findings, conclusions, and recommendations based on information gathered by the committee and the committee's deliberations. Each report has been subjected to a rigorous and independent peer-review process and it represents the position of the National Academies on the statement of task.

Proceedings published by the National Academies of Sciences, Engineering, and Medicine chronicle the presentations and discussions at a workshop, symposium, or other event convened by the National Academies. The statements and opinions contained in proceedings are those of the participants and are not endorsed by other participants, the planning committee, or the National Academies.

For information about other products and activities of the National Academies, please visit www.nationalacademies.org/about/whatwedo.

COMMITTEE ON ASSESSMENT OF TECHNOLOGIES AND APPROACHES FOR REDUCING THE FUEL CONSUMPTION OF MEDIUM- AND HEAVY-DUTY VEHICLES, PHASE TWO

Andrew Brown, Jr., NAE,[1] Delphi Corporation (ret.), Troy, MI, *Chair*
Inês Azevedo, Stanford University, Stanford, CA
Rodica Baranescu, NAE, University of Illinois, Chicago, IL
Tom Cackette, California Air Resources Board (ret.), Sacramento, CA
Nigel Clark, West Virginia University, Morgantown, WV
Ronald Graves, Oak Ridge National Laboratory, Knoxville, TN
Daniel Hancock, NAE, General Motors (ret.), Indianapolis, IN
W. Michael Hanemann, NAS,[2] Arizona State University, Tempe, AZ
Winston Harrington, Resources for the Future, Washington, DC
Gary Marchant, Arizona State University, Tempe, AZ
Paul Menig, Tech-I-M, Sherwood, OR
Amelia Regan, University of California, Irvine, CA (resigned February 2017)
Mike Roeth, North American Council for Freight Efficiency, Fort Wayne, IN
Gary Rogers, Roush Industries Inc., Livonia, MI
Chuck Salter, Independent Consultant (ret.), Chambersburg, PA
Christine Vujovich, Cummins, Inc. (ret.), Columbus, IN
John Woodrooffe, University of Michigan Transportation Research Institute (ret.), Ann Arbor, MI
Martin Zimmerman, University of Michigan (ret.), Ann Arbor, MI

Staff

Elizabeth Zeitler, Study Director, Senior Program Officer, Board on Energy and Environmental Systems (beginning December, 2017)
Martin Offutt, Study Director, Senior Program Officer, Board on Energy and Environmental Systems (until December, 2017)
Dana Caines, Financial Manager, Board on Energy and Environmental Systems
Rebecca DeBoer, Program Assistant, Board on Energy and Environmental Systems
Lanita Jones, Administrative Coordinator, Board on Energy and Environmental Systems
Michaela Kerxhalli-Kleinfield, Research Assistant, Board on Energy and Environmental Systems
Joseph Morris, Senior Program Officer, Transportation Research Board
Janki Patel, Research Associate, Board on Energy and Environmental Systems
E. Jonathan Yanger, Research Associate, Board on Energy and Environmental Systems
James Zucchetto, Senior Scientist, Board on Energy and Environmental Systems

NOTE: See Appendix B, Disclosure of Conflict(s) of Interest.

[1] NAE, National Academy of Engineering.
[2] NAS, National Academy of Sciences.

BOARD ON ENERGY AND ENVIRONMENTAL SYSTEMS

[1] NAE, National Academy of Engineering.
[2] NAS, National Academy of Sciences.

TRANSPORTATION RESEARCH BOARD 2019 EXECUTIVE COMMITTEE

Preface

The fuel consumption and greenhouse gas (GHG) emissions of medium- and heavy-duty vehicles (MHDVs) have become a focus of legislative and regulatory action, starting with the Energy Independence and Security Act of 2007 (EISA), P.L. 110-140. Section 101 of EISA mandated the U.S. Department of Transportation to promulgate fuel consumption standards for MHDVs for the first time. In addition, Section 108 of that same Act required the secretary of transportation to contract with the National Academy of Sciences to undertake a study on the technologies and costs for improving fuel consumption in MHDVs and include a follow-on report at 5-year intervals.

In response to the secretary's request, the National Research Council[1] (NRC) in 2010 completed *Technologies and Approaches to Reducing the Fuel Consumption of Medium- and Heavy-Duty Vehicles,* referred to henceforth as the "NRC Phase One Report." The NRC Phase One Report provided findings and recommendations on the following: the development of a fuel consumption program for MHDVs; metrics for measuring MHDV fuel consumption; availability and costs of various technologies for reducing fuel consumption; potential indirect effects and externalities associated with fuel consumption standards for MHDVs; alternatives for the scope, stringency, certification methods, and compliance approach for the standards; and a suggested demonstration program to validate innovative certification procedures and regulatory elements. Thereafter, in 2011, the National Highway Traffic Safety Administration (NHTSA) and the U.S. Environmental Protection Agency (EPA) issued the "Phase I rule" on fuel consumption and GHG emissions of MHDVs.

In March 2013, the NRC initiated the Phase Two Study, the first 5-year follow-on to the NRC's 2010 Phase One Report, called for in EISA. The NRC Phase Two Study includes the present report, as well as a first report issued in April 2014 titled *Reducing the Fuel Consumption and Greenhouse Gas Emissions of Medium- and Heavy-Duty Vehicles, Phase Two: First Report.* The sponsoring agency had requested the committee provide advice on the then-pending revision to the Phase I rule. At the time of the 2014 NRC Phase Two Study, First Report, NHTSA was revising its regulatory regime with three objectives: (1) reducing in-use emissions of carbon dioxide from MHDVs; (2) reducing in-use emissions of other GHGs from MHDVs; and (3) improving in-use efficiency of fuel use in MHDVs—by driving innovation, advancement, adoption, and in-use balance of technology through regulation. The 2014 report advised on pathways to accomplish this, subject to the following constraints provided by NHTSA: (a) holding life-cycle cost of technology change or technology addition to an acceptable level; (b) holding capital

[1] Effective July 1, 2015, the institution is called the National Academies of Sciences, Engineering, and Medicine. References in this report to the National Research Council are used in a historical context identifying programs prior to July 1, 2015.

cost of acquiring required new technology to an acceptable level; (c) acknowledging the importance of employing a balance of energy resources that offers national security; (d) avoiding near-term, precipitous regulatory changes that are disruptive to commercial planning; (e) ensuring that the vehicles offered for sale remain suited to their intended purposes and meet user requirements; (f) ensuring that the process used to demonstrate compliance is accurate, efficient, and not excessively burdensome; and (g) not eroding control of criteria pollutants or unregulated species that may have health effects.

The present report is the second and final report of the NRC Phase Two Study. At the request of the sponsoring agency, the committee took account of the August 2016 publication of the Phase II rule on fuel consumption and GHGs. The new rule was supported by extensive analysis, and the committee has weighed this in the development of this report, concerned as it is with possible technological and policy developments out to 2030. This longer time frame freed up the committee to envision more dramatic technological developments affecting the engine and vehicle as well as possible changes to the goods movement enterprise and to regulatory and compliance alternatives. At the same time, the committee has considered near-term improvements to the current power plants—to diesel compression-ignition and spark-ignition internal combustion engines—including the possibility of incorporating hybrid energy storage and drives. Many of the constraints the committee noted in the context of its first report in 2014 still prevail: To pick one example, the committee has considered the costs and benefits of compliance with the Phase II rule and looked in some depth at constraints (a) and (b).

The committee is grateful to all of the federal agencies, original equipment manufacturers and suppliers and their respective associations, and nongovernmental organizations who contributed significantly of their time and efforts to this study, either by giving presentations at meetings or by responding to committee requests for information. We acknowledge the valuable contributions of individuals and organizations that provided information and made presentations at our meetings in Appendix C.

As a final check on the quality and objectivity of the study, all National Academies reports whether products of studies, summaries of workshop proceedings, or other documents must undergo a rigorous, independent external review by experts whose comments are provided anonymously to the committee members. The review process is structured to ensure that each report addresses its approved study charge and does not go beyond it, that the findings are supported by the scientific evidence and arguments presented, that the exposition and organization are effective, and that the report is impartial and objective. We wish to thank the following individuals for their review of this report:

Ewa Bardasz, Zual Associates in Lubrication LLC, Mentor, OH
R. Steven Berry, University of Chicago, Aspen, CO
Rob Brenner, Air Policy Office, EPA (Retired), Arlington, VA
Rebecca Brewster, American Transportation Research Institute, Marietta, GA
Alessandro Faldi, ExxonMobil Research and Engineering Company, Irving, TX
Georgios Fontaras, European Commission, Joint Research Center, Ispra, Italy
Arthur Fraas, Resources for the Future, Washington DC
Thomas Jahns, University of Wisconsin–Madison, Madison, WI
John Johnson, Michigan Technological University, Houghton, MI
Jim Kesseli, Brayton Energy, Hampton, NH
Steve Kratzke, National Highway Traffic Safety Administration, Rockville, MD
Rolf D. Reitz, University of Wisconsin–Madison, Madison, WI
Dorothy Robyn, Independent Consultant, Washington, DC
Aymeric Rousseau, Argonne National Laboratory, Argonne, IL

Although the reviewers listed above have provided many constructive comments and suggestions, they were not asked to endorse the conclusions or recommendations nor did they see the final draft of the report before its release. The review of this report was overseen by **ELISABETH DRAKE**, Massachusetts Institute of Technology, and **CHRIS HENDRICKSON**, Carnegie Mellon University, who were responsible for making certain that an independent examination of this report was carried out in accordance with institutional procedures and that all review comments were carefully considered. Responsibility for the final content of this report rests entirely with the authoring committee and the institution.

The MHDVII Committee delivered this report in October 2019. It relies on work conducted through December 31, 2017. Our original intent was to deliver the report by March 2018. However, a prolonged lapse of funding from the study sponsor regrettably resulted in an extended delay in the completion of the committee report. Nevertheless, the committee firmly believes its work is meaningful, and remains current and objective. The report provides findings and recommendations that multiple interested communities will find valuable and actionable in their endeavors. Section 101 of the EISA 2007 directs the U.S. Department of Transportation to commission the National Academies to undertake studies on the technologies and costs for improving fuel consumption in MHDVs at 5-year intervals. Since initiation of the next (third) study phase is imminent, this report also highlights areas, topics, technologies, and issues for future examination. We hope the next committee will consider them in their deliberations.

Contents

Summary

BACKGROUND

Medium- and heavy-duty vehicles (MHDVs) are a significant contributor to energy consumption and greenhouse gas (GHG) emissions in the transportation sector, accounting for approximately 22 percent of U.S. transportation energy consumption.[1] Unlike the light-duty vehicle sector, which has long been subject to corporate average fuel economy standards, the fuel consumption of MHDVs has only recently begun to be regulated at the federal level. This regulation of MHDV fuel efficiency was mandated by Congress in 2007 in the Energy Independence and Security Act of 2007 (EISA), P.L. 110-140. Section 101 of EISA required the U.S. Department of Transportation (DOT) to promulgate fuel efficiency standards for MHDVs. The statute anticipates that these standards would be updated over time and requires DOT to provide 4 years of lead time between promulgation and enforcement of fuel efficiency standards, and also requires a period of 3 years of stability once the standards are in effect. On May 21, 2010, President Barack Obama directed the National Highway Traffic Safety Administration (NHTSA), on behalf of DOT, to issue MHDV fuel economy standards in close coordination with GHG emission standards to be promulgated for the same vehicles by the U.S. Environmental Protection Agency (EPA).

Section 108 of EISA also requires the secretary of transportation to contract with the National Academy of Sciences to undertake a study on the technologies and costs for improving fuel efficiency in MHDVs. Upon completion of the National Research Council (NRC) Phase One Report[2] and its own study of fuel efficiency standards for MHDVs, DOT was instructed to promulgate by rulemaking a fuel efficiency program for MHDVs that is "designed to achieve the maximum feasible improvement" in fuel economy and to "adopt and implement appropriate test methods, measurement metrics, fuel economy standards, and compliance and enforcement protocols that are appropriate, cost-effective, and technologically feasible for commercial medium- and heavy-duty on-highway vehicles and work trucks" (49 U.S.C. § 32902(k)(2)).

Figure S-1 provides a schematic of the different phases of the regulation, the model year (MY) vehicles included, and the types of MHDVs that are considered in the standards.

[1] Davis, S.C., S.W. Diegel, and R.G. Boundy. 2014. *Transportation Energy Data Book: Edition 33*. Oak Ridge, TN: Oak Ridge National Laboratory. March.

[2] National Research Council. 2010. *Technologies and Approaches to Reducing the Fuel Consumption of Medium- and Heavy-Duty Vehicles*. Washington, DC: The National Academies Press.

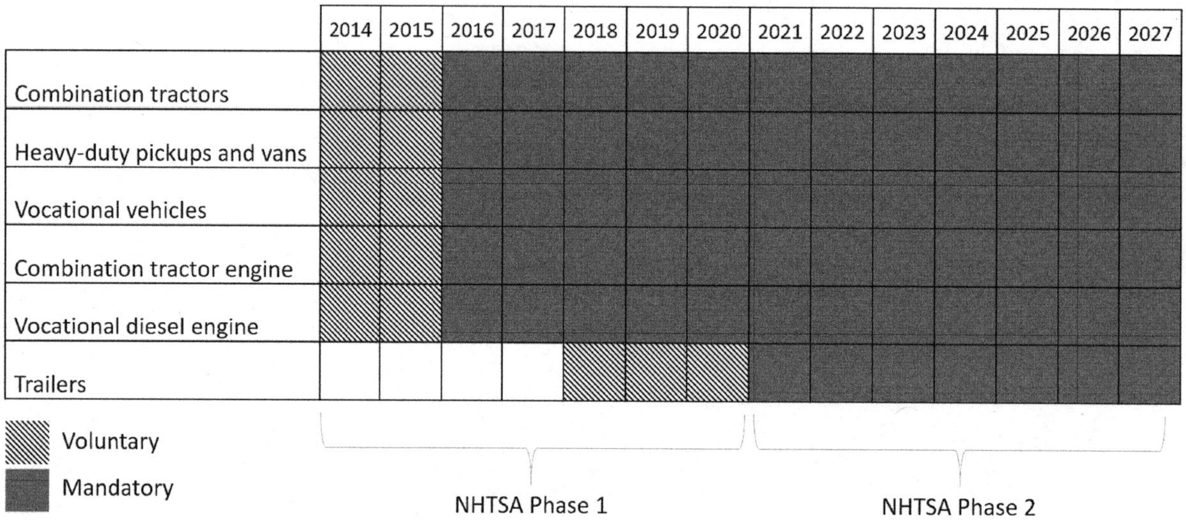

FIGURE S-1 Summary of the different MHDV fuel consumption regulatory phases, vehicle types included, and affected MYs. "Vocational vehicles" includes Classes 2b through 8. Combination tractors are connected to trailers and fall into either Class 7 or Class 8, depending on gross combined vehicle weight. Note: There are instances in which the requirements of the Phase II standards have implications during Phase I, for example, the "pull-ahead" engine standards (see *Federal Register*, Vol. 81, p. 73,498 (81 FR 73498)). NHTSA trailer standards are voluntary until 2021 (81 FR 73643). Phase I continues in effect through 2019 and 2020 (81 FR 73496). Any further revision to the regulatory regime that might occur post-2027 is sometimes referred to as "Phase III." EPA has adopted mandatory GHG requirements that apply during the shaded years for which the fuel economy standards are voluntary.

The committee issued its Phase Two First Report in 2014.[3] The Summary chapter from the 2014 report is provided as Appendix E for reference. It provided guidance for the Phase II rule, which was directed at the post-2018 time frame. The present report is the final report of the NRC Phase Two study and covers a broader range of technologies and issues focused on the 2022 to 2030 time frame.

The Summary captures the overarching report themes of vehicle technology progress, alternative technology and approaches, economic assessment and considerations, future regulatory framework, and interim evaluation. Given the broad scope of the statement of task (SOT), it focuses only on the most significant recommendations. The associated findings are addressed in the individual chapters. Chapter 1 contains a detailed mapping of the SOT items to the report chapters, but briefly, the report is organized with a discussion of MHDV regulation structure and compliance (Chapters 2 and 3), followed by powertrain (Chapter 4) and power demand reduction (Chapter 5) technologies and fuel consumption benefits (Chapter 6), hybrid and electric and battery technologies (Chapters 6–8), freight efficiency (Chapter 9), intelligent transportation systems and automation (Chapter 10), and manufacturing, cost-benefit, and alternative regulatory approaches (Chapters 11–13). A number of appendixes follow with detailed information.

VEHICLE TECHNOLOGY PROGRESS

The energy efficiency of the two widely used conventional engine platforms—compression ignition (i.e., diesel) and spark ignition—can and is still being improved. Diesel engines using the well-known diffusion-dominated compression-ignition (CI) combustion are presently the most efficient engines for MHDVs, with

[3] National Research Council. 2014. *Reducing the Fuel Consumption and Greenhouse Gas Emissions of Medium- and Heavy-Duty Vehicles, Phase Two: First Report.* Washington, DC: The National Academies Press.

progress evident for further reductions in CO_2 and fuel consumption, and NO_x emissions. Supported by the U.S. Department of Energy's SuperTruck initiative, demonstration of 55 percent peak engine efficiency in a research vehicle environment[4] may occur in the next few years. Improvements in waste-heat recovery will be a necessary precursor to reaching such efficiencies with current production engine architectures.

Spark-ignition (SI) engines continue to evolve and improve, with potential to reach more than 40 percent peak brake thermal efficiency while still achieving stringent criteria emissions with relatively low-cost aftertreatment. Even higher efficiency would be achievable in technology pathways where a high fuel octane number is used. Such pathways are well suited to renewable fuels such as ethanol or fuels with low carbon, like natural gas or propane. Based on emerging experimental and modeling results, the expected efficiency of SI engines could exceed the Phase II regulatory requirements, suggesting such engines could attain more stringent regulations depending on the extent that advanced technologies for passenger vehicle engines are adapted to MHDV gasoline engines and based on duty cycle considerations.

The current regulatory structure, which focuses on engine and truck fuel consumption and GHG emission standards, does not appear to have regulatory flexibility to address advanced fuels and methods of improving the efficiency of goods (i.e., freight) movement. For example, increases in heavy-duty vehicle size and weight limits could reduce total system fuel consumption; however, the authority to make such changes lies with Congress.

For Class 8 combination tractor-trailers, a total weight reduction of at least 4,000 pounds appears achievable, which would produce approximately 3 to 4 percent reduction in fuel consumption compared to trucks without lightweighting features, depending on duty cycle. Since most trucks operate at less than maximum weight limit, reduced weight benefits fuel consumption as well as load-specific fuel consumption (LSFC), with the impact of weight reduction greatest in the smaller truck classes.

The committee's technology review, projections, and simulations indicate that, depending on vehicle class and the progress in engine efficiency within each, roughly 12 to 19 percent reduction in LSFC[5] is technologically feasible for MHDVs in the 2030 time frame compared to the committee's 2019 baseline. This expected feasible reduction pertains especially to Class 8 combination vehicles and their drive cycles that consume most of the fuel in the MHDV sector (whereas hybrid vehicles may achieve even greater reductions in fuel consumption in certain duties). Economic feasibility of such an efficiency improvement requires further analyses. This reduction is without the effects of significant changes in fuels or overall operational efficiencies. The committee also conducted simulations under an aggressive set of assumptions for engine efficiency (e.g., 55 percent peak thermal for a combination tractor) and key vehicle parameters,[6] post-2027, and found roughly a 30 percent reduction over the 2019 baseline for the vehicle classes larger than a Class 2b pickup truck. These ambitious CO_2 reductions at the vehicle level will need to be evaluated for sufficiency relative to meeting any national goals and international agreements to which the United States is a party. Substantial reductions greater than 30 percent in CO_2 compared to a 2019 baseline may necessitate approaches beyond the vehicle such as efficiency gains from fuels, operations, and so forth.

Recommendation: NHTSA, in coordination with EPA, should evaluate and quantify the life-cycle GHG emissions and fuel consumption of all fuels and technologies whose use could contribute to meeting a third phase of standards, and take them into consideration in developing a third phase of regulation. It will be critically important to incorporate a life-cycle perspective in those instances where some fuel-technology pathways' life-cycle emissions may lead to an increase, rather than a decrease, in emissions. However, given the resources needed for detailed life-cycle analysis and uncertainty characterization, and given the resource constraints from the agencies, we would recommend assessing carefully which fuel-technology pathways may deserve specific attention in a life-cycle perspective. (**Recommendation 2-7**)

[4] The next phase of the SuperTruck program calls for 55 percent engine efficiency of a Class 8 tractor-trailer with 65,000-pound gross combined vehicle weight at 65 mph. The peak thermal efficiency may be slightly higher.

[5] LSFC is measured in units of volume of fuel consumed per mass of payload per distance traveled (e.g., gallons per 1,000 ton-mile).

[6] Reflecting aggressive improvements in aerodynamic drag, rolling resistance, and mass reduction.

Recommendation: NHTSA, in cooperation with EPA, should establish what MHDV GHG and fuel consumption reductions need to be achieved in the 2030 to 2050 time frame, consistent with national goals and international agreements to which the United States is a party. (**Recommendation 2-2**)

Recommendation: Engine manufacturers, with appropriate engagement of NHTSA and EPA, should allow and promote use of biodiesel and renewable diesel in their engines. NHTSA, in concert with EPA, is encouraged to continue some types of incentives for use of biomass-derived diesel fuel, provided that analyses continue to show overall life-cycle GHG reduction. The agencies are encouraged to seek harmony and synergy between their separate rules for renewable fuels and tailpipe regulations to enable further GHG reduction and petroleum displacement. (**Recommendation 4-7**)

Recommendation: NHTSA and EPA, through their representation in or coordination with the relevant entities within the U.S. Department of Commerce, the U.S. Department of Energy, and the Office of Science and Technology Policy, should prioritize development and application of additive manufacturing, materials joining processes, nanostructured materials, and other yet-to-be-identified promising manufacturing innovations on those parts and systems that offer the greatest potential for fuel consumption and GHG reduction. The progress of these technologies, their prioritization status, and prognosis for effective commercialization within the Phase III regulatory period should be included in the interim evaluation proposed by this report. (**Recommendation 11-5**)

Recommendation: Future MHDV fuel consumption standards should include expectations for weight reduction adoption, with the likely potential for 3 to 4 percent reduction in load-specific fuel consumption at market-acceptable return on investment. (**Recommendation 5-11**)

Recommendation: Government and industry should continue the development of higher-efficiency SI gasoline and natural gas engines for vocational vehicles and should continue to ensure substantial CO_2 and fuel consumption reductions are realized, because (1) higher efficiency appears feasible and (2) market forces may cause a shift toward the SI engines. (**Recommendation 6-2**)

ALTERNATIVE TECHNOLOGY AND APPROACHES

Implementation of new propulsion technologies, new low-carbon fuels, and more efficient freight operations and logistics offers the opportunity to reduce fuel consumption and GHG emissions beyond what is achievable from improving the efficiency of conventional MHDVs—which may be necessary to meet future climate goals. There are numerous alternative-configuration combustion engines. Many of these are merely alternative mechanisms to carry out thermodynamic processes very similar to conventional engines, but there are a few that provide clear enhancements via changing the basic cycles or reducing a prominent loss mechanism (such as heat transfer or friction). Few alternative concepts have reached a scale of development where they could be tested in a MHDV system for performance and emission requirements. With limited data found on alternative configurations, the magnitude of improvement over advanced conventional engines is not readily predictable and will depend on prevailing emission requirements.

For example, progress in understanding and demonstrating the benefits and challenges for kinetics-dominated combustion has continued, but there are still many recipes of the phenomena in flux, depending on fuel properties and stratification modes. Full-scale heavy-duty engines have been successfully demonstrated on test stands, and light-duty vehicle demonstrations have been conducted. A full commercial application is still pending.

Further, waste-heat recovery (WHR) used in Class 8 over-the-road vehicles potentially offers significant cost-effective fuel savings. Expected progress in WHR technology suggests this technology will see increasing penetration in Class 8 combination tractors. There are many approaches to waste-heat conversion to power that can provide up to 4 percent efficiency improvements in modern truck engines. This effect is comparable to other engine improvements in combustion, air handling, friction, etc., and even comparable to many vehicle technology improvements. The WHR systems have proven technically feasible in a research vehicle, yet carry challenges in their weight, packaging, thermal response, and cost. Progress in addressing the challenges is evident. The Regulatory Impact Analysis[7] noted a potential 25 percent penetration in some Class 8 segments by the end of the Phase II rule period.

Hybridization

With respect to hybridization of commercial vehicles, several international light-duty vehicle manufacturers and Tier 1 battery suppliers are projecting costs for plug-in hybrid-electric or battery electric vehicle battery packs that will achieve $120/kWh by 2020 and $100/kWh or less by 2025. At these cost levels, significant market growth in electrified light-duty vehicles can be expected worldwide by 2030. However, due to the duty cycle, service life, and other boundary conditions for commercial vehicles, the battery-system costs are expected to reduce significantly by 2030 but will remain somewhat higher than light-duty battery systems.

Yet, the specific energy, charge and discharge rates, vehicle duty cycles, service life, and reliability requirements for commercial vehicles will limit the potential for direct carryover of battery systems from light-duty vehicles. Additional research and development, above that being done for light-duty applications, is necessary and could delay the market penetration of electrified commercial vehicles. However, international regulatory incentives to promote combustion-free powertrains could support earlier introduction of electrified commercial vehicles than would normally occur in the technology and manufacturing cost-development cycle.

In 2027 and beyond, stop-start technology applications are expected to have payback periods that would make them attractive in many applications to private firms. The greater fuel savings, however, that are possible with the application of mild and strong hybrids are expected to be less attractive to private firms, given their relatively long payback periods. However, technological developments and likely cost evolution and potential life-cycle benefits results suggest that hybridization is a technology that is "next up" for meeting more stringent post–Phase II standards.

Fuel Considerations

Further, petroleum-derived diesel fuel will likely remain the dominant CI engine fuel through the time period of this study (approximately 2030). Regarding changes to diesel fuel to reduce GHG emissions, the most effective measure would be additional use of biomass-derived fuel components (in contrast to changing diesel fuel performance specifications). Renewable diesel fuel (hydrogenation-derived renewable diesel) and biodiesel are both well developed, commercially available, and suitable for this path. They offer overall GHG reduction of up to about 80 percent relative to petroleum-derived diesel.

Since SI engines fueled with gasoline-based fuels are economically attractive for some market segments, GHG reductions with these engines can be enhanced by improving the performance properties of the fuel, such as octane number and maintaining or increasing the content of renewable fuels. The effect of a moderate increase in octane number (research octane number) on efficiency is 5 to 10 percent improvement, whereas the GHG reduction is on the order of 30 percent if renewable ethanol is the high-octane-number blending material.[8,9]

Improved Freight Movement Efficiency

The relatively high cost and long payback periods of "next-up" technologies—in particular, mild and strong hybrids—which will be necessary to meet aggressive reductions in fuel consumption in many of the medium-

[7] NHTSA and EPA. 2016. *Regulatory Impact Analysis, Final Rule for Greenhouse Gas Emissions and Fuel Efficiency Standards for Medium- and Heavy-Duty Engines and Vehicles, Phase II.* Washington, DC: NHTSA and EPA.

[8] Leone, T.G., J.E. Anderson, R.S. Davis, A. Iqbal, R.A. Reese, M.H. Shelby, and W.M. Studzinski. (2015). The effect of compression ratio, fuel octane rating, and ethanol content on spark-ignition engine efficiency. *Environmental Science & Technology* 49(18):10778-10789. doi: 10/1021/acs.est.5b01420.

[9] Theiss, T., T. Alleman, A. Brooker, A. Elgowainy, et al. Summary of High-Octane, MidLevel Ethanol Blends Study. ORNL/TM-2016/42, Oak Ridge National Laboratory, 2016.

and heavy-duty truck segments, suggest that future efforts to increase fuel efficiency should look beyond vehicle technology solutions alone. Consideration should be given to actions that increase the efficiency of the overall freight system, as is described in Chapter 2.

A more integrated approach to systemic improvement in fuel efficiency would, however, require closer cooperation among government agencies than takes place today. A range of opportunities exists to improve energy efficiency and reduce GHG emissions in freight transportation, as detailed in Chapters 9 and 10. For example, modifying truck size and weight standards, facilitating intermodal shipments, and truck platooning could improve the fuel efficiency of freight shipments. Taking advantage of these opportunities will require action and leadership by the federal government, state and local governments (who own and operate the highway system), and private industry.

In considering methods and technologies for improving the efficiency of freight operations, the committee found that higher weight limits and longer combination vehicles could significantly improve productivity and therefore reduce the overall distance traveled in the heavy-vehicle long-haul transportation sector. In addition, the development of freight transfer facilities near urban areas would increase the use of more agile, fuel efficient, and less polluting vehicles for "last-mile" freight movements and would facilitate the early adoption of autonomous vehicles. As such, automated operation that enables truck platooning has the opportunity to save fuel and GHG emissions achieved through decreases in the drag of both trucks.

The commercialization and deployment of advanced technologies, fuels, and freight movement methods that are significantly different from those currently in use may need to start as early as 2030 if ambitious national GHG emissions reduction and fuel-economy goals are established for the 2050 time frame. Because of the long lead time needed to develop such advanced technologies and strategies, planning and preparation for such measures must begin well before the target dates for commercialization and implementation.

Recommendation: NHTSA should coordinate with EPA to identify a portfolio of technologies and options that could achieve national GHG emission and fuel-economy goals for the 2030 to 2050 time frame and to identify the research, planning, and preparation needed to commercialize and implement such technologies and options in the relevant time frame. (**Recommendation 2-3**)

Recommendation: As SI engines continue to be improved, NHTSA and EPA should reassess the future balance in MHDVs between SI and CI, and the reductions in GHG emissions and fuel consumption that might be achieved with a more challenging efficiency requirement for SI engines, including an optimized low-carbon or renewable fuel. (**Recommendation 4-1**)

Recommendation: NHTSA, together with EPA, the Department of Energy (DOE), and the Department of Defense (DOD), should maintain a technology assessment of developments and progress in alternative-configuration engines which evaluates them, analytically and experimentally, against benchmarks and fundamental criteria. (**Recommendation 4-5**)

Recommendation: Industry and government should continue to research, develop, and apply waste-heat recovery (WHR) systems where technical and economic considerations are reasonable to capture this opportunity to reduce fuel consumption. (**Recommendation 4-6**) Since hybrid technologies and WHR potentially could play increasing roles in achieving reductions in fuel consumption in the post-2027 period, developments in the cost and efficiency of these technologies should be monitored and be included in a formal interim review of fuel-consumption standards. (**Recommendation 12-1**)

Recommendation: NHTSA should coordinate with EPA to engage other agencies in the rulemaking process, such as DOE, Federal Motor Carrier Safety Administration, and Federal Highway Administration, who have authorities that can facilitate commercialization of low-carbon fuels and more efficient freight movement methods. NHTSA, in cooperation with EPA, should evaluate how incentives or other regulatory provisions can be incorporated to facilitate implementation of fuels and freight movement approaches that lie outside of NHTSA and EPA authorities. (**Recommendation 2-3**)

Recommendation: NHTSA and the U.S. Department of Transportation should consider urging Congress to adopt size and weight regulations that are in line with those of other North American trade partners. (**Recommendation 9-4**)

ECONOMIC ASSESSMENT AND CONSIDERATIONS

The costs and benefits of specific measures to reduce fuel consumption can be estimated accurately in the near term. The committee had to consider technologies and approaches for medium- and heavy-duty vehicles that could be introduced in model year 2028 or later. Their costs and benefits cannot be estimated with accuracy at this time.

The committee chose to use the NHTSA Lifetime Benefits and Costs Evaluations and the NHTSA and EPA Regulatory Impact Analysis as the foundation of its evaluations. The analysis done by these agencies provided insight into the "marginal cost" and "marginal benefits" of going beyond the rule. The committee assumed that the Phase II rule would take effect and the agencies would be considering a rule to take effect in the 2027 period and beyond. Under this assumption, the required 2.5 percent per annum improvement (needed to comply with the 2027 step of the Phase II rule) would already have been met and further improvements are being considered. In order to meet the higher standard, additional technology will need to be applied to the new vehicle fleet.

In its review of the Regulatory Impact Analysis,[10] the committee found greater reductions are achieved, according to the results shown in Table 10-15 of that document, by the more widespread use of electrification technologies—electronic power steering and accessories, 12-volt stop-start, and strong hybrids. Other technologies either plateau at the high rates, such as eight-speed automatic transmission or turbocharging. Stop-start, mild hybrids, and strong hybrids appear to be a key incremental technology necessary to move beyond the proposed rule and can be used to measure the marginal cost—that is, the incremental cost of increasing fuel economy beyond the proposed rule—as well as the incremental benefit of compliance. Mass reduction and aerodynamic improvements also play a role in moving from 2.5 to 4.0 percent per annum improvement.

The committee's own analysis, referenced in Chapters 6 through 8, generally supports the conclusions of the analysis of NHTSA and EPA. In addition, it finds the following:

- WHR used in Class 8 over-the-road vehicles potentially offers significant cost-effective fuel savings in the post-2027 period. Expected progress in WHR technology suggests that in the post-2027 period this technology will see increasing penetration in Class 8 tractors.
- In 2027 and beyond, stop-start technology applications are expected to have payback periods that would make them attractive in many applications to private firms. The greater fuel savings, however, that are possible with the application of mild and strong hybrids are expected to be less attractive to private firms, given their relatively long payback periods.
- Considering 10-year fuel savings, however, breakeven fuel prices are either below or at the social cost of fuel[11] in many hybrid applications. These results suggest that hybridization is a technology that is "next up" for meeting more stringent post–Phase II standards. Only in Class 2b does the breakeven fuel price for full hybrids exceed the social cost of fuel.
- If the anticipated cost reductions and efficiency gains are in fact consistently achieved, in the post-2027 period increases in fuel efficiency beyond the Phase II rule are likely to be met with increasing applications

[10] NHTSA (National Highway Traffic Safety Administration) and EPA (Environmental Protection Agency). 2016. *Regulatory Impact Analysis, Final Rule for Greenhouse Gas Emissions and Fuel Efficiency Standards for Medium- and Heavy-Duty Engines and Vehicles, Phase II.* Washington, DC: NHTSA and EPA.

[11] The social cost of fuel includes the private cost of the fuel itself as well as its external costs, including the cost to society of carbon emissions (the social cost of carbon), the societal cost of other pollutants, and any national security costs associated with oil consumption. The payback analysis of private firms excludes these external costs in the evaluation of fuel-saving technology. Furthermore, the 10-year fuel savings metric assumes that the factors underlying the relatively short payback periods used by private firms, such as the uncertainties and lack of information—discussed in Chapter 13—will be resolved over time and with more widespread use of the technology.

of hybrid technologies, including stop-start, mild, and strong hybrids. The technology penetration will vary depending on vehicle class and duty cycle.

- The cost of technology tends to fall over time as experience accumulates and firms invest in research and development. Projecting these cost developments is difficult because of the variability in experiences—in the characteristics of specific technologies and in the level of firm investments. In order to capture learning effects, NHTSA and EPA use a complicated algorithm to project how technology costs will evolve over time. While there is ample evidence of cost reduction through learning, the empirical evidence does not support these agencies' complex procedure.

Recommendation: NHTSA and EPA should employ a simpler and more transparent method and rationale on how the unit costs of technologies will evolve, either over time or as a function of the production capacity, in their future assessment. Two options the agencies should consider include (1) assuming specific costs per unit over time for each of the technologies under study, informed by manufacturers' and experts' elicitation of likely costs in the future, and (2) applying simple parametric one-factor learning curves, complemented with a sensitivity analysis, to assess how high or low these learning factors would need to be for a technology to be more economically appealing than another. NHTSA and EPA also are encouraged to do *ex post* assessments of their forecasts of learning effects. (**Recommendation 12-6**)

FUTURE REGULATORY FRAMEWORK

NHTSA and EPA currently lack reliable data on real-world vehicles that can be used to establish a credible regulatory baseline. It is essential for evaluating the effectiveness and success of the regulatory program and identifying future regulatory priorities and directions.

An effective method of determining in-use fuel consumption of trucks, and thereby of determining the overall effectiveness of the regulatory program, is not currently feasible technologically. Such a method is necessary to assess any differential between the LSFC of an individual vehicle model as certified pre-sale using simulation and the LSFC achieved by such vehicles on the road. The data stream this creates will allow NHTSA to identify opportunities to improve the effectiveness and reduce the cost of the program.

In dealing with MHDV fuel consumption and emissions, the question arises of whether to adopt the same regulatory approach for all classes of MHDVs or different approaches for different classes. If one adopts different strategies, there is a risk of artificially distorting owner choices. Because of differential regulatory impacts on vehicle cost or performance, a buyer may shift to a less regulated class, causing a less efficient vehicle to end up in use, thereby frustrating the purpose of the regulation. However, if there is significant heterogeneity across classes, applying the same regulatory strategy to all classes may lead to an outcome determined by the lowest common denominator. In particular, if it becomes possible to measure simultaneously fuel use, vehicle weight, and distance traveled, there is reason to anticipate that telematics systems that permit in-use measurement of vehicle fuel use will be widely adopted in Class 8b vehicles well ahead of other classes, making it more practical to formulate a performance standard for those sectors.

Given that NHTSA's Phase II rule covers model years through 2027, NHTSA and EPA have sufficient time to explore potential in-use compliance concepts and approaches. The available time should allow stakeholder involvement in a deliberative process prior to the issuance of any post-2027 regulations to establish the objectives to be achieved and assess and develop an in-use compliance concept that best meets the objectives.

Recommendation: NHTSA and EPA should commit resources to collecting GHG emission and real-world fuel consumption data for a large and representative national sample of pre-control trucks and for each model year subject to the Phase I and Phase II standards, with priority given to those categories of trucks with the greatest fuel consumption. These data could be used to assess the effectiveness of Phase I and early Phase II standards. If the resources for collecting these data exceed those available to the agencies, the agencies should consider requiring the regulated industry to provide the needed data or fund its collection. (**Recommendation 3-8**)

Recommendation: NHTSA, in coordination with EPA, should develop an effective in-use compliance method that would allow the overall performance of the regulatory program to be quantified, identify whether groups of in-use trucks may not be in compliance, and provide insight into truck operating conditions where GHG emissions and fuel consumption of future trucks could be further reduced. **(Recommendation 3-6)**

Recommendation: NHTSA, in coordination with EPA, should undertake studies and establish a dialog with stakeholders to identify the most feasible method of determining on a continuous basis the effectiveness of the MHDV fuel consumption and GHG emission program, including in-use compliance. **(Recommendation 3-7)**

Recommendation: NHTSA and EPA should also evaluate if actual in-use emissions and fuel consumption could become the principal metric for determining compliance, including the costs, complexities, and benefits of such an approach, and if implemented which regulated entity should be responsible for complying with emissions in use based on considerations of practicality, feasibility, and fairness. NHTSA and EPA should also determine whether a compliance program based on in-use emissions could supplant the need for engine emission test and simulation model-specific standards now in place, and if in-use compliance could be extended for the full life cycle of vehicles. **(Recommendation 3-7)**

INTERIM EVALUATION

The long time period during which NHTSA's Phase II regulations on fuel consumption will be in effect—through model year 2027—creates an opportunity and need for a midcourse evaluation of the preparations for any future regulations. This would evaluate the implementation progress of the Phase II regulations and other developments in technology availability and feasibility. The Phase II standards—covering model years 2019 to 2027—and any follow-on regulations that may then be in the early stages of formulation, both will depend on many assumptions about technology availability, cost, implementation, and effectiveness. An interim evaluation would provide an opportunity to assess those assumptions and projections and to make a determination on whether any adjustments may be appropriate in the preparations for any future rulemaking. This interim evaluation could also coincide with the next National Academies report, required every 5 years, which could feed into that evaluation in the same way the recently published report on light-duty vehicles[12] (LDVs) will contribute to the 2017 midterm evaluation of the LDV standards.

An interim evaluation of its MHDV regulation in the 2021–2022 time period would help improve the latter's overall effectiveness and value contribution. Its primary focus would be on preparations for any future regulations beyond the Phase II standards.

Recommendation: NHTSA, in coordination with EPA, should conduct an interim evaluation of the MHDV program in the 2021–2022 time period, focusing on adjustments to and preparations for future revisions to the regulations. **(Recommendation 2-5)**

The interim evaluation would address the list of tasks enumerated in Chapter 2 and related to the themes of this Summary. All of the report findings and recommendations are discussed thoroughly in each of the chapters. An outline of the report structure, its associated topics, and the committee scope are addressed in Chapter 1.

[12] National Research Council. 2015. *Cost, Effectiveness, and Deployment of Fuel Economy Technologies for Light-Duty Vehicles*. Washington, DC: The National Academies Press. https://doi.org/10.17226/21744.

1

Introduction

Medium- and heavy-duty vehicles (MHDVs) are a significant contributor to energy consumption and greenhouse gas (GHG) emissions in the transportation sector, accounting for approximately 22 percent of U.S. transportation energy consumption (Davis et al., 2014). Unlike the light-duty vehicle sector, which has long been subject to corporate average fuel economy (CAFE) standards, the fuel consumption of MHDVs has only recently begun to be regulated at the federal level. This regulation of MHDV fuel efficiency was mandated by Congress in the Energy Independence and Security Act of 2007 (EISA), P.L. 110-140. Section 101 of EISA required the U.S. Department of Transportation (DOT) to promulgate fuel efficiency standards for MHDVs. The statute anticipates that these standards would be updated over time and requires the DOT to provide 4 years of lead time between promulgation and enforcement of fuel efficiency standards, and also requires a period of 3 years of stability once the standards are in effect.[1]

Section 108 of EISA also requires the secretary of transportation to contract with the National Academy of Sciences (NAS) to undertake a study on the technologies and costs for improving fuel efficiency in MHDVs. Upon completion of the NAS report and its own study of fuel efficiency standards for MHDVs, the DOT was instructed to promulgate by rulemaking a fuel efficiency program for MHDVs that is "designed to achieve the maximum feasible improvement" in fuel economy and to "adopt and implement appropriate test methods, measurement metrics, fuel economy standards, and compliance and enforcement protocols that are appropriate, cost-effective, and technologically feasible for commercial medium- and heavy-duty on-highway vehicles and work trucks" (49 U.S.C. § 32902(k)(2)).

The key events in implementing the EISA requirement for MHDV standards are shown in Figure 1-1. The NAS, through its operating arm the National Research Council (NRC), issued a report on March 31, 2010, titled *Technologies and Approaches to Reducing the Fuel Consumption of Medium- and Heavy-Duty Vehicles* (henceforth the "NRC Phase One Report") (NRC, 2010). On May 21, 2010, shortly after the release of the NRC report, President Barack Obama directed the National Highway Traffic Safety Administration (NHTSA), on behalf of the DOT, to issue MHDV fuel economy standards in close coordination with GHG emissions standards to be promulgated for the same vehicles by the EPA (White House, 2010). Given the connection between fuel economy and GHG emissions, a coordinated approach to fuel economy and GHG standards would reduce regulatory costs and

[1] These timing requirements only apply to the National Highway Traffic Safety Administration (NHTSA) MHDV fuel efficiency standards, not to the accompanying Environmental Protection Agency (EPA) MHDV CO2 standards, which are subject to the Clean Air Act rather than EISA requirements.

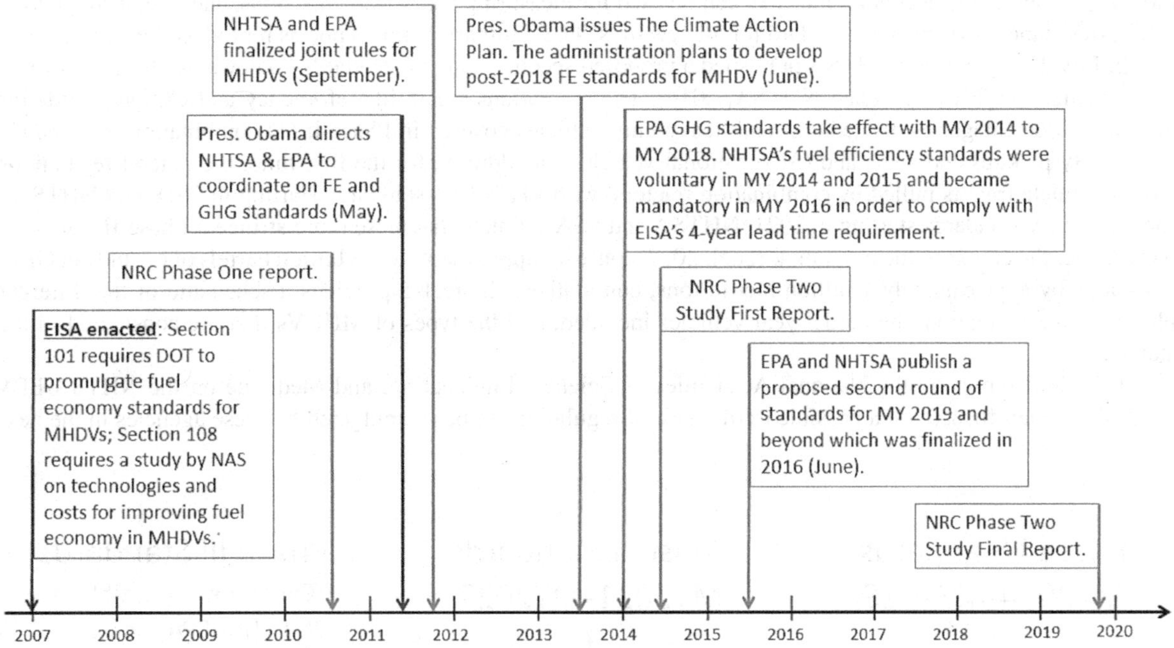

FIGURE 1-1 Timeline of key events in MHDV regulation.
NOTES: Third NRC report data collection ended in 2017 but was not published until 2019. DOT = Department of Transportation, EISA = Energy Independence and Security Act, EPA = Environmental Protection Agency, GHG = greenhouse gas, MHDV = medium- and heavy-duty vehicle, MY = model year, NAS = National Academy of Sciences, NHTSA = National Highway Traffic Safety Administration, NRC = National Research Council.

burdens and minimize inconsistent regulatory requirements by allowing manufacturers to build one set of vehicles to comply with both sets of standards.

On September 15, 2011, NHTSA and EPA finalized joint Phase I rules to establish a comprehensive Heavy-Duty National Program to reduce GHG emissions and fuel consumption for on-road medium- and heavy-duty vehicles. NHTSA adopted final fuel consumption standards under its statutory authority provided by EISA, and EPA adopted carbon dioxide emission standards under its Clean Air Act authority. The coordinated rules were tailored to the same three regulatory categories of heavy-duty vehicles: (i) combination tractors, (ii) heavy-duty pickup trucks and vans, and (iii) vocational vehicles. The EPA GHG emission standards, which were not subject to the EISA 4-year lead-time requirement, commenced with model year (MY) 2014, whereas NHTSA's fuel efficiency standards were voluntary in MYs 2014 and 2015 and became mandatory in MY 2016, in order to comply with EISA's lead-time requirement (EPA and NHTSA, 2015).

In June 2013, President Barack Obama issued The President's Climate Action Plan, which states that the administration plans to work with stakeholders "to develop post-2018 fuel economy standards for heavy-duty vehicles to further reduce fuel consumption through the application of advanced cost-effective technologies and continue efforts to improve the efficiency of moving goods across the United States" (White House, 2013).

As NHTSA and EPA began working on a second round of standards, the NRC (2014) issued another report—*Reducing the Fuel Consumption and Greenhouse Gas Emissions of Medium- and Heavy-Duty Vehicles, Phase Two: First Report*—authored by the present committee and providing recommendations for the Phase II standards. Key recommendations of that report were for NHTSA and EPA to consider various suggested improvements in the certification and compliance procedures using the Greenhouse Gas Emissions Model (GEM), to adopt fuel efficiency and GHG emission standards for natural gas engines and vehicles, to regulate trailers, to develop and monitor a baseline for fleet emissions that can be used for measuring real-world emissions, and to consider a low-rolling-resistance performance standard for all new tires. The summary of this report is reproduced in Appendix F.

In July 2015, EPA and NHTSA published a proposed second round of standards—known as Phase II—which were finalized in 2016 (EPA and NHTSA, 2016). The regulations create fuel efficiency and CO_2 standards for the same three categories of medium- and heavy-duty vehicles covered in Phase I: (1) combination tractors, (2) heavy-duty pickups and vans, and (3) vocational vehicles. In addition, for the first time, the federal regulations would regulate trailers pulled by combination tractors, with EPA's CO_2 standards starting in 2018 and NHTSA's fuel efficiency standards starting in 2021. NHTSA and EPA estimate that these more stringent Phase II standards would, over the course of the program through 2027, conserve approximately 1.8 billion barrels of oil and cut GHG emissions by approximately 1 billion metric tons, cumulative. Figure 1-2 provides a schematic of the different phases of the regulation, the model year vehicles included, and the types of MHDVs that are considered in the standards.

This third report by the National Academies of Science, Engineering, and Medicine on the NHTSA/EPA MHDV program focuses on a possible third phase of regulations to be promulgated by these agencies in the next

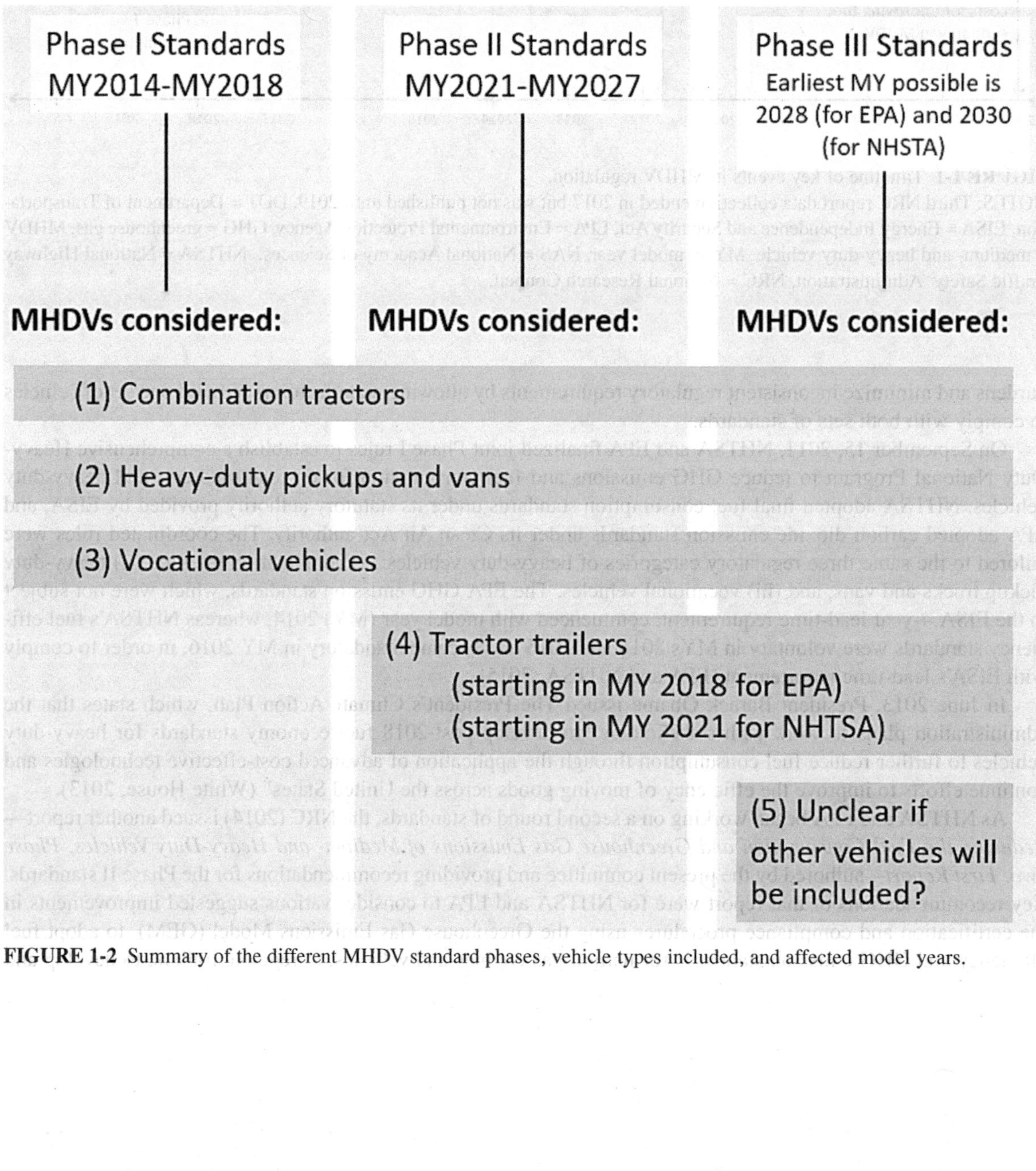

FIGURE 1-2 Summary of the different MHDV standard phases, vehicle types included, and affected model years.

decade (i.e., in the 2020s). The timing for implementation of a third phase of standards would be affected by several factors that are different for fuel efficiency and GHG emissions. Under EISA, NHTSA must provide 4 years of lead time and 3 years of stability when adopting or revising fuel efficiency standards. These discrete timing provisions do not apply to EPA's regulatory authority under the Clean Air Act (CAA), which specifies that standards "shall take effect after such period as the Administrator finds necessary to permit the development and application of the requisite technology, giving appropriate consideration to the cost of compliance within such period" (CAA §202(a)(2)). It is therefore possible for EPA to implement its GHG standards before NHTSA can enforce its fuel efficiency standards. The Phase I standards reflected these different requirements with EPA implementing GHG standards with the 2014 models, while NHTSA's fuel efficiency standards became mandatory in 2016. The Phase II standards will not be fully implemented until 2027. EPA's third set of GHG standards could begin with the 2028 model year provided they were adopted early enough to provide adequate lead time. NHTSA's fuel efficiency standards could not go into effect until 2030 due to the statutory 3-year stability requirement. However, to date both standards have been designed to be met with a single engine and vehicle design. Thus the implementation date for the third set of standards will likely be established by the EPA GHG standards, and could be as early as the 2028 model year.

The committee's work has focused on addressing the technological and regulatory issues related to future regulatory activities regarding MHDV fuel efficiency and GHG emissions, including a potential third phase of rulemakings in the future. The committee's work was directed at, and guided by, the committee's statement of task reproduced in Box 1-1. Table 1-1 shows which elements of the statement of task are addressed by which chapters.

REFERENCES

Davis, S.C., S.W. Diegel, and R.G. Boundy. 2014. *Transportation Energy Data Book: Edition 33.* Oak Ridge, TN: Oak Ridge National Laboratory. March.

EPA (Environmental Protection Agency) and NHTSA (National Highway Traffic Safety Administration). 2011. Greenhouse gas emissions standards and fuel efficiency standards for medium- and heavy-duty engines and vehicles. *Federal Register* 76:57106-57513 (September 15).

EPA and NHTSA. 2015. Greenhouse gas emissions and fuel efficiency standards for medium- and heavy-duty engines and vehicles—Phase 2. *Federal Register* 80:40138-40765 (July 13).

EPA and NHTSA. 2016. Greenhouse gas emissions and fuel efficiency standards for medium- and heavy-duty engines and vehicles—Phase 2. *Federal Register* 81:73478-74274 (October 25).

NRC (National Research Council). 2010. *Technologies and Approaches to Reducing the Fuel Consumption of Medium- and Heavy-Duty Vehicles.* Washington, DC: The National Academies Press.

NRC. 2014. *Reducing the Fuel Consumption and Greenhouse Gas Emissions of Medium- and Heavy-Duty Vehicles, Phase II: Report 1.* Washington, DC: The National Academies Press.

White House. 2010. Improving Energy Security, American Competitiveness and Job Creation, and Environmental Protection through a Transformation of our Nation's Fleet of Cars and Trucks. Office of the Press Secretary. May 21.

White House. 2013. The President's Climate Action Plan. Available at https://www.whitehouse.gov/sites/default/files/image/president27sclimateactionplan.pdf (accessed August 27, 2019).

BOX 1-1
Statement of Task

The committee will give priority in its analytic work to assessing the current medium- and heavy-duty corporate average fuel economy (CAFE) program broadly while making recommendations for improvements to future programs. The areas to be assessed should include rulemaking assumptions, analytical methods used, and values used in the analyses. The committee will also produce an updated assessment of technologies and programs for reducing the fuel consumption of medium- and heavy-duty vehicles and reassess the technologies analyzed in the National Research Council (NRC) report, *Technologies and Approaches to Reducing the Fuel Consumption of Medium- and Heavy-Duty Vehicles* (2010), providing updated estimates of the cost, potential efficiency improvements, and barriers to commercial deployment of technologies that might be employed in model years 2022 and beyond. The committee will investigate developments that have occurred and technologies that have entered the market since the initial report was issued, especially any new technologies that may become important by 2030. In particular, the committee will address the following tasks:

1. Review and contrast the final rule for fuel efficiency and greenhouse gas emissions standards for medium- and heavy-duty engines and vehicles for model years 2014-2018 with the recommendations offered in the NRC 2010 report. Identify the potential implications and gaps in the U.S. regulatory process for medium- and heavy-duty vehicles as they pertain to the engine manufacturers, trailer manufacturers, final stage manufacturers, and complete vehicles. As part of this activity, explore regulatory options for trailers from a total vehicle perspective. Contrast the U.S. regulatory approach to that in other parts of the world, notably Europe, Japan, China, and India.
2. Analyze and provide options for improvements to the certification and compliance procedures for medium- and heavy-duty vehicles, including the use of representative test cycles and simulation using the Environmental Protection Agency's Greenhouse Gas Emissions Model (EPA's GEM).
3. Review updated baseline information on the medium- and heavy-duty truck fleet, including combination tractors and trailers, as well as the methodology for providing on-road information on fuel consumption necessary to inform baseline standards.
4. Examine advanced gasoline engine technologies, including the ability of those engines to meet load demands, the impact of those engines on cost, the need for aftertreatment systems, and their market acceptability. Consider the impacts of possible alternative fuels on these technologies. Technology assessments shall include, but not be limited to, engine technologies such as waste-heat recovery, low-temperature combustion, downsizing, and others.
5. Examine diesel emission control systems, including the capabilities of emission control systems to meet current and possible future criteria pollutant emissions standards, the impacts on fuel consumption attributed to meeting emissions standards, and the fuel characteristics needed to enable low-emissions diesel technologies. Also consider the impacts of possible alternative fuels on these technologies.

6. Examine electric powertrain technologies, including the capabilities, limitations, and cost of hybrids, plug-in hybrids, battery electric vehicles, and fuel cell vehicles. Specific areas of interest include electric hybrids, hydraulic hybrids, and other hybrids; all electric vehicles; stop-start systems; implementation strategies; total capital and operating costs; modeling and simulations as a tool; the certification process; and emissions compliance.

7. Examine battery technologies including an examination of the cost, performance, range, durability (including performance degradation over time), and safety issues related to lithium-ion and other possible advanced energy storage technologies that are necessary to enable plug-in and full-function electric vehicles.

8. Examine vehicle technologies such as mass reduction, aerodynamic drag improvements, automatic tire inflation systems, improved transmissions, improved efficiency of accessories, fans, and water pumps, and other approaches. Evaluate the potential for reducing vehicle mass including technologies such as materials substitution, and the use of new vehicle, structural, system, or component designs.

9. Evaluate intermodal and intelligent systems for potential fuel consumption benefits, including a survey of the current fleet communication systems (vehicle to vehicle, vehicle to infrastructure), existing barriers to implementation, and future technologies.

10. Review the potential impacts of fuel-consumption-reduction technologies on medium- and heavy-duty vehicle safety including aerodynamic components, wide-based wheels and tires, tire pressure monitoring and automatic inflation systems, hybridization and alternative fuels, combination vehicles and higher gross vehicle weight ratings, lightweighting, idle reduction and stop-start, and others.

11. Provide an analysis of how fuel efficient technologies may be practically integrated into manufacturing processes and how such technologies are likely to be applied in response to requirements for reducing fuel consumption. Also provide an analysis of how technology implementation is likely to impact capital equipment and engineering, research, and development costs, the certification process, and emissions compliance.

12. Examine the costs, cost multipliers, and benefits that could accompany the introduction of technologies for reducing fuel consumption. Consider, to the extent possible, initial and life-cycle costs, including such aspects as operation, maintenance, insurance, and other factors. Become informed of and consider the activities of an ongoing study for National Highway Traffic Safety Administration (NHTSA) by the National Research Council on the assessment of technologies for improving fuel economy for light-duty vehicles to assess whether the analysis of that study with regards to incremental indirect costs and the impacts of learning on costs are relevant for technologies used for medium- and heavy-duty vehicles.

13. To the extent possible, address uncertainties and perform sensitivity analyses for the fuel consumption and cost-benefit estimates and provide guidance to NHTSA on improving its uncertainty analyses given the relatively long time frame for these future estimates.

14. Write and provide to NHTSA, the Congress, and the public a final report documenting its conclusions and recommendations.

TABLE 1-1 Relationship Between Statement of Task Items and Report Chapters

Report Chapters	Statement of Task
1. Introduction	14
2. Setting the Stage: Regulatory Horizons, Challenges, and Influences	2, 3
3. Certification, Compliance, and Enforcement	1, 2, & 3
4. Powertrain Technologies	4, 6, & 8
5. Technologies for Reducing the Power Demand	8 & 10
6. Projected Benefits of Technologies on Fuel Consumption	4, 5, & 12
7. Hybrid and Electric Powertrain Technologies	6
8. Battery Technologies	7
9. Operational Efficiency	9 & 10
10. Intermodal ITS and Safety	9
11. Manufacturing Considerations	11
12. Cost-Benefit Analysis	12 & 13
13. Alternative Regulatory Approaches	1
Appendixes	1-14

2

Setting the Stage: Regulatory Horizons, Challenges, and Influences

The fuel consumption and greenhouse gas (GHG) emissions of medium- and heavy-duty vehicles (MHDVs) are affected by the intersection of a variety of technological factors, market forces, regulatory programs, and other government and private-sector policies and initiatives. Before examining specific technologies in the main body of this report, this chapter provides the relevant regulatory, policy, and market background and context for the present study. It first describes the future prospects and projections for the regulation of MHDV fuel consumption and GHG emissions in the period from 2030 to 2050 when potential future Phase III standards might be implemented (Section 2.1). It then addresses some key gaps and challenges of the National Highway Traffic Safety Administration (NHTSA)/Environmental Protection Agency (EPA) program (Section 2.2). Other regulatory programs directly or indirectly affecting MHDV fuel consumption and GHG emissions are summarized next (Section 2.3).

2.1 FUTURE REGULATORY PATHWAYS

This report focuses on a potential future Phase III set of NHTSA/EPA MHDV standards (i.e., likely affecting model year (MY) 2028 at the earliest for EPA, and MY 2030 at the earliest for NHTSA). These vehicles will in many cases remain in operation through the middle of the century. Because this report is addressing technologies and strategies for reducing fuel consumption and GHG emissions from MHDVs that will be fully implemented more than a decade into the future, we necessarily take a prospective and horizon-scanning approach in this analysis. Over the time period in which Phase III standards might be adopted and implemented, many important changes and advancements are likely in vehicle technology and costs, transportation systems and planning, fuel availability and price, other measures and policies affecting MHDV fuel economy and GHG emissions, and in the economic and political circumstances of the nation and world. Thus, the forward-looking analysis in this report will obviously necessitate various projections, assumptions, and uncertainties, which will be identified throughout this report.

This section proceeds in three parts. First, projections of future fleet characteristics are discussed, focused on the likely time period for adoption and implementation of the Phase III standards (Section 2.1.1). Next, the potential pathways for reducing fuel consumption and GHG emissions over this time period are assessed (Section 2.1.2). Finally, the timing opportunities and challenges in planning for the Phase III rulemaking based on these analyses are discussed (Section 2.1.3).

2.1.1 Fleet Characteristics During the Phase III Standards Time Horizon

The direction and stringency of a future set of Phase III standards will be dictated by the intersection of technological feasibility, market demands, economic trends, and the policy objectives set by the federal government for reducing fuel consumption and GHG emissions from MHDVs. However, the uncertainties regarding the future characteristics of the MHDV fleet are significant. The extent to which the fleet characteristics are likely to change is difficult to predict. Changes in the fleet may result from changes in the demand for goods and services; the competing alternatives available to provide those goods and services; fuel availability and fuel prices; future standards and regulations; and other policies and exogenous factors such as changes in the economy or change in fuel prices. Even in the absence of policy and regulatory changes, projections on the composition of the fleet fall short of being accurate. In addition, uncertainty in national and international policies creates further difficulties in providing projections.

To illustrate this uncertainty see Figure 2-1, which affects the transportation sector overall. Every year, the Energy Information Administration (EIA) produces estimates of future energy consumption for the next several decades. Seven of these annual estimates, made in calendar years 1985 to 2015, are shown. Also shown is the actual U.S. transportation energy consumption for this period. As the data show, the EIA estimates of future transportation energy consumption often deviate substantially from actual consumption, illustrating the difficulty of making accurate long-term projections. These deviations between projections and actual consumption result from two kinds of uncertainties: (i) uncertainties relating to structural changes in energy markets such as the 1970s OPEC embargo and (ii) fluctuations in supply, demand, and prices within existing structural markets.

The most recent EIA Annual Energy Outlook (AEO; EIA, 2015) projects that the number of vehicle miles traveled by freight trucks (> 10,000 pounds) and commercial light trucks will continue to increase, from about 284 and 72 billion vehicle miles, respectively, in 2015 to about 400 and 105 billion vehicle miles traveled, respectively, in 2040 (see Figure 2-2). AEO 2015 also includes projected fuel efficiency (in miles per gallon) for freight trucks

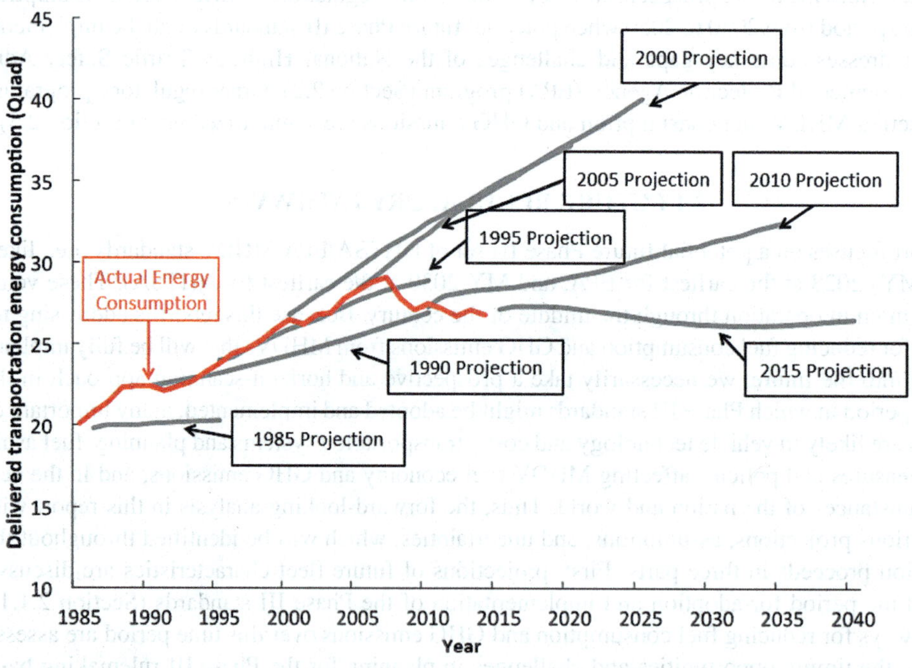

FIGURE 2-1 Actual delivered energy consumption for the U.S. transportation sector from 1985 to 2015, and several projections by the Energy Information Administration made in calendar years 1985, 1990, 1995, 2000, 2005, 2010, and 2015. The EIA produces these projections every year, but we are only showing those for every fifth year.
NOTE: Plot produced by Evan Sherwin and Inês Azevedo using data from the AEO's retrospective analysis.

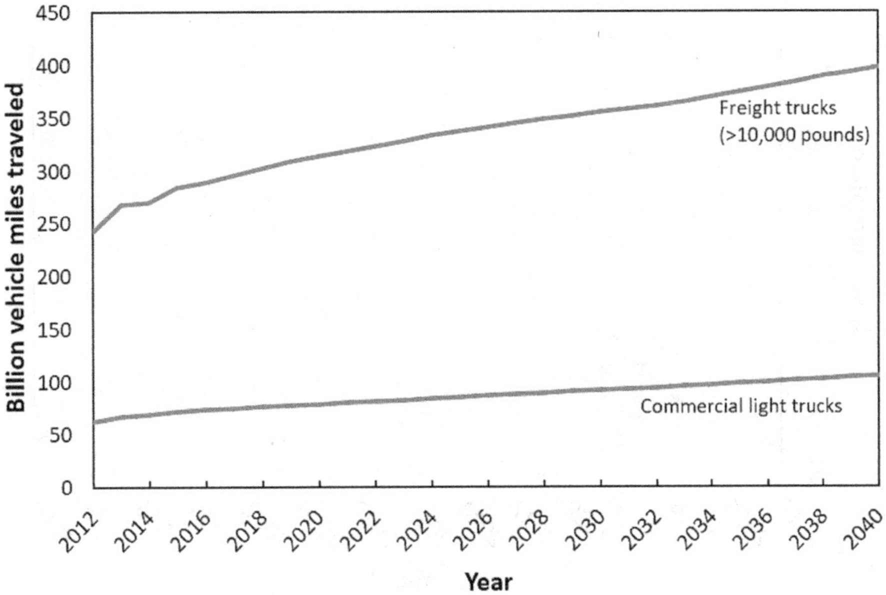

FIGURE 2-2 Annual Energy Outlook (2015) projections for number of vehicle miles traveled by freight trucks (> 10,000 pounds) and commercial light trucks.
NOTE: Plot produced by the committee using data from the AEO 2015.

(> 10,000 pounds), which is expected to increase to 7.7 mpg in the late 2030s and then plateau or even slightly decrease (see Figure 2-3). This projection reflects the implementation of the Phase I MHDV fuel efficiency and GHG standards, but not the Phase II or subsequent standards or any other future standards (EIA, 2015, p. 10).

Fuel consumption from freight trucks (> 10,000 pounds) is also expected to increase according to the EIA AEO's most recent report, from about 2.8 million barrels per day oil equivalent in 2015 to about 3.4 in 2040, a 21 percent increase (see Figure 2-4). Again, this projection assumes implementation of the Phase I MHDV standards, but not any subsequent potential standards.

Finding: Technological, market, economic, and policy variables make future projection of fuel consumption and GHG emissions by MHDVs highly uncertain. This can result from both structural changes within energy markets and price volatility and fluctuations within market structures.

Recommendation 2-1: Given the uncertainties in projecting future MHDV fuel consumption and GHG emission trends, periodic or ongoing assessments should be made by NHTSA, in cooperation with EPA, to evaluate the implementation of existing fuel efficiency and GHG emission standards for MHDVs, and to adjust planning for future standards.

2.1.2 Assessing the GHG Reductions and Fuel Savings Potential

2.1.2.1 International Studies and Insights

Fuel savings and GHG reductions from transport are each affected and informed by international climate agreements and voluntary goals such as the Kyoto Protocol and the goals adopted in the 21st Conference of Parties in Paris. For example, the European Union adopted aggressive targets under its "20-20-20" policy of reducing greenhouse gas emissions by 20 percent relative to 1990 and achieving this with 20 percent renewable energy and 20 percent improvement in energy efficiency by 2020 (European Commission, 2008) and is now moving to even

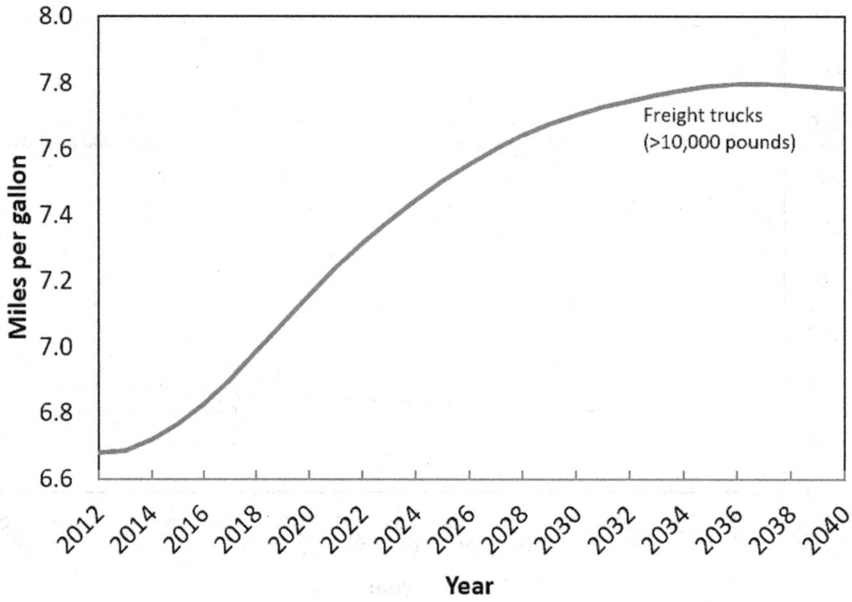

FIGURE 2-3 Annual Energy Outlook (2015) projections for fuel efficiency in miles per gallon for freight trucks (> 10,000 pounds).
NOTE: Plot produced by the committee using data from the AEO 2015.

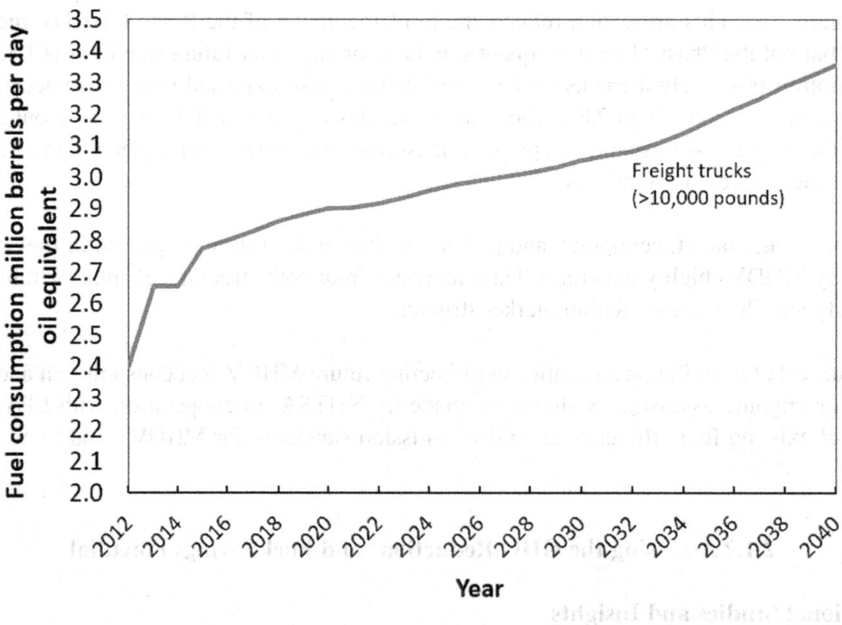

FIGURE 2-4 Annual Energy Outlook (2015) projections for fuel consumption for freight trucks (> 10,000 pounds) in million barrels per day of oil equivalent.
NOTE: Plot produced by the committee using data from the AEO 2015.

more ambitious goals. Large GHG reductions in the transportation sector may include strategies and technologies such as vehicle electrification, fuel switching to biofuels derived from biomass, etc. (IEA, 2015).

Meeting large GHG reduction goals at a national or regional level will require a portfolio of strategies in all sectors, including those outside of transportation such as electricity, buildings, and industry, where mitigation strategies may be the most cost effective. Several studies have considered pathways to stabilize global temperature rise at 2 degrees Celsius (°C) above pre-industrial levels and made varied observations about the abatement of GHGs that might be achievable in different sectors. For example, the Organisation for Economic Co-operation and Development's (OECD's) International Energy Agency (IEA; 2010), found that most of the abatement occurs in the electricity sector—a finding shared by Energy Modeling Forum 24 (Clarke et al., 2014, p. 21) and by the Transportation Research Board (2011). In transportation, the IEA's (2015) scenarios, for example, showed electrification in light-duty vehicles to exceed 40 percent of new cars by 2040. The Global Energy Assessment (GEA, 2012) evaluated a group of energy efficient pathways consistent with 2-degree stabilization that envisaged increased technical efficiency of all transportation coupled with mode shifting to mass transit for personal transportation in developed countries by 2050 and the increase in goods movement is fulfilled mostly by rail transport. The Intergovernmental Panel on Climate Change (IPCC, 2014, p. 24), estimated that, on a global basis, GHG emissions in the transportation sector could be reduced 15 to 40 percent below business-as-usual values by 2050 using a mix of technological and behavioral changes (such as mode shifting), transit-oriented development, and other considerations. Taptich et al. (2015) assessed the potential for GHG intensity reductions for passenger and freight transportation that would be achievable in 2030 and 2050 through the adoption of alternative technologies, mode switching, and electrification across eight regions of the world. The authors argue that non-OECD countries have lower reduction potential than OECD countries for all modes except oceangoing vessels, and that the reduction of GHG intensity occurs more slowly for freight transportation when compared to passenger transportation.

Though the studies do not converge on a particular technology or fuel pathway, all studies find that to achieve deep GHG reductions transformation in the energy system must occur well before 2050, anticipated and driven by large investment in the preceding decades. The Deep Decarbonization Pathways Project (DDPP) of the sustainable Development Solutions Network study finds that incremental investments in transportation of a few hundred billion dollars per year would be needed, peaking in about 2040 before declining as the new technologies became cheaper (DDPP, 2015). The lack of convergence in the results of these models could be due to uncertainty surrounding both the readiness of the various technologies and their potential to abate GHG emissions (Clarke et al., 2014, p. 29).

2.1.2.2 National Studies and Insights

At the national level, several studies have examined emissions reduction potential for the U.S. MHDV sector. The U.S. contribution to the DDPP (2015) examined possible pathways to reduce the aggregate emissions of the U.S. energy system by 80 percent below 1990 levels by 2050. For heavy-duty diesel engines, the study's modeling assumptions led to scenarios where drop-in replacement biomass-derived fuels dominated, and other scenarios found substantial deployment of heavy-duty vehicles using hydrogen or natural gas. All cases that met the 80 percent economy-wide reduction target showed substantial improvement of 50 percent or greater in heavy-duty vehicle fuel economy. The DDPP scenarios for heavy-duty vehicle (HDV) emissions reduction as part of a national, economy-wide 80 percent GHG emissions reduction strategy are shown in Figure 2-5.

In economy-wide scenarios, MHDV emissions do not generally fall to 80 percent of 1990 levels by 2050, with other sectors taking on larger reductions. If the U.S. government were to set a goal of 80 percent GHG emissions reduction in the MHDV sector specifically,[1] rather than as an economy-wide goal, even more aggressive measures and strategies for MHDV emissions reduction would likely be needed. One study of freight transported by MHDVs in California found that GHG emissions could be reduced in that subsector by 80 percent by 2050 from a 1990 baseline by implementing zero emission vehicle and fuel technologies at an aggressive rate (Nahlik et al., 2015),

[1] This discussion of an 80 percent emissions reduction goal in the MHDV sector is provided for illustrative purposes only, without implying endorsement of any specific goal, a matter for policy makers and beyond the scope of this report.

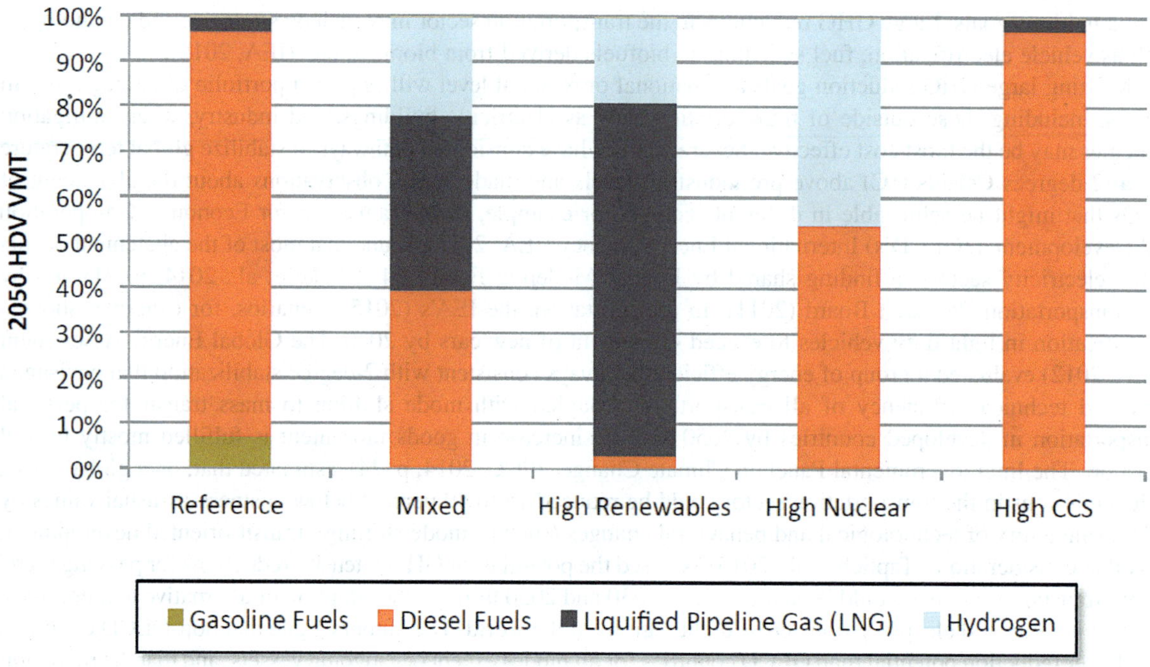

FIGURE 2-5 Summary of fuel use in the HDV sector for various economy-wide emissions reduction scenarios as part of a nationwide strategy to achieve 80 percent reductions in national GHG emissions by 2050 relative to 1990. The scenarios labeled Reference, Mixed, High Renewables, High Nuclear, and High CCS are descriptors of a scenario's distinguishing features in its electricity generation strategy. The colors indicate the projected fuel use mix for HDVs, given a modeled energy system with the labeled electricity generation mix. CCS = carbon capture and sequestration as applied to electricity generation emissions. SOURCE: Williams et al. (2014).

but that those reductions would be overwhelmed by increased freight traffic, leading to a net increase in GHG emissions under a scenario involving incremental improvements in existing technologies.

Other scenario analyses have found a similar need for aggressive technology shifts and timelines to meet a goal of 80 percent reduction in MHDV emissions by 2050 (Fulton and Miller, 2015; Sharpe, 2013). The results of the Fulton and Miller scenario analysis for the United States (extrapolated from a model for California) are shown below (Figure 2-6). According to the authors, a substantial growth is expected in MHDV carbon dioxide (CO_2) emissions due to increased goods movement activity, which makes achieving steep reductions in emissions by 2050 very challenging. To achieve an 80 percent reduction in CO_2 emissions from 2010 levels, two scenarios were constructed. One included a near complete turnover of the MHDV fleet to electric and hydrogen fuel cell trucks using very low carbon energy sources. A second scenario included a mix of these technologies (roughly 50 percent of new truck sales by 2050) plus widespread use of very low carbon biodiesel fuel (60 percent blend in conventional diesel).

Yang et al. (2009) assessed whether California could achieve a reduction in transportation GHGs by 80 percent below 1990 levels by 2050. They found that the goal could be achieved with a concerted effort that would include changes in travel behavior, in vehicles and fuels, and argued for a portfolio approach. Morrow et al. (2010) used the National Energy Modeling System (NEMS) from the EIA to assess the implications of several policy scenarios (economy-wide CO_2 prices, fuel taxes, fuel economy standards, purchase tax credits for new vehicle purchases, and combinations of these policies). They concluded that a GHG emissions reduction goal 14 percent below 2005 levels by 2020 cannot be achieved. In Table 2-1 we summarize the studies reviewed, the climate mitigation goals considered, and the suggested alternatives for emissions reductions.

These findings suggest that additional approaches beyond improving the efficiency of conventional combustion engines, trucks, and trailers will be necessary in the 2030-2050 time frame if 80 percent reductions in GHG

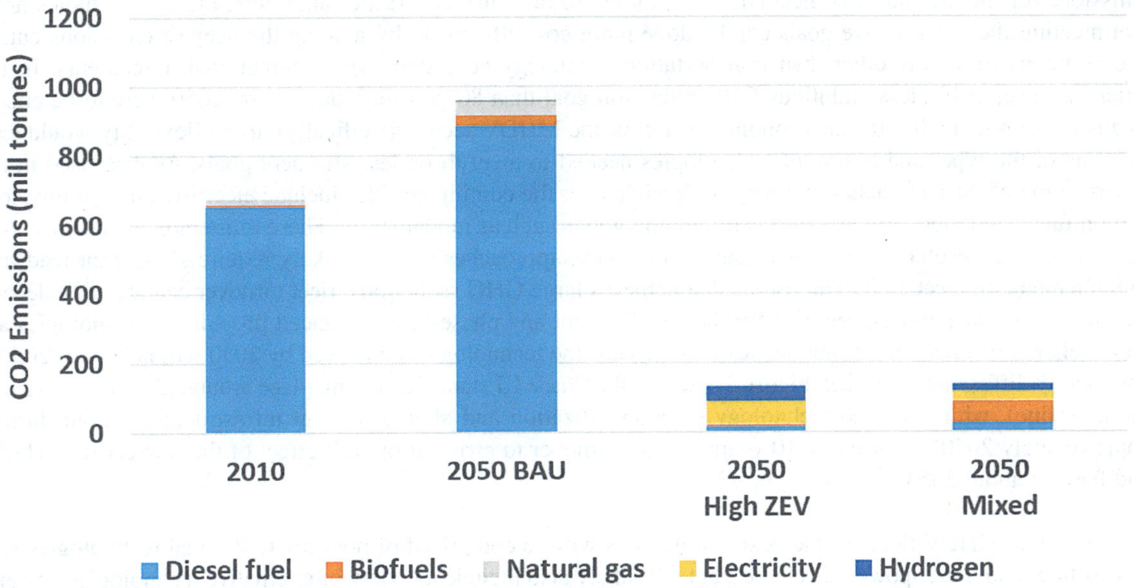

FIGURE 2-6 Projected well-to-wheels MHDV CO_2 emissions in 2010 and in a 2050 business-as-usual (BAU) case, along with two scenarios that achieve a modeled 80 percent reduction of MHDV CO_2 emissions for the United States by 2050. SOURCE: Adapted from Fulton and Miller (2015) by Lew Fulton for the committee.

TABLE 2-1 Summary of Studies Reviewed That Included Large GHG Emissions Reductions Goals for MHDVs in the United States or a Region Within the United States

Author (Date)	GHG Reductions Goal	Technologies Selected
Deep Decarbonization Pathways Project (DDPP, 2015)	Reduce emissions in the U.S. energy system by 80% below 1990 levels by 2050.	Different solutions: (i) drop-in replacement biomass fuels for diesel engines; (ii) deployment of vehicles using H or natural gas.
Nahlik et al. (2015) (California only)	2050 goals of reducing MHDV GHG emissions by 80% from a 1990 baseline.	Zero emission vehicles and fuel technologies would need to be implemented at a very aggressive rate.
Fulton and Miller (2015); Sharpe (2013)	80% reduction in CO_2 emissions from 2010 levels by 2050.	Two solution scenarios were constructed: (i) a near complete turnover of the MHDV fleet to electric and hydrogen fuel cell zero emission trucks; (ii) a mix of these technologies (roughly 50% of new truck sales by 2050) plus widespread use of very low carbon biodiesel fuel (60% blend in conventional diesel).
Yang et al. (2009) (California only)	Reduction in transportation GHGs by 80% below 1990 levels by 2050.	Goal could be achieved with a concerted effort that would include changes in travel behavior, in vehicles and fuels, and argued for a portfolio approach.
Morrow et al. (2010)	GHG emissions reduction goal 14% below 2005 levels by 2020.	Several policy scenarios (economy-wide CO_2 prices, fuel taxes, fuel economy standards, purchase tax credits for new vehicle purchases, and combinations of these policies). They conclude that the goal cannot be achieved with these policies.

emissions for the medium- and heavy-duty sector are to be achieved. At the same time, the above studies noted that meeting these aggressive goals can be done more cost effectively by making the deepest emissions cuts in energy-intensive sectors other than transportation, a strategy consistent with international agreements. In this latter instance, or if a less ambitious GHG reduction goal than 80 percent reduction by 2050 were to be chosen as a national goal (either for the economy overall or the MHDV sector specifically), more flexibility would exist in terms of the types and timing of technologies needed to meet those less stringent goals. As discussed in this report, some of the approaches that may be feasible over the coming decades include electrified drivetrains, low-carbon fuels, and more efficient means of moving goods such as modal shifts. These more radical changes could be described as revolutionary, rather than evolutionary, approaches and will likely require significant lead time to implement. To meet a 2050 milestone that achieves large GHG reductions, fleet turnover can absorb a decade, and another decade may be required for the development and phase-in of advanced propulsion technologies and new fuel and/or freight movement infrastructure. Thus the technologies conceived by 2030 will take until 2050 to completely diffuse into the fleet. Figure 2-7 shows the Phase III standards taking place around 2030 (shown above the blue line), with advanced technology commercialization and alternative fuel infrastructure rollout through approximately 2040, followed by 10 years of fleet turnover to arrive at the full effect of the low-carbon vehicles and fuels at about 2050.

Finding: The MHDV fleets of the next two decades will be comprised of not only traditional technologies such as gasoline- and diesel-powered vehicles, but will likely also include a variety of alternative technologies powered by a variety of energy sources.

2.1.2.3 Public Investment in Energy Research and Development

The public-sector energy research enterprise has relied on a portfolio approach that includes investment in a range of technologies. These portfolios can be explicitly designed to be robust (i.e., to provide technology options) across a variety of possible future states of the world (e.g., one that is carbon constrained or has very high oil prices) as a hedge against the uncertainty of long-range planning (NRC, 2007). Operationalizing a portfolio approach can be challenging, but it requires extending decision points as far into the future as possible without sacrificing the overall objective in order to maximize the number of feasible options that can be considered and developed for as long as possible. Other factors complicate the task of projecting the future incumbent technologies. For

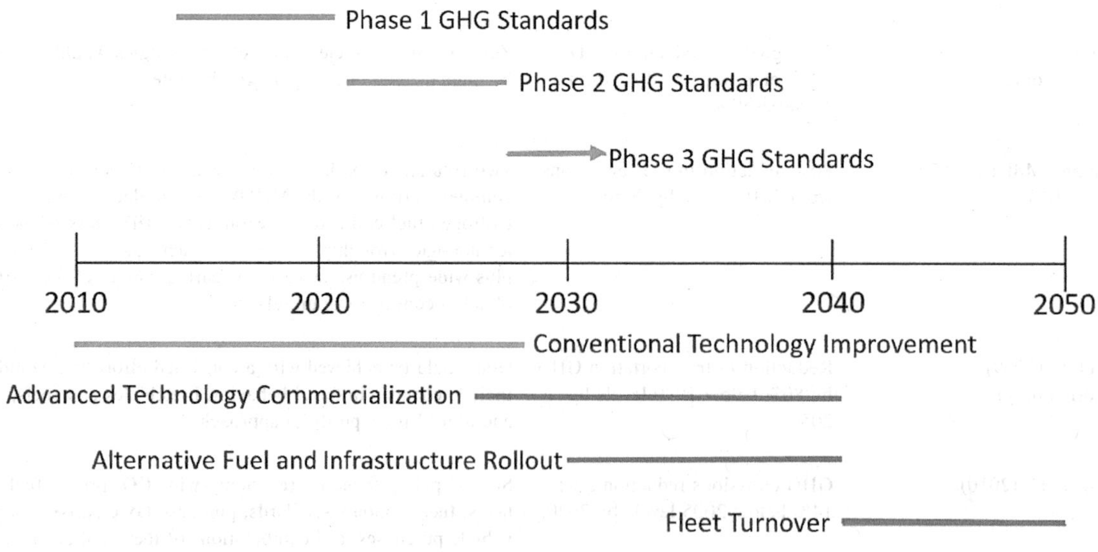

FIGURE 2-7 Example of timeline to populate the fleet with efficient, low-carbon technologies by 2050, given the existing and next phases of the GHG standards, technology development, commercialization, and fleet turnover periods.

example, attractive technology options can emerge from innovation pathways unforeseen today (NRC, 2009). In addition, present-day estimates may not show a clear winner when, for example, the cost of electricity from several technologies is compared (NRC, 2009, p. 57). Historically, such portfolios yield a mix of market-ready technologies with realized economic and environmental benefits as well as so-called knowledge benefits derived from research results that did not lead to a market-ready technology (NRC, 2001). The portfolio approach obviates the need to choose one technology in which to invest.

The history of technology innovation and policy provides several lessons with respect to the long lead time required to fully implement advanced technologies, fuels, and freight efficiency for the technology pathways that may be considered in a third phase of rulemaking:

First, developing a portfolio of technologies will increase the likelihood that the most effective advanced technologies, fuels, and freight movement approaches—capable of meeting longer-term goals—will be available to support compliance with a Phase III rulemaking. The goal here is not to have the government "pick winners," which often is unsuccessful (Marchant, 2014; Nelson and Langlois, 1983) given the many challenges and uncertainties in predicting future technology pathways and timelines. Rather, the goal should be for both government and the private sector to play an active role in ensuring a robust portfolio of technology options is available for consideration, which usually does not happen without active nourishment of the most promising options by both the private and public sectors (Jaffe et al., 2005).

The investment needed to develop such a portfolio of technology options and deploy those that are market ready might be less effective were there to be over-investment in technologies that provide only modest energy consumption and GHG reductions beyond what can be achieved by conventional technologies (Nahlik et al., 2015). In addition, the opportunity cost of misdirected investment could slow progress in the more advanced technologies that in some cases might achieve larger GHG reductions and could adversely affect economic competitiveness. Once made, these investments are unlikely to be pushed aside in favor of a more promising advanced technology and/or fuel option that could achieve greater reductions in fuel consumption and GHG emissions, and decision makers would need to deal with a long-lasting and potentially suboptimal infrastructure.

The timing of the investment in the portfolio is also critical: If this investment waits until after a Phase III rulemaking has commenced, the available technology choices may not evolve beyond the mix of diesel and natural gas combustion engines and trucks and tractors that is currently available. The timeline for technological readiness and implementation of additional options—including advanced technologies, fuels, and freight movement approaches—will accordingly be delayed, beyond 2050.

Second, proactive government involvement to provide clearer signals about future environmental performance requirements is essential to drive private innovation in the direction of longer-term solutions and technologies (Norberg-Bohm, 1999).

Third, limiting the scope of the third phase of rulemaking to consider only those technologies and approaches for which NHTSA and EPA have direct authorities (e.g., standards for engines, trucks, and conventional fuels) will limit the opportunities to achieve the maximum feasible and cost-effective GHG and fuel consumption reductions. For example, involvement of agencies with authorities or influence over goods movement (e.g., Federal Highway Administration [FHWA]) or alternative fuels (Department of Energy [DOE]) in the development of the rulemaking could identify credits or incentives to be implemented in parallel with the rulemaking that would provide performance-based rewards to industry for achieving the best-performing advanced technologies, fuels, and goods movement methods.

Finding: Implementation of new propulsion technologies, new low-carbon fuels, and more efficient freight operations and logistics may offer the opportunity to reduce GHG emissions beyond what is achievable from improving the efficiency of combustion engine MHDVs.

Recommendation 2-2: NHTSA in cooperation with EPA should establish what MHDV GHG and fuel consumption reductions need to be achieved in the 2030 to 2050 time frame, consistent with any national goals and international commitments the United States has adopted.

Finding: The commercialization and deployment of new propulsion technologies, low-carbon fuels, and more efficient freight movement methods that are significantly different from those currently in use may need to start as early as 2030 if ambitious national GHG emissions reduction and fuel economy goals are established for the 2050 time frame. Because of the long lead time needed to develop such advanced technologies and strategies, planning and preparation for such measures must begin well before the target dates for commercialization and implementation.

Finding: The current regulatory structure which focuses on engine and truck fuel consumption and GHG emission standards does not appear to have regulatory flexibility to address new fuels and methods of improving the efficiency of goods movement. For example, increases in MHDV size and weight limits could reduce fuel consumption; however, the authority to make such changes lies with another agency not part of this rulemaking process.

Recommendation 2-3: NHTSA should coordinate with EPA to identify a portfolio of technologies and options that could achieve national GHG emission and fuel economy goals for the 2030 to 2050 time frame and to identify the research, planning, and preparation needed to commercialize and implement such technologies and options in the relevant time frame. NHTSA should also coordinate with EPA to engage other agencies in the rulemaking process, such as DOE, the Federal Motor Carrier Safety Administration (FMCSA), and FHWA, who have authorities that can facilitate commercialization of low-carbon fuels and more efficient freight movement methods. NHTSA in cooperation with EPA should evaluate how incentives or other regulatory provisions can be incorporated to facilitate implementation of fuels and freight movement approaches that lie outside of NHTSA and EPA authorities.

2.1.3 Timing: An Opportunity and a Challenge for the Phase III Rulemaking

The long time period until Phase III standards might be proposed, adopted, and implemented leads to two considerations. First, the situation is very different than the NHTSA and EPA's Phase I and Phase II rulemaking, in which the agencies noted they lacked time and resources to address some fundamental issues that therefore had to be reserved for future rulemakings.[2] The long interim period provides a window of opportunity to undertake much of the preparation, groundwork, and identification and mitigation of obstacles for some fundamental and broader strategies for effectively controlling GHG emissions and fuel consumption by MHDVs. These issues may include, but are not limited to, development of an effective in-use compliance strategy, a methodology for analyzing and comparing life-cycle GHG emissions of different fuel options, a strategy for integrating fuel economy and emission standards with other policies and methods for reducing GHG emissions and fuel consumption (e.g., intermodal shifts, weight and height regulations, carbon tax, etc.), and a technological roadmap for deployment of vehicles that can potentially have near-zero emissions such as electric and fuel cell vehicles. In the Phase I (EPA and NHTSA, 2011) and Phase II (EPA and NHTSA, 2016) rulemakings, the agencies identified a number of such future issues that they did not have sufficient time and resources to address in those rulemakings, including addressing innovative and integrated designs of tractor-trailer combinations, analyzing the life-cycle consequences of different fuel options, and designing a comprehensive in-use compliance and enforcement program.

Second, the long time period during which NHTSA's Phase II regulations on fuel consumption will be in effect creates an opportunity and need for a midcourse evaluation of the preparations for any future regulations. This would evaluate the implementation progress of the Phase II regulations and other developments in technology availability and feasibility. The Phase II standards—covering model years 2021 to 2027—and any follow-on regulations, which may then be in the early stages of formulation, both will depend on many assumptions about technology availability, cost, implementation, and effectiveness. An interim evaluation would provide an opportunity to assess those assumptions and projections, and to make a determination whether any adjustments may be

[2] For example, in the Phase I rulemaking NHTSA and EPA stated that they would reserve consideration of requirements applicable to trailers to a future rulemaking. In the Phase II rule, the agencies postponed consideration of better integrated tractor-trailer design changes and life-cycle analysis of natural gas and other alternative fuels.

appropriate in the preparations for any future rulemaking. This interim evaluation could also coincide with the next National Academies of Sciences, Engineering, and Medicine report, required every 5 years.

An interim evaluation of its MHDV regulation in the 2021-2022 time period would help improve the latter's overall effectiveness and value contribution. Its primary focus would be on preparations for any future regulations beyond the Phase II standards.

Finding: NHTSA and EPA have a decade between the promulgation of the second and implementation of the third set of MHDV standards to develop more fundamental approaches and strategies for transforming the MHDV sector, such as planning for the expanded use of low-carbon fuels, development of zero emission technologies, and approaches to increasing the efficiency of moving goods. Such approaches could form a more comprehensive and effective program of reducing fuel consumption and GHG emissions from MHDVs.

Recommendation 2-4: To avoid the problems of having inadequate time and resources to address fundamental issues such as life-cycle analysis and technological, economic, and environmental analysis, in-use compliance and enforcement, and more integrated and comprehensive fuel efficiency and GHG emissions reduction policies, NHTSA in cooperation with EPA should be proactive in fully utilizing the 10-year window before a possible third phase of standards is implemented to sponsor and conduct studies, analyses, pilot projects, workshops, and collaborative efforts, including discussions with other relevant government agencies, to advance, anticipate, and address opportunities and challenges for long-term strategies for achieving more ambitious emissions and fuel consumption reductions in the 2030s and beyond.

Finding: The Phase II regulations, and any post-2027 regulations then being contemplated, will depend on many assumptions about technology availability, cost, implementation, and effectiveness. An interim evaluation would provide an opportunity to assess those assumptions and projections, and to make a determination whether any adjustments may be appropriate in the preparations for future rulemakings. This interim evaluation would also coincide with the next National Academies report, required every 5 years.

Recommendation 2-5: NHTSA, in coordination with EPA, should conduct an interim evaluation of the MHDV program in the 2021-2022 time period, focusing on adjustments to and preparations for future revisions to the regulations.

2.2 KEY REGULATORY GAPS AND CHALLENGES

In addition to the timing issues discussed above, several complexities and challenges in regulating MHDVs result in inherent gaps and limitations for any regulatory program that seeks to regulate such a diverse set of vehicles. Complexities include defining fuel economy for regulatory purposes; the effects of exogenous factors such as volatility in fuel prices; the need to consider vehicle components other than the engine and powertrain since these affect fuel consumption and GHG emissions; the significance of vehicle weight and length for the payload carried (and thus the freight efficiency); the variety and diversity of vehicle types and functions that complicate manufacturing, regulation, and enforcement; the diversity in the size and capabilities of companies in the MHDV sector; differences in the results of life-cycle analyses of different fuel options; challenges in ensuring effective certification and in-use compliance; and miscellaneous other factors affecting fuel consumption and greenhouse gas emissions.

2.2.1 Defining Fuel Economy

The first complexity is in defining fuel economy for MHDVs. The terms fuel economy and fuel consumption are both used to show the efficiency of fuel use in vehicles. Fuel economy is a measure of how far a vehicle will go with a gallon of fuel and is expressed in miles per gallon (mpg). Fuel consumption is the inverse measure—the amount of fuel consumed in driving a given distance—and is measured in units such as gallons per 100 miles.

Fuel consumption is a physical measure and is useful because it can be related directly to the goal of decreasing the amount of fuel required to travel a given distance. A key concept for MHDVs is load-specific fuel consumption (LSFC) (see NRC, 2010, p. 25). The primary purpose of most medium- and heavy-duty vehicles is to deliver freight or passengers (the payload). An extreme example illustrates the problem: A simple way to reduce the fuel consumption of a truck is to leave the cargo on the loading dock. This approach, however, ignores the purpose of these vehicles. Therefore, the way to represent an attribute-based fuel consumption metric is to normalize the fuel consumption to the payload that the vehicle hauls. Adding payload to a vehicle increases fuel consumption, but the higher payload actually improves the efficiency of the vehicle (in terms of LSFC). The Phase I rule employs the U.S. customary system of units, and LSFC is expressed as gallons per 1,000 ton-miles.

2.2.2 Volatile Fuel Prices

A second complexity is the uncertainty and volatility in fuel prices. Fleet owners will make purchasing and operating decisions based on expectations of what fuel prices are likely to be in the future. However, accurate projections of oil prices have proven to be difficult. For example, every year, the EIA of the Department of Energy produces a set of projections of energy quantities using their NEMS, a general equilibrium model. Figure 2-8 shows the projections for oil prices in the dashed grey lines that were produced by the EIA in its AEOs over time, for projections made since year 1985. The actual oil price is shown by the black line. All values are in constant dollars. Two messages come across: (1) oil prices are hard to predict and (2) there is substantial volatility in the prices, and that volatility seems to have increased in recent years.

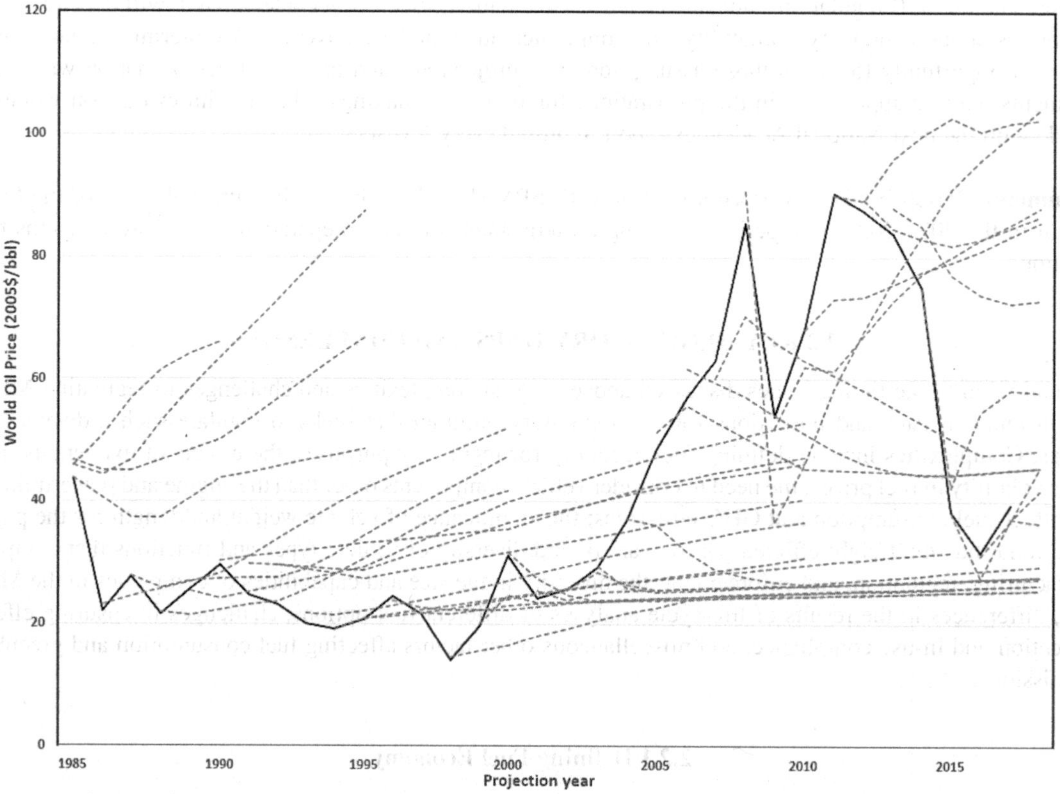

FIGURE 2-8 Actual (black line) and projected oil prices in constant dollars from 1985 to 2018. Each dashed line corresponds to a set of oil price projections from the reference case outputs in the Annual Energy Outlook produced by the EIA.
SOURCE: Compiled by Evan Sherwin and Inês Azevedo using data from EIA's retrospective analysis: https://www.eia.gov/outlooks/aeo/retrospective/ (accessed March 4, 2019).

One potential strategy currently being discussed is the use of natural gas for transportation. In that regard, it is worth pointing out that the same issues mentioned for oil—volatility and difficulty in projecting the prices—apply to natural gas prices, as shown in Figure 2-9.

2.2.3 Factors Other than Engine/Powertrain Affecting Fuel Consumption/GHG Emissions

Improving the efficiency of the engine and drivetrain is an important approach to reducing fuel consumption and CO_2 emissions; however, a third complexity is that many factors impact fuel consumption, aside from engine and powertrain operation. Of the fuel consumed by a Class 8 tractor-trailer moving at constant highway speed, roughly 40 percent provides tractive power to overcome road load—aerodynamic drag and rolling resistance. The National Research Council (NRC, 2014) found that reducing tractor and trailer aerodynamic drag and tire rolling resistance were viable methods of reducing fuel consumption and CO_2 emissions.[3] The remainder of the energy is

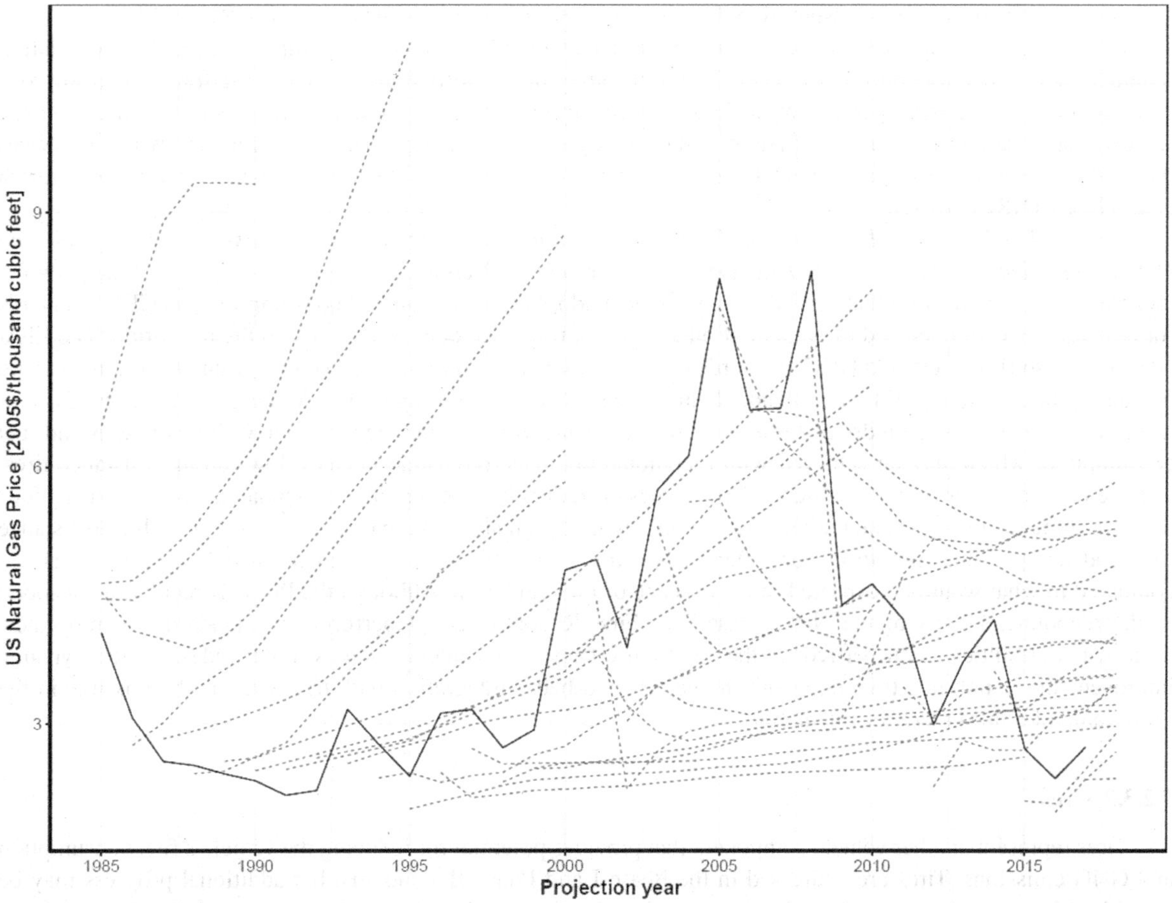

FIGURE 2-9 Actual (black line) and projected natural gas prices in constant dollars from 1985 to 2018. Each dashed line corresponds to a set of natural gas price projections from the reference case outputs in the Annual Energy Outlook produced by the EIA.
SOURCE: Compiled by Evan Sherwin and Inês Azevedo using data from EIA's retrospective analysis: https://www.eia.gov/outlooks/aeo/retrospective (accessed March 4, 2019).

[3] Note the energy use of a lower-speed vocational (urban) truck will differ, with stop-and-go driving and tire rolling resistance more important than aerodynamics.

consumed by accessories, parasitic loads, and friction and other inefficiencies in the engine and drivetrain (NRC, 2014, p. 68). This subsection briefly identifies potential opportunities to reduce fuel consumption and GHG emissions using strategies other than those focused on the engine and drivetrain. More detailed analysis of some of these strategies is provided later in this report, and a full discussion of vehicle characteristics is presented in Chapter 5.

2.2.3.1 Trailers

Trailers provide an opportunity to reduce CO_2 emissions and fuel consumption. As of 2009, there were about 5.7 million trailers commercially registered in the United States (Roberts, 2016). About half are 53-foot or longer dry van or refrigerated van trailers. The remainder is composed of shorter van trailers, or specialty trailers such as tankers or flat beds. The use of boat tails, fairings, and side skirts has the potential to reduce fuel consumption caused by dry van trailers by up to 10 percent (NRC, 2010, pp. 99-100). Use of low-rolling-resistance tires on a tractor and trailer can reduce fuel consumption by about 3 percent, and tire pressure monitoring and inflation systems can help ensure fuel consumption does not increase due to tire under-inflation (NRC, 2014).

California has adopted GHG standards for trailers, which now are being implemented. This regulation requires tractor operators pulling trailers (53-foot or longer dry or refrigerated vans) to use trailers that improve fuel economy of the tractor-trailer by at least 5 percent compared to a trailer without aerodynamic features. Observations made in California and Arizona showed a greater proportion of trailers using aerodynamic devices than did observations made in Oregon, Texas, Michigan, Pennsylvania, and Maryland, where travel to California is less likely (NRC, 2014, p. 83).[4]

The 2014 NRC report, *Reducing the Fuel Consumption and Greenhouse Gas Emissions of Medium- and Heavy-Duty Vehicles, Phase 2: First Report* (the first report of National Academy of Sciences "Phase Two"), included a recommendation that the federal government adopt by regulation fuel consumption and GHG standards for new van (box) trailers, and to evaluate if other types of trailers should be included in the regulation. The EPA/NHTSA Phase II rule adopted GHG standards for new trailers beginning in 2018 (fuel consumption standards are mandatory beginning in 2021). The standards increase in stringency through 2027. Manufacturers of new 50-foot and longer box trailers (both dry and refrigerated) must achieve the greatest reduction in GHG emissions and fuel consumption, which may be achieved with a combination of aerodynamic devices, low-rolling-resistance tires, and tire monitoring or inflation systems. Other types of trailers have less stringent standards based on difficulties in using some technologies, and others may comply with design standards based on use of low-rolling-resistance tires and tire pressure monitoring or inflation systems. CO_2 emissions reductions range from 2 to 9 percent compared to what would be expected in the absence of the regulation. Although the Phase II regulations respond to the recommendations of the NRC Phase Two First Report to adopt performance standards for most new trailers, opportunities to further reduce the energy use caused by trailers, such as trailer redesign, aerodynamic improvements to non-box trailers, weight reduction, and better integration with tractors, may be possible in the time frame of a third phase of rulemaking.

2.2.3.2 Tires

Tires represent another vehicle component that provides potential for reducing the vehicle's fuel consumption and GHG emissions. Tires are addressed in the Phase I and Phase II standards, but additional progress may be possible. Most manufacturers of truck, tractor, and trailer tires produce low-rolling-resistance (LRR) models for all wheel positions. Many of these are verified by the EPA SmartWay program to reduce tire rolling resistance by at least 15 percent, which equates to about a 3 percent improvement in fuel economy for a tractor-trailer. The best-in-class tires reduce rolling resistance by 30 percent.

Manufacturers reported in 2013 that the majority of new tractors and trailers are equipped with LRR tires. Greater use of LRR tires can be expected as new truck manufacturers rely on them to help comply with the federal

[4] The Oregon survey was taken on I-84, which is an east-west highway that connects Portland to Idaho and points east, not toward California.

fuel economy standards. Currently, about three of every four new tires sold are for replacement, representing one of the few technologies available for improving fuel efficiency of existing vehicles. However, most of these replacement tires are not LRR, and less than half of the tire models available are SmartWay verified (NRC, 2014).

To ensure the fuel economy benefits of both new and replacement tires are maximized, the committee in its 2014 report (NRC, 2014) recommended that NHTSA should further evaluate and quantify the rolling resistance of new tires, especially those sold as replacements. If additional cost-effective fuel savings can be achieved, the report recommended that NHTSA and EPA adopt a LRR performance standard for all new tires sold, including those sold as replacements. In the Phase II rule, the agencies did not adopt a LRR performance standard for all new tires. The agencies expect that LRR tires will likely be used as a means of complying with the GHG emission and fuel consumption standards for new trucks and trailers. However, this does not address the key fact that more replacement tires are sold each year than tires sold for installation on new trucks and trailers, and given the incremental higher costs of LRR tires there remains a lack of incentive for those who only own trailers to invest in fuel-saving technologies such as replacement tires. Thus the committee continues to believe in the appropriateness of its previous recommendation that NHTSA and EPA evaluate the cost effectiveness of low-rolling-resistance tires, especially replacement tires, and consider adopting a LRR performance standard if indicated (NRC, 2014, p. 80). More detailed analysis of the potential fuel consumption savings from tires is provided in Section 5.3 in Chapter 5.

2.2.3.3 Cabin Air Conditioner Refrigerant

Most vehicle air conditioners leak refrigerant during their life, either by slow leakage through hoses or seals, or from the failure of the various components.[5] Refrigerants can also be released from collisions and during servicing and scrappage if not properly reclaimed.

The refrigerant currently in service for truck cabin conditioning is HFC-134a, which has a 100-year global warming potential (GWP) of 1,430.[6] Truck and trailer refrigerant units (reefers) also use R-404A which has a still higher GWP of 3,922. The Phase II rule included a minimum leakage standard for Class 2b to Class 8 vocational trucks, beginning in 2021.

Other factors may favor a move away from HFC-134a to lower-GWP refrigerants in MHDVs. EPA has finalized a rule delisting approval of HFC-134a in light-duty vehicles by MY 2021 (EPA, n.d.).[7] The European Commission has eliminated the use of refrigerants having GWP above 150 effective January 1, 2017, in new cars and vans. The European Commission reviewed and reaffirmed the directive[8] after one manufacturer and Germany's Federal Motor Transport Authority (KBA) identified a risk of flammability of HFO-1234y, a substitute for HFC-134a, in head-on collisions.[9]

Finding: NHTSA and EPA have adopted requirements in the Phase II rule that will reduce fuel consumption and/or GHG emissions caused by nonengine components such as trailers, tires, and air conditioning. However, fuel consumption and GHG emission reductions beyond those adopted in Phase II appear possible.

Finding: The use of refrigerants in light-duty vehicles for cabin air conditioning has moved to refrigerants having lower GWP than HFC-134a. The use of such refrigerants may be feasible in medium- and heavy-duty vehicles.

Recommendation 2-6: The committee recommends that each of these nonengine components be further evaluated for feasibility and cost effectiveness during the period prior to issuing a Phase III proposal. Specifically, to achieve

[5] This discussion of hydrofluorocarbons (HFCs) affects the GHG emissions of a MHDV but not its fuel consumption.

[6] GWP is defined as the ratio of the time-integrated radiative forcing of a gas to that of an equal mass of CO_2. The period of time integration is user defined.

[7] Protection of Stratospheric ozone: Change of listing status for certain substitutes under the significant new alternatives program. *Federal Register* 80:42870.

[8] See http://europa.eu/rapid/press-release_MEMO-14-50_en.htm (accessed December 1, 2019).

[9] See http://europe.autonews.com/article/20130808/ANE/130809861/german-officials-provide-mixed-ruling-on-honeywell-refrigerant (accessed December 1, 2019).

greater reductions in fuel consumption and GHG emissions, the design of trailers, and better integration with tractors, should be evaluated; and the rolling resistance of new tires, especially those sold as replacement tires, should be quantified and evaluated. If additional cost-effective fuels savings can be achieved, NHTSA should adopt a regulation establishing a low-rolling-resistance performance standard for all new tires designed for tractor and trailer use.

2.2.3.4 Length and Weight

Truck size and weight federal policy was enacted to preserve the infrastructure and ensure that heavy vehicles could operate on the road system within the load-carrying capacity of roads and bridges. It also ensured that vehicle size was controlled to allow operation within geometrical constraints dictated by bridge height, lane widths, and curve radius constraints, which limit vehicle maneuverability. Since the policy limits vehicle length and weight, it de facto limits options to improve vehicle productivity. U.S. federal truck size and weight policy has been frozen since 1982. Meanwhile, other countries have reformed their policies, yielding improved freight efficiency and resulting in substantial fuel emissions reductions for a given freight task. It is estimated that the reduction in vehicle miles traveled (VMT) and CO_2 emissions attributable to potential size and weight policy reform is in the range of 10 to 20 percent (Woodrooffe et al., 2009). A 10 percent reduction in VMT translates to an annual reduction of approximately 11 billion liters of diesel fuel and 28 million metric tons CO_2 and has an estimated safety benefit of about 330 fewer fatalities and over 4,200,000 fewer injuries due to reduced vehicle exposure. The annual value of these savings is approximately $16 billion (Woodrooffe, 2016). Heavier and longer vehicles could affect and be affected by road infrastructure, although the Federal Highway Administration recently completed a truck size and weight study (FHWA, 2015) that examined how selected heavier and longer vehicles would influence infrastructure life. The study found that the infrastructure costs for most vehicles examined were minimal and there was no significant infrastructure risk identified. This issue of size and weight and its relationship to reducing fuel consumption and GHG emissions is discussed further in Chapter 9.

2.2.4 Complexity and Variety of Vehicle Types

A fourth complexity for MHDV regulations is the variety of vehicle types and duty cycles. Passenger vehicles are relatively uniform, varying mainly by weight, footprint, and frontal area (aerodynamic drag), and primarily perform movement of people. Most passenger vehicle models are produced in relatively high annual volumes. By contrast, medium- and heavy-duty (MHD) trucks vary greatly in weight, size, power, and drag. A local delivery truck and an 18-wheel tractor-trailer share much less in common than do a subcompact car and a large sport utility vehicle. The MHDV on-road vehicle industry is composed of a variety of vehicles with missions spanning small business and utility service to delivery and commercial cargo trucks (see Figures 2-10 and 2-11). The complexity and variety of vehicle types and tasks makes all aspects of the industry more complicated, from manufacturing to regulation to enforcement.

Most MHDVs have historically been powered by diesel engines; however, the impact of environmental regulation compounded by high fuel costs has caused the evolution of a more complex set of technologies designed for these vehicles in recent years. At the lighter end of the weight class for these vehicles, gasoline- and diesel-fueled heavy pickups compete based on fuel economy and vehicle price. At the heavier end of the weight class, favorable pricing of natural gas versus diesel fuel has brought about an increase of natural gas–powered vehicles in the MHDV markets. Compounding this is the need for a myriad of engine, transmission axle, tire, and programmable electronic features that satisfy the specific work each vehicle is designed to achieve. With the rapid development of electronics, many MHDVs in the late 2020s and 2030s will also be equipped with features that selectively power vehicle functions. Finally, the role of the operator cannot be discounted in how effective reductions in fuel consumption can be. Whether benefits and/or incentives are available to the driver will play a significant role in whether the vehicle is driven in the most efficient manner. All this will produce a complex mix of technology features and behaviors which will play a role in the control of GHG emissions and improve fuel economy in the future. Chapter 9 explores these considerations further.

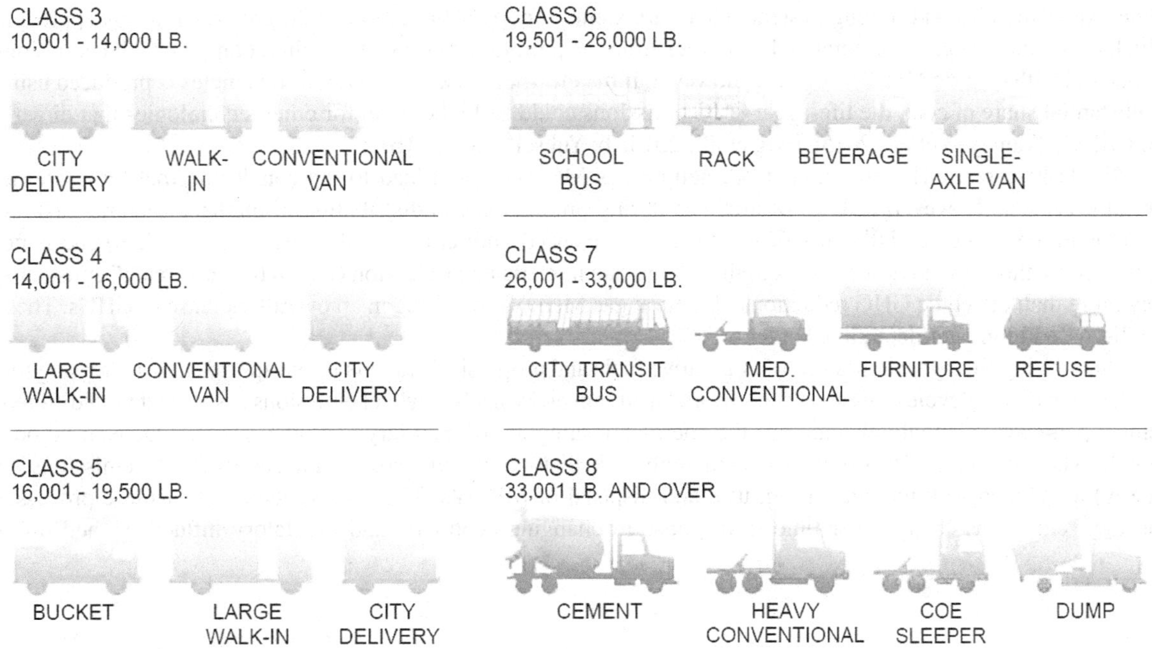

FIGURE 2-10 Size and weight classes of medium- and heavy-duty vehicles (Class 3 and above). The NHTSA regulation also covers so-called 2b vehicles having gross vehicle weights of 8,501 to 10,000 pounds.
SOURCE: National Petroleum Council (2012).

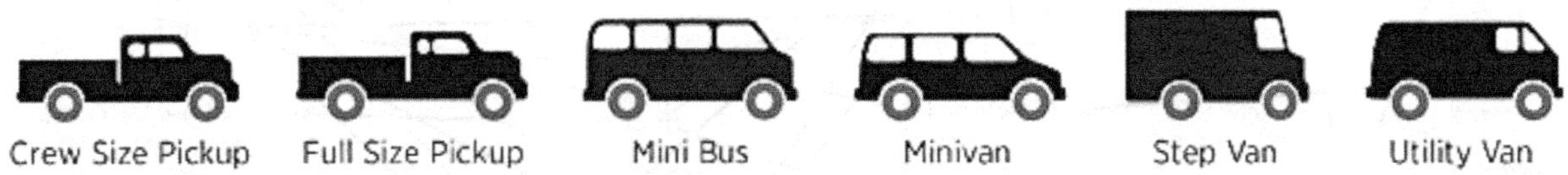

FIGURE 2-11 Class 2 vehicles: The heavier vehicles in this class comprise "Class 2b" and have gross vehicle weights of 8,501 to 10,000 pounds. These latter vehicles are subject to the NHTSA MHDV regulations.
SOURCE: AFDC (2012).

2.2.5 Large Versus Small Companies

A fifth complexity the regulation of medium- and heavy-duty vehicles must address is the range and variation of the size of companies potentially subject to or otherwise affected by the regulation. Especially in the fleet ownership sector of the industry, companies will range from large commercial fleet owners such as Walmart, FedEx, and UPS, with substantial regulatory resources and expertise, to small (one- to three-truck) family businesses with much more limited resources and expertise for regulatory compliance. The Phase I standards exempted small businesses from compliance with the standards, but the Phase II standards will no longer exempt small businesses. However, NHTSA and EPA have provided additional flexibility to small businesses subject to the Phase II standards, notably small business trailer manufacturers, such as providing a 1-year delay for compliance and relaxing certification and reporting requirements. Going forward, continued flexibility and accommodation is important for small businesses subject to the standards, as well as small businesses that may be affected by the regulations (e.g., small business fleets that purchase new trucks).

2.2.6 Life-Cycle Analysis of Vehicle/Fuel Systems

The importance of life-cycle analysis, particularly for GHG emissions, is another complexity MHDV efficiency regulations must address. The current regulatory design only includes GHG emissions from the tailpipe and to some extent the air conditioning system. There are some important limitations with undertaking this approach, which may lead to negative unintended outcomes from the policy. For example, the direct tailpipe GHG emissions associated with electric MHDVs are zero; however, if the electricity used to power such vehicles is produced using a substantial share of coal, the life-cycle GHG emissions could be higher than for other technologies that directly emit GHGs (Tamayao et al., 2015; Tong et al., 2015a,b; Yuksel et al., 2016).

Similarly, direct GHG emissions from natural gas MHDVs could lead to the conclusion that these are net emissions savers. However, leakage of methane during production and distribution of methane (natural gas) has been identified as a potentially significant source of GHGs (Brandt et al., 2014; Tong et al., 2015a,b) and is not addressed by the current regulations. A high leakage rate in methane production (i.e., in the "upstream" emissions) may more than offset the GHG reductions of natural gas MHDVs, resulting in an overall increase in GHGs. These are discussed in greater depth in Chapter 6.

Quantifying life-cycle emissions of alternative fuels and propulsion systems is one approach to addressing this regulatory gap. Life-cycle emissions can include upstream and wheel-to-wheel emissions associated with different vehicle pathways. It should also include the energy consumption of auxiliary and other power sources used on a vehicle. The advantage of this type of system analysis is that—in its full scope—it accounts for the emissions (or energy) associated with manufacturing, use, and disposal of MHDVs. The disadvantage is that these processes and their emissions change over time in response to changing economic and regulatory influences, and differ

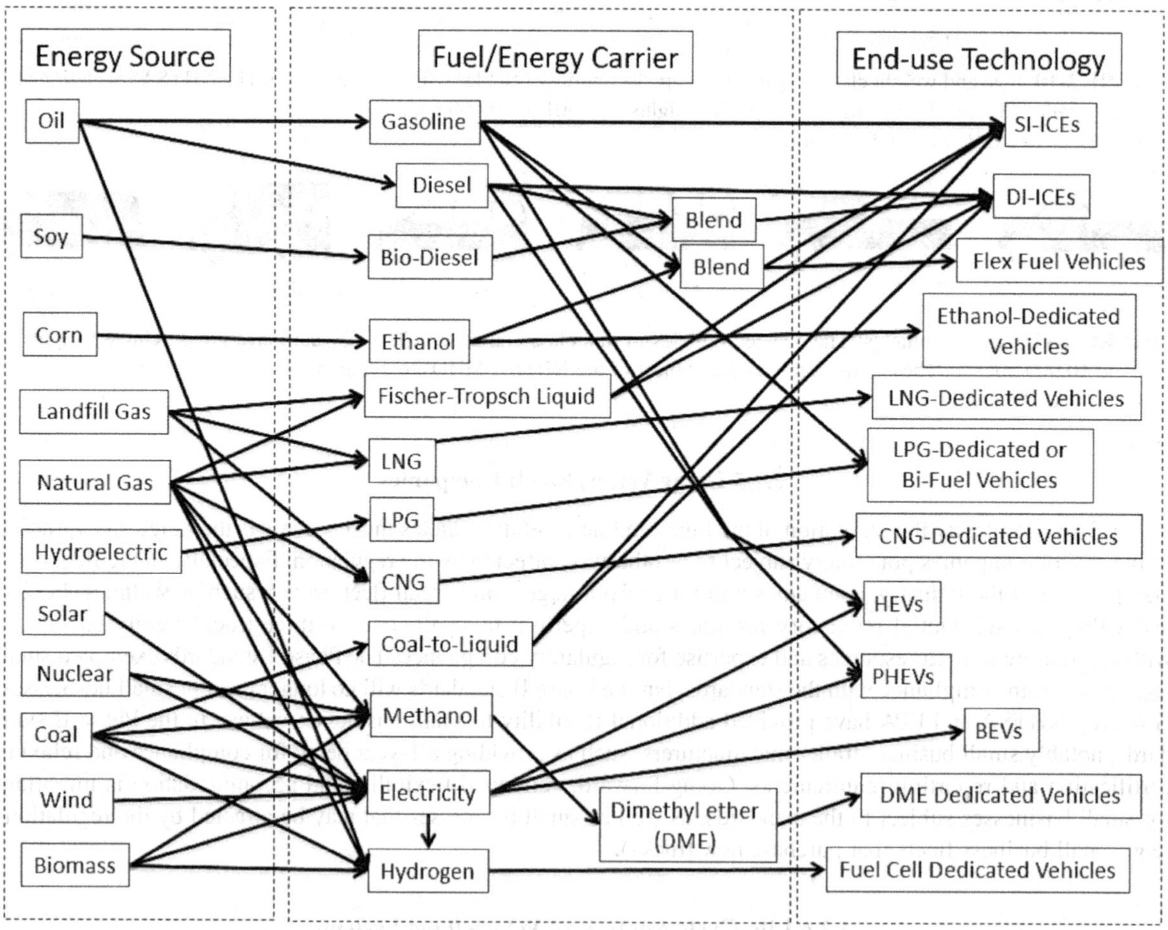

FIGURE 2-12 Summary of potential fuel and technology pathways for MHDVs including energy source, energy carrier, and end-use vehicle.

regionally, which may make policy development and implementation more difficult. The effect of accounting for life-cycle GHG emissions and fuel use should be quantified, including uncertainties to determine if accounting for life-cycle emissions is necessary to create a level playing field for emerging fuels and technologies or for incorporation into future regulatory design is necessary.

Figure 2-12, reproduced from the committee's 2014 First Report, provides a framework to assess potential fuel pathways to serve different MHDVs.

Finding: New fuel and power source options (including hybrid-electric systems and auxiliary power sources) may be needed to achieve significant further progress in reducing fuel consumption and GHG emissions from MHDVs over the next couple of decades. In addition to direct fuel consumption and emissions from the vehicle, these fuel options will likely differ significantly in upstream and nontailpipe emissions. These nontailpipe and upstream emissions may have an impact on program benefits achieved.

Recommendation 2-7: NHTSA, in coordination with EPA, should evaluate and quantify the life-cycle GHG emissions and fuel consumption of all fuels and technologies whose use could contribute to meeting a third phase of standards, and take them into consideration in developing a third phase of regulation. It will be critically important to incorporate a life-cycle perspective in those instances where some fuel-technology pathways' life-cycle emissions may lead to an increase, rather than a decrease, in emissions. However, given the resources needed for detailed life-cycle analysis and uncertainty characterization, and given the resource constraints from the agencies, we would recommend assessing carefully which fuel-technology pathways may deserve specific attention in a life-cycle perspective.

2.2.7 Certification and Real-World Compliance

For passenger vehicles, EPA requires the vehicle manufacturer to perform pre-sale emission and fuel economy testing of the completed vehicle to validate that it is capable of meeting emission and fuel economy standards. The test cycles are designed to mimic real-world operation. During the vehicle life following sale, EPA performs testing of many models using the same test cycles to determine if the vehicles continue to comply in-use with standards. Models that do not may be recalled. EPA also performs extensive testing using a broader range of test cycles, including use of on-vehicle portable emission measurement equipment, to determine the emission performance of the vehicle fleet as a whole, which provides data that allow an assessment of regulatory program effectiveness. Together these testing requirements help ensure (1) new vehicles, before they are sold, are capable of meeting standards; (2) vehicles meet standards during most of their operational life, and if they do not, remedial actions are taken; and (3) the overall regulatory program is effectively reducing emissions and fuel consumption (NRC, 2015). The test results may also suggest ways the effectiveness of the regulations can be increased, and the cost of compliance reduced. These compliance programs have identified many in-use light-duty vehicles with high emissions, for which design changes were made through recalls, and also identified opportunities to improve the effectiveness of the regulations.

For MHDV Class 4 and above, the current regulation for GHG emissions and fuel economy, when compared to the program for passenger vehicles, falls considerably short of providing an assurance of compliance and program effectiveness. Pre-sale certification emission testing applies to the engine, but not the vehicle. Certification of the vehicle is accomplished using a simulation model with no direct vehicle GHG emission or fuel consumption testing. Once vehicles are sold, in-use testing is sparse and limited to on-vehicle emission measurements made during random truck operation. There are generally no state and emissions inspections of the sort used for light-duty vehicles (LDVs). This approach does not lend itself to determining in-use compliance since the on-road testing differs from the certification engine test and the cycles used in the simulation model, and it is the on-road testing rather than the certification testing that provides the necessary real-world measurement of emissions. Removing engines and testing them in the same manner as pre-sale certification can be done, but it is far too costly and time consuming to be a viable and comprehensive method of determining in-use compliance. It also remains unclear if an on-road vehicle test that suggests the vehicle is exceeding standards can be used as a basis for enforcement given the vehicle was originally certified using a simulation model rather than an emission test.

This regulatory gap is discussed in more detail in Chapter 3, Certification, Compliance, and Enforcement, and some possible alternative approaches to ensuring compliance and measuring program effectiveness are identified that should be evaluated by NHTSA and EPA prior to proposing a third phase of regulations.

2.2.8 Regulatory Baselines and Metrics

The overall objective of the MHDV fuel efficiency and GHG emissions reduction regulations is to achieve real-world benefits in reducing fuel consumption and GHG emissions in the MHDV sector. As explained in the committee's 2014 report (NRC, 2014, Chapter 4), a reliable baseline of pre-regulatory fleet characteristics, miles driven, fuel consumption, and emission data is critical for measuring real-world impacts. As explained in the 2014 report, NHTSA and EPA currently lack a reliable source of data on real-world vehicles that can be used to establish a regulatory baseline. Phase II is based on an assumption of compliance with Phase I standards as the baseline for the Phase II regulations, but this modeled baseline is not connected to or validated by real-world measurements and thus does not provide a valid baseline that can be used to measure real-world performance. The nearly decade-long gap between the Phase II and a possible "Phase III" rulemaking provides an opportunity for NHTSA and EPA to consult with other governmental and private entities to coordinate efforts to obtain real-world vehicle data that can be used to establish a reliable baseline.

Finding: NHTSA and EPA currently lack reliable data on real-world vehicles that can be used to establish a regulatory baseline. While engine certification data are available and useful, they do not address the need for on-road real-world truck data. A reliable regulatory baseline based on real-world data is essential for evaluating the effectiveness and success of the regulatory program, and identifying future regulatory priorities and directions.

Recommendation 2-8: NHTSA and EPA will have almost a decade between the Phase II and a possible "Phase III" rulemaking in which to coordinate with other governmental and private entities to collect real-world vehicle data that can be used to establish a reliable baseline. The agencies should commit resources to collecting real-world fuel consumption and GHG emissions data from a robust and representative sample of pre-control trucks and for each model year subject to Phase I and Phase II standards. These data can be used to establish a regulatory baseline that can be used to evaluate program effectiveness and future regulatory priorities.

2.2.9 Other Factors Affecting Fuel Efficiency

There are a number of other factors that can affect the fuel efficiency and GHG emissions performance of a MHDV. For example, vehicle speed is an important factor, with higher vehicle speeds resulting in greater fuel consumption needed to transport a given load a specified distance. As a result, many fleet owners are now installing speed limiters on their vehicles to require lower speeds and save money through fuel conservation. Another important factor is driver behavior, as inefficient driver behavior (e.g., nonoptimal gear shifting or excessive idling) can result in significant fuel wastage. Many vehicle owners are now using telematics and surveillance technologies to monitor driver behavior, creating sensitive tensions between optimal performance and driver privacy. Yet another important factor is road condition and traffic management, all of which can also dramatically affect fuel efficiency. Because these and other factors have a significant effect on vehicle fuel efficiency and GHG emissions, it is important to consider how fuel efficiency and GHG emission regulations will interact with or be affected by these factors.

Some additional market and technology factors, some of which are discussed in more detail in the following chapters, include (i) economic trends such as construction and manufacturing levels; (ii) the availability of industry incentive programs such as tax credits for electric vehicles and other energy-conserving technologies (Diamond, 2009); (iii) the development and deployment of connected and automated vehicle systems; (iv) modal shifts between trucks, ships, rail, and water vehicles; (v) changes in transportation planning and traffic management; (vi) emerging technologies such as 3D printers and unmanned aerial vehicles that may affect transportation and delivery systems (Basulto, 2014; Harford, 2013; Vance, 2010); and (vii) the expansion of the Panama Canal and other infrastructure projects.

2.3 OTHER REGULATORY PROGRAMS

A variety of other regulatory programs at the state, federal, and international levels seek to control directly the fuel consumption and GHG emissions of MHDVs, as does the NHTSA/EPA program. These other programs, summarized below, may provide useful lessons and synergies for the NHTSA/EPA program. In addition to these regulatory programs that directly regulate the fuel conservation or GHG emissions of MHDVs, there are a number of other regulatory programs, also discussed below, that indirectly affect fuel consumption and GHG emissions associated with the MHDV sector.

2.3.1 Other Programs Directly Regulating MHDV Fuel Efficiency and GHG Emissions

2.3.1.1 States

Under federal clean air law, only California can adopt its own emission limits for new vehicles, which must be at least as stringent as the EPA standards. For new heavy truck engines, the California Air Resources Board (CARB) required stricter emission standards for criteria air pollutants than EPA until the 1990s, when CARB and EPA agreed to jointly adopt uniform standards that meet both federal and California's urban air pollution control needs. This continued with the adoption of the Phase I MHDV GHG standards, and the agencies have worked cooperatively in developing the Phase II GHG standards, which were adopted in 2016. In addition, California is the only state to regulate certain truck operators by requiring large trailers to have improved aerodynamic performance, which reduces tractor fuel use, and requiring use of low-rolling-resistance tires.

Other states are preempted from adopting their own vehicle emission standards under the Clean Air Act, but section 177 of the statute allows other states under certain circumstances to adopt and enforce the California emission standards for LDVs or MHDVs. Nine states (mostly in the Northeast) have used this statutory provision to require California certification of new MHDVs sold in their states. Because the California and federal criteria emission standards for MHDVs have been essentially the same since the 1990s, and EPA was first to adopt MHDV GHG standards, most of these states have not updated their regulations to require California-certified heavy trucks, but the authority to require sale of California vehicles remains should it be needed.

2.3.1.2 Other National Programs

A number of other nations have adopted their own MHDV fuel efficiency or GHG regulations. The *European Union* (EU) has been an international leader in regulating overall GHG emissions but has not yet adopted standards for fuel efficiency or GHG emissions specifically for MHDVs. The Commission of the European Union has developed a computer simulation tool called VECTO to calculate CO_2 emissions from new MHDVs. The Commission issued a strategy for low-emission mobility which indicates the EU will need to introduce measures to actively curb CO_2 emissions from lorries (trucks), buses, and coaches, as other parts of the world have. The Commission will speed up analytical work and public consultation to prepare proposed standards (European Commission, 2016).

Japan in 2005 was the first country to establish fuel economy standards for new heavy-duty trucks and buses over 3.5 tonnes. The standards apply only to diesel-fueled trucks and buses, and must be met on average for a vehicle class (typically weight based). The standards are set by the best-performing vehicle per class in a base year (2002), and are enforceable beginning in 2015, providing a 12 percent reduction in fuel consumption from new vehicles. Compliance is based on engine dynamometer testing and a simulation model with actual engine and drivetrain inputs and standard assumptions for typical driving cycle, drag, rolling resistance, and weight. The concept, called Top Runner, is that all trucks and buses of a given class should be able to perform equal to or better than the best in class given time for redesign. The 10-year lead time acknowledged that standards for NO_x and $PM_{2.5}$ were also strengthened in 2009. No in-use compliance program is specified (Olivares and Wagner, 2014a).

China has adopted fuel economy standards for heavy trucks and buses over 3.5 tonnes, fueled by diesel or gasoline. The second-phase standards went into effect in July 2015 and vary by weight, and five truck or bus types. Each truck must meet its appropriate standard. The standard is expected to reduce fuel consumption by about 11 percent compared to the first phase of standards. Compliance is based on a chassis emission test for a base model representing the worst-case truck in a class, with a simulation model used to determine fuel consumption for closely related variants to the base model. The standards are less stringent for gasoline trucks. No in-use compliance program is specified (ICCT, 2014).

Canada has adopted GHG standards nearly identical to those of the United States, beginning with 2014 models. Vehicles certified to U.S. standards are deemed in compliance with Canadian standards (Olivares and Wagner, 2014b).

India in February of 2016 proposed to adopt the EuroVI standards and procedures for criteria pollutants, with implementation beginning with 2020 models (Olivares et al., 2016). India has been evaluating the GHG and fuel consumption of heavy-duty trucks and is currently developing fuel efficiency standards to address the unique vehicles and driving conditions there (ICCT, 2014).

2.3.1.3 Voluntary Programs

The U.S. EPA initiated a public-private partnership in 2004 called SmartWay®, the objective of which is reducing GHG and pollutant emissions resulting from movement of freight across a firm's supply chain (EPA, 2017).[10] The program works with corporate partners and recognizes those fleet operators and logistics firms that make voluntary reductions in their trucking emissions (Berth, 2014). SmartWay partners agree to provide data on their operations, which are input into standardized tools to produce a measurement of environmental efficiency, such as grams pollutant per ton-mile. These benchmark results are compared to similar categories of freight movement, for example, dry vans, and ranked in quintiles. Fleet operators and shippers use the results to improve their efficiency, identify green options, and achieve recognition. SmartWay also includes a Technology Program that develops test procedures and verifies the performance of technologies, equipment, and strategies that save fuel and reduce emissions.

With the implementation of mandatory GHG emission and fuel economy standards, beginning with the 2014 models, many of the fuel-saving features for new tractors encouraged by the SmartWay program will be used on new tractors in order to meet government standards. This includes low-rolling-resistance tires and idle-reduction technologies. The government regulation effectively supplants this portion of the (voluntary) SmartWay program for new trucks. However, the trailer standards in the Phase II rule would not be fully implemented until 2027, and more importantly would affect only new trailers, and not those already in use. Thus SmartWay can continue to encourage improvements to existing tractor-trailer combinations.

2.3.2 Other Regulatory Programs That Indirectly Affect Fuel Consumption and GHG Emissions

Many other regulatory programs indirectly affect fuel consumption and GHG emissions of MHDVs and may intersect with or otherwise affect the NHTSA and EPA MHDV standards. Such programs include the regulations of tailpipe emissions from vehicles of both so-called criteria pollutants and of carbon dioxide, vehicle safety regulations, and fuel programs.

2.3.2.1 NO_x Emissions and Regulation

Efforts to improve fuel efficiency of diesel engines can be impacted by requirements for reducing emissions of the ozone precursor NO_x. The committee received testimony from CARB that the need for significant additional NO_x reductions in California requires an integrated approach in future GHG and NO_x emissions reduction regulation (Sax et al., 2014). Analyses by CARB and the South Coast Air Quality Management District indicate that the greater Los Angeles area needs additional NO_x emissions reduction by 2023 and 2031 to meet federal ozone ambient air quality standards. Figure 2-13 indicates that compared to 2023 NO_x emissions that reflect only current regulatory requirements, NO_x emissions must be reduced by another 50 percent in 2023 and by 65 percent before

[10] This program is discussed in more detail in the committee's Phase Two First Report (NRC, 2014).

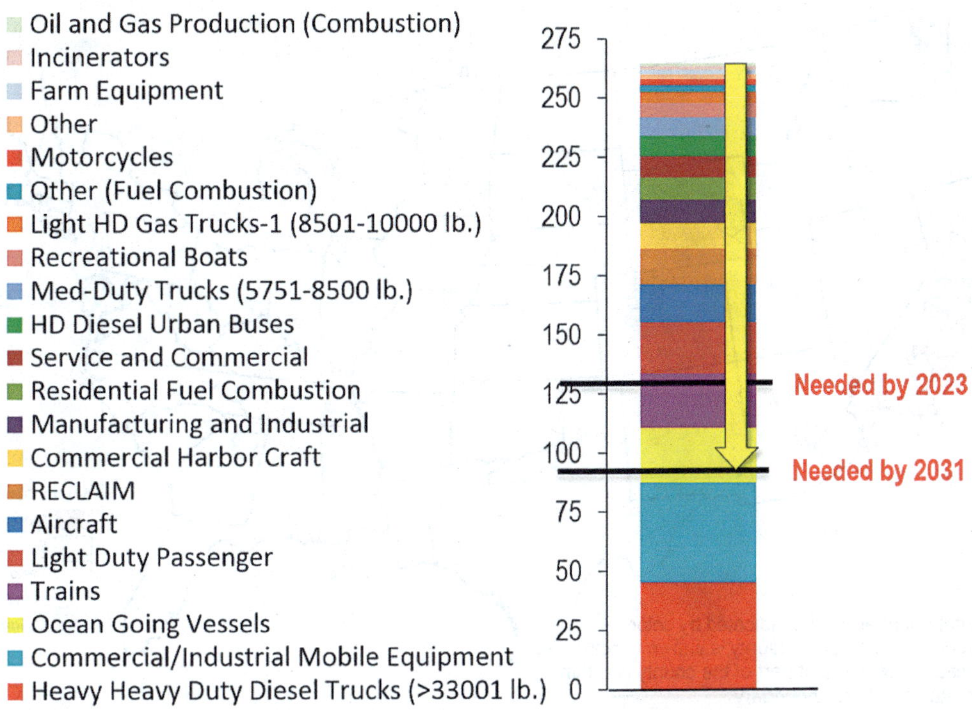

FIGURE 2-13 NO_x emission reductions by different sources needed in the greater Los Angeles area to meet the federally mandated ozone ambient air quality standard by the 2032 deadline, compared to required NO_x emissions in 2023 without new requirements.
SOURCE: SCAQMD (2016).

2032. Also shown in the figure is that on-road heavy-duty diesel trucks are the largest source of NO_x emissions. Thus it will be necessary to further reduce their emissions. Eleven state and regional air pollution agencies, most from outside of California, have petitioned EPA for tighter NO_x standards for MHD engines.[11] In December 2016, EPA responded to the petition indicating it would initiate a rulemaking to propose revisions to the on-road heavy-duty control program to reduce NO_x emissions, starting with the 2024 model year. Many regions across the country in addition to California are also facing challenges in complying with the national ambient air quality standard for ozone (see Figure 2-14). Additional reductions in NO_x emissions from MHDVs would likely be welcomed by many of these states to help them meet the ozone standard in their region.

CARB and the South Coast Air Quality Management District have begun the process of determining how to achieve the additional NO_x reductions needed from diesel trucks. The agencies are sponsoring studies of approaches to optimize selective catalytic reduction control efficiency, and to achieve lower emissions from natural gas vehicles. In the California Sustainable Freight Action Plan issued in July 2016, CARB laid out its time frame for adopting a more stringent NO_x standard for new MHDVs. CARB states it will work with and/or petition EPA to adopt a nationwide low-NO_x engine standard by 2019 to be implemented with the 2024 models. CARB has stated that, were EPA not to act, it would adopt a low-NO_x standard for new engines sold in California, beginning with 2023 models. While EPA did not propose or adopt a more stringent NO_x standard for new MHD engines, in the preamble to the 2016 Phase II rule EPA states it plans to engage with stakeholders to discuss opportunities to develop more stringent standards to reduce NO_x from heavy-duty on-highway engines through a coordinated effort with CARB. In addition, it should be noted that the need or desire of some states to achieve greater reductions in ground-level

[11] Petitioners represent four state agencies (CT, DE, NH, and WA) and seven local or regional agencies (South Coast [CA], Bay Area [CA], Akron [OH], New York City [NY], Pima County [AZ], Puget Sound [WA], and Washoe County [NV]).

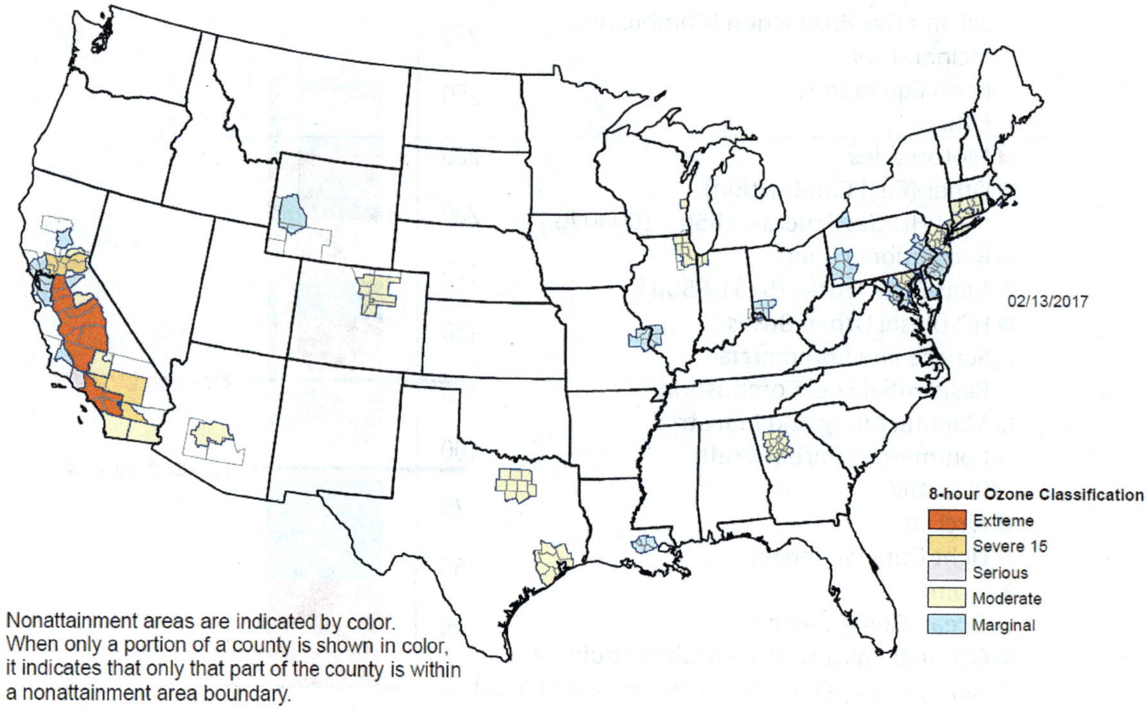

FIGURE 2-14 Ozone nonattainment areas across the continental United States.
SOURCE: EPA (2016).

ozone precursors or fine particulates may be the driver for those states to opt into the California program, which would also include the California GHG emission standards.

Finding: CARB has made a finding that additional reduction of NO_x emissions is needed to attain the federal ambient air quality standard for ozone in California. Other state and local air quality agencies from several parts of the country have also indicated a desire for additional NO_x reductions. In California, emissions of NO_x are primarily from mobile sources, and on-road heavy-duty diesels are one of the largest sources. These factors suggest that the current NO_x standard for MHD engines may be tightened either by California or on a national level.

Recommendation 2-9: Because reductions of NO_x may increase fuel consumption and emissions of CO_2, it is important that EPA and NHTSA thoroughly evaluate the effect of current and emerging technologies on this relationship, and determine an optimum approach for addressing this issue in time for consideration as part of a third phase of regulatory development.

2.3.2.2 Climate Change Regulation

President Barack Obama issued a Climate Action Plan in June 2013 (White House, 2013). In addition to the adoption of GHG emission standards for new passenger vehicles and MHDVs, the plan directed EPA to issue GHG emission limits for new and existing power plants, which were finalized in August 2015. The Climate Action Plan sets a goal of reducing annual U.S. GHG emissions in the range of 17 percent below 2005 levels by 2020. In a subsequent agreement with China, a goal of reducing by 2025 U.S. GHG emissions by 26 to 28 percent below 2005 levels was announced. This goal was submitted as the United States' intended Nationally Determined Contribution as part of the 21st Conference of Parties under the United Nations Framework Convention on Climate

Change held in Paris in December 2015 (White House, 2015). On June 1, 2017, President Trump announced his intention to withdraw the United States from the Paris Agreement, and the new administration has not yet set any specific goals for GHG emission reductions.

At the state level, California requires GHG emissions to be reduced to 1990 levels by 2020, and 40 percent below this level by 2030. This is being accomplished by source-specific regulatory action, such as the GHG standards for new cars and trucks, renewable fuel requirements for in-state power plants, the Low Carbon Fuel Standard for transportation fuels, GHG reduction targets to be achieved by local governments through improved planning, recovery of hydrofluorocarbons, and a cap-and-trade program which generates funds that are used to achieve reductions in GHG emissions. Relevant to this report, cap-and-trade funds of about $150 million yearly are being allocated to support commercialization of near-zero-emission MHDVs. The results from these pilot programs will demonstrate if new technologies can be commercially viable under normal use and, if they are, can provide data valuable to the development of a third phase of MHDV regulations.

Many city and county governments have also taken actions to reduce GHG emissions. For example, the C40 Cities Climate Leadership Group, with over 70 megacity members, has committed to over 8,000 actions with the goal of reducing more than one gigaton of CO_2 emissions by 2020.

Finding: A growing number of actions, commitments, and goals of national, regional, and local governments are supporting the reduction of GHG emissions from energy use including MHDVs.

2.3.2.3 Natural Gas and Hydraulic Fracturing Regulation

The rapid increase in hydraulic fracturing in the United States over the past decade has greatly expanded the availability, and reduced the cost, of natural gas produced from shale rock, potentially increasing the use of natural gas as a fuel in the MHD trucking sector. The economic feasibility of natural gas production through hydraulic fracturing is sensitive to fuel prices, and the recent decline in oil prices has suppressed fracturing activity, but this form of energy production is capable of rapidly expanding again as fuel prices increase (Mills, 2016). Federal and state regulation of hydraulic fracturing can also limit activity in this area. Regulation can focus on a variety of different elements, including limits on discharges to groundwater, surface waters, and air, restrictions on the chemicals used in the fracturing process, requirements to disclose information, or outright prohibitions (Richardson et al., 2013). In addition to environmental concerns, some jurisdictions are also considering restrictions to protect against the possibility that hydraulic fracturing may contribute to seismic activity. Finally, a number of lawsuits have been filed against companies engaged in hydraulic fracturing seeking monetary damages for alleged personal injury and property damages as well as injunctive relief (Hall, 2012). Notwithstanding such legal risks and economic costs, hydraulic fracturing will continue to play an important role in U.S. energy production, and hence MHDV fuel options and costs, for many decades to come.

2.3.2.4 Infrastructure and Road Regulation

Infrastructure and road regulation pertaining to trucks is governed by the Federal Highway Administration, by individual state governments, and by municipalities in the form of size and weight regulations, seasonal axle weight limits, road classification, lane limits, high occupancy vehicle and truck lanes, traffic calming, weigh in motion, and road pricing. In addition, in August 2016, NHTSA proposed to require all newly manufactured U.S. trucks, buses, and other vehicles with a gross vehicle weight rating more than 26,000 pounds to come equipped with speed limiting devices. These regulatory programs and their potential for reducing MHDV fuel consumption and GHG emissions are discussed in greater detail in Chapter 9.

2.3.2.5 Safety Regulations

Commercial vehicle safety regulations at the federal level are governed by two separate agencies. NHTSA controls regulations governing "new vehicles." These regulations are generally performance based and deal with occupant safety, vehicle component performance such as brakes and rear underride systems, and conspicuity. The FMCSA safety regulations focus on vehicle operations. They cover, develop, and enforce data-driven regulations that balance motor carrier (truck and bus companies) safety with efficiency; harness safety information systems to focus on higher-risk carriers in enforcing the safety regulations; target educational messages to carriers, commercial drivers, and the public; and partner with stakeholders including federal, state, and local enforcement agencies, the motor carrier industry, safety groups, and organized labor on efforts to reduce bus- and truck-related crashes.

As discussed in greater detail in Chapter 9, some fuel efficiency technologies may create safety trade-offs, although these trade-offs are not as obvious and controversial as have been the case for LDV fuel economy standards (NRC, 2002). Some fuel efficiency measures, such as lower speed levels, may have beneficial safety implications for MHDVs. Other technologies, such as alternative fuels and low-rolling-resistance tires, could potentially have negative safety trade-offs. NHTSA has comprehensively examined these positive and negative safety trade-offs with fuel efficiency regulations in a report produced as part of the Phase II regulatory proceeding (NHTSA, 2015).

2.3.2.6 Fuel Regulations and Government Programs

EPA has adopted a renewable fuel standard for transportation which is being phased in through 2022. The standard requires specified volumes of renewable transportation fuel, principally ethanol and biodiesel, in four categories that have varying lower life-cycle GHG emission footprints compared to petroleum fuels. The 2013 requirement for biomass-based diesel, one of the four categories, is 1.13 percent of total diesel fuel volume (1.28 billion gallons), and this fuel must have a carbon footprint at least 50 percent lower than petroleum-derived diesel. In 2014, 1.63 billion gallons of biomass-derived diesel were produced. On November 30, 2015, EPA adopted a requirement that 2.0 billion gallons of biomass-based diesel be produced in 2017 (EPA, 2015).

California adopted in 2009 a Low Carbon Fuel Standard for transportation fuels. The regulation requires that each producer or importer of gasoline or diesel fuel for use in California gradually reduce the life-cycle GHG emissions of its products by 10 percent by 2020. Compliance can be achieved by directly reducing the fuel's carbon intensity, or by buying credits generated by third-party producers or importers of lower-carbon fuels. Credits generated to date are 60 percent from ethanol, 28 percent from biodiesel, 10 percent from natural gas (28 percent less CO_2 per unit enthalpy than diesel), and 2 percent from electricity, and the available credits exceed the current year regulatory requirement by about 65 percent. Almost 170 biofuel facilities have registered to provide low-carbon fuels to California (CARB, 2014). The state has invested \$60 million in supporting the development of advanced biofuel production facilities. Beginning in January 2015, transportation fuels have been added to California's cap-and-trade program.

These government programs are encouraging, incentivizing, and in some cases requiring the use of cleaner substitutes for petroleum diesel fuels, and a nascent market for their use is developing. Alternative fuels that can be used in diesel-like engines, but require engine modification and new fueling infrastructure, are also being evaluated for commercial viability.

Another approach to reducing GHG emissions from MHDVs is to produce a fuel from renewable and other sources such that the fuel has a much smaller well-to-tank carbon footprint. For example, biomethane can be produced from waste products. Similarly, dimethyl ether (DME) can be efficiently produced from waste products and combusted in a diesel cycle engine with a very low net production of CO_2 relative to standard diesel fuels (Semelsberger et al., 2006).

Synthetic diesel fuel can also be made from hydrocarbon sources such as natural gas, or coal, which would reduce conventional diesel consumption while avoiding the need for engine modifications and new fuel infrastructure. This approach can reduce our dependence on petroleum, but even with carbon capture and storage, well-to-wheels GHG emissions from indirect liquefaction of coal are roughly the same as that of petroleum fuels from conventional reservoirs due to high energy use and resulting GHG emissions during fuel production (NRC, 2013). Conversion of natural gas to petroleum-like liquid fuel, without carbon capture and sequestration, also results in GHG emissions similar to conventional petroleum fuels. Vocational trucks and buses can be powered by electricity or hydrogen fuel cells with much lower CO_2 emissions, depending on how the electricity or hydrogen is produced.

Finding: Improvements in the efficiency of diesel and gasoline engines, trucks, and tractor-trailers have been the focus of regulatory efforts to reduce fuel use and GHG emissions. However, if the practical application of these approaches is inadequate to achieve desired GHG reductions, other alternative fuels such as natural gas, propane, DME, biomethane, and renewable low-carbon diesel may offer opportunities for further reductions in GHG emissions. Battery electric and fuel cell electric powertrains could also provide additional GHG reduction in urban trucks.

2.4 REFERENCES

AFDC (Alternative Fuels Data Center). 2012. Types of Vehicles by Weight Class. Department of Energy (DOE). Available at http://www.afdc.energy.gov/data/10381 (accessed December 1, 2019).

Basulto, D. 2014. How 3D printing could transform Amazon and online shopping. *Washington Post*, March 13.

Berth, D. 2014. EPA recognizes logistics firms for reducing emissions. *Transport Topics* September 23.

Brandt, A.R., G.A. Heath, E.A. Kort, F. O'Sullivan, G. Pétron, S.M. Jordaan, P. Tans, et al. 2014. Methane leaks from North American natural gas systems. *Science* 343(6172):733-735.

CARB (California Air Resources Board). 2014. Low Carbon Fuel Standard Re-Adoption Concept Paper. Available at http://www.arb.ca.gov/fuels/lcfs/lcfs_meetings/030714lcfsconceptpaper.pdf (accessed December 16, 2019).

Clarke, L.E., A.A. Fawcett, J.P. Weyant, J. McFarland. V. Chaturvedi, Y. Zhou. 2014. Technology and emissions reductions goals: Results of the EMF 24 modeling exercise. *Energy Journal* 35(Special Issue 1):9-31.

Diamond, D. 2009. The impact of government incentives for hybrid-electric vehicles: Evidence from US states. *Energy Policy* 37(3):972-983.

EIA (U.S. Energy Information Administration). 2015. Annual Energy Outlook 2015 (AEO2015). Washington, DC. DOE/EIA-0383(2015).

EPA (Environmental Protection Agency). 2015. Renewable fuel standard program: Standards for 2014, 2015, and 2016 and biomass-based diesel volume for 2017. *Federal Register* 80(239):77420 (December 14).

EPA. 2016. 8-Hour Ozone Nonattainment Areas (2008 Standard). Retrieved from https://www3.epa.gov/airquality/greenbook/map8hr_2008.html (accessed September 20, 2019).

EPA. 2017. About SmartWay. Available at http://www.epa.gov/smartway/about/index.htm (accessed August 11, 2017).

EPA. Undated. Refrigerant Transition & Environmental Impacts. Available at https://www.epa.gov/mvac/refrigerant-transition-environmental-impacts (accessed April 5, 2017).

EPA and NHTSA (National Highway Traffic Safety Administration). 2011. Greenhouse gas emissions standards and fuel efficiency standards for medium- and heavy-duty engines and vehicles: Final rule. *Federal Register* 76(179):57106-57513 (September 15).

EPA and NHTSA. 2016. Greenhouse gas emissions and fuel efficiency standards for medium- and heavy-duty engines and vehicles—Phase 2. *Federal Register* 81:73478-74274 (October 25).

European Commission. 2008. Communication from the Commission: Energy efficiency: Delivering the 20% target (COM/2008/0772). November 13.

European Commission. 2016. Communication from the Commission to the European Parliament, the Council, the European Economic and Social Committee and the Committee of the Regions: A European Strategy for Low Emission Mobility. Available at http://ec.europa.eu/transport/themes/strategies/news/doc/2016-07-20-decarbonisation/com(2016)501_en.pdf (accessed December 1, 2019).

FHWA (Federal Highway Administration). 2015. Highway Safety and Truck Crash Comparative Analysis Report. U.S. DOT (prepublication version).

Fulton, L., and M. Miller. 2015. Strategies for Transition to Low-Carbon Emission Trucks in the United States. White Paper from the Sustainable Transportation Energy Pathways Program at University of California, Davis, and the National Center for Sustainable Transportation. June.

GEA (Global Energy Assessment). 2012. *Global Energy Assessment—Toward a Sustainable Future*. Cambridge, UK: Cambridge University Press; Laxenburg, Austria: International Institute for Applied Systems Analysis.

Hall, K.B. 2012. Hydraulic fracturing litigation. *For the Defense*, January, p. 66.

Harford, T. 2013. How delivery drones could transform the world. *Financial Times*, December 5.

ICCT (International Council on Clean Transportation). 2014. Final Phase Two China Fuel Consumption Standard for Heavy-Duty Motor Vehicles (May 2014). Available at http://www.theicct.org/sites/default/files/publications/ICCTupdate_China-Phase2_june2014.pdf (accessed December 16, 2019).

IEA (International Energy Agency). 2010. World Energy Outlook 2010. Paris: Organisation for Economic Cooperation and Development.

IEA. 2015. *World Energy Outlook Special Report on Energy and Climate*. Paris: Organisation for Economic Co-operation and Development.

IPCC (Intergovernmental Panel on Climate Change). 2014. *Climate Change 2014: Synthesis Report*. Contribution of Working Groups I, II and III to the Fifth Assessment Report of the Intergovernmental Panel on Climate Change, Core Writing Team, R.K. Pachauri and L.A. Meyer (eds.). Geneva: IPCC, 151 pp.

Jaffe, A.B., R.G. Newell, and R.N. Stavins. 2005. A tale of two market failures: Technology and environmental policy. *Ecological Economics*, *54*(2-3):164-174.

Marchant, G.E. 2014. Complexity and anticipatory socio-behavioral assessment of government attempts to induce clean technologies. *UCLA Law Review* 61:1858-1894.

Mills, M.P. 2016. After the carnage, shale will arise again. *Wall Street Journal*, January 18.

Morrow, W.R., K.S. Gallagher, G. Collantes, and H. Lee. 2010. Analysis of policies to reduce oil consumption and greenhouse-gas emissions from the US transportation sector. *Energy Policy* 38(3):1305-1320.

Nahlik, M.J., A.T. Kaehr, M.V. Chester, A. Horvath, and M.N. Taptich. 2015. Goods movement life cycle assessment for greenhouse gas reduction goals. *Journal of Industrial Ecology* 20(2):317-328.

Nelson, R.R., and R.N. Langlois. 1983. Industrial innovation policy: Lessons from American history. *Science* 219(4586):814-818.

NHTSA. 2015. *NHTSA Study: Review and Analysis of Potential Safety Impacts and Regulatory Barriers to Fuel Efficiency Technologies and Alternative Fuels in Medium- and Heavy-Duty Vehicles*. Washington, DC: NHTSA.

Norberg-Bohm, V. 1999. Stimulating "green" technological innovation: An analysis of alternative policy mechanisms. *Policy Sciences* 32(1):13-38.

NPC (National Petroleum Council). 2012. Advancing Technologies for America's Transportation Future. Washington, DC: U.S. Department of Energy.

NRC (National Research Council). 2001. *Energy Research at DOE: Was It Worth It? Energy Efficiency and Fossil Energy Research 1978 to 2000*. Washington, DC: National Academy Press.

NRC. 2002. *Effectiveness and Impact of Corporate Average Fuel Economy Standards*. Washington, DC: National Academy Press.

NRC. 2007. *Prospective Evaluation of Applied Energy Research and Development at DOE (Phase Two)*. Washington, DC: The National Academies Press.

NRC. 2009. *America's Energy Future: Technology and Transformation*. Washington, DC: The National Academies Press.

NRC. 2010. *Technologies and Approaches to Reducing the Fuel Consumption of Medium- and Heavy-Duty Vehicles*. Washington, DC: The National Academy Press.

NRC. 2013. *Transitions to Alternative Vehicles and Fuels*. Washington DC: The National Academies Press.

NRC. 2014. *Reducing the Fuel Consumption and Greenhouse Gas Emissions of Medium- and Heavy-Duty Vehicles, Phase 2: First Report*. Washington DC: The National Academies Press.

NRC. 2015. *Cost, Effectiveness and Deployment of Fuel Economy Technologies for Light-Duty Vehicles*. Washington DC: The National Academies Press.

Olivares, E., and V. Wagner. 2014a. Japan—Heavy-Duty: Emissions (last updated 2014). Available at http://www.transportpolicy.net/index.php?title=Japan:_Heavy-duty:_Emissions (accessed September 23, 2019).

Olivares, E., and V. Wagner. 2014b. Canada: Heavy-Duty: GHG (last updated 2014). Available at http://transportpolicy.net/index.php?title=Canada:_Heavy-duty:_GHG (accessed September 23, 2019).

Olivares, E., V. Wagner, M. Morgan, and E. Foukes. 2016. India: Heavy-Duty: Emissions (last updated 2016). Available at http://transportpolicy.net/index.php?title=India:_Heavy-duty:_Emissions (accessed September 23, 2019).

Richardson, N., M. Gottlieb, A. Krupnick, and H. Wiseman. 2013. The State of State Shale Gas Regulation. Resources for the Future. May. Available at http://www.rff.org/files/sharepoint/WorkImages/Download/RFF-Rpt-StateofStateRegs_Report.pdf (accessed December 16, 2019).

Roberts, D. 2016. 1.8 million American truck drivers could lose their jobs to robots. What then? *Vox*. August 3.

Sax, T., S. Yoon, C. Misra, J. Collins, J. Herner, D. Hawelti, S. Lemieux, et al., California Environmental Protection Agency Air Resources Board. 2014. An Integrated Approach to Heavy Duty GHG and NO_x Control. Testimony presentation to the National Research Council's Committee on Committee on Assessment of Technologies and Approaches for Reducing the Fuel Consumption of Medium- and Heavy-Duty Vehicles, Phase Two, Ann Arbor, MI, July 29.

SCAQMD (South Coast Air Quality Management District). 2016. Petition to EPA for Rulemaking to Adopt Ultra-Low NO_x Exhaust Emission Standards for On-Road Heavy-Duty Trucks and Engines. Available at http://www.aqmd.gov/docs/default-source/default-document-library/news-docs/nox-petition-to-epa-june-2016.pdf?Status=Temp&sfvrsn=2 (accessed September 23, 2019).

Semelsberger, T.A., R.L. Borup, and H.L. Greene. 2006. Dimethyl ether (DME) as an alternative fuel. *Journal of Power Sources* 156(2):497-511.

Sharpe, B. 2013. Examining the Costs and Benefits of Technology Pathways for Reducing Fuel Use and Emissions from On-Road Heavy-Duty Vehicles in California. Report No. UCD-ITS-RR—3-17. Davis, CA: University of California, Davis.

Tamayao, M., J. Michalek, C. Hendrickson, and I.L. Azevedo. 2015. Regional variability and uncertainty of electric vehicle life cycle CO_2 emissions across the United States. *Environmental Science & Technology* 49(14):8844–8855.

Taptich, M.N., A. Horvath, and M.V. Chester. 2015. Worldwide greenhouse gas reduction potentials in transportation by 2050. *Journal of Industrial Ecology* 20:329-340.

Tong, F., I.L. Azevedo, and P. Jaramillo. 2015a. A comparison of life cycle greenhouse gases from natural gas pathways for light-duty vehicles. *Energy and Fuels* 29(9):6008-6018.

Tong, F., I.L. Azevedo, and P. Jaramillo. 2015b. A comparison of life cycle greenhouse gases from natural gas pathways for medium and heavy-duty vehicles. *Environmental Science & Technology* 49:7123-7133.

Transportation Research Board. 2011. *Policy Options for Reducing Energy Use and Greenhouse Gas Emissions from U.S. Transportation: Special Report 307*. Washington, DC: The National Academies Press. https://doi.org/10.17226/13194.

Vance, A. 2010. 3-D printing spurs a manufacturing revolution. *The New York Times*, September 13.

White House. 2013. The President's Climate Action (June). Available at https://obamawhitehouse.archives.gov/sites/default/files/image/president27sclimateactionplan.pdf (accessed December 16, 2019).

White House. 2015. U.S. Cover Note, INDC and Accompanying Information, available at https://www4.unfccc.int/sites/submissions/INDC/Published%20Documents/United%20States%20of%20America/1/U.S.%20Cover%20Note%20INDC%20and%20Accompanying%20Information.pdf (accessed March 31, 2015).

Williams, J.H., B. Haley, F. Kahrl, J. Moore, A.D. Jones, M.S. Torn, and H. McJeon. 2014. *Pathways to deep decarbonization in the United States: The U.S. report of the Deep Decarbonization Pathways Project of the Sustainable Development Solutions Network and the Institute for Sustainable Development and International Relations.* San Francisco, CA: Energy and Environmental Economics, Inc.

Woodrooffe, J. 2016. Opportunity cost for society related to US truck size and weight regulation: Freight efficiency. *Transportation Research Record: Journal of the Transportation Research Board* 2547:25-31.

Woodrooffe, J., B.M. Belzowski, J. Reece, and P.F. Sweatman. 2009. Analysis of the Potential Benefits of Larger Trucks for U.S. Businesses Operating Private Fleets. Prepared for the National Private Truck Council. University of Michigan Transportation Research Institute. May.

Yang, C., D. McCollum, R. McCarthy, and W. Leighty. 2009. Meeting an 80% reduction in greenhouse gas emissions from transportation by 2050: A case study in California. *Transportation Research Part D: Transport and Environment* 14(3):147-156.

Yuksel, T., M-A. Tamayao, C. Hendrickson, I.L. Azevedo, and J. Michalek. 2016. Effect of regional grid mix, driving patterns and climate on the comparative carbon footprint of gasoline and plug-in electric vehicles in the United States. *Environmental Research Letters* 11(4):044007.

3

Certification, Compliance, and Enforcement

3.1 INTRODUCTION

As with any regulatory program for mobile sources, the issues of certification, compliance, and enforcement are critical to the success of the program. For medium- and heavy-duty vehicles (MHDVs), designing and implementing effective certification, compliance, and enforcement is particularly challenging given the nature of this sector, including the multiple components of a truck that could be subject to certification, compliance, and enforcement measures (e.g., whole vehicle, engine, powertrain, trailer, etc.), the diversity of potential testing regimes (engine certification, whole vehicle simulation, in-use enforcement, etc.), varying duty cycles, and the variety and heterogeneity of the vehicles subject to any regime.

The committee and its predecessor (the "NRC Phase One" committee) have provided extensive comments on certification, compliance, and enforcement in their first two reports (NRC, 2010, 2014). This chapter expands on those previous findings and recommendations by addressing the following topics. Section 3.2 of this chapter provides an overview of the certification, compliance, and enforcement approaches adopted by the National Highway Traffic Safety Administration (NHTSA) and the Environmental Protection Agency (EPA) in the Phase I standards and Phase II standards. Section 3.3 provides our findings and recommendations with respect to improvements and challenges with the current pre-sale certification and compliance models and tools used by NHTSA and EPA in their rulemakings. Section 3.4 addresses in-use compliance and enforcement.

3.2 SUMMARY OF CERTIFICATION APPROACHES IN PHASE I AND PHASE II RULES

Perhaps the most anticipated aspects of the Greenhouse Gas Emissions and Fuel Efficiency Standards for Medium- and Heavy-Duty Engines and Vehicles—Phase II rules (EPA and NHTSA, 2016) are the changes to the certification structure. Considerable changes have been made in the Phase II Greenhouse Gas Emissions Model (GEM) for heavy-duty vehicles. In general, the Phase II GEM features more manufacturer-specified inputs and fewer default inputs compared to Phase I GEM, and more real-world route choices versus previous standardized routes. An engine test remains an element of the certification of the vehicles, and the fuel maps used in the engine certification can be directly used by the manufacturer for use in GEM modeling. This section summarizes the major changes made to the certification regulations in the Phase II rule for MHDVs. The section is divided by regulated classes: first, heavy-duty vehicles; second, vocational vehicles; and third, trailers for heavy-duty tractors.

3.2.1 Heavy-Duty Combination Tractors

Certification in the Phase I rule was intentionally designed to be readily implementable. NHTSA and EPA wanted to quickly achieve a means to reduce greenhouse gas (GHG) emissions and improve vehicle efficiency without requiring the regulated manufacturers to invest sizable capital for test equipment and new technologies which were not currently available. The certification structure for Phase I has two major components. One is an engine dynamometer certification test to measure CO_2, CH_4, and N_2O emissions in grams per brake-horsepower-hour, which is aligned with what has been used for criteria pollutants for many years. (Emissions of CH_4 and N_2O are relatively low compared to CO_2; however, they are measured because on an equal-mass basis they have global warming potentials that are 28-36 and 265-280 times greater than CO_2, respectively [EPA, 2017]). The other is the use of the GEM, which simulates vehicle fuel efficiency and CO_2 emissions over three distinct driving cycles for prescribed vehicle configurations and weights, resulting in a single output expressed in fuel consumed in gallons per 1,000 ton-miles and an equivalent CO_2 output expressed as grams per ton-mile. This certification strategy was expedient at the time, but it is representative of vehicle performance over only certain routes and confines the inputs for the simulation model to minimize complexity—and therefore precision—of the measured emissions and efficiency of the certified vehicles.

After the promulgation of the Phase I rules, the comments from industry and recommendations in the NRC Phase One Report (NRC, 2010) encouraged NHTSA and EPA to improve the representativeness of GEM in the ensuing rulemaking. The committee in its GEM analysis encouraged these agencies to include more route choices. It argued in favor of more manufacturer-specific inputs in lieu of the default attributes for the vehicle in GEM and made specific recommendations for refinements to the GEM inputs.

Some original equipment manufacturer (OEM) presenters to the committee recommended that EPA consider eliminating the engine test aspect of the certification strategy and rely only on the full vehicle simulation, which would include CO_2 results from the use of a specified engine map. In general, the Phase II certification approach has not deviated from the original Phase I test strategy: the engine test for CO_2 emissions remains, as does the GEM simulation for the vehicle. The certification structure for the vehicle (GEM), however, has been modified considerably to address changes in expected available technology and improvements to the robustness and fidelity of the tests and simulation model.

NHTSA and EPA propose numerous changes in response to the recommendations made by the Committee in its first report (2014). Those recommendations are summarized below.

- Recommendation 3.2: GEM 3 should be compatible with manufacturer's order entry system.
- Recommendation 3.3: Output from GEM 3 should include graphs and other presentation methods to help manufacturers gain greater insight into actions to improve GHG emissions and fuel efficiency.
- Recommendation 3.4: Develop a mechanism to determine accurate tire rolling resistance, and store it for public availability.
- Recommendation 3.5: Improve methods for determining aerodynamic performance to be more representative of real-world experience.
- Recommendation 3.6: Configure test cycles and routes to ensure that products are not optimized for the test, but optimized for real-world fuel economy and CO_2 reduction.
- Recommendation 3.7: Consider allowing OEMs to replace fixed code with their own validated models of components or systems which represent real-world operation.
- Recommendation 3.8: Determine whether the steady-state speed-torque test is sufficient to capture all efficiency measures taken (e.g., downsizing engine is an approach that GEM would not properly address).
- Recommendation 3.9: Measured values for vehicle weights should replace fixed values.
- Recommendation 3.10: GEM should accommodate the power-to-weight ratio of a vehicle if GEM permits the benefit accrued from integrating the powertrain and aftertreatment systems with interactive controls.
- Recommendation 3.11: Revise cycles and schedules to replicate over-the-road activities without becoming cumbersome.
- Recommendation 3.12: Determine whether GEM can be used to derive the fuel economy benefits from all trailers in combination with tractors.

NHTSA and EPA addressed all the recommendations made by the committee in its 2014 report; however, with regard to Recommendation 3.12, the agencies were unable to justify the regulation of all trailers (see the subsection "Trailers" below).

Another important change in the Phase II rule is the application of useful life. EPA has always stipulated that engines or vehicles certified for criteria pollutant emissions would meet the standard through a "useful life" period, and that the technology required to meet those certified values cannot be disassembled from the certified engine configuration during the useful life period. The Phase I fuel economy regulations promulgated by NHTSA (EPA and NHTSA, 2011) did not include useful life requirements. EPA applied the same useful life periods for other criteria pollutant emissions to the Phase I CO_2 standard for MHDVs. In the Phase II regulations, NHTSA incorporates the EPA useful life periods, and EPA has changed the useful life of the light heavy-duty engine category from 110,000 miles to 150,000 miles to reflect today's experience while keeping useful life numbers for the other categories the same as in Phase I. The useful life compliance period for trailers is 10 years (see Table 3-1).

In addition to useful life compliance, the Phase II regulations include a production audit of five vehicles per manufacturer per year. These vehicles will be tested on a chassis dynamometer. NHTSA and EPA concede this is not a confirmatory test to certify results; however, these results can be used to refine a production audit process in the future.

As a final enforcement practice, EPA states in the Phase II rule that it will continue to conduct selective enforcement audits with the manufacturers and evaluate in-use compliance by testing Phase II in-use engines for CO_2 compliance. More detailed analyses of the GEM refinements for vehicle certification are included in other sections of this report.

3.2.2 Vocational Vehicles

The manufacture of the vocational vehicle is generally shared between the chassis manufacturer, which produces an incomplete or stripped chassis, and the final stage manufacturer, which assembles the body of the vehicle that accommodates the load and work-specific equipment. In some cases, the entire vehicle is built by a specialty manufacturer. The responsibility for certifying these vehicles falls to the chassis manufacturer in both Phase I and Phase II regulations. In Phase I, NHTSA and EPA considered only the vehicle weight class in grouping vehicles for certification: light heavy-duty (Class 2b through Class 5), medium heavy-duty (Class 6 and Class 7), and heavy heavy-duty (Class 8).

The NRC Phase One Report expressed concerns that test cycles and standards should reflect the performance of vocational vehicles over varied duty cycles. In response to this concern, NHTSA and EPA have widened the categories under which vocational vehicles are certified. In the Phase II rule, the three vocational vehicle weight classes are further divided into distinct categories reflecting driving patterns: regional, multipurpose, and urban. The CO_2 certification test cycle weighting factors are adjusted to reflect these driving cycles. Certain exemptions for emergency vehicles will apply.

TABLE 3-1 Phase II Useful Life Compliance Periods by Vehicle Class and Trailers

Light heavy-duty	150,000 miles
Medium heavy-duty	185,000 miles
Heavy heavy-duty	435,000 miles
Trailers	10 years

SOURCE: EPA and NHTSA (2016, pp. 74256, 74262).

3.2.3 Trailers

The Phase I rule did not regulate trailers. In its Phase Two First Report, the committee stated clearly its belief that trailers could and should be regulated. It pointed to the success of the EPA voluntary SmartWay program as a sound foundation for efficiently enforcing a Phase II performance standard nationwide which could result in anywhere from 3 to 11 percent improvement in fuel economy from aerodynamic treatment to the trailer (NRC, 2010).

The committee's Phase Two First Report also made recommendations regarding trailers for the agency to consider in its Phase II rulemaking. These recommendations have generally been included in the Phase II rule:

- Recommendation 6.1: Require all new 53-foot and longer dry and refrigerated box vans to meet performance standards that will reduce fuel consumption and CO_2 emissions.
- Recommendation 6.2: Consider including other trailers such as pups, flatbeds, and container vans if practicable.
- Recommendation 6.3: Define an optimum full vehicle test and validate it against real-world experiences. NHTSA should assess if adding yaw loads would provide significantly increased value to the calculated C_d of the vehicle. NHTSA should also compile test data and fuel economy results and make it available to the public.
- Recommendation 6.4: NHTSA should prepare a white paper clarifying the minimum performance needed from a safety perspective.

For most of the dry box, refrigerator box trailers, and pup vans between 28 and 53 feet the agency was able to justify applying performance standards in the Phase II rule. Technologies to be applied to achieve these standards include a variety of aerodynamic features and low-rolling-resistance tires, primarily carried over from EPA's SmartWay program, and the California Air Resources Board's Greenhouse Gas Reduction Regulation. The advantage these programs provide is that these features have already been road tested and verified by the SmartWay Technology Program.

For some box trailers with field equipment prohibiting aerodynamic treatment to either end of the trailer, referred to in a category called non-aero box vans, NHTSA and EPA have concluded that the low-rolling-resistance tires and the use of automatic tire inflation (ATI) will be applicable. A list of heavy hauler tractor-trailers and specialty field trailers are exempt from the trailer standards.

Table 3-2 provides EPA's summary of the major technical input changes which have been incorporated in GEM from the Phase I rule to the Phase II rule.

3.3 PRE-MARKET CERTIFICATION

3.3.1 Introduction

Pre-market certification has been favored for decades in the automotive world and may be directly based on test results (crash-worthiness, criteria pollutant emissions, ad noise) or through component specifications (lighting and braking systems), which in turn requires compliant component design or verification testing. Usually there is an assumption that the test or specification addresses real-world use sufficiently well that no subsequent compliance procedure is needed. Vehicle recalls may arise when premature component failures or a specific design flaw or malfunction exceed a threshold for concern, but these are not part of an organized after-sales measurement campaign. In-use enforcement of vehicle performance is complex because both the manufacturer and the owner may be implicated in deviations from an expected norm.

Several examples of in-use compliance exist. As one example, there is a program in the United States that examines in-use heavy-duty vehicle NO_x and particulate matter (PM) emissions using on-board equipment, which originated following the realization that certain diesel engine manufacturers controlled the engines so that brake-specific NO_x was substantially higher under steady load over the road than during the transient FTP pre-market certification. As another example, dynamometer-based inspection and maintenance programs were used to control

TABLE 3-2 GEM Input Comparisons by Category—Phase I and Phase II

Category	GEM for Phase I	GEM for Phase II
Cycles	Time-based cycles California Air Resources Board (CARB) Transient. 55 mph cruise. 65 mph cruise. EPA Federal Test Procedure (FTP) – Vocational.	Distance-based cycles. CARB Transient – steady-state and transient cycle average fuel map required. 55 mph cruise with road grade and flat road effects; steady-state and transient cycle average fuel map optional. 65 mph cruise with road grade and flat road effects; steady-state and transient cycle average fuel map optional. EPA FTP – Vocational with adjusted composite weighting.
Engine	EPA default steady-state fuel map.	Manufacturer provided steady-state fuel map. Additional idle points. Optional powertrain test map instead of fuel map.
Transmission	EPA default gears and shift strategy.	Manufacturer-specified number of gears and gear ratios. EPA default shift strategy. Optional powertrain test.
Axle	EPA default.	Manufacturer specified axle ratio. Optional manufacturer axle efficiency table.
Vehicle Mass	EPA default curb weight and payload. Curb weight reduction and payload increase for material substitution in tractors, component tables for high-strength steel and aluminum.	EPA default curb weight and payload. Curb weight reduction and payload increase for material substitution in tractors and vocational. Added component table for thermoplastics.
Aerodynamics	C_d bins based on measured C_dA.	C_dA bins based on measured C_dA with adjusted boundary limits. Manufacturer required minimum coast-down testing validation.
Tires	Manufacturer-specified C_{rr}.	Manufacturer-specified C_{rr}.
Trailer (box and partial box)	EPA non-aerodynamic trailer default.	Trailer includes skirt, low-rolling-resistance tires and automatic tire inflation (ATI) systems. Revised test procedure to account for average wind yaw.
Trailer (non-aero box and non-box)	Not included.	Trailer includes low-rolling-resistance tires and ATI system.
Yes/No technologies ("pulldown")	Vehicle speed limiter (tamper-proof). Hotel idle reduction for tractors (tamper-proof).	Vehicle speed limiter (tamper-proof). Hotel idle reduction for tractors (tamper-proof). Workday idle for vocational (tamper-proof). 4x2, 6x2 rear axles (including switchable 6x4 to 6x2 axles). Low-friction axle lubes. Auto tire inflation. Auto/Automatic Manual/Dual Clutch transmission for tractors. Predictive cruise control for tractors. High-efficiency AC for tractors. Electric power steering or cooling pumps.

SOURCE: Derived from email to committee member by EPA.

in-use automobile emissions in air districts that exceeded national ambient air quality standards. These have been replaced in many cases by reliance on the vehicle on-board diagnostics, a practice which is shown to have flaws. Safety checks and annual inspections of trucks, buses, and automobiles represent in-use compliance practice. California has a diesel exhaust smoke opacity test as a surrogate for transient PM production. However, in all of these cases, the in-use standards are less stringent than the pre-market standards because it is acknowledged that the performance is hard to measure accurately in the field, that components can deteriorate, and that the in-use operation or test circumstances may be substantially dissimilar from the original certification test. Thresholds also allow for accuracy of field grade measurement equipment. In this way, most in-use approaches do not represent a high standard and would not be suitable on their own for controlling vehicle design through pre-market certification.

An obverse view is that in-use performance represents the real truth, and that formalized pre-market testing fails to control adequately for all circumstances because it must be limited in nature. For example, real-world automobile fatality statistics cannot be predicted by crash-worthiness ratings alone (Elliott, 2009). Designing toward the compliance test exacerbates this disconnection. Where the design includes controls or components that can behave in a nonlinear fashion, a vehicle may prove to be highly fuel efficient during the pre-market certification test, but less so when exploring other operating regimes.

Vehicle models or simulations may be used to predict fuel use. The model employs maps or modules to describe major components in the drivetrain (engine and transmission) or the interaction of components with the outside world (tires, surfaces, and gravitational force). Internal to the model, variables such as torque or speed are exchanged between component modules, and engine fuel flow yields the numerator in the overall efficiency metric. The model must include the controller logic used in the actual vehicle and must simulate the behavior of a driver. By employing a vehicle simulation, rather than one or two prescribed tests, current and proposed MHDV pre-market certification approaches provide the benefit of determining the fuel efficiency over a variety of operating schedules and permit cost-effective examination of vehicles with different combinations of components. However, the simulation accuracy relative to the real world is no better than the representativeness that the component maps in the simulation would imply. Those component maps or modules are themselves obtained from test procedures, and these procedures cannot cover all operational circumstances.

It is argued below that simulation represents a more versatile approach than whole vehicle testing and is a reasonable short- and mid-term approach. However, the growing need to reduce fuel use in all types of operation, over a full vehicle life span, and in an increasingly interconnected world may require in-use methods, as discussed below. Ultimately the problem must be addressed as a system optimization of a society of interacting vehicles and control devices.

3.3.2 Engines

3.3.2.1 Engine Dynamometer Testing Adequacy

At present, MHDV fuel efficiency is determined by simulation and is determined separately from criteria pollutant values, which are still engine based.[1] However, the engine fuel efficiency is regulated in addition to the MHDV fuel efficiency and is determined using the same well-established test procedure that is used for criteria pollutant standards. Heavy-duty engine dynamometer testing procedures are well established because they have been employed for three decades to quantify the emissions of regulated criteria pollutants from engines during EPA and California Air Resources Board (CARB) certification. The procedure defines not only the dynamometer measurement of load and speed, but also the way the exhaust is captured, diluted, and analyzed. In 2007 the reduction of PM from diesel engines to levels that require exhaust filtration was accompanied by new procedures that sought a higher level of accuracy from the measurement process (EPA, 2005). Engine fuel consumption and carbon dioxide production can be measured independently during the certification process and can be compared with one another by using fuel composition data.

[1] Some Class 2b vehicles are vehicle tested.

The variability in test run inputs, and the inability of dynamometer tests to assess all possible MHDV operation modes, is a challenge to engine fuel consumption estimation and testing of individual components. Cycle-averaged dynamometer brake-specific fuel consumption measurements in a single laboratory typically are repeatable from hot run to hot run with a precision of a fraction of a percent, but the permissible accuracy of fuel flow measurements, engine speed measurements, and engine shaft torque measurements alone suggests that the brake-specific fuel consumption could vary by about 2 percent between laboratories.[2] It is therefore difficult to determine the relative merits of certain technology choices by differencing total fuel consumption between laboratories and complete engines, since many technology improvements are claimed to have an efficiency improvement smaller than the intertest variability. This is even more true because fuel consumption variability is within a narrow window limited by thermodynamic considerations. If an assumption is made that engines can reduce their fuel consumption by at most 10 percent (a brake efficiency improvement of about 5 percent), then the potential measurement variability would be as much as one-fifth of that improvement range. However, there is no clear pathway for substantial improvement in this accuracy.

The full scope of all possible real-world operational modes is addressable using any given test procedure. The efficiency of modern engines with sophisticated controls is strongly influenced by speed, load, and transient behavior. To evaluate the real-world benefits of engine technology improvement in reducing fuel use, it is important that the engine is exercised in a way that mimics real-world use adequately. Unfortunately, engine speed, load, and dynamic behavior in real-world operation vary widely due to both vehicle application and operational circumstances. The controlled approach that offers relatively high accuracy and very high precision in the test cell does not permit the evaluation of the engine's efficiency during extremes of temperature, humidity, and altitude. The approach also cannot account for the variable cooling power demands placed on the engine as climate varies or road grade changes. Also, it is far more difficult to measure instantaneous fuel consumption than cycle-averaged fuel consumption accurately. If it is measured using carbon dioxide instantaneous measurement, the combustion products are delayed and diffused over time by the exhaust and measurement systems (Madireddy and Clark, 2006). Therefore, there is difficulty in employing test cell instantaneous fuel economy data to verify models used to predict instantaneous efficiency of an engine or vehicle under transient operation. The more divorced a test procedure becomes from the real-world operation, the more the engine is likely to be configured to the test rather than the end use. Fuel price and technology cost changes will influence this gap through the design and calibration processes.

The test procedure has the most stringent requirements for criteria pollutant emissions and for fuels used, neither of which reflect real-world performance. The engine must exhibit the highest-criteria pollutant reduction stringency during the dynamometer test, because in-use compliance is inherently more relaxed in restricting the pollutant level. In some cases there are identifiable causes for fuel use deviations between test cell and real-world emissions: with current-technology diesel engines these include inactive selective catalytic reduction (SCR) operation, the need for diesel particulate filter (DPF) regeneration in cold weather, and variable back pressure from the DPF. Some proposed technologies, such as waste-heat recovery, depend on engine load history and ambient temperature for performance: their benefits are hard to quantify generically. The test cell procedure also specifies the fuel to be used in the vehicle. This assists with repeatability, but real-world diesel energy content, and hence either mass or volume of fuel used, vary widely (Bacha et al., 2007) and differ from the test fuel. For example, the specific gravity of No. 2 diesel fuel can vary from 0.82 to 0.88 at 15.6°C, and this affects heating value.

Dynamometer fuel efficiency measurements and real-world fuel efficiency will vary in sympathy when the dynamometer test schedule simulates torque and speed values that are encountered on the road. An engine that has multiple applications, such as line haul operation and transient urban operation, and engines paired with boost-interrupting transmissions cannot be correctly simulated with a single test cycle. Current measurements average fuel consumption over the operation period in the test cell for a legacy transient test with a high proportion of operation near rated speed, and a set of steady-state points. The cycle-averaged value from the transient test or a weighting of the point efficiencies can be used directly as a standard for engine efficiency. The current approach in GEM uses individual set-point efficiencies to map the steady-state efficiency to predict engine efficiency during

[2] The standard estimate of the error for torque can be as high as 2 percent of the maximum torque, with additional statistical limitations for torque calibrations shown in 40 C.F.R. Part 1065.307, Table 1.

vehicle modeling. However, measurement and digital post-processing methods already exist to determine the instantaneous, transient fuel consumption of the engine and to relate this to the instantaneous torque and speed and past history of commands given to the engine controller. With these data, one is able to reconstruct the efficiency of an engine over a wide range of behaviors or cycles during a simulation, although still under constraint of the test cell conditions. Cold-start operation in the test cell can add to the sophistication of these predictions. In this way, the engine efficiency is more realistically modeled in a range of applications with little additional effort during the test cell procedures. Test cell in combination with modeling rewards care in matching the engine and transmission in cases where the engine and transmission are not tested as a combined powertrain unit.

Measured instantaneous fuel consumption data may be used in both whole vehicle modeling to test against a vehicle standard or in simulation of engine cycles to determine compliance of engines with efficiency standards for a variety of vocations. Ultimately, one may see the test cell as an input measurement device for engine and vehicle performance simulation, rather than a technique for gathering cycle-averaged data used in direct comparison against a standard. Current use of steady-state points to produce an efficiency map in GEM is a first step toward such a goal but is insufficient for future use, because the steady-state efficiency and transient efficiency may not be the same, even at the same instantaneous torque and speed. One cause is due to the varied intake and exhaust pressures that may occur at the same torque and speed when the turbocharger is rising or falling in speed, or changing in geometry. Another cause is that the injection control strategy may vary because in-cylinder versus SCR NO_x reduction strategies change during transient operation. Although the argument can be extended that test cell measurements with varying ambient conditions will extend the applicability of the data for real-world operation, these measurements would be costly and onerous.

3.3.2.2 Harmony with Criteria Emissions

A trade-off exists between engine efficiency and engine-regulated (criteria) emissions in most cases, except where reductions in internal energy losses benefit both. Maintaining a strong linkage between regulated emissions and efficiency measurements is important, first because both should reflect reality, and second because it is counterproductive to develop engines that are efficient in one zone of operation and clean in another zone. Currently both are determined in the test cell, using the same transient test and steady-state test points. Whereas instantaneous fuel efficiency and instantaneous pre-aftertreatment criteria pollutants can be measured in a test cell, both urea SCR and the DPF to reduce PM make it difficult to associate tailpipe or stack criteria pollutant levels directly and repeatedly with engine operation.

3.3.2.3 Powertrain Measurement

Integrated powertrain measurement has been presented as an important alternative to engine testing in cases where the engine and transmission have been closely integrated and are controlled as one unit. This is a viable alternative to separate engine and transmission characterization, with subsequent combination of the engine and transmission within a vehicle simulation. In the future it is likely that the engine, transmission, and aftertreatment may be optimized and commonly controlled to the point that operation of the engine alone will not be possible. Characterization of the whole powertrain package is also important in that it will encourage designs with the benefits of a high level of integration. Nevertheless, it is essential that the test cell characterization of the powertrain provides no advantage to the powertrain results other than to acknowledge the benefits of the integrated design and control in the resulting data. Noting the variety of vocations of trucks, it is difficult to select a powertrain test that will compare equitably with separate engine and transmission tests for a range of real-world cycles. Powertrain tests, and tests of a mule (powertrain plus drive axle), are philosophically as valid as engine tests in achieving vehicle efficiency goals, but will require very careful consideration to establish an equivalency. As with engine tests, a powertrain test will have difficulty in providing data on instantaneous tailpipe or stack emissions of criteria pollutants.

In the case of hybrid drivetrains, the instantaneous energy use by the engine does not necessarily correspond proportionally to the energy delivered by the vehicle to the road, nor do the criteria pollutant emissions. Energy may be drawn from, or stored in, an electric, hydraulic, or mechanical storage system following the direction of a control algorithm. It is possible, from measurement or combined measurement and control simulation, to create a hybrid powertrain model, but the instantaneous fuel demand depends on history of operation as well as the current vehicle activity. Even when a complete hybrid drivetrain is not used, some technology approaches such as the supply of hotel loads through electro-turbo compounding will not allow the engine to be considered in isolation from other integrated components.

3.3.2.4 Findings and Recommendations

Finding: A new technology or design improvement is likely to cause incremental reductions in engine fuel use that challenge the accuracy of the test cell in measuring the difference in fuel consumed and shaft power absorbed. The variability of measurement of fuel use in a test cell is a substantial fraction of the foreseeable reductions in engine fuel use in the future, but there is no clear pathway to reduce this variability substantially.

Finding: Current test cell procedures do not measure engine behavior over all circumstances encountered in the real world, nor are the test schedules weighted to match all real-world combinations of torque, speed, and transient behavior. Even at the same instantaneous torque and speed, steady-state engine efficiency does not necessarily represent the efficiency during transient operation. However, measurement and data processing techniques exist to yield instantaneous fuel consumption models that can be used in the simulation of fuel use over unseen cycles.

Recommendation 3-1: Steady-state maps should be replaced with an advanced predictive fuel flow model, derived from transient test cell data, for use in transient engine, powertrain, and vehicle simulations.

Finding: Currently engines are characterized for both fuel use and regulated pollutant production in the same test cell. If fuel efficiency is determined using simulation and an engine model derived in the test cell, the test circumstances for fuel efficiency and criteria pollutants will differ.

3.3.3 Whole Vehicle Modeling

3.3.3.1 Adequacy and Possible Improvements

Whole vehicle modeling offers substantial advantages over whole vehicle testing in that vehicle design changes and operational differences may be examined rapidly and inexpensively. In addition, a limited period of whole vehicle testing, whether on road or off road, usually fails to capture a range of real-world effects. However, substantial effort is needed to ensure that the model has desired accuracy in predicting real-world performance. The software platforms needed for accurate simulation are already established, but reaching this goal demands complete and sophisticated input data, and the inclusion of all significant components and effects.

Environments that are suited to graphical programming and model-assisted design and optimization are available to support vehicle simulation. Many software packages, private and commercial, are based on these environments and offer predetermined architectures, structures, and information flow to support simulation. The GEM tool is one such package. Errors associated with digital approximation are typically small, and the accuracy of the simulation depends almost entirely on the representativeness of the modules describing each component.

Meeting future fuel efficiency standards will require the use of a combination of strategies and components that reduce fuel use each by a fraction of a percent. As a result, simulation must be able to deal with these small quantities accurately. It will not be satisfactory to develop correction factors for simulation based on whole vehicle tests, because these factors will vary from test schedule to test schedule.

Few components can be modeled so simply that they just require a fixed efficiency, although this is often used today. As an example, an axle incorporating reduction and a differential have bearing losses, rubbing losses, and stirring losses, and the loss model can change with reversal of rotation and torque. Moreover, the viscous losses depend on the lubricant temperature, which in turn depends on ambient temperature and recent history of operation.

Some components can be modeled acceptably with a steady-state map for efficiency, but complex assemblies such as automatic transmissions, engines, and powertrains need acknowledgment of transient behavior.

As an example of the need for an accurate transient model, consider a modern engine that employs exhaust-gas recirculation, injection management strategy, and urea SCR to achieve low NO_x emissions. The strategy is usually different during transient operation than steady-state operation, and the instantaneous efficiency may be lower. Use of a steady-state map would deny the distinction between use of cruise control and a driver who dithers the pedal. Once these engines incorporate systems such as waste-heat recovery, the model will also need to include a factor to describe recent history of operation. Far more complexity is introduced if the model is expected to perform over a wider weather envelope than the present-day temperature window.

The driver model used in the simulation can affect predictions. Real-world drivers differ in gear choice, shifting style, lane control, and pedal operation. A simulation should employ the style of a typical driver so that the true benefits of driver aids, cruise management, automated shifting, and look-ahead technologies, can be quantified.

Hybrid drivetrains are a major pathway to vocational efficiency improvement. It is necessary to simulate hybrid vehicles that include a hybrid drivetrain, rather than relegate them to a physical test requirement, to allow equitable comparison with conventional vehicles and to examine their benefit over a variety of test schedules. These test schedules must be selected with great care, because even the rearrangement of microtrips or duration of grades within a schedule can affect hybrid vehicle performance.

In addition to selecting test schedules that are representative to use in MHDV simulations, the power-to-weight ratio must also be considered carefully, so as not to drive the market toward over- or under-powering of trucks. This involves not only the mass assigned to the truck in the simulation, but also the payload granted to the truck if ton-mile reference is used.

Currently, simulations such as GEM provide an attractive opportunity to determine the fuel efficiency of a range of vehicle designs over different routes. However, the inputs to GEM are relatively simple, such as a single rolling resistance value for tires, and a wind-averaged drag coefficient that is not speed dependent. The engine maps are also steady-state based. When the demand for greater fidelity and the ability to include more innovations increases, GEM will become more complex, and the component performance verification will become critical and substantial. As vehicles become progressively more complex, integrated, controlled, and connected, simulation with sufficient accuracy and attention to detail may become impossible. Not only will the simulation itself become difficult, but the compliant and well-defined collection of component input data will demand excessive resources. It is likely that simulation will remain a strong mid-term solution. However, post-sale in-use data collection, either with or without a level of pre-market simulation, is a likely long-term solution. Therefore, in the future, in-use measurement may replace simulation as a result of the design and operational complexity of MHDVs.

3.3.3.2 Specific Findings and Recommendations

Finding: Accurate simulation of MHDVs relies on the accuracy of the component maps or modules within the vehicle simulation model. Components cannot be modeled accurately using a steady-state approach or limited independent variables if their behavior is complex or depends on operating history of the component or other components.

Recommendation 3-2: Transient maps or advanced models, as opposed to fixed efficiency or steady-state maps, should be employed for components with complex behavior within a vehicle simulation unless it is ensured that the simpler models support acceptable accuracy.

Finding: Selections of test schedules and test weights for a simulation will affect results substantially, and hybrid vehicles will be more sensitive to the ordering of events within a test schedule. The choice of the baseline driver model will also affect simulation results both absolutely and relative to simulations that include driver aids or vehicle automation.

Recommendation 3-3: The baseline driver model used in MHDV simulations should be chosen to represent average driver behavior, so that benefits of driver training, automation, and driver aids are realistically modeled.

3.3.4 Components

3.3.4.1 Tires

Tire rolling resistance represents the substantial portion of the road load for most vocational vehicles, for example, transit buses, package delivery vehicles, and refuse haulers. For vocational vehicles, it is as important to characterize the tire losses as accurately as the engine efficiency. For line haul, rolling losses are still a substantial factor.

Currently tire inputs to the GEM consist of an effective rolling diameter and rolling loss coefficient. The tire's effective rolling diameter varies over its lifetime, with a change on the order of 2 percent. If modeling is based on wheel rotational speed, the real ground speed will be affected, and for reduced diameter the required engine torque required to propel the vehicle will be reduced. If ground speed is used as a reference, gearing will change over the tire's lifetime, and engine speed will be affected. In addition, longitudinal slip will cause tires to have a larger number of rotations per unit distance under positive torque, and a smaller number of rotations per mile under negative torque (e.g., during braking). Longitudinal slip of drive wheels will increase as a smaller proportion of the load is on those drive wheels. These effects may appear proportionally small, but so are the fuel efficiency gains needed to meet certification.

Tires do not maintain the same rolling resistance over their lifetime. A new tire generally exhibits the highest rolling resistance, with some of the resistance attributed to tread design. The resistance then declines as the tire ages to a baseline value that is more indicative of the casing materials and design. Tires with the same rolling resistance when new may trend to different values of rolling resistance when worn. Low-rolling-resistance tire casings may be regrooved or retreaded, with further change in rolling resistance.

Tire rolling resistance is affected by the type of road surface, which may affect a truck owner's choice of tires. Furthermore, tire rolling resistance is not accurately described as a constant fraction of the vertical load on the tire. It varies with that load, with the speed of operation, with temperature, and with tire pressure.

If modeling is employed solely for certification against a standard, the use of a new-tire simple rolling resistance coefficient and rolling diameter may be a sufficient insurance of performance. However, for comparisons of models to real-world operation, diameter change, rolling resistance change, speed and load coefficient dependency, pavement surface effects, and longitudinal slip are all factors that can impact fuel use. An evaluation is needed to see whether use of average values for these parameters is sufficient, and a method is needed for determining or measuring these average values. Furthermore, the method of measuring rolling resistance must yield values that represent real-world use on average pavement, and must be consistent between laboratories and of high quality.

Recommendation 3-4: For certification against a standard using vehicle simulation, it is essential that the rolling diameter and rolling resistance tests are standardized to meet the same data quality level as engine tests, and that they yield results that are sufficiently representative for modeling average real-world performance. For comparisons of simulation data with real-world operation, tires should be treated with a more sophisticated model than a combination of rolling diameter and rolling resistance.

3.3.4.2 Aerodynamics

Aerodynamic losses are the dominant power demand in high-speed trucking and require the same degree of accuracy as a model input as does engine efficiency. However, it is difficult to measure the product of aerodynamic drag coefficient and frontal area (C_dA) reliably and repeatedly. Methods include computational modeling, scaled wind-tunnel testing, coast-down testing, and constant-speed testing. With on-road or track testing methods, the aerodynamic loss must be decoupled from rolling losses, and yaw angle of the wind is hard to control and measure. Moreover, yaw angle causes side thrust on a truck, which is reacted by the steering angle, which in turn causes elevated tire rolling losses. In a track test these additional losses would be interpreted as aerodynamic yaw angle effects, whereas in a wind tunnel or simulation, they may be neglected. With wind-tunnel testing, scaling issues exist, ground effects may not be well modeled, and simulation of yaw in a tunnel requires correction for lateral force on the truck. Computational methods have difficulty in modeling vehicles without employing calibration from base cases, and require high computing power to support fine grid sizes.

Frontal C_dA does not necessarily reflect performance under yaw conditions, and C_dA does vary as a function of vehicle speed. It is also possible to have the same relative wind speed and yaw angle at different wind conditions and truck speed, so that both truck speed and wind speed (relative to ground speed) and the wind direction (relative to the truck direction) are all important in computing instantaneous drag. Two trucks with the same frontal C_dA value may differ in their values under yaw. Just as grade has been added to time-speed schedules for simulations, so could wind speed and direction to challenge the vehicle performance without simply using an average value. Inability to measure C_dA in an accurate and representative fashion is a substantial barrier to accurate simulation. Additional discussion of real-world conditions that impact aerodynamic drag can be found in the section on the SmartWay program in Chapter 5.

Engines are acknowledged to have efficiencies of the order of 40 percent, so that about 60 percent of the fuel energy is lost. The power supplied by the engine is used primarily to overcome road load. During low-speed operation tire rolling resistance dominates the road load—an effect which increases as tires wear and their performance degrades—and at high speed the aerodynamic losses are dominant. In this way engine inefficiency, tire losses, and aerodynamic losses are all important contributors to fuel use. All merit the same level of attention and measurement accuracy in the quest to improve vehicle fuel efficiency.

Finding: Regulation has focused more upon engine efficiency compliance than separate tire or aerodynamic compliance, even though all three systems are associated with the same order of magnitude of energy in truck operation. If engines are regulated for fuel efficiency in addition to regulation of the whole MHDV, then there is a parallel argument that tires and the aerodynamic body might also be regulated separately in addition to the engine and vehicle. If engines are regulated for fuel efficiency in a manner that acknowledges their complexity, then there is a parallel argument that tires and the aerodynamic body might also be regulated with the same acknowledgment of complexity including the degradation of tire performance as they wear.

Recommendation 3-5: It is essential that the determination of C_dA is standardized to meet the same data quality level and repeatability as engine tests, and that the values of C_dA are available separately at different yaw angles for use in advanced simulations. Where the simulation is used to compare with real-world activity, dependence of C_dA on vehicle speed should be measured or simulated.

3.4 IN-USE COMPLIANCE AND ENFORCEMENT

3.4.1 Need for In-Use Compliance and Enforcement

The purpose of pre-sale certification is to demonstrate that the design of each engine and vehicle model is capable of meeting applicable emission and fuel consumption standards and requirements such as on-board diagnostics (OBD). The purpose of post-sale evaluation of performance is twofold. The first objective is to verify that under real-world conditions vehicles continue to be in compliance with emission and fuel consumption requirements throughout their statutory useful life. Should a specific vehicle model have in-use emissions or fuel consumption that exceeds standards, or have a systematic design defect, NHTSA and EPA may require the specific model vehicle to be recalled and modified to reduce emissions or fuel consumption. The second objective of post-sale evaluation is to provide insight into whether real-world emissions and fuel consumption of the fleet as a whole are consistent with expectations based on regulatory standards. If they are not, NHTSA and EPA may make changes to the pre-sale test and certification procedures so that future engine and vehicle designs perform on road as they do in certification testing.

The importance of a post-sale verification of emission and fuel consumption standards compliance has been demonstrated for light-duty vehicles. Several examples follow:

- In the 1980s and 1990s, in-use compliance testing by EPA of groups of similar passenger vehicles found widespread noncompliance with criteria emission standards. The cost and bad publicity of the many resulting recalls resulted in vehicle manufacturers improving the reliability of emission control systems. The result is that few criteria emission recalls are occurring today, contributing to lower emissions.
- Random testing of in-use passenger vehicles revealed the inadequacy of the certification emission test to represent many modes of operation, and resulted in addition of new pre-sale tests that better predicted actual on-road emissions and fuel consumption.
- Hyundai/Kia recently paid a large fine for not following the proper procedures in determining light-duty fuel economy, which was revealed by in-use compliance testing. This finding resulted in relabeling vehicles not yet sold with lower fuel economy values, and the companies sent refund checks to owners to compensate for the misleading high ratings.
- The International Council on Clean Transportation recently reported on the well-known discrepancy between the certified level of fuel consumption and the measured on-road fuel consumption of in-use passenger vehicles. In Europe the certification test results under-predicted real-world fuel consumption of 2001 models by 8 percent. The surprising result was that the amount of under-prediction has grown year by year, as more stringent GHG emission standards were implemented, to 38 percent for 2013 models (Tietge et al., 2015). In-use testing revealed this trend and may provide the basis for correcting the shortfall with improved certification processes.

These examples from light-duty vehicles demonstrate the importance of an in-use compliance program. While some important lessons for a potential MHDV in-use compliance program can be gained from the extensive experience with the light-duty program, it is important to note that in-use compliance for MHDVs differs from passenger vehicles. For passenger vehicles, the complete vehicle is tested pre-sale for both criteria emissions and fuel consumption. The shortfall between the certification test and real-world fuel consumption is corrected in the fuel economy label attached to each new passenger vehicle. The pre-sale test can be easily replicated in use by borrowing vehicles from owners and testing them at certified laboratories. Emissions and fuel consumption can also be measured on road using portable emission measurement systems. These approaches allow vehicle models with high emissions, or faulty emission control system designs, to be identified, which can lead to corrective action.

For the larger MHDVs,[3] only the engine receives a pre-sale emission and fuel consumption test in the Phase I and Phase II rules. The fuel consumption of the complete vehicle in these larger truck classes is determined by the GEM simulation model, based on user inputs including characteristics of the vehicle such as the drag coefficient or rolling resistance, and, in the Phase II rule, on engine fuel maps and torque curves. No actual measurement of fuel consumption of the vehicle being certified is performed. The pre-sale certification approach for MHDVs thus limits how an in-use compliance program is structured in several ways:

- Determining if the engine is in compliance with fuel consumption (or criteria emission) standards requires the engine be removed from a truck and tested on an engine dynamometer. The cost and time involved are very high, and procurement of in-use trucks for engine removal is difficult. Thus such testing is limited compared to testing passenger vehicles.
- Determining if the complete truck (Class 4 and above) is in compliance is limited to verifying the inputs to the GEM model, since there is no pre-sale vehicle GHG emission or fuel consumption test to duplicate in use. At best, in-use testing using on-truck emissions or fuel consumption measurement devices can identify suspect groups of vehicles with high GHG emissions or fuel consumption, but regulatory action to remedy the problem, such as recall, is likely limited to engine removal for testing, or validation testing of inputs to the GEM simulation model.

[3] Class 2b and 3 vehicles are usually certified using a chassis dynamometer.

Table 3-3 compares the current certification and in-use compliance requirements for passenger vehicles and MHDVs.

For these reasons, an in-use compliance program for MHDVs patterned after the in-use compliance program for passenger vehicles is not likely to be practical or successful in providing confirmation that MHDVs are meeting standards and requirements. Thus, a different approach is needed to provide assurance of in-use compliance.

Finding: An effective method of determining in-use compliance of trucks, important for the overall effectiveness of the regulatory program, does not currently exist. One is necessary to ensure that individual vehicle models are meeting fuel consumption and GHG emission standards (as well as criteria emission standards), and to identify opportunities to improve the effectiveness and reduce the cost of the program.

Recommendation 3-6: NHTSA, in coordination with EPA, should develop an effective in-use compliance method that would allow the overall performance of the regulatory program to be quantified, identify whether groups of in-use trucks may not be in compliance, and provide insight into truck operating conditions where GHG emissions and fuel consumption of future trucks could be further reduced.

3.4.2 Approaches to Determine In-Use Compliance for MHDVs

In general, the objective of an in-use compliance program should be to determine if real on-road emissions and fuel consumption match the results expected from the regulatory standards and requirements. It should also identify groups of trucks that appear to have higher emissions and fuel consumption than expected, in order to implement corrective action. In a broader sense, it should provide data that suggest how the regulatory program could be made more effective (i.e., the concept of continuous improvement).[4] Depending on the comprehensiveness and effectiveness of the in-use compliance program, it may complement, or could replace, the pre-sale certification program. Several in-use compliance concepts that meet some or all of the above objectives are discussed next.

TABLE 3-3 Comparision of Pre- and Post-Sale Compliance Requirements for Passenger Vehicles and MHDVs

Pre- or Post-sale?	Type of Verification	Passenger Vehicle and Class 2b/3 Trucks	HD Engine[a]	HD Truck[b]
Pre-sale	Emission tests[a] for certification	Yes	Yes	No
Pre-sale	Simulation model for certification	No	No	Yes
Post-sale	Emission test (to verify in-use compliance with standards)	Yes	No[c]	No
Post-sale	Emission test (on-road emissions to measure program effectiveness)	Yes	No[c]	No

[a]Criteria and GHG gases, which allow calculation of fuel consumption.
[b]Fuel consumption and GHG only. Class 4 vehicles and larger.
[c]Quantity insufficient to determine overall compliance or program effectiveness.

[4] See Recommendation 2-5 on the need for NHTSA to conduct an interim assessment of the regulations.

3.4.2.1 On-Board Diagnostics and Warranty Reporting

Modern passenger vehicles are required to have an on-board diagnostic system that is closely correlated to criteria emissions such that a warning light and diagnostic code is set whenever emissions increase by about 50 percent above the emission standard. In the past few years, MHDVs have also become subject to OBD requirements, although the sensitivity of the OBD system to identify problems with particle filters and exhaust catalysts on diesel engines currently lags far behind the effectiveness of gasoline vehicle OBD. It may be possible to develop OBD monitoring for MHDVs that is more closely tied to GHG emissions and fuel consumption, and is sufficiently comprehensive.[5] If successful, the OBD data could be transmitted electronically from the truck to the truck manufacturer and/or NHTSA and EPA, to assess in-use performance of GHG emissions and fuel consumption, as well as criteria air pollutants, both for individual trucks and groups of trucks as a whole.

Engine manufacturers are currently required to report warranty repairs to NHTSA and EPA, and these reports have been used to identify systemic defects in engines. This reporting system could be strengthened to more directly become a means of recalling engines (and could be extended to complete trucks) that show a pattern of emissions and fuel-related component failures, especially for components whose current OBD monitoring capabilities are limited. The emission warranty period, currently only 100,000 miles, would need to be extended as well, perhaps to the useful life (as much as 435,000 miles), since 100,000 miles is often driven in the first year of a Class 8 truck's life.

A combined OBD and warranty reporting program could serve as a somewhat limited in-use compliance program. It would meet the objective of identifying groups of engines that are exceeding emission and fuel consumption standards, and cause their repair. However, it would not alone provide a quantitative assessment of on-road emissions, nor would it identify the need to make program improvements that would address high off-cycle emissions or fuel consumption. These limitations exist because pre-sale development of the OBD monitoring system is primarily tied to sensor signals generated during a standard test procedure, not to the direct measurement of emissions or fuel consumption.

3.4.2.2 In-Use Test Cycle

NHTSA and EPA could adopt an in-use test cycle representative of real-world driving that could be easily replicated on an in-use truck equipped with standard portable emission measurement instruments. The cycle could be added to the GEM. The regulation would specify that in-use compliance will be based on the results of on-road tests from a sample of in-use trucks. The tests could be conducted under any conditions, such as temperature or load, which the trucks might normally experience. It would be the truck manufacturer's responsibility to ensure that its trucks comply with the standards under a broad range of conditions for the specified in-use test cycle. This approach would partially meet the first objective of determining real-world emissions and fuel consumption (outlier trucks such as those operating at extreme conditions or cycles would not be reflected in on-road emission assessments). It could also identify groups of similar trucks with high emissions and fuel consumption, but may not identify specific areas for program improvement since testing would still involve a single defined test cycle.

3.4.2.3 Compliance Based on Actual On-Road Emissions

Modern trucks use sensors and an on-board computer to determine the needs of the engine, such as the amount of fuel needed to achieve required performance. Since fuel consumption is one of the metrics of the standards adopted by NHTSA and EPA, it is plausible that fuel use, or possibly directly measured CO_2 if a low-cost CO_2 sensor could be developed, could be collected from every truck on a real-time basis, and stored and periodically reported to the agencies. Other information such as speed can also be collected and, for many fleet operators, is currently reported electronically for operational analysis. A regulatory requirement that such data be reported would meet the objectives of quantifying real-world emissions and fuel consumption for comparison to what was

[5] For a discussion of the technical issues in collecting real-world fuel consumption data from OBD systems, see NRC (2015, p. 349).

expected from the adopted standards and requirements, and could also identify groups of trucks that appear to be performing abnormally. It might also provide insight into how to improve the regulatory program, such as by identifying modes of operation where GHG emissions and fuel consumption are higher than expected, which could lead to improvements in the pre-sale certification protocol.

A more significant change could be to require the truck builders to be responsible for the on-road GHG emissions and fuel consumption of the trucks they sell. Truck manufacturers could select a standard they will meet, subject to a regulatory maximum, with in-use compliance based on on-road data subsequently collected on their trucks. Truck manufacturers know how most of their trucks will be operated (based on their sophisticated simulation models) and can reflect in their selection of a standard the fraction of vehicles that operate at loads, speed, altitudes, etc., that differ significantly from the average. This approach provides a greater risk of noncompliance to the truck manufacturer than does the current certification approach that has minimal risk once the vehicle is sold. However, if in-use compliance is based on a large group of their vehicles, such as all of a manufacturer's vocational trucks, those trucks that are outliers may not have a large effect on the average. To further reduce the risk to the truck manufacturer, the initial years of implementation could be phased in with minimal penalties until experience with the in-use data is gained. Furthermore, NHTSA and EPA could impose monetary nonconformance penalties, as currently set forth in the Clean Air Act, in lieu of the threat of a recall, especially where defective design of components is not involved.

This approach significantly differs from current in-use compliance approaches. It places the responsibility for actual in-use emissions and fuel consumption on the truck manufacturer, increases the risk of noncompliance because responsibility would include all operational conditions, some of which may not be precisely known by the manufacturer, and raises numerous issues which need to be assessed, such as knowing the load on the vehicle, processing large amounts of data, and many more. However, it is the only approach that fundamentally meets the criterion of providing assurance that actual, real-world emissions and fuel consumption are achieved, and thus is included in this list of possible in-use compliance approaches.

Table 3-4 summarizes how each possible in-use compliance approach meets the three objectives listed at the beginning of this section.

The in-use compliance programs discussed above are concepts designed to illustrate the degree of change and need for new thinking in developing an effective in-use compliance program for MHDVs. This is in contrast to the current, limited in-use compliance program which is based in part on the program structure used for passenger vehicles. As the color coding in Table 3-4 illustrates, the program concept involving the most significant change meets all the objectives of an in-use compliance program, whereas the OBD/warranty reporting concept, which relies on improving existing regulatory provisions, only addresses one objective.

Finding: With at least 5 years available before a notice of proposed rulemaking (NPRM) for possible Phase III standards might be issued, EPA and NHTSA have sufficient time to explore potential in-use compliance concepts and approaches. The available time should allow stakeholder involvement in a deliberative process, before the Phase III NPRM is issued, to establish the objectives to be achieved, and assess and develop an in-use compliance concept that best meets the objectives. The method should provide sufficient granularity to offer insight on how

TABLE 3-4 Objective-Based Comparison of Various Approaches to Determining Compliance

	Objectives		
In-use compliance approach	Determine on-road emissions	ID models out of compliance	ID program improvements
OBD + warranty data reports			
In-use truck compliance test cycle			
Data from all on-road trucks			

NOTE: Light gray, meets objective; medium gray, partially meets objective; black, does not meet objective.

the GHG and fuel consumption reduction program can be made more effective, and to identify potential groups of trucks that appear to be failing to achieve standards, and why.

Recommendation 3-7: NHTSA, in coordination with EPA, should undertake studies and establish a dialog with stakeholders to identify the most feasible method of determining on a continuous basis the effectiveness of the MHDV fuel consumption and GHG emission program, including in-use compliance. The agencies should also evaluate if actual in-use emissions and fuel consumption could become the principal metric for determining compliance, including the costs, complexities, and benefits of such an approach, and if implemented which regulated entity should be responsible for complying with emissions in use based on considerations of practicality, feasibility, and fairness. NHTSA and EPA should also determine whether a compliance program based on in-use emissions could supplant the need for engine emission test and simulation model specific standards now in place, and if in-use compliance could be extended for the full life cycle of vehicles.

3.4.3 Program Effectiveness

The OBD and in-use compliance test cycle approaches discussed above do not alone provide for a comprehensive database on which to assess program effectiveness and identify ways of improving program effectiveness. However, these two approaches could be supplemented if NHTSA and EPA undertook a comprehensive and continuous program of collecting actual on-road truck GHG and fuel consumption data. Such a program would have to test a large sample of trucks under varying conditions each year, and include pre-control vehicles for a baseline, as well as vehicles subject to the Phase I and Phase II standards and requirements. The committee has recommended in prior reports (NRC, 2010, 2014) the importance of establishing baselines of fleet fuel consumption and GHG emissions to allow assessment of program effectiveness. When combined with an in-use compliance program more removed from actual on-road emissions, more of the objectives could be met.

Finding: Comprehensive and ongoing testing of in-use MHDVs can provide data necessary to evaluate the effectiveness of the MHDV program and to help develop an effective in-use compliance program.

Recommendation 3-8: NHTSA and EPA should commit resources to collecting real-world GHG emission and fuel consumption data for a large and representative national sample of pre-control trucks and for each model year subject to the Phase I and Phase II standards, with priority given to those categories of trucks with the greatest fuel consumption. These data could be used to assess the effectiveness of the Phase I and early Phase II standards. If the resources for collecting these data exceed that available to the agencies, the agencies should consider requiring the regulated industry to provide the needed data or fund its collection.

3.4.4 Individual Vehicle Compliance

Many states require passenger vehicles to undergo periodic (usually biennial) smog inspections as a condition of re-licensing. Older cars without advanced OBD typically get a dynamometer exhaust test for criteria pollutants. Newer models get an electronic OBD check.

Typically, larger trucks are not included in these periodic tests, for a number of reasons which include the logistics of inspection due to the size of trucks, and the interstate travel and multistate registration of Class 8 highway trucks and trailers which complicates state enforcement based on licensing. With OBD available on most 2013 and newer diesel MHDV engines, roadside OBD inspections of trucks (such as at weigh stations near urban areas) are possible. These OBD inspections would be directed at identifying vehicles with excessive fuel consumption rates and/or high CO_2 emissions. Roadside inspections could also verify that truck and trailer aerodynamic devices are present and low-rolling-resistance tires are being used (when the Phase II trailer regulation is implemented).

Finding: Periodic inspections of the OBD system on 2013 and newer in-use MHDVs could identify individual vehicles with excessive fuel consumption and/or high CO_2 and criteria emissions.

Recommendation 3-9: NHTSA and EPA should evaluate the on-road data they are collecting (see Recommendation 3-8) and determine if there is a need for periodic inspections of individual MHDVs to ensure fuel consumption and GHG emissions are not excessive (and as a co-benefit criteria emissions remain low). If a need exists, the agencies should develop recommended inspection practices and work with states to establish inspection programs.

3.5 REFERENCES

Bacha, J., J. Freel, A. Gibbs, L. Gibbs, G. Hemighaus, K. Hoekman, J. Horn, et al. 2007. Diesel Fuels Technical Review. Chevron. Available at https://www.chevron.com/-/media/chevron/operations/documents/diesel-fuel-tech-review.pdf (accessed December 16, 2019).

Elliott, H. 2009. Why small cars won't keep you safe. *Forbes.* http://www.forbes.com/2009/04/13/car-crash-tests-lifestyle-vehicles-crash-test-report.html (accessed December 16, 2019).

EPA. 2005. Engine-Testing Procedures. *Federal Register* 70:40419-40612 (July 13).

EPA. 2017. Understanding Global Warming Potentials. Available at https://www.epa.gov/ghgemissions/understanding-global-warming-potentials (accessed December 16, 2019).

EPA and NHTSA (National Highway Traffic Safety Administration). 2011. Greenhouse gas emissions standards and fuel efficiency standards for medium- and heavy-duty engines and vehicles. *Federal Register* 76:57106-57513 (September 15).

EPA and NHTSA. 2016. Greenhouse gas emissions and fuel efficiency standards for medium- and heavy-duty engines and vehicles—Phase 2. *Federal Register* 81:73478-74274 (October 25).

Madireddy, M.R., and N.N. Clark. 2006. Sequential inversion technique and differential coefficient approach for accurate instantaneous emission measurement. *International Journal of Engine Research* 7(6):437-446.

NRC (National Research Council). 2010. *Technologies and Approaches to Reducing the Fuel Consumption of Medium- and Heavy-Duty Vehicles*. Washington, DC: The National Academies Press.

NRC. 2014. *Reducing the Fuel Consumption and Greenhouse Gas Emissions of Medium- and Heavy-Duty Vehicles, Phase 2. First Report*. Washington, DC: The National Academies Press.

NRC. 2015. *Cost, Effectiveness and Deployment of Fuel Economy Technologies for Light-Duty Vehicles*. Washington, DC: The National Academies Press.

Tietge, U., N. Zacharof, P. Mock, V. Franco, J. German, A. Bandivadekar, N. Ligterink, and U. Lambrecht. 2015. From Laboratory to Road: A 2015 Update of Official and "Real World" Fuel Consumption and CO_2 Values for Passenger Cars in Europe. ICCT. Available at http://www.theicct.org/sites/default/files/publications/ICCT_LaboratoryToRoad_2015_Report_EEnglis.pdf (accessed December 16, 2019).

4

Powertrain Technologies

4.1 INTRODUCTION

The powertrain is the engine plus the drivetrain which kinematically couples it to the wheels. The internal combustion engine (ICE) is the dominant prime mover for medium- and heavy-duty vehicles (MHDVs). For definition, ICEs encompass all configurations of piston engines, both reciprocating piston and rotary, and even gas-turbine vehicle engines. Four-stroke (or four-cycle) reciprocating piston engines are the prevailing configuration, and the Environmental Protection Agency (EPA) and the National Highway Traffic Safety Administration (NHTSA) in forming the Phase II rule have not anticipated a change. A key question the National Academies of Sciences, Engineering, and Medicine committee is reviewing is whether there will be developments or circumstances that would cause a shift from diesel to other engine types. Additionally, the committee is addressing whether there are emerging technology developments that would be game changers for diesels as well as other engine categories. Examples would include new materials and manufacturing methods, and new chemistries or catalysts for after-treatment systems. Changes in emissions regulations and fuels could also cause a shift in greenhouse gas (GHG) potential or market preference among engine configurations. In addition to discussing improvements in the efficiency of the prime mover—whether diesel, gasoline, or other concept—this chapter discusses the transmission and drivelines and the changes in those technologies. (Electric drive, for which centralized electric power generating stations are the prime mover, and hybrid configurations are discussed in Chapter 7.)

Internal combustion engines using reciprocating pistons can be described as using any of three overall categories of combustion (Figure 4-1):

1. Spark-ignition (SI) engines where the combustion process is dominated by flame propagation in a mostly premixed fuel-air mixture;
2. Compression-ignition engines where the combustion is dominated by diffusion burning with start of combustion strongly influenced by timing of fuel injection; and
3. Compression ignition where chemical kinetics dominate the combustion process. Examples here would include homogenous charge compression ignition (HCCI) and partially premixed combustion—often called low-temperature combustion because dilution by air or exhaust gas is usually significant.

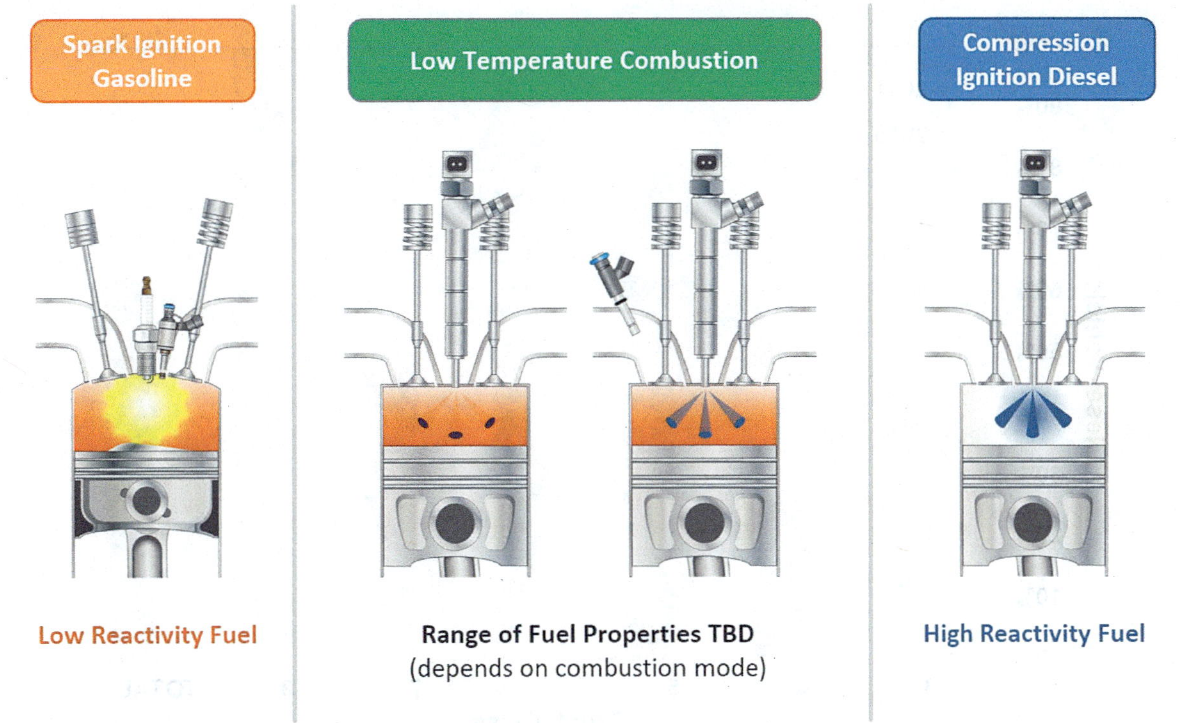

FIGURE 4-1 Categories of combustion with associated fuel type.
SOURCE: Wagner (2015).

Gas turbine combustion engines use a quite different combustion system and are reviewed in a separate section. External combustion engines, such as Stirling, Rankine, and Brayton closed-cycle heat engine cycles, are presently not used in road vehicles as prime movers, but find application as waste-heat recovery systems.

4.2 MARKET TRENDS IN ENGINE USE

Diesel engines continue to be dominant in Class 8 trucks and recently gained market share in Class 6. In Classes 3 through 5, diesels have declined in use in the past few years (see Figure 4-2) in favor of gasoline and natural gas (NG) engines. Note, however, that NG vehicles represented less than 10,000 of a total of about 160,000 sales of Class 4 through 7 medium-duty (MD) trucks in 2013 (NGVAmerica, 2014; ORNL, 2016.

4.3 SPARK-IGNITION ENGINES

Since spark-ignition (SI) engines have different efficiency characteristics and fuel requirements than compression-ignition (CI) engines, their GHG signature is also different. The committee assessed whether there are technical and economic factors that could alter the current engine preferences across Class 2b through 8 vehicles, as well as assessing the potential efficiencies and emissions of SI engines.

We see several factors emerging that could cause an increase in use of SI engines for GHG benefits:

- The initial cost and operating costs of the control technology for criteria emissions is much less for SI than for CI engines.

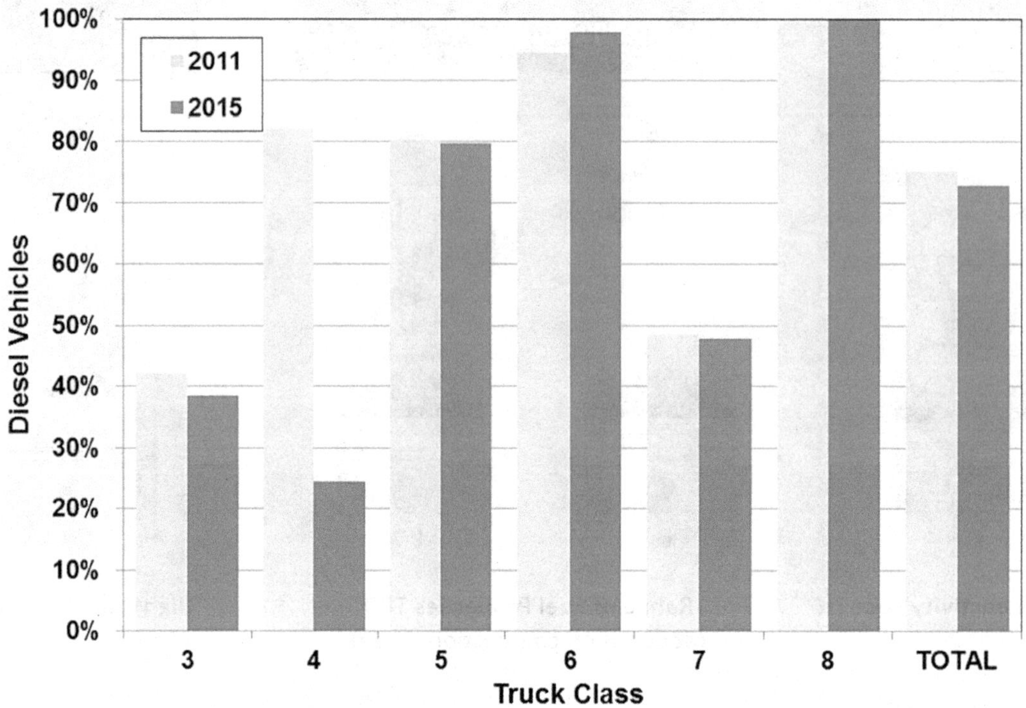

FIGURE 4-2 Shares of diesel truck sales by class, 2011 and 2015.
SOURCE: ORNL (2017).

- Lower-carbon fuels available at large scale (natural gas, ethanol, and propane) are well suited for SI engines.[1]
- The cost of fuels for SI engines is expected to remain below that for CI fuels (diesel) for considerable time due to large world demand for diesel fuel.
- The efficiency of SI engines is continuing to be pushed to higher levels as progress is made to improve practically every feature of an engine's energy balance: overall cycle efficiency (increased compression ratio and exhaust energy use), heat transfer losses, pumping losses, mechanical friction, and auxiliaries.

The National Research Council (NRC) light-duty vehicle report (NRC, 2015a) presents a thorough review of SI engine fundamentals, as did the NRC Phase One Report (NRC, 2010), which will not be repeated here. The committee attempted to identify plausible 2025 SI engine efficiencies to scale engine maps for whole vehicle studies to allow comparison of different engines and fuels in similar vehicles. Three sources or methods were used:

1. Review of modern engine test results for peak and part load efficiency, and extrapolating progress. For example, the Cummins ETHOS engine achieved a brake efficiency near 43 percent on high-ethanol-content fuel (Perfetto and Geckler, 2014) (Figure 4-3), and Toyota has reported results for 2016 production engines of 40 percent (GreenCar and SAE). Department of Energy (DOE)-sponsored projects have yielded multicylinder experimental engines that have surpassed 42 percent efficiency (Chrysler) (DOE, 2013). A deep literature survey of SI engine and fuel trends revealed a continuing increase in compression ratio (Figure 4-4), which is a fundamental parameter governing engine efficiency (Splitter et al., 2016). In the

[1] Of these fuels, propane is used in the smallest quantity for transportation. Yet 0.4 billion gallons are used annually in engines, approximately equal to the energy used by 1 million electric cars.

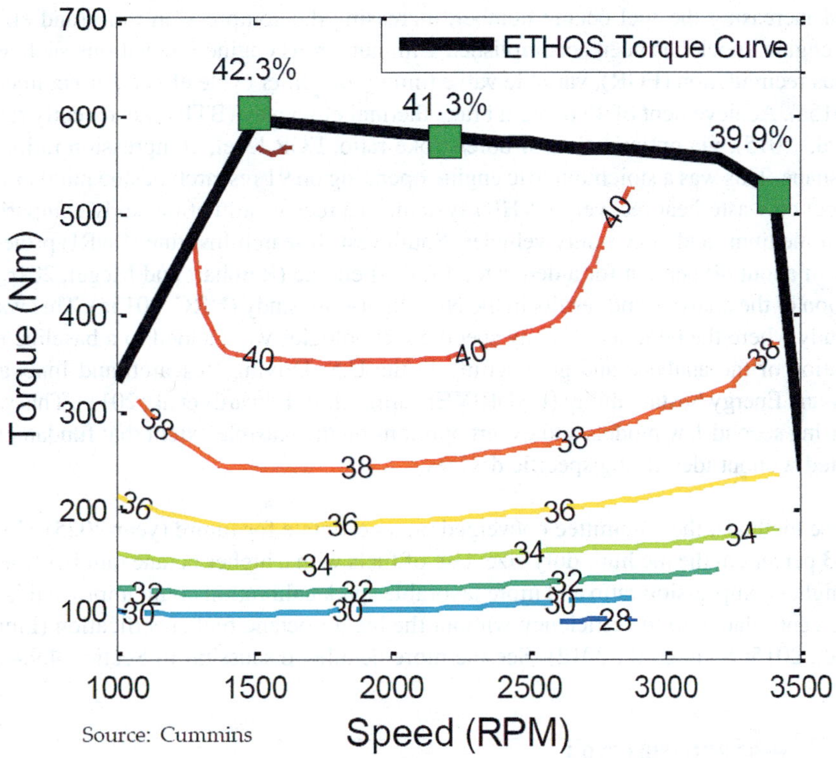

FIGURE 4-3 Cummins ETHOS engine project achieved greater than 42 percent engine brake thermal efficiency with E85 fuel.
SOURCE: Perfetto and Geckler (2014).

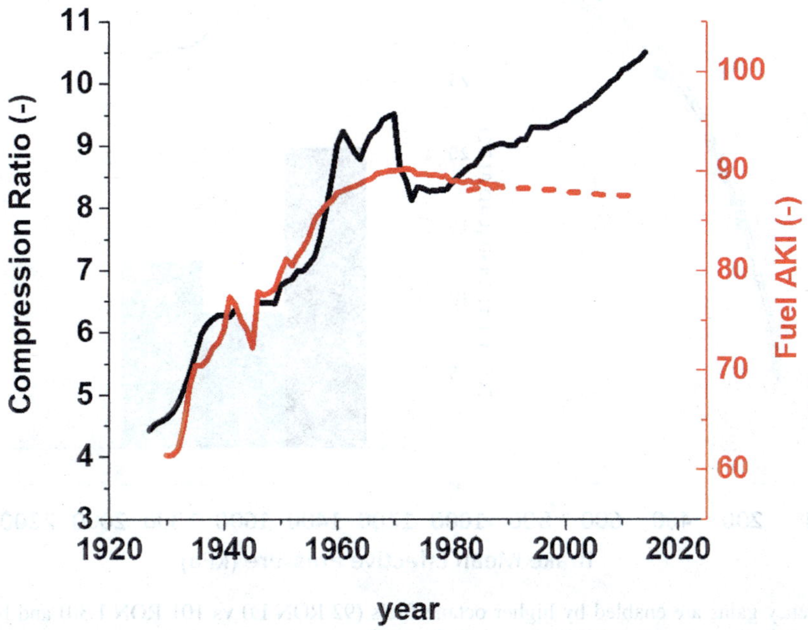

FIGURE 4-4 Technology allowed increased compression ratio (black) even through years of stagnant fuel octane number (red).
SOURCE: Splitter et al. (2016).

absence of increasing the fuel octane number, increasing the compression ratio (and efficiency) without incurring engine knock has been accomplished with numerous engine innovations such as knock sensors, exhaust-gas recirculation (EGR), variable valve timing for Miller cycle effects, direct injection, and others (NRC, 2015a). Achievement of 45 percent brake thermal efficiency (BTE) was recently reported by Honda (Ikeya et al., 2015), via optimization of bore/stroke ratio, EGR level, compression ratio, and combustion chamber shape. This was a stoichiometric engine operating on 91 research octane number (RON) fuel, with boosting but no waste-heat recovery (WHR) system. In a recent simulation study comparing SI and diesel engines for medium- and heavy-duty vehicles, Southwest Research Institute (SwRI) projected a best point efficiency of about 40 percent for a dedicated EGR SI engine (Reinhart and Megel, 2016).

2. Consultation of the analysis and results in the NAS light-duty study (NRC, 2015a). This was a "ground-up" type of study where the benefits of many specific technologies were added to a baseline engine.
3. Consideration of the analysis and goal setting in the U.S. Driving Research and Innovation for Vehicle Efficiency and Energy Sustainability (USDRIVE) partnership (Edwards et al., 2011). This was a "top-down" analysis using second-law models and expert opinions on the feasible extent that fundamental losses could be mitigated without identifying specific devices.

Based on these methods, the committee converged on an estimate for future (year-2025) SI engine peak efficiency of about 43 percent in the medium-duty size. Use of fuels with a higher octane number (the example here is RON) to enable higher compression ratio and more favorable spark calibration could improve this by an additional approximate 5 percent relative to the efficiency without the higher octane fuel specification (Figure 4-5) (Leone, 2015; Sluder et al., 2015; Stein et al., 2012). See the more detailed discussion in Section 4.9.4 in Chapter 4. In

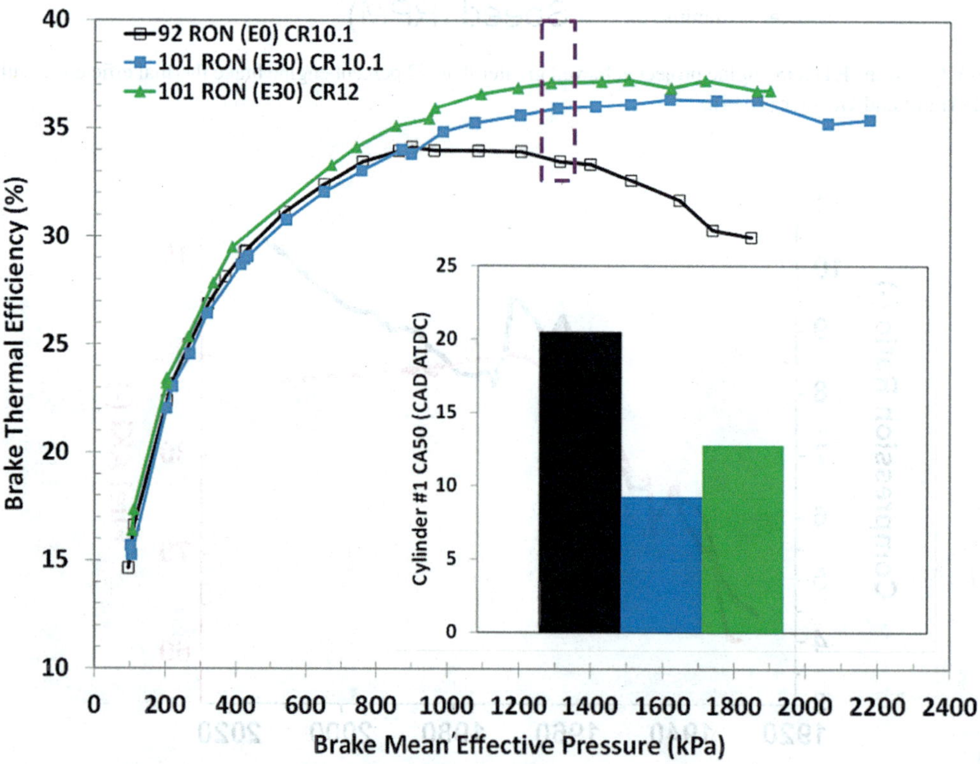

FIGURE 4-5 Efficiency gains are enabled by higher octane fuels (92 RON E0 vs 101 RON E30) and higher compression ratio (CR 10.1 vs CR 12). Note the 101 RON fuels have earlier 50 percent burn completion (see inset, blue and green bars).
SOURCE: Sluder et al. (2015).

comparison, NHTSA and EPA adopted 660 grams CO_2 per horsepower-hour (g CO_2/hp-h) as a 2010 baseline[2] for a vocational truck gasoline engine and placed a standard of 627 g CO_2/hp-h for 2016-2017. These equate to 28 to 30 percent efficiency over the cycle. For these SI engines available for medium- and heavy-duty vehicles, NHTSA and EPA are not requiring more aggressive CO_2 targets beyond 2017. The CO_2 requirements for full vehicles with SI engines (large pickup trucks, for example), are reported by the original equipment manufacturers (OEMs) to be difficult to achieve, presenting a paradox to be studied. Reasons for the paradox in efficiency are discussed in the following paragraphs.

With increasingly stringent regulations for criteria emissions in passenger vehicles, the SI engine system has achieved remarkably low emissions of hydrocarbons (HCs), CO, NO_x, and particulate matter (PM). Developments of the three-way catalyst (TWC), engine fuel systems, and electronic controls have been key enablers. Tier 3 emissions requirements will also bring lower-sulfur gasolines to assist performance and longevity of the aftertreatment systems. If future regulations were to require lower NO_x (for example by 90 percent as California has suggested), there is little doubt that the SI engine and TWC could meet this requirement, and probably at affordable cost. For example, citing "lower acquisition cost and simpler maintenance due to the complexities that diesel buses face in meeting emissions standards," Navistar and Power Solutions International (PSI) recently introduced a gasoline engine for school buses to complement their propane and natural gas SI engine offerings (PSI, 2016). If SI engines for MHDVs adopt gasoline direct fuel injection, there could be a challenge in meeting PM standards unless gasoline particle filters are applied.

In the Phase II rule (EPA and NHTSA, 2016), the agencies would retain the Phase I fuel consumption/GHG standards for SI engines instead of adopting technology-forcing standards. NHTSA and EPA developed the baseline fuel consumption and CO_2 emissions for the gasoline engines from manufacturer-reported CO_2 values used in the certification of non-GHG pollutants.[3] The baseline engine for the analysis was developed from a model year 2011 engine, because this is the most current information available. The average CO_2 performance of the heavy-duty gasoline engines considered (all naturally aspirated V-8 without cam phasers or variable valve timing [VVT])[4] was 660 gram per brake horsepower-hour (g/bhp-hour). For the Phase I standards—reaffirmed for the Phase II standards through 2027[5]—NHTSA and EPA settled on an engine CO_2 standard of 627 g/hp-h,[6] a 5 percent reduction in fuel consumption from the baseline, or approximately 30 percent BTE. This modest stringency was partly predicated on the fact that there are relatively few loose SI engines sold on the vocational market compared to whole vehicles. Hence any technology forcing is done in the whole vehicle standards. This could possibly further improve the cost competitiveness of SI engines and increase the incentive for the MHDV sector to shift away from more efficient diesels. In NHTSA's fuel efficiency technology study Report #1 (Reinhart, 2015), the gasoline baseline engine was a 6.2-liter (L) V-8 based on an OEM product and consistent with the discussion above. Interestingly for the Phase II rule, NHTSA and EPA selected a 6.8-L Ford V-10 engine to build a fuel map for use in the Greenhouse Gas Emissions Model for vocational vehicle simulation. This engine fuel map emulated a 627 g/bhp-h CO_2 performance.

SI engines today are not the dominant engine in vehicles with high annual vehicle miles traveled due to their inferior efficiency compared to diesel (compression-ignition) engines. The lower efficiency of SI engines is due to the following:

- Lower compression ratio, mostly due to combustion knock limits, which is related to gasoline octane number.
- Throttling (pumping) losses at part load. Since SI engines must operate at stoichiometric air/fuel ratio for the TWC system (14.7:1 air:fuel), to reduce load they must reduce the air flow, which induces pumping losses. Boosting, downsizing, and EGR have all contributed to reducing pumping losses over the past 15 years.

[2] 81 *Federal Register* 57234.

[3] 81 *Federal Register* 57234. "The baseline gasoline engines are all stoichiometric port fuel injected V-8 engines without cam phasers or other variable valve timing technologies."

[4] 81 *Federal Register* 57234.

[5] 81 *Federal Register* 83555.

[6] 76 *Federal Register* 57235 and 40 CFR 1036.108(a)(1)(i).

- SI engines, again needing to operate near lambda = 1, where the ratio of air to fuel is chemically correct for complete combustion, also have a less favorable ratio of specific heat in combustion products, thus reducing the work extraction (compared to diesel engines that run fuel-lean).

As mentioned, the technologies in development to reach 42 percent peak efficiency, and to approach this level across the most used operating areas of the engine map, have been covered in other reports. For reciprocating engines, the more important innovations include the following:

- A wide range of boosting and downsizing technologies, which may call for a fuel anti-knock upgrade (this includes electric assist boosting);
- Optimized and high combustion heat release rate within noise and structural limits;
- Friction reduction via materials, designs, and lubricants;
- Heat transfer reduction via low-temperature combustion and thermal management designs;
- Lean burn (needs further development in enabling technologies such as ignition and catalysts);
- Other means of reducing pumping losses (variable valve actuation [VVA], variable displacement);
- Atkinson, Miller over-expansion systems;
- Waste-heat recovery;
- In-cylinder reforming (for heat recovery and lean-limit extension) (Chadwell et al., 2014); and
- Variable compression ratio.

One would expect the advantage of diesels to be more pronounced in duty cycles requiring much part-load operation and idling. Computer simulations comparing projected 2025 diesel and SI engines in vocational vehicles did not show a dramatic advantage for the diesel. This is discussed further in Chapter 6 in the section "Class 6 Delivery Truck." In spite of the increasing efficiency of SI engines and affordability of emission controls, the committee recognizes that SI engines would need to meet durability expectations for an overall attractive cost of ownership. The industry has advanced considerably in its ability to accurately model and simulate pressure and temperature stresses imposed on engine component parts and subsystems, and their related effect on duty-cycle-based durability. When SI engines are adapted for commercial vehicle applications, these advanced analyses are used to identify many upgrades compared to light-duty vehicle use, which are intended to address the higher durability requirements. Improved materials, more robust structural designs, and recalibration to manage peak cylinder pressures are examples. While it is true that such SI engines would likely not achieve the high durability of diesel engines if these modern analysis tools are not used, or the results not heeded in creating robust redesign of critical parts and interfaces, the lower overall production cost of SI engines, combined with the much less expensive exhaust aftertreatment system, can lead to similar total cost of ownership.

Finding: Spark-ignition engines continue to evolve and improve, with potential to reach well over 40 percent peak brake efficiency while still achieving stringent criteria emissions with relatively low-cost aftertreatment. Even higher efficiency would be achievable in technology pathways where high fuel octane numbers were used. They are well suited to renewable fuels such as ethanol or fuels with lower carbon like natural gas or propane. Based on emerging experimental and modeling results, the expected future efficiency of SI engines appears well beyond the Phase II regulatory requirements depending on the extent that advanced technologies for passenger vehicle engines are adapted to MHDV gasoline engines.

Recommendation 4-1: As SI engines continue to be improved, NHTSA and EPA should reassess the future balance in MHDVs between SI and CI, and the reductions in GHG and fuel consumption that might be achieved with a more challenging efficiency requirement for SI engines, including an optimized low-carbon or renewable fuel. An assessment on the extent that high-efficiency technologies emerging in light-duty SI engines can be adapted to MHDVs should be carried out and could be part of an "interim evaluation" for MHDVs.

4.4 COMPRESSION-IGNITION-DOMINATED ENGINES

Compression-ignition engines (diesel engines) are the most prevalent MHDV power plant type, especially for Class 7 and 8 vehicles. This section discusses compression-ignition engine technology, emissions aftertreatment, and total costs. The compression-ignition combustion process is dominated by diffusion burning, with a start of combustion strongly influenced by fuel injection timing. Diesel engines generically exhibit high efficiency by virtue of

- Higher compression ratio than typical SI engines,
- Overall lean operation for more favorable gamma and less heat loss,[7] and
- Unthrottled operation.

The fundamentals of compression-ignition engines have been covered well in the NRC light-duty vehicle study (NRC, 2015a) and other reports and are not repeated here. Their continuous development has made diesels the most efficient transportation power plant available today, with a predominant position in heavy-duty applications. New developments in diesel engines are driven forward and demonstrated through the 21st Century Truck Partnership (21CTP), which seeks to implement and bring to market new transportation power plants through government-industry partnership. The 21CTP uses a coherent and convergent approach to improve engines, emission control and fuels for thermal efficiency, reduced petroleum dependency, lower exhaust emissions, lower cost, and improved safety. Related efforts in the Department of Energy programs include a long-standing catalysis/emission forum (Crosscut Lean Exhaust Emissions Reduction Simulation [CLEERS]), fuel effects studies such as Fuels for Advanced Combustion Engines and more recently fuel-engine optima, and combustion research and development (R&D) collaboratives. These efforts support diesel engines' continuing ability to meet requirements for criteria emissions as well as fuel efficiency and GHG emissions.

The 21st Century Truck Partnership projects like SuperTruck (ST) demonstrate the possibilities in record engine efficiency. The recently completed phase of the SuperTruck program saw several teams achieve engine efficiencies reaching 50 percent BTE in real-world packaging and on-road conditions (see NRC, 2015b). Concepts were defined to achieve 55 percent BTE in these projects. In the second phase of SuperTruck, teams are challenged to demonstrate 55 percent (Funding Opportunity Announcement [FOA] issued in March 2016).[8] Some engine concepts may utilize features of low-temperature combustion, which is described below in the section "Kinetics-Dominated Combustion Engines."

Other concepts include improved combustion, as embodied in the technologies listed in Figure 4-6. The optimized bowl, injector, and heat transfer efforts represent combustion improvement. The team is performing simulation and experiments with the objective of shortening the combustion interval as much as possible to maximize the work from expansion and minimize heat loss. As shown in Figure 4-6, the objective is to gain approximately three percentage points of improvement in engine BTE through this combustion improvement. Details about the engine technologies used by the four SuperTruck engine teams are given in detail in the NRC (2015b) report, *Review of the 21st Century Truck Partnership, Third Report.* Noteworthy is waste-heat recovery used and considered by all four groups, either with Rankine cycle or turbocompound plus Rankine cycle. The engine efficiency improvements have been met while meeting emissions regulations.

An important aspect of diesel engines and their efficiency is meeting criteria emissions. Because of their lean burn and variable air/fuel ratio over their operating map, diesel engines require a very different type of emission control aftertreatment system than stoichiometric SI engines. Before 2010, diesel emissions mostly used in-cylinder NO_x controls. That year marked a transition to the synergistic integration of engine and fuel combustion development, with the capabilities of the aftertreatment systems. The use of selective catalytic reduction (SCR) on medium- and heavy-duty diesel engines became widespread in 2010 to meet the 2010 standard that required

[7] Lower temperatures and leaner mixtures within the cylinder result in larger values of gamma, the ratio of specific heats of the gases in the cylinder during the closed portions of the engine's cycle ($\gamma = c_p/c_v$). A larger average γ results in more work being extracted during the closed-cylinder portion of the engine's mechanical cycle, leading to a decrease in the amount of useful energy leaving the system in the exhaust.

[8] The next phase of the SuperTruck program calls for 55 percent engine efficiency of a Class 8 tractor-trailer with 65,000-pound gross combined vehicle weight traveling at 65 mph. The peak thermal efficiency may be slightly higher.

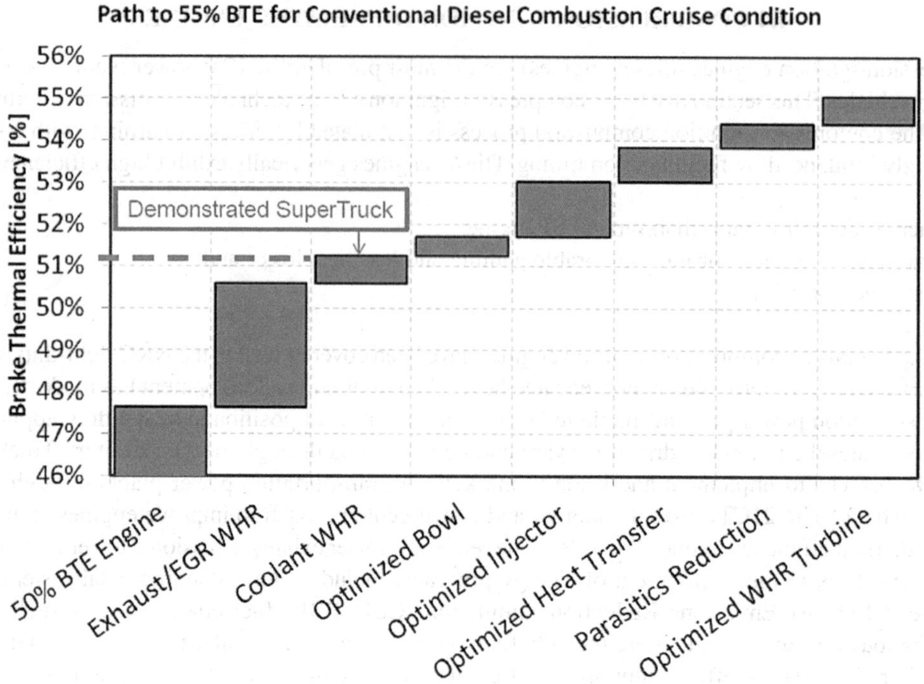

FIGURE 4-6 Cummins-Peterbilt SuperTruck team's projected incremental gains to reach 55 percent BTE from its current 50 percent BTE engine.
SOURCE: Koeberlein (2014).

tailpipe NO_x emissions of 0.2 g/bhp-h. The introduction of SCR allowed higher engine-out emissions of NO_x, and an associated 5 percent reduction in fuel consumption of the 2010 engines compared with the 2007 engines. However, a trade-off required an addition of about 2 percent (per unit diesel fuel uses) of the urea solution diesel exhaust fluid required by the SCR system (Cummins Filtration, 2009; Prevost/Volvo, 2010).

Costs of the criteria emissions aftertreatment systems for MHDVs are important factors in the choice of power plant. As described above, the prevailing system is urea SCR for NO_x, in combination with a diesel oxidation catalyst and diesel particulate filter (DPF) since 2010. The content and complexity of these systems result in substantially higher cost than the TWC used with SI engines. The International Council on Clean Transportation (ICCT) reported the added cost of diesel emission controls (direct manufacturing costs) as about $7,000 for meeting the 2010 U.S. standards and Euro VI (Figure 4-7).

Presentations at the Technology and Maintenance Council meeting in 2012 showed cost increases considerably higher than this, but the increases likely included more than just the aftertreatment system.[9] In contrast, Martec reviewed light-duty vehicle diesel costs, including the NRC 2015 Light Duty Report (NRC, 2015a), and found the costs in that report to be overestimated (Martec Group, 2016). Martec concluded that progress in diesel emission control technology, along with reductions in platinum group metal costs, would enable very cost-effective diesel engine systems for many light-duty vehicles. These findings as generally applicable to heavy-duty (HD) diesels and an indicator of probable success for HD diesels meeting future NO_x targets cost effectively compared to alternatives. The added initial cost and maintenance cost of diesel aftertreatment is certainly a parameter to watch regarding the market share of diesel versus SI engines in the future.

The Phase II proposed rule (EPA and NHTSA, 2015) assumed no change in regulation for engine NO_x output, yet there is an interest and investigation regarding more stringent NO_x standards by the California Air Resources

[9] Fleet Experiences with EPA 2010 Compliant Diesel Engines, Technology & Maintenance Council S3 Engines Study Group Annual Meeting February 23, 2012.

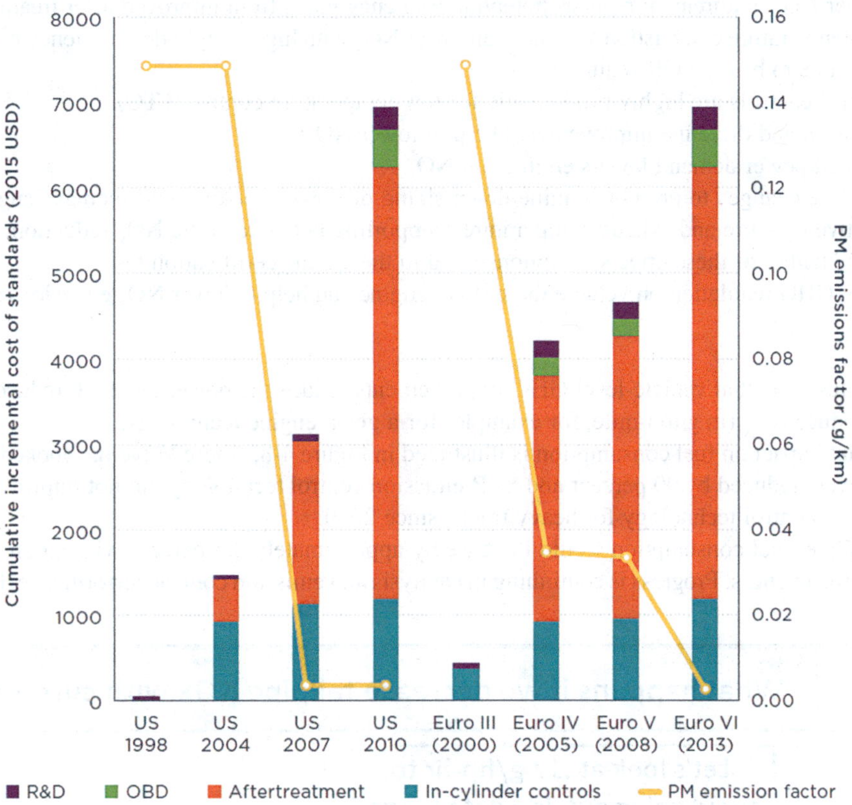

FIGURE 4-7 Estimated overall costs of emission controls for diesel engines for meeting U.S. standards through 2010, and EU standards to 2013, subdivided by costs of various aspects of technologies for reducing emissions.
SOURCE: ICCT (2016).

Board (CARB). CARB has stated its intent to lower NO_x as much as 90 percent below current (0.2 g/hp-h) standard, i.e., to 0.02 g. In December 2016, EPA issued a memorandum indicating that work would be begin on regulations to further reduce NO_x from HD diesels with a target date of 2024 for all 50 states. This follows much public discussion and technology assessments especially in California about the need for stricter emissions regulations and petitions to EPA from California municipalities to act. Proposed regulations in China (Beijing 6 and China 6) stand to be the most stringent standards in the world in the 2017-2019 time frame (Johnson, 2016). The increasing stringency internationally provides strong additional incentive to improve the technology and increases confidence in success. In the next few pages of this report, data are presented which are encouraging that diesels could comply with new standards without loss of efficiency, but there will undoubtedly be additional cost.

Achieving lower NO_x levels could increase the cost of the diesel engine system and perhaps reduce efficiency (increase fuel consumption). In a presentation to the committee, Dr. Anthony Greszler[10] effectively summarized the NO_x-fuel consumption trade-off as follows:

- Trade-off between NO_x and engine efficiency has been long established:
  - In-cylinder NO_x reduction is highly detrimental to diesel cycle combustion efficiency.
  - Improvements in aftertreatment can lower NO_x (at potentially high cost) or contribute to better efficiency at current NO_x levels.

[10] Anthony Greszler, Volvo Trucks North America, "Impact of NO_x Control on Heavy Duty Vehicle Efficiency," presentation to the committee, July 30, 2014, Ann Arbor, MI.

- Any lower NO_x requirement reduces potential efficiency gains from improved after treatment:
  - Low-temperature combustion schemes can lower NO_x with high in-cylinder efficiency but large pumping losses due to heavy EGR required.
- Real-world controls are highly problematic for low-temperature control (LTC).
- Vehicle level and driveline improvements help to lower NO_x:
  - Reduced power demand lowers engine-out NO_x.
  - Driveline changes to promote engine down-sizing or down-speeding will increase engine brake mean effective pressure and exhaust temperatures supporting better catalytic NO_x reduction.
  - Note: Neither of these effects is demonstrated in the engine certification test.
- Focusing GHG regulation on vehicle rather than engine can help to lower NO_x emissions at any regulated NO_x standard.

A key conclusion is that vehicle-level GHG improvements reduce the power demand, reduce fuel consumption, and thus reduce NO_x (in g/ton-mile, for example) for a given engine technology.

The potential impact on fuel consumption is illustrated in Figure 4-8, where Volvo has looked at a case where NO_x standards were reduced by 90 percent and SCR emission control technology did not improve. SCR has been the dominant NO_x control technology for heavy trucks since 2010.

In this example, fuel consumption would increase by approximately 3.5 percent without an improvement in aftertreatment effectiveness. Progress is continuing in catalysis and emission control performance, however. Per the

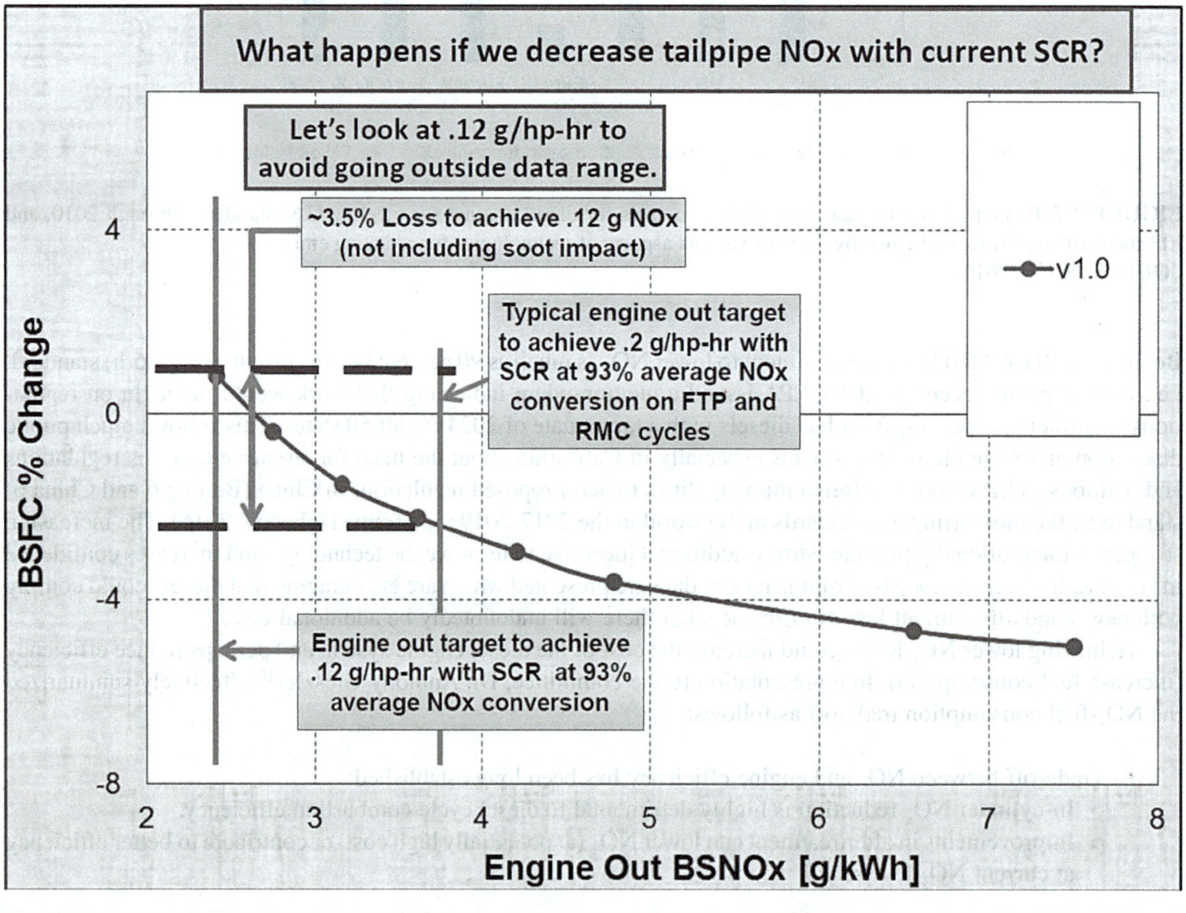

FIGURE 4-8 Trade-off between NO_x and fuel consumption in a case where SCR aftertreatment performance is not improved.
SOURCE: Greszler (2014).

Manufacturers of Emission Controls Association (MECA), "Every generation of SCR systems has led to improvements in catalyst conversion." Further findings of MECA reveal that improving integration of emission control approaches is yielding simultaneous reductions of both CO_2 and NO_x, far from the experiences of a decade ago. Figure 4-9 shows certification data from 2002-2015, indicating that CO_2 and NO_x emissions have not necessarily been a trade-off since the introduction of SCR aftertreatment in 2010, except that there is additional cost of urea consumption and added component features or technology content. The trend is very clear toward simultaneous reduction of NO_x and CO_2 for diesel compression-ignition engines.

The graph from MECA refers to future enabling technologies for reduced NO_x. Many of the technologies are reviewed by Johnson (2014) and Johnson and Corning Incorporated (2016) and include SCR catalyst systems with improved low-temperature performance, SCR+DPF (SCRF Johnson Matthey TM), SCR combined with NO_x storage catalysts, and importantly new calibrations to enable faster catalyst warmup. Non-urea forms of delivering ammonia continue to be studied and developed. Sax et al. (2014) reviewed the California low-NO_x program for the committee as well. Research is targeting high NO_x conversion at 150 degrees Celsius (°C). For reaching a cycle standard of 0.02 g/hp-h NO_x, what is needed, per MECA, is

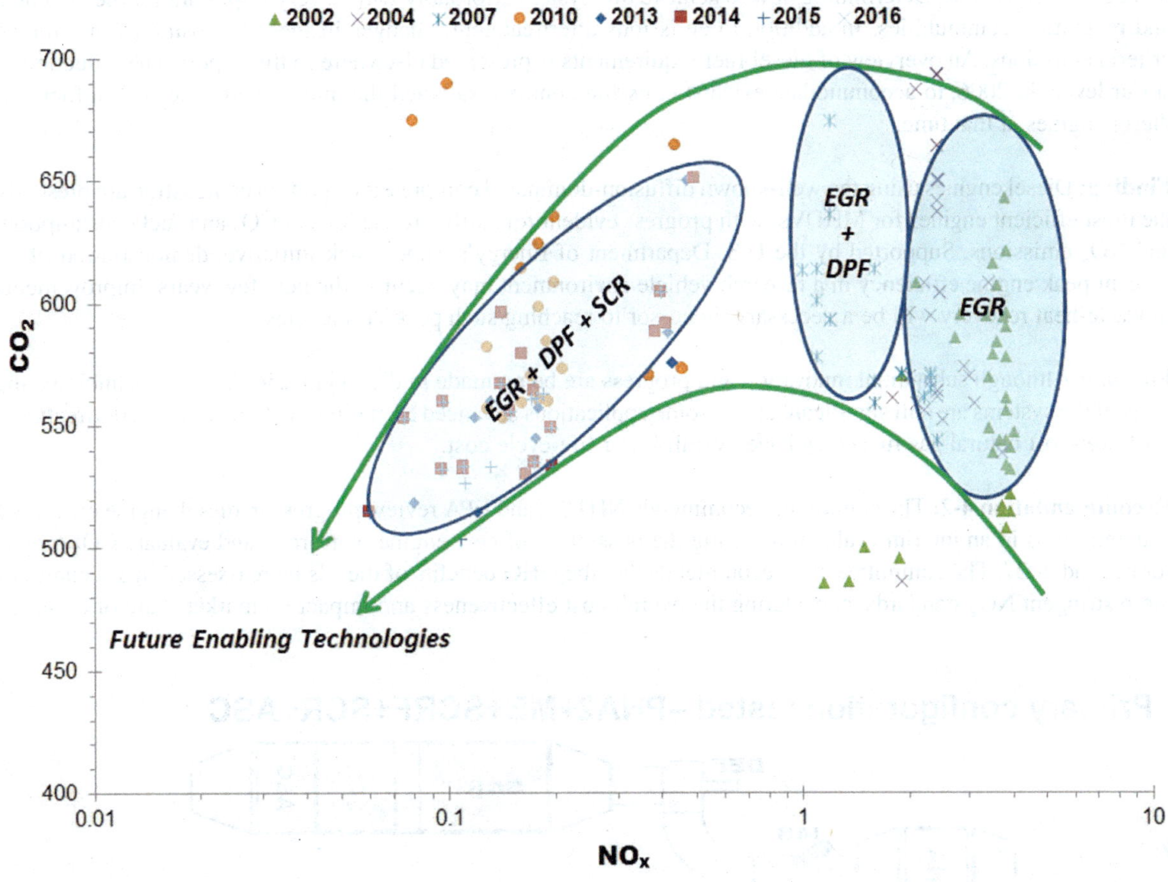

FIGURE 4-9 CO_2 and NO_x emissions from certification data over 14 years reveal progress and advantages of engine/aftertreatment development and integration. Moving from EGR, to EGR + DPF, to EGR + DPF + SCR showed orders of magnitude reduction in NO_x and constant or decreasing CO_2.
SOURCE: Tenneco, from data compiled by Dr. Benjamin Shade. Copyright 2015, Tenneco Automotive Operating Company, Inc. for MECA (2015).

- 95 percent efficiency for cold start,
- 99 percent for hot start, and
- 99 percent on steady-state cycle.

Low-temperature performance is an area of ongoing emphasis in the partnerships between DOE and industry as evidenced in the CLEERS consortia. The approach includes development of improved catalyst formulations with higher Cu content (Gao et al., 2015), developing improved understanding and models of ammonia absorption (Pihl et al., 2015), and development of improved protocols for characterizing catalysts (Rappé et al., 2016). A recent announcement by Continental regarding close-coupled oxidation catalysts is a further example of how the diesel aftertreatment systems are continuing to improve (ET Auto, 2016).

CARB sponsored research at SwRI to explore paths to reduce NO_x by an order of magnitude below the current standard of 0.2 g/hp-h. Many configurations were examined, and recently the configuration shown below in Figure 4-10 was found to achieve a NO_x level of 0.012 g/hp-h on the composite test cycles. It utilizes a passive NO_x adsorber, mini-burner, SCR and SCR on filter, plus an ammonia slip catalyst (Sharp, 2016).

The impact on CO_2 (fuel consumption) was reported as a 1.4 percent increase over the base engine. These results are encouraging that the CI engine's efficiency can be mostly preserved with measures installed to greatly reduce NO_x emissions. Determining how to achieve this system affordably will be very important for the technical and regulatory communities. In addition to emissions aftertreatment, changes in fuel composition can improve criteria emissions. An overview of diesel fuel requirements is presented elsewhere in this report. The reduction of sulfur levels in 2006, to accommodate exhaust emission controls, satisfied the major change needed in fuels for diesel engines at that time.

Finding: Diesel engines using the well-known diffusion-dominated compression-ignition combustion are presently the most efficient engines for MHDVs, with progress evident for further reductions in CO_2 and fuel consumption, and NO_x emissions. Supported by the U.S. Department of Energy's SuperTruck initiative, demonstration of 55 percent peak engine efficiency in a research vehicle environment may occur in the next few years. Improvements in waste-heat recovery will be a necessary precursor to reaching such peak efficiencies.

Finding: Although substantial innovation and progress are being made in diesel emission control technology, the costs of the systems are still significant and in some applications advanced SI engines with lower-cost aftertreatment (or lower-cost natural gas fuel) may have overall lower life-cycle cost.

Recommendation 4-2: The committee recommends NHTSA and EPA review progress in diesel engine efficiency and emissions in an interim evaluation, noting the possibility of new engine platforms, and evaluate CO_2 targets for beyond 2027. The committee also recommends that the GHG benefits of diesels be reassessed in a scenario of more stringent NO_x standards, considering the overall cost effectiveness and impact on market share of diesels.

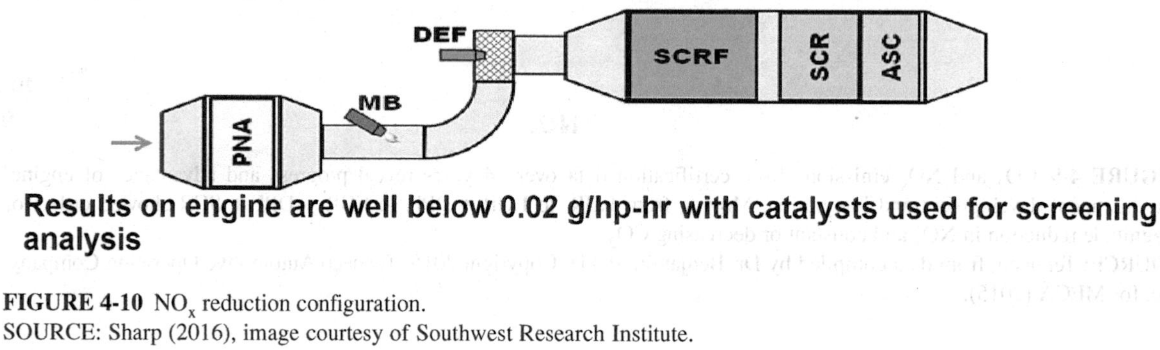

FIGURE 4-10 NO_x reduction configuration.
SOURCE: Sharp (2016), image courtesy of Southwest Research Institute.

4.5 KINETICS-DOMINATED COMBUSTION ENGINES

Kinetics-dominated combustion processes are another opportunity to improve powertrain efficiency. Kinetics-dominated combustion processes have generally fallen under the term LTC, also referred to as HCCI, premixed charge compression ignition (PCCI), and reactivity controlled compression ignition (RCCI), and by many other names. LTC is dominated by chemical kinetics and is largely flameless and is controlled by the staged, autoignition combustion of dilute mixtures created by excess air and/or EGR). Recent assessments of fuel effects for LTC identified approximately 13 documented variants of LTC. An abbreviated spectrum is illustrated below in Figure 4-11, showing the key distinguishing features of fuel injection timing, degree of mixing, and fuel properties (Dempsey et al., 2016).

Kinetics-dominated low-temperature combustion is an area of research that has been pursued in recent decades and has been addressed in numerous publications including in recent National Academies reports reviewing DOE's 21st Century Truck Partnership to which the reader is referred for additional details (NRC, 2008, 2012, 2015b). The following brief discussion follows from those reports.

The basic concept of LTC, in its various forms (Figure 4-11), is to carry out the combustion process in an ultralean or dilute mixture, the temperature of which is low enough to forestall the fixation of nitrogen and oxygen, i.e., with the intent to reduce NO_x formation to levels capable of meeting prevailing emission standards without aftertreatment, and whose lean composition avoids the formation of particulate matter. Equally significant, there are features of LTC that also can enhance engine efficiency and reduce GHGs. The fundamental technical issue or barrier to LTC combustion in diesel (and SI) engines is that there is no longer a source of positive ignition, which is easily controlled. Although stable control can be achieved at a steady-state set point, difficulty arises in

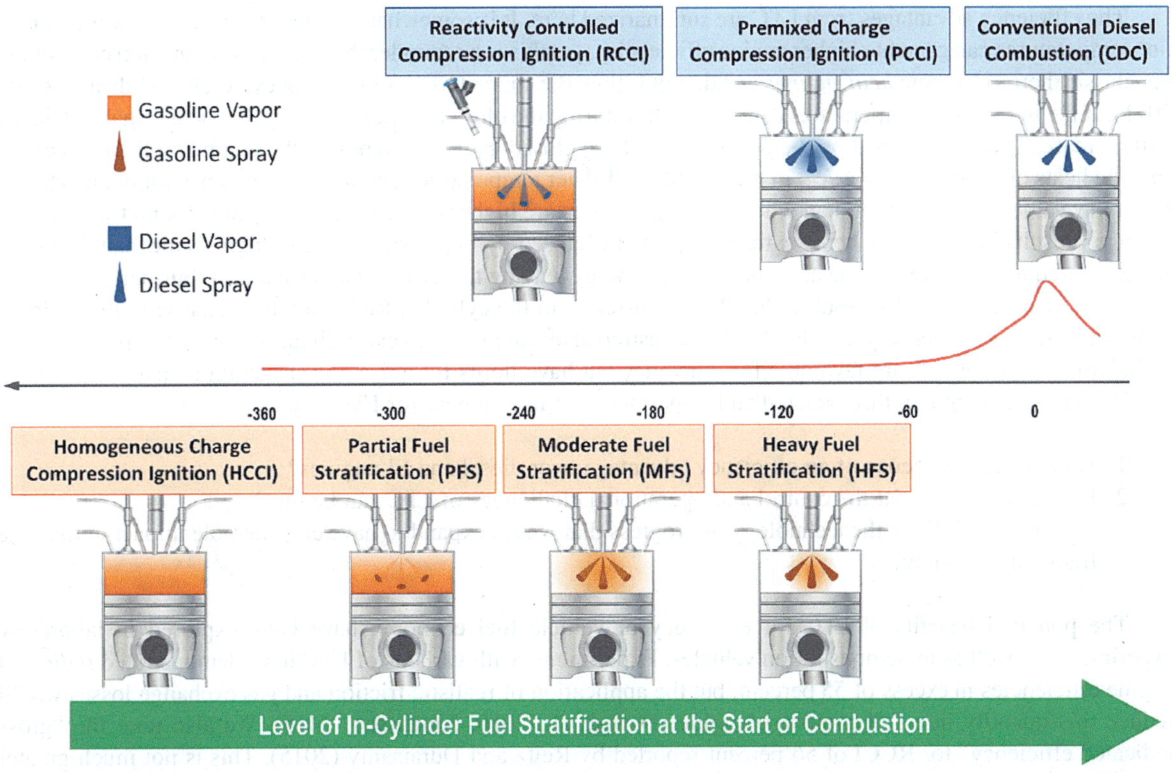

FIGURE 4-11 Landscape of low-temperature combustion modes according to fuel stratification, fuel type, and timing of fuel injection into the combustion chamber.
SOURCE: Dempsey et al. (2016).

establishing control over a broad operating envelope and during rapid load transients. The fuel and air are premixed before entry into the cylinder, or fuel is injected very early in the case of HCCI to achieve homogeneity. In the case of PCCI, fuel injection takes place later in the compression stroke, so there is considerable stratification of the fuel and air. In either case, the mixture is so dilute that combustion has to occur through the natural chemical kinetic mechanisms between the fuel and the surrounding air and residual gases (i.e., burned gases containing nitrogen, oxygen, CO_2, and water [H_2O] as major species) as the mixture is compressed to higher temperatures during the compression stroke of the engine. This reliance on gas-phase chemical kinetics results in major control and operational issues because the kinetic rates do not scale with engine speed (measured in revolutions per minute [RPM]), are variable at different loads, are dependent on engine thermal condition and prior-cycle events, and are a function of the fuel composition, which can vary across a significant range (cetane number in the United States, for example, typically ranges from 40 to 54, and biodiesel content varies from 0 to 11 percent; Infineum, 2014) and still be acceptable for traditional diesel combustion.

A method of overcoming the difficulty of managing fast transients in an engine and expanding the engine operating envelope involves the use of two fuels of differing reactivity introduced into the cylinder in a controlled sequence. The resulting RCCI combustion varies the ratio of fuels fed to the cylinder depending on load required (Reitz and Duraisamy, 2015). This provides more rapid and simple control than the use of variable engine geometry, boost, exhaust gas recirculation, and intake temperature control. However, it has the disadvantage of a design with two separate fuels on board, with no guarantee that these fuels will be used in the same ratio in different vehicle activities and applications. As described in most literature, RCCI uses port injection for the low-reactivity fuel (often gasoline) and direct injection for the high-reactivity fuel (often diesel fuel). A recent development utilizes direct injection for both fuels and is termed direct dual-fuel stratification (DDFS). Comparison of DDFS to other LTC modes has been described by Wissink and Reitz (2015). On-board reforming to control fuel properties in real time has also been explored (Posada et al., 2007).

The efficiency advantages from LTC are summarized here. It is somewhat counterintuitive that lower combustion temperatures can generate higher engine efficiencies, yet this is a reminder that internal combustion engines are not classic thermodynamic *heat engines* and do not follow the characteristics of Carnot cycles, etc. (Edwards et al., 2011). If the in-cylinder temperatures can be kept low during the closed portion of the cycle, the thermal efficiency will increase. This is explained by the dependence of the closed-cycle efficiency of the engine on the ratio of the specific heats of the gases ($\gamma = c_p/c_v$) in the cylinder.[11] Lower temperatures and leaner mixtures within the cylinder result in γ values that are larger than when the temperatures are higher or the mixtures are stoichiometric. A larger average γ results in more work being extracted during the closed-cylinder portion of the engine's mechanical cycle, which subsequently decreases the amount of useful energy leaving the cylinder in the exhaust. Furthermore, lower in-cylinder temperatures also result in less heat transfer from the cylinder. Reduction of radiation to the cylinder walls also increases efficiency. Finally, LTC combustion also can produce very high rates of pressure rise and peak cylinder pressures, which are favorable for efficiency but have limits for noise and structural loads.

With this background, three related and important questions appear for LTC modes:

1. Is there clear evidence of an efficiency advantage over diesel and SI engines?
2. Is there evidence that the control and operational challenges of LTC can be met?
3. Is there an uptake of the technology in progress that would expand to have considerable impact in the time frame of this study?

The potential benefits of LTC for efficiency or vehicle fuel economy have been explored in laboratory experiments as well as in demonstration vehicles. Experiments with variants of LTC have demonstrated *indicated* engine efficiencies in excess of 55 percent, but the application of realistic friction and gas exchange losses would reduce this quantity in a full prototype engine (Daw et al., 2010; Johansson, 2015). We also note the "gross indicated efficiency" for RCCI of 56 percent reported by Reitz and Duraisamy (2015). This is not much greater than expected for peak efficiency of a diesel engine on a test stand. This same review describes experiments with

[11] C_p, specific heat at constant pressure; C_v, specific heat at constant volume.

higher compression ratio and no piston cooling that stretched gross efficiency toward 60 percent. Additionally, there are experimental results combined with vehicle simulation, as well as data from prototype vehicles. Oak Ridge National Laboratory (ORNL) used a multicylinder RCCI engine to develop engine maps that were then applied to a simulation of a passenger vehicle (Curran et al., 2015). The key results are shown below in Figure 4-12, where we see RCCI enabling 8 to 15 percent higher fuel economy than a baseline diesel, and even much higher improvement compared to a gasoline port fuel injection engine. As we note elsewhere, RCCI is challenged by HC emissions, and additional aftertreatment penalties are acknowledged but not included in this study.

An RCCI hybrid research vehicle was assembled by a University of Wisconsin team and evaluated in chassis dynamometer experiments at Ford and ORNL. Importantly the vehicle with the RCCI engine successfully negotiated the Federal Test Procedure (FTP) cycles, but its lack of optimization and high losses in the drivetrain inhibited demonstration of a fuel economy advantage (Hanson et al., 2015).

Again, for a light-duty vehicle application, Delphi has taken gasoline direct injection compression ignition (GDCI) engine development to a vehicle demonstration (Sellnau et al., 2014, 2015). GDCI is characterized by multiple injections and significant stratification and high compression ratio (approximately 15). Initial multicylinder engine development achieved efficiency of about 38 to 39 percent on 91 RON gasoline, approaching the best production diesel engines and somewhat better than SI engines. The vehicle demonstration project, with Hyundai as partner, "culminated in a vehicle level demonstration of 39 percent combined City+Highway fuel economy improvement with a warmed-up engine relative to a 2009 port fuel injected baseline vehicle also run warm" (Confer and Delphi Advanced Powertrain, 2015). A follow-on project is addressing emission control development for GDCI.

LTC processes depend on the ignition quality of the fuel used in their operation. If a new fuel specification were needed, most likely the market introduction and uptake of LTC engines would be slowed. Fuels with high cetane numbers (CNs) such as those typically used for diesel fuels have significant low-temperature reactivity and thus ignite easily at modest overall engine compression ratios. These low compression ratios limit the maximum thermal efficiency obtainable from such cycles. Fuels with lower cetane, or higher octane, ratings have a smaller amount of low-temperature reactivity and thus require higher compression ratios to obtain ignition. Research activity includes investigating the optimization of fuel composition for such approaches to obtain combustion at appropriate compression ratios and engine operating temperatures. Volatility of the fuel also plays a role in

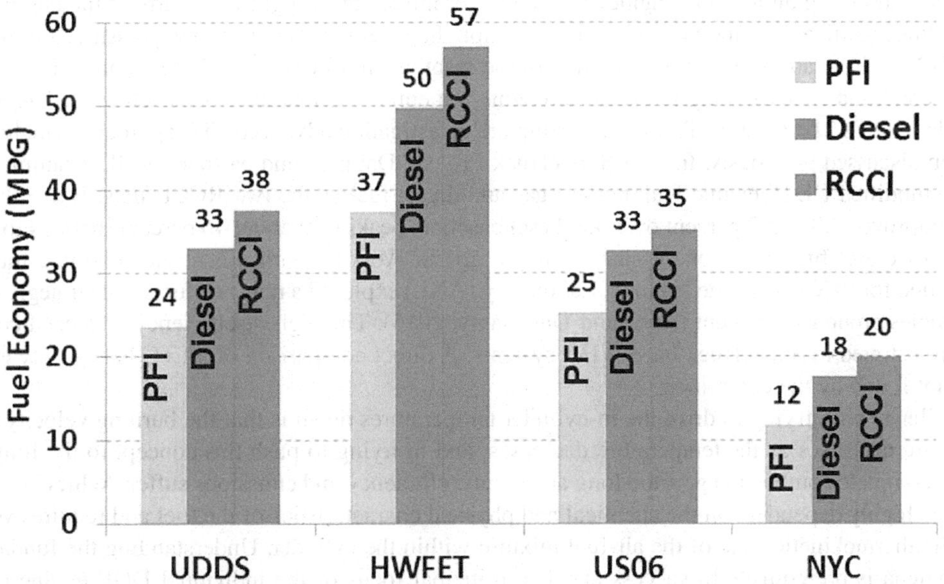

FIGURE 4-12 Drive-cycle fuel economy comparison of low-temperature combustion schemes: port fuel injection (PFI), diesel, and multimode RCCI.
SOURCE: Curran et al. (2015).

establishing the ignition timing. It is difficult to prepare completely homogenous gaseous mixtures with heavy fuels, and ignition may start in a preferred region for a mixture that is inhomogeneous on either a large or small length scale.

Furthermore, if LTC is ultimately able to operate successfully at the maximum engine load, the lean or dilute nature of this combustion process will require significant increases in engine maximum cylinder pressures or engine displacement to achieve output similar to those of engines presently used in highway trucks. If these changes in engine cylinder pressure requirements or displacement are introduced into the market, these product introductions would require major changes in engine materials and design and require a research and development cycle of over 5 years to demonstrate appropriate engine life targets. Higher cylinder pressures may also introduce significantly higher friction losses, which should be considered. It is possible to operate an engine with a novel combustion regime at light loads and using conventional diesel injection at high loads, but the emissions spectrum then demands the same aftertreatment found on current production engines.

DOE, through its 21st Century Truck Partnership and USDRIVE Partnership, has funded much of the research on LTC within the national laboratories, industry, and academia. The goals of the DOE laboratory and university research programs are to develop the fundamental understanding and tools necessary to facilitate such a high level of engine combustion control. Numerous engine producers are involved in research projects and the fuel companies have also joined in understanding the effects of fuel composition on advanced combustion strategies (NRC, 2012, 2015a). Critical aspects of how HCCI and diesel LTC combustion progress, how their heat release rate and combustion phasing can be controlled, the sources of HC and carbon monoxide (CO) emissions when the LTC approaches are pushed to limits of operation, and fuel effects on LTC are being unraveled. Advanced combustion regimes have been explored using fuels other than diesel. Although blends of naphtha or fuel cuts lying between diesel and gasoline are suited to the evaporation and ignition requirements, some may represent a safety hazard in the fuel tank ullage, and they do not have a current distribution infrastructure. The Coordinating Research Council has led research on the properties of both diesel and gasoline fuels for alternative combustion mode research, with support from DOE (Kim et al., 2012; Zigler et al., 2012). More recently DOE has revisited this key issue of fuels for LTC via a newly planned Fuel-Engine Optimization program (Farrell, 2015).

Higher efficiencies using LTC in heavy-duty truck engines have also been shown in the laboratory with examples described below. Implementation of diesel LTC approaches has begun in heavy-duty diesels for a portion of the fuel burned during almost any engine cycle and over moderate- to light-load parts of the engine operating range, providing significant engine-out emissions reduction. In general, higher injection pressure, multipulse injection, and EGR use have allowed a greater fraction of the reactive mixtures during diesel combustion to be pushed toward LTC conditions, contributing to the lower engine-out emissions that have been achieved. The SuperTruck projects under DOE's 21st Century Truck Partnership are investigating advanced LTC approaches including RCCI that has been discussed previously. In their SuperTruck Project, Daimler, and partner ORNL, examined dual fuel (RCCI) in a modified DD15 engine with natural gas and diesel fuel as the two RCCI fuels. Brake thermal efficiency was improved 3.8 to 5.7 percent over the diesel baseline, peaking at about 46 percent, with 74 to 86 percent replacement of diesel fuel on an energy basis (Singh, 2015). Working with a Navistar modified multicylinder engine modified for RCCI, Argonne National Laboratory (ANL) explored a range of fuels and strategies and found peak efficiencies around 43 percent (Reitz and Duraisamy, 2015). The highest-efficiency advanced conventional engines demonstrated in SuperTruck have WHR systems. A direct comparison of a LTC heavy-duty engine with WHR was not found by the committee.

The challenge with trying to drive the in-cylinder temperatures down is that the burning velocity of the fuel and air mixture decreases as the temperature decreases, and in trying to push this concept to the limit, the time necessary to complete combustion gets too long and engine efficiency and emissions suffer. Achieving this type of combustion is highly dependent on the chemical and physical characteristics of the fuel and requires very precise control of the thermokinetic state of the air-fuel mixture within the cylinder. Understanding the fundamentals of these phenomena is prerequisite to success and is a principal focus of the individual DOE engine combustion research projects in 21CTP.

If successfully achieved, LTC strategies yield higher closed-cycle efficiency that minimizes heat loss from the cylinder and exhibits low NO_x and particulate emissions. This is shown in the plot of fuel-air equivalence ratio versus temperature (Figure 4-13), which shows that low-temperature combustion operates in a regime that has both low soot and NO_x. At the same time, LTC introduces concern about unburned HC and CO emissions. Consequently, it is likely that aftertreatment will still be required, if not for PM and NO_x, then for HC and CO, and the aftertreatment systems will most likely need to operate at lower temperatures than current systems today.

A recent presentation by DOE to the National Academies Committee on Review of the Research Program of the U.S. DRIVE Partnership noted that LTC is still a long-range research endeavor with uncertainty about its application and commercial success.[12] It was noted that LTC faces a number of challenges and that the following issues need to be addressed:

Challenges facing the success of LTC operation include

- Combustion noise (American Council of Engineering Companies guidelines) and combustion robustness,
- High engine-out HC and CO emissions,
- Transient combustion control, and cold and transient emissions,
- Wider speed and load operating region, and
- Cost and complexity.

In summary, the understanding and demonstration of LTC efficiency advantages has become clearer, as have the fuel requirements, and also the challenges to application in commercial engines. The efficiencies of advanced conventional diesel and spark-ignited engines are also being improved, relegating the kinetics combustion modes to the research status or moderate portions of engine operation maps for the time being. Research programs are in place to continue the exploration and development of kinetics-driven combustion. Vehicle simulations with experimental LTC engine maps, as well as full light-duty (LD) vehicle tests have shown moderate *potential* fuel consumption benefits if practical emissions and operational challenges are overcome.

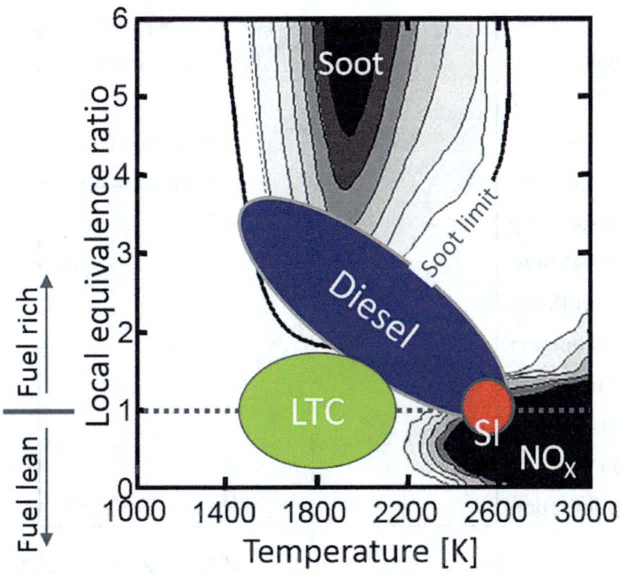

FIGURE 4-13 Fuel-air equivalence ratio versus temperature, showing areas of increasing soot and NO_x emissions, and the operating regimes of diesel combustion, spark ignition (SI), and low-temperature combustion (LTC).
SOURCE: DOE (2012).

[12] Solomon, A., and K. Howden. 2016. Advanced Combustion and Emission Control Technical Team. Presentation to the National Academies Committee on Review of the Research Program of the U.S. DRIVE Partnership, Phase 5, April 20, Washington, DC.

Finding: Progress in understanding and demonstrating the benefits and challenges for kinetics-dominated combustion has continued, but there are still many recipes of the phenomena in flux depending on fuel properties and stratification modes. Full-scale HD engines have been successfully demonstrated on test stands, and LD vehicle demonstrations have been conducted. A full commercial application of LTC was not found by the committee.

Recommendation 4-3: The status of LTC should be followed and reassessed as part of an interim evaluation. The government-industry R&D portfolio should include determination of overall GHG reduction via LTC in comparison to advanced conventional combustion engines, including the applicability of WHR to both engine systems. The market uptake of LTC should be tracked and reported. This comparison should consider overall cost effectiveness including aftertreatment requirements.

4.6 NATURAL GAS ENGINE UPDATE

In the committee's first report (NRC, 2014) natural gas engine availability and performance characteristics were reviewed, including discussion of potential efficiency-improving technologies. Cummins has continued work on their market well-received ISX12-G natural gas SI stoichiometric/cooled EGR engine, and recently EPA certified this engine for the next EPA GHG standard with compliance required in January 2017.

Cummins has identified technologies they believe will enable efficiency development toward EPA's Phase 2 standards, proposed to be effective in 2021. These are listed in Figure 4-14.

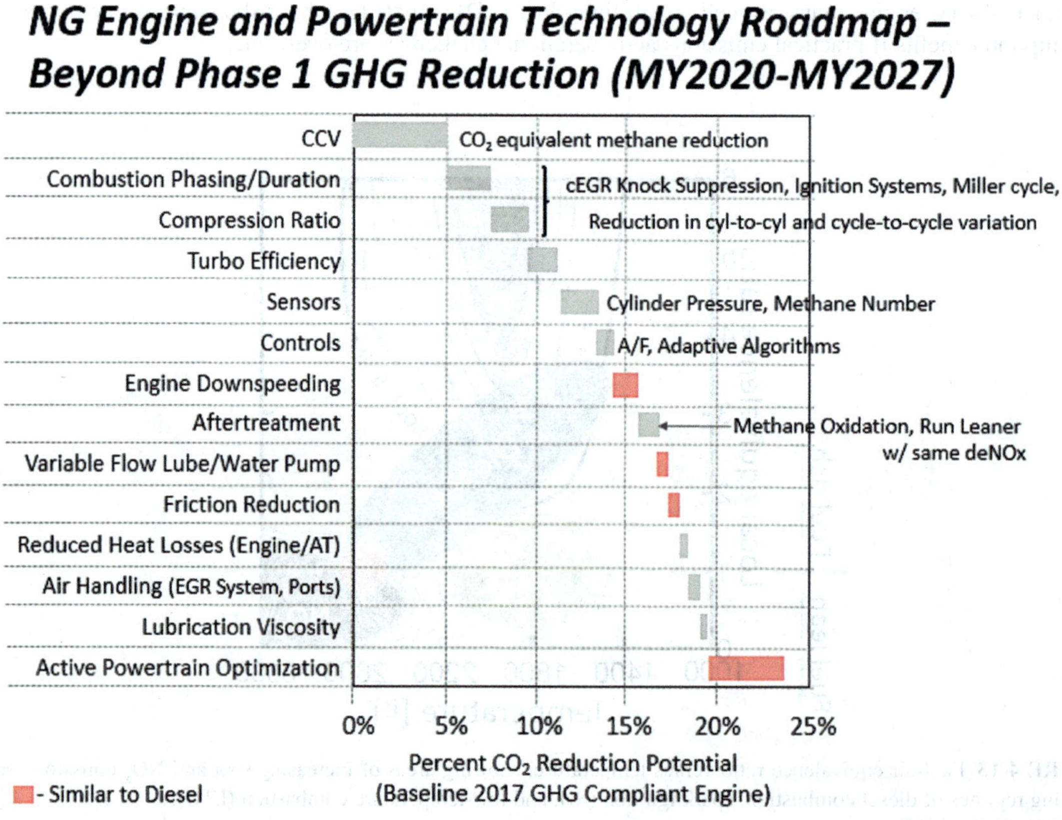

FIGURE 4-14 Cummins engine technologies for advanced NG fueled engines. Their CO_2 reduction forecast is for 20 percent or more, including a notable contribution from avoiding methane loss through the crankcase (CCV, closed crankcase ventilation).
SOURCE: Stanton (2014).

4.6.1 Comparing Diesel and Natural Gas Engines for Reduced NO_x and GHG

A presentation by Cummins to a 2015 CARB Symposium on Phase II GHG provided valuable forward-looking GHG and criteria emission forecasts both for baseline diesel and derivative NG fueled engines (Eckerle, 2015). Figure 4-15 from their presentation used their expected 2017-2018 results as a base case and looked to the late-2020s time frame for potential results from intense research and development. The small (3 percent) quantitative separation between the two CO_2 base cases implies that the combined effects of the NG engine efficiency gap[13] and the methane tailpipe leakage basically eliminate the low-carbon advantage of natural gas.

This result is similar to the 5 percent NG net improvement reported for the ISX 12 Diesel and NG-2013 FTP data from Figure 5-5 of the NRC (2010). These two reports together suggest that technology development will likely reduce GHGs from both the SI natural gas and DI diesel engines, but the net CO_2 differences are not be expected to be near the 25 percent carbon differences of these fuels.

Note that the current diesel behavior (black line) "anchors" the graph with its 2017-2018 NO_x at 0.2 g/hp-h and the 0 percent CO_2 change point. Applying new diesel and new natural gas technologies drives the expected NO_x lower along with simultaneous GHG (CO_2) reductions as shown. Both late-2020s regions center at about a 10 percent CO_2 reduction from their respective bases with the identified mix of technologies.

Figure 4-15 also illustrates a potential solution for application in ozone nonattainment regions, another CARB objective. That NO_x solution favors the NG, with the currently imagined new technologies for both engines.

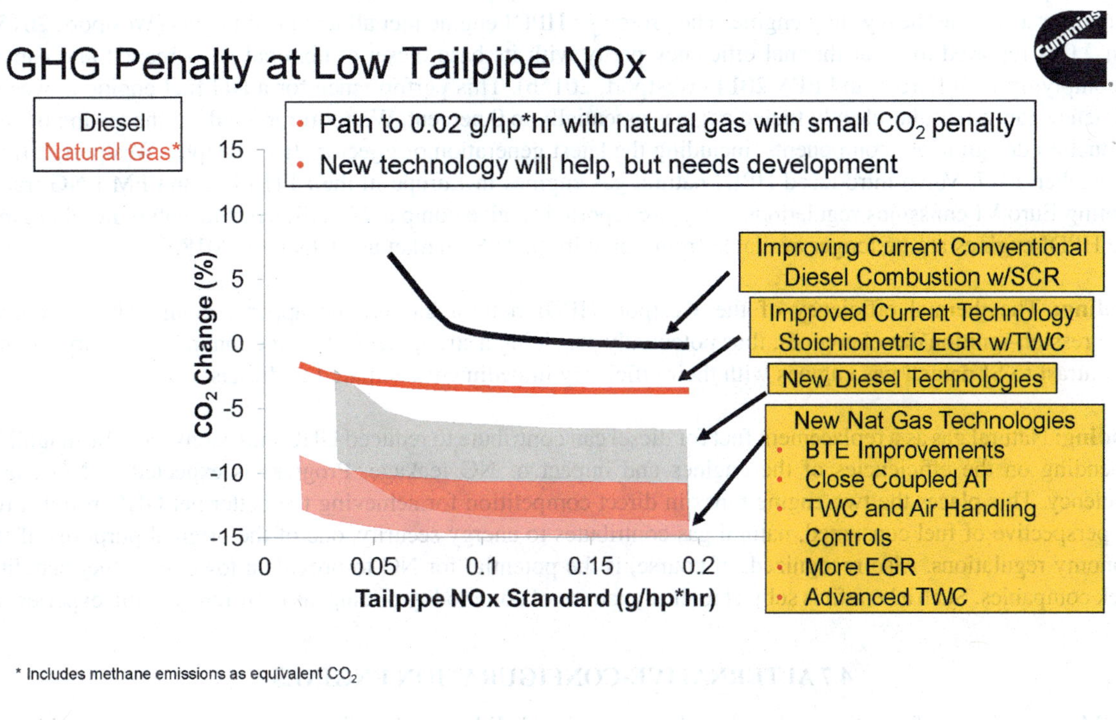

FIGURE 4-15 Projected impacts on CO_2 emissions from heavy-duty truck diesel and natural gas engines with the addition of technologies designed to lower NO_x emissions.
SOURCE: Eckerle (2015).

[13] NG spark-ignition engines have features that make them less efficient than compression-ignition diesel fuel engines. The NG engines have lower compression ratio, operate at stoichiometric fuel-air mixture, and use throttling for load control, all tending to reduce efficiency compared to diesels. NG engines that use high-pressure direct injection (DI) are closer to diesel engine efficiency.

Finding: There are engine research, development, demonstration, and deployment opportunities both for advanced diesel-fueled and NG-fueled heavy-duty truck engines, which can contribute to GHG (CO_2) and fuel consumption reduction in truck transportation as recently demonstrated and forecast.

Recommendation 4-4: NHTSA should coordinate with other federal agencies to define the necessary research, development, and demonstration projects focused on NG engine thermal efficiency to achieve near-diesel parity, and to manage tank-to-wheels methane leakages in combustion processes as well as crankcase intrusion.

4.6.2 High-Pressure Direct-Injection Natural Gas Engines

Natural gas engines offer opportunities for improved efficiency and lower GHG emissions. In this NG fuel delivery system, the high-pressure-cylinder injected gas is combustion ignited by an appropriately timed pilot quantity of diesel fuel. In the committee's first report (NRC, 2014), this technology was reviewed and it was reported that the Westport high-pressure direct injection (HPDI) was discontinued in the fall of 2013. This natural gas fuel delivery process is now in a Generation 2 design configuration. This evolution by the manufacturer is to optimize system mechanics, hydraulics, electronics, and cost, seeking particularly to overcome the cost obstacle of their first-generation system. In Chapter 6, results are presented for simulation of an advanced HPDI engine in a Class 8 vehicle, where the engine map was approximated based on the 2010 Westport HPDI. This simulation found near parity on CO_2 with a highly advanced diesel engine, and near parity on thermal efficiency with the committee's baseline 2019 diesel.

Westport announced that it has completed an engineering program with Daimler AG to develop the Gen 2.0 system for a Daimler heavy-duty engine. The prototype HPDI engine met all technical targets (Westport, 2015a). Gen 2.0 is reported to be at thermal efficiency parity with its base diesel at nominal 44 percent[14] and capable of complying with Euro 6 and EPA 2014 (Westport, 2015b). This performance for a NG fuel engine moves the efficiency gap versus diesel-only fueled engines essentially to 0 percent. Westport reported engine testing of early production design intent components, including the latest generation of injectors from Delphi (Westport, 2015c). In October 2017, Volvo introduced HPDI natural gas engines in Europe in their FH LNG and FM LNG trucks, meeting Euro VI emissions regulations. They are reported to give comparable efficiency to Volvo diesel engines. The HPDI engines are no longer available from Volvo in the U.S. market as of January 2018.[15]

Finding: The thermal efficiency of the Westport HPDI natural gas engine appears comparable to current compression-ignition diesel engines, thus potentially providing a strong net CO_2 reduction in heavy-duty engines, in contrast to SI natural gas engines with their efficiency impediment compared to diesels.

Finding: Natural gas as a replacement fuel for diesel can contribute to reduced GHG emissions with the magnitude depending on the efficiencies of the engines and impact of NG leakage. Progress is expected in NG engine efficiency. This places the two engine types in direct competition for achieving the better net GHG result. From the perspective of fuel consumed, natural gas contributes to energy security, one of the original purposes of fuel economy regulations. Also recognized, of course, is the potential for NG to provide a lower-cost fuel benefit to truck companies. Such a benefit is subject to the dynamics of commodity pricing, as is currently being experienced.

4.7 ALTERNATIVE-CONFIGURATION ENGINES

Many engine configurations other than the conventional slider-crank reciprocating piston system (with four-stroke cycle) have been investigated and are even presently being explored. A partial list is shown below:

[14] See http://www.westport.com/is/core-technologies/hpdi-2 (accessed December 18, 2019).
[15] See https://www.dieselnet.com/news/2017/10volvo.php (accessed December 18, 2019).

- Variants of reciprocating piston engines
  - Two- and six-stroke reciprocating piston engines (Blain, 2015; Kierstein, 2014).

    Two-stroke spark-ignition and diesel engines have been used in a number of on-highway vehicles and remain popular in handheld equipment and applications where power/weight and simplicity are imperative. New developments or concepts have appeared. The emissions characteristics are generally challenging. A profound advantage in fuel efficiency does not appear likely for MHDV uses, but see the discussion on opposed piston engines below.

    Six-stroke engines have been analyzed and tested wherein an additional compression and power stroke is added to recover residual heat after the initial power stroke. The residual heat can be captured by adding water or additional fuel after the initial power stroke. In the case of water injection, steam is created that expands to create work during the second power stroke. If fuel is injected, endothermic reforming can occur that results in fuel with added energy content and hydrogen content which enhances combustion. In some respects the six-stroke concept is waste-heat utilization. Unlike the split-cycle concepts, there need not be transfer of the working fluid between multiple cylinders (Conklin and Szybist, 2010; Kothari et al., 2014).
  - Piston engines with staged compression and expansion cylinders ("split cycle"). Multiple expansion cylinders may be used. There have been many embodiments of this approach with generally the same barrier of pressure (exergy) losses as the working fluid is transferred between expansion cylinders. The concept is directionally advantageous and similar to the effect of an Atkinson configuration, but the expansion ratio can be much greater. Hence these engines in concept can remove some of the thermodynamic constraints of the traditional slider-crank mechanism (Clarke and O'Malley, 2013; General Motors, 2012; Ricardo, 2013).
  - Opposed piston engines are in development, some based on designs used decades ago in various applications (Huo et al., 2015; Naik et al., 2015; Pirault and Flint, 2009). These engines typically use a two-stroke cycle, meaning combustion and a power stroke every revolution. Elimination of the cylinder head eliminates a heat-loss path, and the configuration can be conducive to lower mechanical friction. Since the power density is inherently higher, the overall fuel/air ratio on each combustion event can be leaner on average, which helps the efficiency. The geometry of two pistons moving in the same cylinder creates a unique cylinder volume-crank rotation profile that influences heat release. Two organizations, Achates and Ecomotors, are noted here as having demonstrated opposed piston engines of significant size, yet with distinct designs. These are among few unconventional engines that have progressed to multicylinder testing.

    Achates has published test data and analyses of their engine compared to a conventional diesel and estimated greater than 15 percent improvement in fuel economy over a representative range of conditions including transient operation. Engine-out emissions have been reported but as of 2016 there has not been a full demonstration of meeting 2010 regulated emissions levels. Engine-out levels would dictate the need for about 95 percent effective aftertreatment (Sharma and Redon, 2016). A peak efficiency of 43 percent was reported in comparison to the 6.7-L diesel at 41 percent. The advantage was attributed to the mechanisms cited above. Further improvements are expected that would increase the advantage, but the baseline diesel engine is improving as well. Achates modeling results show a path to 55 percent peak brake efficiency in a heavy-duty scale even without waste-heat recovery systems (Abani et al., 2016). Achates as well as AVL were awarded contracts from the U.S. Army Tank Automotive Research Development and Engineering Center to develop and test demonstrator engines for military vehicles. Cummins is a partner to Achates. In addition, the Advanced Research Projects Agency-Energy (ARPA-e) awarded a development contract to Achates in 2015 to develop a gasoline-fueled compression-ignition engine for medium-duty vehicles. Opposed piston engines have their specific challenges in emission controls, lubricant consumption, and thermal management, which are being addressed (Redon, 2011).
  - Axial configuration engines where the crankshaft is replaced by an alternative device such as a swash plate. The result is usually a barrel-shaped engine. There are many configurations.

- Linear alternators using a reciprocating free-piston combustion process. There are many attempted versions of this type of engine that directly generates electricity from reciprocating movement of a free piston within electric coils. The Department of Energy supported a project in this area for many years at Sandia National Laboratory. In concept, very high compression ratios can be used for high efficiency. A recent concept has been publicized as derived by Toyota, and a development project sponsored by ARPA-e was recently started at West Virginia University (WVU) for stationary power. There is a body of work at WVU dating back over 15 years on this topic (Atkinson et al., 1999; Robinson, 2015).
- Other positive displacement devices. There are many mechanical systems that can produce a power cycle based on compression, combustion, and expansion.
- Rotary/eccentric piston engines such as the Wankel and Liqui-Piston. This category includes other cam-like piston motions.
- Rotating vanes such as in the TurbX engine (Arimilli et al., 2003). Other than the Wankel rotary engine, no alternatives to the reciprocating piston/crankshaft configuration have been widely used for on-highway vehicles. Most alternative mechanical configurations produce power with processes that are thermodynamically similar to conventional engines, whereas to gain a notable advantage they would need to produce an over-expansion effect (Atkinson cycle), reduce heat transfer, reduce mechanical friction, enable high dilution or lean burn, etc.

- Gas turbine/Brayton engines. Gas turbines have been explored for use in trucks and cars for many decades and have appeared in truck demonstrations again in the past few years.[16] Gas turbines have benefits in power density and low emissions (they run very lean, so NO_x and particle emissions are relatively low) and are adaptable to many fuels. Gas turbine engines are used in combined-cycle electric power plants and have achieved approximately 60 percent efficiency at the multimegawatt scale; however, such configurations are generally considered impractical for vehicles. Marine diesel piston engines of 10-19 MW exceed 50 percent efficiency. The turbine engine is still the choice for powering the U.S. Army M1 Abrams main battle tank, due to this engine's high power-to-weight ratio. However, the efficiency of this engine was not a primary military specification. Its engine peak efficiency is far below that of the diesel alternatives. They are generally less fuel efficient than piston engines. Turbines rely on using high combustion temperatures to achieve high efficiency, but the continuous combustion occurs at lower temperature than the intermittent combustion in piston engines. The Brayton cycle turbines also require use of highly effective regenerative heat exchangers to deliver competitive efficiency. In spite of progress in compact heat exchangers, the recuperators are large and a packaging challenge. In 2005, Volvo summarized the development and demonstration of a gas turbine for heavy trucks (Figure 4-16), achieving low emissions but with peak efficiencies only around 40 percent (Olsson et al., 2005). In the present committee study, information from Brayton Energy indicated peak efficiency potential for their system approaching 50 percent employing an advanced ceramic turbine. However, in a full system demonstration the efficiency peaked at about 38 percent (DeLizo, 2015).

Simulated vehicle fuel consumption is described in Chapter 6. If regulatory agencies were to require further reduction in NO_x and PM emissions from medium- and heavy-duty vehicles, the turbine engine may receive further investigation. The California Energy Commission in conjunction with Kenworth, ICR Turbine Engine Corporation, and Brayton Energy conducted a demonstration project using a turbine engine in a Class 8 truck in a series hybrid configuration running on natural gas. The vehicle was operated on a chassis dyno and test track, gaining valuable experience and data (DeLizo, 2015). Walmart, Peterbilt, Capstone, Wrightspeed, FedEx, and others are exploring turbines in hybrid powertrains as well, but there are limited data available from these projects (Cobb, 2014; Karagozian, 2013; Senefsky, 2003). Walmart is unique in this field in that they have favored the gas turbine parallel or mechanical drive over the more common series hybrid electric. Many vehicle integrators have utilized the Capstone C30 and C65 in vehicle demonstrations. These demonstrations usually showed low emissions, but

[16] See https://www.braytonenergy.net/brayton-presenting-at-mit-center-for-center-for-advanced-nuclear-energy-systems-canes/ (accessed December 18, 2019).

FIGURE 4-16 Volvo T300 gas turbine engine for heavy duty trucks, circa 1998.
SOURCE: Olsson et al. (2005).

efficiency was quite low, generally well below that of today's SI engines. Low efficiencies are bound by uncooled metal turbine alloys and have low shaft conversion efficiency from the high-frequency alternator generator, combined with the battery and electrical powertrain losses. Finally, the very large thermal inertia of the recuperator results in wasted fuel for those series hybrid applications where the engine is frequently cycled on and off. In November 2016, Wrightspeed and the Ratto Group introduced a refuse hauler powered by a gas turbine generator range extender hybrid, reported to achieve 70 percent reduction in fuel use in the refuse hauler operation compared to a conventional powertrain.[17] In this special application, diesel engines are not able to meet emission standards and thus the alternative engine in the hybrid is SI. The single-shaft turbine generator is 80 kW but with the batteries and electric motors the max power is closer to 300 kW. It uses an intercooler and recuperator, and is capable of operating on diesel fuel, propane, or natural gas. The turbine is in a dedicated range extender configuration and does not follow load (it is either running near peak efficiency or is off). The gas turbine in this system has a best efficiency of about 32 percent, hence its principal advantage in the application is low emissions.

The key question to address is whether there will be developments or circumstances that would cause a shift from diesel to other engine types. Additionally, are there technology developments emerging that would be game changers for diesels as well as other engine categories? Examples would include new materials and manufacturing methods and new chemistries or catalysts for aftertreatment systems. Changes in emissions regulations and fuels could also cause a shift in GHG potential or market preference for engine configurations.

[17] See http://www.wrightspeed.com/releases/wrightspeed-ratto-group-deploy-first-route-enabled-refuse-truck (accessed December 18, 2019).

Finding: There are numerous alternative-configuration combustion engines. Many of these are merely alternative mechanisms to carry out thermodynamic processes very similar to conventional engines, but there are a few that provide clear enhancements via changing the basic cycles or reducing a prominent loss mechanism (such as heat transfer or friction). Few alternative concepts have reached a scale of development where they could be tested in a MHDV system for performance and emission requirements. With limited data found on alternative configurations, the magnitude of improvement over advanced conventional engines is not readily predictable and will depend on prevailing emissions requirements.

Finding: Since so few new concepts pass fundamental tests of merit, and even fewer scale to meaningful power levels, independent benchmark tests would be reasonable endeavors.

Recommendation 4-5: NHTSA, together with EPA, DOE, and the Department of Defense, should maintain a technology assessment of developments and progress in alternative-configuration engines which evaluates them, analytically and experimentally, against benchmarks and fundamental criteria.

4.8 WASTE-HEAT RECOVERY

4.8.1 Introduction

Approximately 50 percent of the fuel energy consumed by typical diesel engines in Class 8 line-haul truck applications is lost through heat transfer and unused exhaust energy and fails to perform useful work. In the example of Figure 4-17 engine heat rejection accounts for 24 percent of the fuel consumed, and exhaust sensible heat accounts for another 26 percent. Technologies and techniques exist or are under development that can recapture a portion of this rejected heat and help reduce vehicle emissions as well as reduce fuel consumption. In these cases, consideration of the second law of thermodynamics dictates how much of that heat can be converted to useful work. Hence in Figure 4-17 Cummins distinguishes between high- and low-quality heat where the higher temperature energy is the greater resource to produce additional shaft work.

Many options exist at least conceptually to utilize these sensible heat streams from the exhaust, EGR coolers, water jackets, and exhaust, with examples shown in Table 4-1.

Several recent review papers (Bhatnagar and Owende, 2015; Liang et al., 2015; Nadaf and Gangavati, 2014; Orr et al., 2016) have covered these very numerous options in some detail. Many WHR systems are investigated or utilized in nonmobile applications, for example, in combined heat-power systems. The ARPA-e Generators for Small Electrical and Thermal Systems program (GENSETS) includes projects on direct energy conversion for small combined heat and power systems. Vehicle applications bring significant challenges such as weight, performance, packaging, and reliability. In general the WHR systems are most effective at high engine loads and in fact offer little benefit in urban driving cycles (Reinhart, 2015). Transient response is a major issue or barrier as well. The Phase II GHG Regulatory Impact Analysis (RIA) includes a summary of NHTSA and EPA's perspective on WHR and concludes that there is high potential yet limited commercialization before 2027 because of the need for further engineering development (EPA, 2015a, p. 2-11).

In particular the organic Rankine cycle (ORC) system has seen considerable progress and adaptation to on-road trucks. An ORC system was demonstrated on a tractor-trailer truck in the 1970s (Patel and Doyle, 1976), and this type of system was selected again in a recent set of advanced vehicle studies. An example project is reviewed here.

4.8.2 Application of Waste-Heat Recovery in SuperTruck and 21st Century Truck Partnership

All of the teams in the DOE SuperTruck Project included waste-heat recovery approaches to improve engine efficiency. The teams chose bottoming cycles and turbocompounding systems in various configurations, but not any of the direct energy conversion methods seen in Table 4-1. The SuperTruck engine systems with WHR are described in the NRC (2015b) report *Review of the 21st Century Truck Partnership, Third Report*. As a represen-

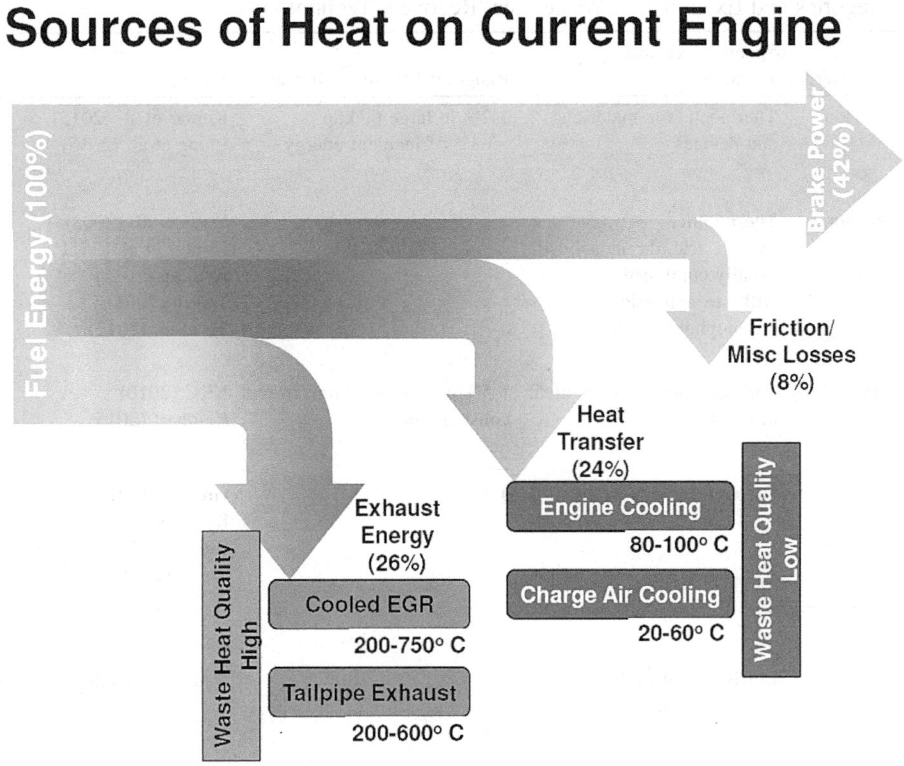

FIGURE 4-17 Sources of heat in engines, including exhaust energy, heat transfer, friction and losses, and brake power. SOURCE: Eckerle (2013).

tative, road-tested example, the Cummins WHR system is shown in Figure 4-18. This system is packaged to fit within the vehicle underhood and underbody space, indicative of overall progress in the technology.

The relative contribution by WHR to engine efficiency gains is shown in Figure 4-17, in comparison to other improvements being introduced by Cummins. Overall, WHR increased the system efficiency (i.e., decreased the fuel consumption) by about 6 to 7 percent.

4.8.3 Continued Development of the Organic Rankine Cycle

In recognition of the need for reducing cost and complexity of WHR systems, the DOE-sponsored project with Eaton, AVL, and Deere is noted in Figure 4-19 and uses a Roots-type expander in the ORC system. Most of the development work has been performed with ethanol as the working fluid.

The system in simulation has a 3 to 6 percent benefit over a wide range of engine speeds and loads, and can be declutched if necessary to avoid parasitic losses. The roots machine operates at a lower speed and lower pressure and can tolerate two-phase flow for potentially improved reliability.

Each of the SuperTruck teams had its own design of ORC and associated WHR components. As in this example of the Roots expander, the committee expects continued refinement in ORC systems.

TABLE 4-1 Categories and Examples of Waste-Heat Recovery Technology

Category of Heat Recovery	Specific Example Technologies	Range of Efficiency Impact	Source
Direct energy conversion	Thermoelectric materials and devices	1-2% in large pickup ~3.3% of incident energy	Kumar et al. (2013) Liang et al. (2015)
Direct energy conversion, additional examples	Thermionic, pyroelectric, thermoacoustic, usually combined with piezoelectric, thermophotovoltaic		Fraas et al. (2003) Hunter et al. (2011) Keolian (2011) Wekin (2008) Yu et al. (2012)
Mechanical recovery	Turbocompound mechanical or electrical	1-5% overall reduction in fuel consumption	NRC (2010) Reinhart (2015)
Heat engine cycle	Organic Rankine cycle	4-8%	Greszler (2012) Koeberlein (2013) Reinhart (2015) Singh (2011)
Heat engine cycle	Refrigeration cycle Stirling engine		Nadaf and Gangavati (2014)
Thermochemical recuperation	Fuel reforming, hydrogen production	10-13%	Daw et al. (2010) Liang et al. (2015) Szybist et al. (2017) Tartakovsky Leonid et al. (2014)

4.8.4 Summary

Many examples of engines with waste-heat recovery have been analyzed, constructed, and tested. Most experiences have been with turbocompounding, Rankine bottoming cycles, and thermoelectric generators; yet many more heat-to-power technologies exist and may see further development. The existence of waste-heat streams throughout the industrial and power generation sectors will likely force technology development at least as much as in vehicles (Elson et al., 2015). In recent projects with modern engines, the WHR systems decrease fuel consumption by 3 to 8 percent. Cost and packaging and transient response are significant challenges to widespread adoption (Delgado and Muncrief, 2015). Costs of WHR systems are discussed in Chapter 12.

Finding: There are many approaches to waste-heat conversion to power that can provide 3 to 8 percent efficiency improvements in modern truck engines. This effect is comparable to other engine improvements in combustion, air handling, friction, etc., and even comparable to many vehicle technology improvements. The WHR systems are clearly technically feasible, yet carry challenges in their weight, packaging, thermal response, and cost. Progress in addressing the challenges is evident. NHTSA and EPA's RIA noted a potential 25 percent penetration in some Class 8 segments by the end of the Phase II rule period.

Recommendation 4-6: Industry and government should continue to research, develop, and apply WHR systems where technical and economic considerations are reasonable to capture this opportunity to reduce fuel consumption.

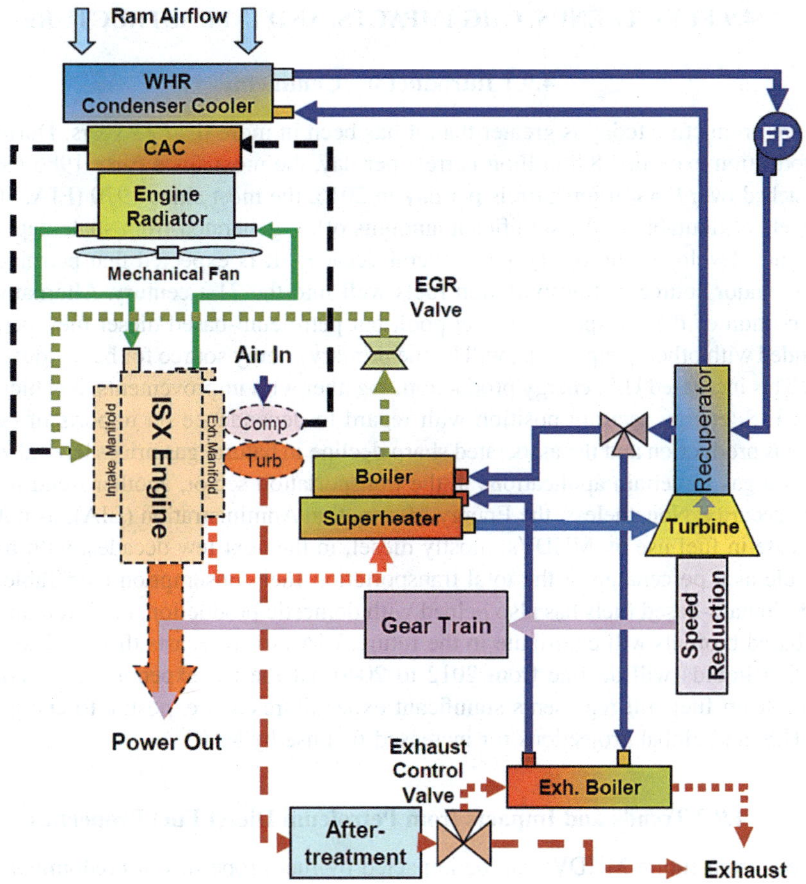

FIGURE 4-18 Cummins waste-heat recovery system.
SOURCE: Koeberlein and Cummins Inc. (2014).

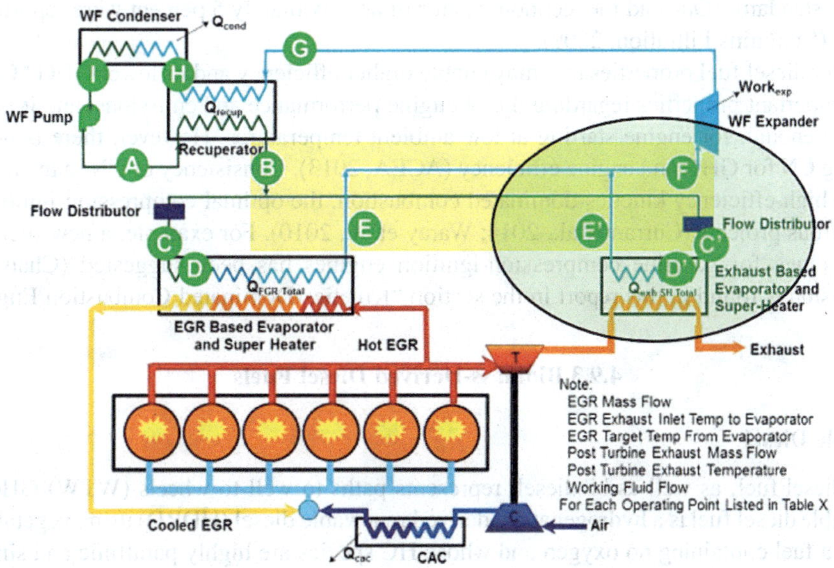

FIGURE 4-19 Organic Rankine cycle waste heat recovery system.
SOURCE: Subramanian (2014).

4.9 FUEL TRENDS, GHG IMPACTS, AND INFRASTRUCTURE

4.9.1 Introductory Comments

U.S. petroleum production today is greater than it has been in more than 25 years. During September 2014, daily U.S. oil production exceeded 8.8 million barrels per day, the most since early 1986 (Shenk, 2014). Crude oil production reached over 9.4 million barrels per day in 2015, the most since 1970 (EIA, 2016). The increased production is largely attributable to the significant amounts of oil generated from shale deposits using advanced extraction techniques developed during the past several decades. It is expected that petroleum-based fuels will continue to be the major source of transportation fuels well into the 21st century. Alternative fuel sources will contribute some portion of the transportation fuel pool, but petroleum-based diesel fuel, perhaps with modified properties or blended with other components, will be the primary energy source for heavy-duty trucks (particularly Classes 7 and 8). This increased U.S. energy production, together with improvements in vehicle fuel consumption, has improved the U.S. energy security position with regard to dependence on imports of petroleum. Increased domestic natural gas production and the associated sharp decline in natural gas prices have also stimulated interest in the use of natural gas in certain applications in the transportation sector, another trend that can contribute to improving energy security. Nonetheless, the Energy Information Administration (EIA), as noted above, forecasts a significant increase in fuel use by MHDVs, mostly diesel, in the next few decades, with natural gas playing a relatively small role as a percentage of the total transportation fuel consumption (see Table A2 of EIA, 2014). Increasing use of biomass-based fuels has also helped with domestic production, but it remains to be seen to what extent cellulose-based biofuels will contribute in the future. EIA also forecasts that the U.S. net import share of petroleum and other liquids will decline from 2012 to 2040 but the net expenditures in constant 2012 dollars will increase. Petroleum fuel still represents significant expenditures and exposure to energy security concerns, especially given U.S. and global projections for increased fuel use by MHDVs.

4.9.2 Trends and Impacts from Petroleum Diesel Fuel Properties

GHG emissions from future MHDVs can be impacted by fuel properties of predominantly petroleum fuels, most notably by addition of renewable fuel content, but possibly by other changes as well. The most impactful fuel change in on-road GHG emissions in recent years was the reduction of diesel fuel sulfur in 2006 that enabled highly effective NO_x aftertreatment devices such as urea SCR introduced in many 2010 on-highway engines. Many engine OEMs were able to recalibrate engines for higher fuel efficiency (and engine-out NO_x) and still meet the tailpipe emissions standards. On-road fuel economy gains of approximately 5 percent were reported, compared to pre-2010 engines (Cummins Filtration, 2009).

Are there other diesel fuel properties that may enable higher efficiency and/or lower GHG? Cetane number is one of the most important properties regarding diesel engine performance and emissions, and it is very important for CN to be high enough for engine starting at low ambient temperatures. However, there is not a compelling case for increasing CN for GHG and engine efficiency (ACEA, 2013). Consistency of CN in the market is desired.

For enabling high-efficiency kinetics-dominated combustion, the optimal compression-ignition fuel is being explored in numerous projects (Curran et al., 2014; Warey et al., 2010). For example, a new fuel in the naphtha/gasoline boiling range for gasoline compression-ignition engines has been suggested (Chang et al., 2012). Additional discussion is found in this report in the section "Kinetics-Dominated Combustion Engines."

4.9.3 Biomass-Derived Diesel Fuels

4.9.3.1 Renewable Diesel

Renewable diesel fuel, as well as biodiesel, represents paths to well-to-wheels (WTW) GHG reduction for MHDVs. Renewable diesel fuel is a hydrogenation-derived renewable diesel (HDRD) using vegetable oil or animal fat that results in a fuel containing no oxygen and whose HC species are highly paraffinic and similar to those in petroleum diesel fuel. See Knothe (2010) for a comparison between biodiesel and HDRD. A comparison of key properties of HDRD to other fuels is shown in Table 4-2.

TABLE 4-2 Comparison of Properties of Three Diesel Fuels for MHDVs: No. 2 Ultralow-Sulfur Diesel Petroleum, FAME Biodiesel, and Renewable Diesel

Property	No. 2 Petroleum ULSD	Biodiesel (FAME)	Renewable Diesel
Carbon, wt%	86.8	76.2	84.9
Hydrogen, wt%	13.2	12.6	15.1
Oxygen, wt%	0.0	11.2	0.0
Specific gravity	0.85	0.88	0.78
Cetane number	40-45	45-55	70-90
T_{90}, °C	300-330	330-360	290-300
Viscosity, mm^2/s @ 40°C	2-3	4-5	3-4
Energy content (LHV)			
Mass basis, MJ/kg	43	39	44
Mass basis, BTU/lb	18,500	16,600	18,900
Vol. basis, 1000 BTU/gal	130	121	122

NOTE: FAME, fatty acid methyl ester; LHV, lower heating value; ULSD, ultralow-sulfur diesel.
SOURCE: Robbins et al. (2011).

HDRD fuel has a very high CN, suggesting it could be used as a blending stream for conventional fuels. There are additional paths of creating renewable diesel fuel which fall under a range of biomass-to-liquid (BTL) processes. Biomass pyrolysis is one example. These paths typically result in a drop-in petroleum-like fuel. CARB studies show that renewable diesel can reach up to 70 percent greenhouse gas reduction compared to petroleum diesel, depending on feedstock and processing methods (Propel Fuels, 2015).

HDRD producers include Neste Oil, Phillips 66, Petrobas, and UOP (NREL and AFDC). Phillips 66 began commercializing renewable diesel in 2006, and similarly did Neste Oil in Europe. Neste Oil is the largest producer of HDRD and now exports to many world regions including North America. Since HDRD is compatible with petroleum diesel and infrastructure, there is no blending limit per se as long as the finished fuel meets ASTM specifications. As an example, NExBTL is the commercialized brand of HDRD by the Finnish oil and refining company Neste Oil.[18] Unblended NExBTL meets the requirements set by the European pre-standard CEN TS 15940. Fuel blends meet the European diesel fuel standard EN 590. Neste Oil was recently cited as the supplier of renewable diesel for vehicles in the City and County of San Francisco, California, and Oakland, California. The feedstock for NExBTL is vegetable oil and waste animal fats; however, fuel quality is equal to the synthetic Fischer-Tropsch BTL and gas-to-liquids diesel fuels (EPA, 2010). Neste Oil says the production cost of NExBTL is mainly driven by utility costs and the price of hydrogen.

4.9.3.2 Biodiesel

Biodiesel is a mix of fatty acid methyl esters (FAMEs) produced through a chemical process called transesterification whereby the glycerin is separated from the fat or vegetable oil. The process leaves behind two products: methyl esters (the chemical name for biodiesel) and glycerin (a valuable byproduct usually sold to be used in soaps and other products). The biodiesel industry has steadily grown over the past decade, with commercial production facilities from coast to coast. The industry reached a key milestone in 2011 when it crossed the billion gallon production mark for the first time. It repeated that in 2012 with more than 1.1 billion gallons and set a new record in 2013 by producing nearly 1.8 billion gallons, according to EPA figures. The industry's total production significantly exceeded the 2013 biodiesel requirement under the Federal Renewable Fuel Standard and was enough to fulfill the majority of the Advanced Biofuel requirement. Biodiesel content in the U.S. diesel supply has a mean

[18] Neste US Inc. 2015. NEXBTL Renewable Diesel, October.

value of 2 to 3 percent with a maximum of 11 percent (Infineum, 2014). B5 is the most prominent blend level. At these blend levels, biodiesel is compatible with the existing fuel-dispensing infrastructure, but pipeline movement of biodiesel is still limited and hindered due to potential mixing with jet fuel if moved in the same pipeline.

Issues with use of biodiesel are mostly well understood, and there is high importance that retail biodiesel comply with ASTM specifications. In past years, a considerable fraction of biodiesel did not meet some of the key specifications, but this has improved greatly (Alleman, 2013; Lopes et al., 2014). The manufacturers of MHDV engines typically approve up to either B5 (e.g., Daimler, DD-15, PACCAR) or B20 (Cummins, Volvo). The National Biodiesel Board maintains a list of approved biodiesel content.[19] The total biomass-based diesel volume of 1.8 billion gallons is primarily biodiesel but also includes renewable diesel.

Biodiesel reduces greenhouse gas emissions by 20 percent to about 80 percent (depending on fuel source and analysis) (EPA, 2009; Pradhan et al., 2012), as well as reducing PM, CO, and HC emissions (EPA, 2002). Biodiesel usually has a CN number well over 40 and typically higher than petroleum diesel fuel. A number of tests of biodiesel in pre-2010 (no aftertreatment on engines) reported increases in NO_x on the order of 2 to 5 percent for B20, although the results varied across engine platforms and duty cycles. The EPA 2002 citation contained an often-cited figure showing a small NO_x increase with biodiesel blends. More recent studies with newer engines generally conclude that biodiesel does not increase NO_x with statistical significance (Lammert et al., 2012; McCormick et al., 2006). A deep review published by SAE (Robbins et al., 2011) concluded that "NO_x effects are small and variable."

4.9.3.3 Summary

Biomass-derived diesel fuels are commercially developed and in widespread use in the U.S. diesel fuel pool. The use of these fuels represents a pathway to reduced GHG emissions (WTW) for MHDVs, with the magnitude depending on the blend level and production process, of course. The impacts on criteria emissions and CO_2 at the vehicle level are dependent on blend levels, baselines of comparison, and specific vehicles (CRC, 2014). EPA identified biobased diesel fuels as meeting at least a 50 percent GHG reduction criterion compared to petroleum as required in the Renewable Fuel Standard (EPA, 2010). Independent studies (Huo et al., 2009) have indicated similar levels of GHG reduction.

Finding: Petroleum-derived diesel fuel will likely remain the dominant CI engine fuel through the time period of this study (approximately 2030). Regarding changes to diesel fuel to reduce life-cycle GHG, an effective measure would be additional use of biomass-derived fuel components (in contrast to changing diesel fuel cetane number). Renewable diesel fuel (HDRD) and biodiesel are both well developed, commercially available, and suitable for this path. They offer overall GHG reduction of up to about 80 percent relative to petroleum-derived diesel.

Recommendation 4-7: Engine manufacturers, with appropriate engagement of NHTSA and EPA, should allow and promote use of biodiesel and renewable diesel in their engines. NHTSA, in concert with EPA, is encouraged to continue some types of incentive for use of biomass-derived diesel fuel because of the overall GHG reduction. The agencies are encouraged to seek harmony and synergy between their separate rules for renewable fuels and tailpipe regulations to enable further GHG reduction and petroleum displacement.

4.9.4 Trends and Impacts from Gasoline Fuel Properties

As noted in the engine section of this report, a significant number of MHDVs are gasoline fueled; over half of Class 3 and 4, for example, are nondiesel, although the overall fuel consumed in these classes is far less than in Class 8 trucks. Recent news noted that about half of buyers of Ford Class 3 to 5 trucks are opting for SI engine power (Government Fleet News, 2014). It is likely that advanced SI engine technologies like boosting and downsizing will find greater use in these classes of trucks, which will close the fuel economy advantage of diesels. The SI engines offer lower initial cost and at least for now a lower fuel price. Two aspects of SI engine fuels' reduction of GHG emissions from MHDVs are noted here:

[19] See http://biodiesel.org/using-biodiesel/oem-information/oem-statement-summary-chart (accessed December 30, 2018).

- Recent studies have suggested that overall WTW GHG emissions in future SI vehicles can be reduced with higher-octane-number gasoline, especially if the engines or vehicles are optimized for the higher-octane fuels. There is even a small benefit in the legacy vehicles if higher-octane fuels are used (Leone et al., 2015; Theiss et al., 2016).
- The magnitude of the GHG reduction is generally greater if renewable, high-octane blending materials (ethanol, butanol, etc.) are used to achieve the higher anti-knock performance (Leone, 2015; Theiss et al., 2016). Higher blend levels of biofuels in gasoline, for example, higher than approximately 25 percent, may require changes in refueling infrastructure, but this is being addressed in new dispenser installations ongoing at present. As of 2017, about 40 percent of gasoline dispensers sold in the United States were compatible with E25, largely because a major supplier, Wayne Fueling, is making all of their new dispensers E25 compatible (Renewable Fuels Association, 2016). Many underground tanks and lines are also already compatible with higher blends, up to E100 (Moriarty et al., 2014). Such higher blends may offer overall GHG emissions on a WTW basis. According to recent studies by three DOE National Laboratories (Theiss et al., 2016):

> The experimental and analytical results of this study considered together show that High Octane Fuels (HOF), specifically mid-level ethanol blends (E25-E40), could offer significant benefits for the United States. These benefits include an improvement in vehicle fuel efficiency in vehicles designed and dedicated to use the increased octane. The improved efficiency of 5-10% could offset the lower energy density of the increased ethanol content, resulting in volumetric fuel economy parity of E25-E40 blends with E10. Most of the flex-fuel vehicles on the road today would be expected to have faster acceleration using HOF, which offers a marketing opportunity in the near term. Furthermore, dedicated HOF vehicles would provide lower well-to-wheel GHG emissions from a combination of improved vehicle efficiency and increased use of ethanol. If ethanol were produced using cellulosic sources, GHG emissions would be expected to be up to 30% lower than those from E10 using conventional ethanol and gasoline. Refinery modeling suggests that refiners could use higher levels of ethanol to meet potentially high market shares of HOF.

Hence, unlike diesel fuel, there appear to be GHG benefits in SI engine fuels by future improvements in performance properties such as octane number.

Tier 3 rules for criteria emissions include a further reduction in gasoline fuel sulfur level to 10 ppm starting in 2017. This is expected to extend performance of aftertreatment devices, reduce criteria emissions from in-use as well as Tier 3 emission vehicles, and bring the U.S. fuel sulfur level in harmony with California, Japan, and Europe.[20] The reduced sulfur could possibly enable U.S. commercialization of lean-burn SI engines which offer part-load efficiency benefits compared to stoichiometric operation engines.

In addition to the importance of sulfur level and octane number, attention is drawn to the impact of gasoline properties on PM emissions from SI engines. Modern direct-injection SI engines emit higher levels of PM than premixed charge engines. EPA Tier 3 standards are calling for a 70 percent reduction in PM emissions from light-duty cars and trucks, to 3 mg/mi; 8 mg/mi from Class 2b trucks; and 10mg/mi from Class 3 trucks. Over the past few years, strong correlations were developed between certain gasoline fuel properties (principally volatility and double-bond-equivalent) and PM emissions, as described in the literature on PM index (PMI). Retail gasolines across the United States were found to create a fourfold variation in PM emissions from typical vehicles. Honda et al. developed a predictive index for PM emissions based on several key fuel properties that appears to be valid for many fuels and engine types (Aikawa and Jetter, 2013; Aikawa et al., 2010). The general trends are as follows:

- Fuels with components of low vapor pressure increase PM.
- Fuels with high content of double carbon bond equivalent (DBE) HCs increase PM.

[20] Emission and Fuel Standards 2014. 40 CFR Parts 79, 80, and 85. April.

- High levels of highly reactive HCs like indene tend to increase PM.
- Oxygenates such as ethanol and methanol tend to decrease PM when merely added to a base gasoline (splash blending). Match blending (for constant volatility) with ethanol can actually cause higher PM emissions. Results are variable in the literature.

The PMI is expressed as a function of the vapor pressure of fuel components and the DBE.[21] The DBE has been found to correlate well with PM mass and number for many gasoline vehicles, although EPA studies found some vehicles that did not follow the correlation (Butler et al., 2015). The committee notes a PMI of some type as a possible future feature of gasoline specifications.

Finding: In consideration that SI engines fueled with gasoline-based fuels are economically attractive for some market segments, GHG reductions with these engines can be enhanced by improving the performance properties of the fuel such as octane number and maintaining or increasing the content of renewable fuels. The effect of a moderate increase in octane number (RON) on efficiency is 5 to 10 percent improvement, whereas the GHG reduction can be up to 30 percent if renewable ethanol is the high-octane-number blending material (Leone, 2015; Theiss et al., 2016).

Recommendation 4-8: NHTSA, in coordination with EPA, should consider and further analyze scenarios where SI engines and gasoline-like fuels play a larger role in freight movement. The regulatory environment they develop should enable and encourage adoption of the best technologies for gasoline SI engines into MHDVs, including higher octane number in the fuels since it is shown as an overall GHG benefit.

4.9.5 Natural Gas

Natural gas and fuels derived from natural gas were covered in depth in the committee's first report (NRC, 2014). A key finding noted lower engine efficiency with NG and the body of analysis estimating that methane leakage tended to offset the inherent GHG benefits of the low-carbon fuel. The load-specific CO_2 emissions from NG engines are generally 5 to 20 percent less than those from comparable diesel engines. NG's inherent GHG benefit by virtue of its low carbon content (approximately 28 percent) is partially negated by lower efficiency in currently available engines and the higher GHG impact of methane emissions (NRC, 2014, p. 65). The committee recommended further study of leakage and related methane emissions to quantify the trade-offs, and will review here several studies released in 2015.

Also in the first report, the committee addressed the "new" interest in natural gas derived principally from increased domestic NG extraction owing primarily to process developments permitting major volumes of gas removal from tight shale deposits in the vast Utica and Marcellus shale region (the "fracking" process). This NG availability has driven its price down to extremely low levels compared to diesel, whose prices had been significantly increasing over the prior decade.

Nearly simultaneously, the petroleum exploration and extraction industries, especially in the United States and Canada, have significantly reduced the North American petroleum dependencies on other world regions. And, now, depressed diesel fuel prices resulting from a world petroleum glut with Organization of the Petroleum Exporting Countries–controlled pricing finds September 2015 natural gas at a small diesel energy equivalent price differential (Q3-2015 Diesel, \$2.62; NG, \$2.35) (ACT, 2015), effectively resulting in an unacceptable payback period, since NG fuel consumption is exacerbated by its poorer engine efficiency.

Progress in engine efficiency continues for both CI and SI engines using a variety of fuels, as well as new analysis of methane leakage impacts. CARB is furthermore studying reduction of NO_x emissions standards by

[21] The double bond equivalent is also called the degree of unsaturation. In the chemical structure, each pi bond will generate one DBE. The DBE can also be calculated with the following formula, where C is the number of carbon atoms, H is the number of hydrogen and halogen atoms, and N is the number of nitrogen atoms:

$$\text{DBE} = C - \frac{H}{2} + \frac{N}{2} + 1 .$$

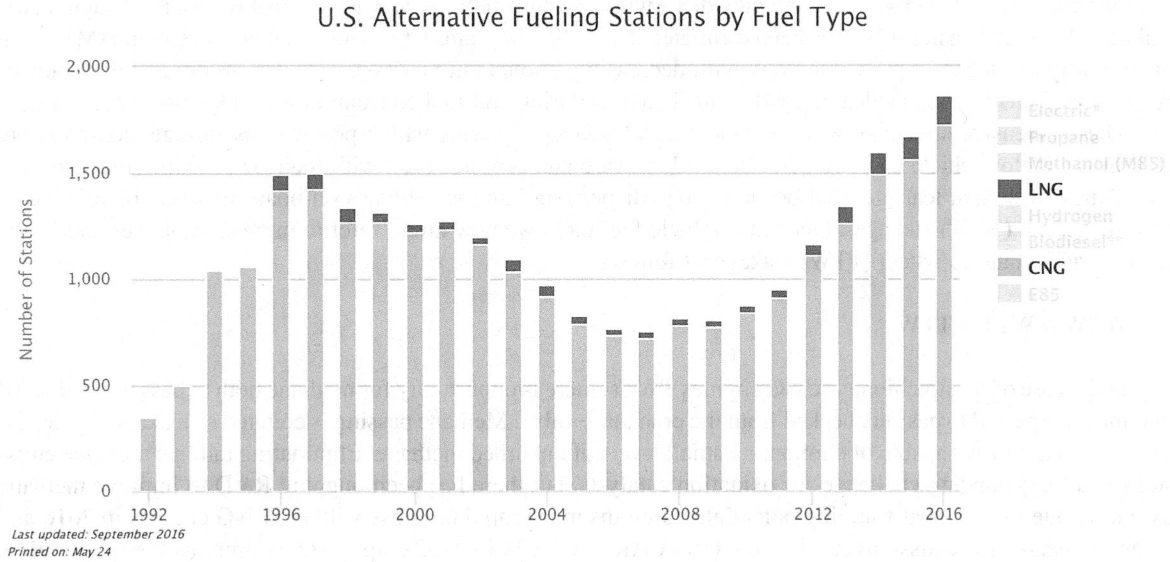

FIGURE 4-20 U.S. LNG and CNG alternative fueling stations in the United States from 1992-2016.
SOURCE: Committee generated using Department of Energy Alternative Fuels Data Center widget titled "U.S. Alternative Fueling Stations by Fuel Type."

approximately 90 percent, which could change the GHG comparison of NG to diesel fuel. New WTW studies of NG have been conducted (Carnegie Mellon, ICCT, and ANL). Sales of NG-fueled MHDVs increased more than 25 percent in 2014 compared to 2013 (NGV America, 2014), although still less than 1 percent of NG is used for vehicle fuel. EIA forecasts large growth in NG usage, especially for freight vehicles (EIA, 2014), yet this still accounts for only 7 percent of heavy vehicle fuel use in 2040. These projections are generally more conservative than the National Petroleum Council study of 2012. The number of liquefied natural gas (LNG) and compressed natural gas (CNG) refueling stations has continued to grow since 2009—about 948 and 76, respectively[22]—and has surpassed the peak number from nearly 20 years ago (Figure 4-20). Over half of these stations are available to the public. Providers of CNG tanks have also made progress in reducing the price of that component.

Some manufacturers of HD NG engines suspended development of those programs in late 2014 and early 2015 when the return-on-investment driver for their product was drying up. Only Cummins-Westport, who introduced their ISX 12L-G in mid-2013, continued forward. Certain financing companies have offered truck companies incentives to continue their purchases (ACT, 2015). Nevertheless, the potential reduction in operating cost still holds great potential for this industry whenever diesel prices return to a reasonably stable and commercially favorable (for NG) differential.

4.9.5.1 Impacts of Natural Gas Leakage on Overall GHG Benefits of Natural Gas–Fueled MHDVs

Recent studies on natural gas supply chain leakage have estimated that relatively low levels of methane emissions could detract from the GHG reduction benefit that might otherwise accrue from the use of this fuel relative to diesel. The United States, through its network of 555,364 producing gas wells[23] and 305,000 miles of

[22] U.S. Department of Energy, Alternative Fuels Data Center. "Natural Gas Fueling Station Locations." Available at http://www.afdc.energy.gov/fuels/natural_gas_locations.html (accessed March 30, 2017).

[23] Energy Information Administration, February 28, 2017, "Number of Producing Gas Wells." Available at https://www.eia.gov/dnav/ng/ng_prod_wells_s1_a.htm (accessed March 30, 2017).

NG pipeline[24] in the lower 48 states, already deals with complex methane leakage control issues. For example, the leakage of domestic natural gas has been estimated at roughly 10 grams CO_2-equivalent per megajoule (MJ), with the majority occurring in production and with decreasing amounts in processing, transmission, and distribution. Another 6 grams CO_2-equivalent per MJ is attributed to flaring and fuel consumption (Venkatesh et al., 2012).

These extensive upstream facilities link to local delivery systems which provide gas storage, compression, and, in the motor vehicle case, vehicle fueling hardware and process, each with their own sealing mechanisms. All of these upstream leakages that occur through dispensing into the vehicle's on-board tank can be referred to as well-to-tank (WTT) leakages. Once in a vehicle fuel tank, we now might refer to methane leakages and losses as comprising tank-to-wheel (TTW) leakage. It follows:

$$WTW = WTT + TTW$$

In the case of SI, stoichiometric NG engines TWCs, there is opportunity for methane both to pass uncombusted into the tailpipe and to pass unchecked from the crankcase valve. Methane passing is counted as leakage. Typically, the TWC is not fully capable of eliminating small levels of unburned methane. Eliminating tailpipe methane emissions is a long-standing challenge for oxidation catalysts, but there has been ongoing R&D to improve methane oxidation due to its global warming potential. Cummins made rapid progress with their NG engines in 2016 and received "near-zero" emissions certification from CARB and EPA for 0.02 g/hp-h NO_x (Cummins Westport, 2016). Methane slip has been reduced by about 70 percent by closing the crankcase ventilation and more effective catalysts.

A recent study (Clark et al., 2017b) of pump-to-wheels (PTW) emissions for heavy-duty vehicles found total methane emissions for an assumed scenario and vehicle mix to be 1.33 percent by mass of the fuel used. Both high-pressure direct-injection CI and stoichiometric SI engines, and both CNG and LNG fuels and fueling stations, were addressed in this study. Engine crankcase emissions represented about 39 percent of the losses, so that the latest engines with closed crankcases would reduce losses to 0.81 percent of fuel used. A second paper (Clark et al., 2017a), which used the data to project future scenarios, suggested that for 2035 methane PTW losses could range from 0.15 to 0.67 percent. Alvarez et al. (2012) computed that 2.4 percent of U.S. gas production was lost due to leaks and venting from production to local distribution, based on an EPA estimate.

A further study, by the ICCT, published an analysis of leakage (Delgado and Muncrief, 2015). The results are summarized in Table 4-3.

4.9.5.2 Methane Leakage and CO_2 Emissions: Scenarios and Pathways (Delgado and Muncrief, 2015)

ICCT conducted and published a study on the potential GHG benefits of NG vehicles by considering various scenarios of market penetration, leakage levels, global warming potential (GWP), and engine efficiency gap relative

TABLE 4-3 Summary of Low and High Estimates of Well-to-Wheels Methane Loss, as a Percentage of Fuel Delivered

Source	Reference Low Estimate	Reference High Estimate
Well to pump	1.4	2.6
Fueling station	0.3	0.3
Tailpipe	0.1	0.9
Crankcase (or HPDI venting)	0.4	0.8
TOTAL	2.2	4.6

SOURCE: Delgado and Muncrief (2015).

[24] Energy Information Administration, 2008, "Estimated Natural Gas Pipeline Mileage in the Lower 48 States, Close of 2008." Available at https://www.eia.gov/pub/oil_gas/natural_gas/analysis_publications/ngpipeline/mileage.html (accessed March 31, 2017).

to diesels. Their primary tool in the analysis was the Argonne National Laboratory Vision Model. The scope of the analysis is strictly limited to heavy-duty vehicles, so nothing can be concluded for light-duty passenger vehicles. Class 3 to 6 natural gas vehicles and Class 7 and 8 straight units were assumed to use SI engines and CNG as fuel supply, and Class 7 and 8 combination trucks were assumed to use HPDI engines and LNG as fuel supply (Delgado and Muncrief, 2015).

The ICCT scenarios for natural gas vehicles (NGVs) market share estimates were described in their text and are consistent with forecasts reported in the committee's first report. Their scenarios assume the market share reaches the selected penetrations in 2025 and beyond. The Vision Model's resulting benefits (positive or negative) are achieved in 2040, where 2040 is imagined as a mature fleet. The model reports a GHG fleet inventory for the scenarios and compares each to a reference case with a trivial 1 percent NGV market share. "The Reference case assumes that the new vehicle fleet sees efficiency improvements of 3.5 percent per year [in vehicle fuel efficiency], and that the share of natural gas–fueled trucks remains at about 1 percent of the market, with methane leakage assumed to be 1.1 percent and natural gas trucks continuing to have a 10 percent efficiency gap (or loss) versus diesel" (Delgado and Muncrief, 2015, p. 17).

The reference case benefits are driven by the assumed 3.5 percent per year fleet efficiency improvements (for 2018 through 2025). That assumed annual improvement is driven by Phase II regulations and is supported by industry forecasts documented in the ICCT reference. Selected scenarios representing a range of WTW cases and a range of engine efficiency gap cases are reported in Table 4-4.

These results at a 10 percent efficiency gap indicate that zero leakage is needed for NG vehicles to reduce GHG emissions below the reference case, and that the loss of benefits is potentially very large (39 percent) compared to the overall reference Phase II benefits. Note that the committee's current analysis presented in Chapter 6 of this report shows an efficiency gap of over 15 percent between a modern diesel and a SI-NG engine. ICCT further examined the impact of increasing NG engine efficiency, such as HPDI engines, so that the gap was 0 to 15 percent. The ICCT results are shown in Table 4-5, where the leakage was fixed at the reference case (latest EPA estimate).

The average NGV efficiency gap needs to be under 10 percent to avoid a worse (negative) global result than the reference case (1.0 percent market share [MS]). The source analysis did not describe the technologies to achieve

TABLE 4-4 Estimated Sensitivity of Emissions of NG Vehicles (Life-Cycle Basis) with Varying Well-to-Wheels Leakage Rates

No.	Name	NGV Market Share	WTW Leakage	Methane GWP	NGV Efficiency Gap	GHG Benefits (millions of metric tons CO_2e)	Percentage Change from Reference
R	Reference (Phase II)	1%	1.1%	28	10%	144.9	-
1	High MS, no leak, 100 y	20%	0.0%	28	10%	151.7	5%
2	Mid MS, no leak, 100 y	10%	0.0%	28	10%	147.1	2%
3	High MS, mid leak, 100 y	20%	2.2%	28	10%	143.1	-1%
4	Mid MS, mid leak, 100 y	10%	2.2%	28	10%	142.7	-2%
5	Mid MS, high leak, 100 y	10%	4.6%	28	10%	137.9	-5%
6	High MS, high leak, 100 y	20%	4.6%	28	10%	133.6	8%
7	Mid MS, mid leak, 20 y	10%	2.2%	84	10%	130.5	-10%
8	High MS, mid leak, 20 y	20%	2.2%	84	10%	119.3	-18%
9	Mid MS, high leak, 20 y	10%	4.6%	84	10%	115.1	-21%
10	High MS, high leak, 20 y	20%	4.6%	84	10%	89.1	-39%

NOTE: CO_2e, carbon dioxide equivalent; GHG, greenhouse gas; GWP, global warming potential; MS, market share; NGV, natural gas vehicle; WTW, well to wheels. Annual benefits are for 2040.
SOURCE: Delgado and Muncrief (2015).

TABLE 4-5 Estimated Sensitivity of Emissions of NG Vehicles (Life-Cycle Basis) with Varying Efficiencies Gaps

No.	Name	NGV Market Share	WTW Leakage	Methane GWP	NGV Efficiency Gap	GHG Benefits ($MMTCO_2e$)	Percentage Change from Reference
R	Reference (Phase II)	1%	1.1%	28	10%	144.9	-
11	High MS, low leak, 100 y, 0% gap	20%	1.1%	28	0%	155.6	7%
12	High MS, low leak, 100 y, 5% gap	20%	1.1%	28	5%	151.7	5%
13	Mid MS, low leak, 100 y, 0% gap	10%	1.1%	28	0%	149.0	3%
14	High MS, low leak, 100 y, 10% gap	20%	1.1%	28	10%	147.4	2%
15	Mid MS, low leak, 100 y, 5% gap	10%	1.1%	28	5%	147.1	2%
16	Mid MS, low leak, 100 y, 10% gap	10%	1.1%	28	10%	144.9	0%
17	High MS, low leak, 100 y, 15% gap	20%	1.1%	28	15%	142.5	-2%
18	Mid MS, low leak, 100 y, 15% gap	10%	1.1%	28	15%	142.4	-2%

NOTE: GHG, greenhouse gas; GWP, global warming potential; $MMTCO_2e$, million metric tons carbon dioxide equivalent; MS, market share; WTW, well to wheels. Annual benefits are for 2040.
SOURCE: Delgado and Muncrief (2015).

1.1 percent leakage level. Indeed, it cannot be achieved without the kinds of leakage reduction processes EPA has proposed for the oil and natural gas industry sectors (EPA, 2015b).

Finding: Both WTW methane leakages and the NG-to-diesel thermal efficiency gap are critical to NG-fueled HD trucks achieving the previously anticipated CO_2 benefit compared to legacy diesels. Even at an optimistic efficiency gap of 10 percent, the upstream leakages potentially negate the benefits of NG low carbon content. The committee's review of the literature has found the NG engine efficiency gap over most drive cycles to be over 15 percent. WTT leakages appear to hold the most importance, and EPA is proposing significant emphasis on industry actions to reduce the upstream methane leaks. NG SI combustion engines will likely always have some thermal efficiency deficit with comparable model year diesel; this is due to inlet air throttling which degrades thermal efficiency. Among available technologies, natural gas HPDI engines are expected to enable the greatest reduction of the efficiency gap.

Finding: Methane leakage to the engine crankcase will persist more significantly in SI versus DI engine combustion systems. This is due to leakage of the cylinder contents into the crankcase during the relative compression strokes, and where in SI there is high methane content. Yet in SI, the crankcase content might be positively ventilated to the vehicle exhaust and catalytically combusted there.

NG engines designed for direct injection of the fuel are found to have much reduced methane tailpipe and methane crankcase leakages, based on their fuel delivery and lean combustion characteristics.

Recommendation 4-9: Given the significance of WTW methane leakages and losses on GHG reduction, NHTSA and EPA should monitor progress on all of those errant upstream methane emissions and use a regulatory approach that accounts for them along with the tailpipe GHG emissions.

4.9.5.3 Analysis of the Report *Life Cycle GHG for Natural Gas Pathways in MHDVs, Including Leakage Effect* (Tong et al., 2015)

In its first report (NRC, 2014), the committee noted multiple pathways to use natural gas in MHDVs, including direct combustion (as in the just-reviewed ICCT study), conversion to electricity, and conversion to other liquid fuels such as Fischer-Tropsch liquids, dimethyl ether, and ethanol. Carnegie Mellon University (CMU) researchers (Tong et al., 2015) analyzed multiple pathways of using natural gas for road transportation and compared them for different types of MHDVs. Life-cycle analyses were performed on a comprehensive set of natural gas–derived fuels, engine technologies, and vehicle types, and most importantly included the impacts of methane leakage. We summarize the results of this study and suggest the reader review the publication for detailed analyses and conclusions.

Shown below in Figure 4-21 are the different pathways examined. As shale gas is an increasing percentage of the NG produced in the United States, it was assumed to be the primary source in the analysis. Carnegie Mellon researchers defined the baseline fuel pathways as conventional gasoline (Class 2b) and conventional (ultralow-sulfur) diesel (Class 3 through 8 MHDVs). There are four stages in the life cycle of the pathways: feedstock (natural gas) production and transport, transportation fuel production, transportation fuel delivery, and vehicle use. The analysis estimates of GHGs, CO_2, methane, and N_2O are converted to CO_2 equivalents for the GWP values.

Table 4-6 considers seven types of MHDVs and their interaction between fuel pathways, vehicle engine technologies, and types MHDVs with various parameters. The detailed assumptions and estimates are provided in the Carnegie Mellon reference. The primary functional metrics are (1) vehicle distance traveled (gCO_2-e/km) and freight-distance moved (gCO_2-e/km-metric-ton).

To account for a potential bias in methane leakage rate estimates, and also to account for choices of GWPs with different integration periods,[25] four scenarios were considered: (1) a baseline methane estimate with 100-year GWP (baseline scenario), (2) a baseline methane estimate with 20-year GWP, (3) a pessimistic estimate with 100-year GWP, and (4) a pessimistic estimate with 20-year GWP.

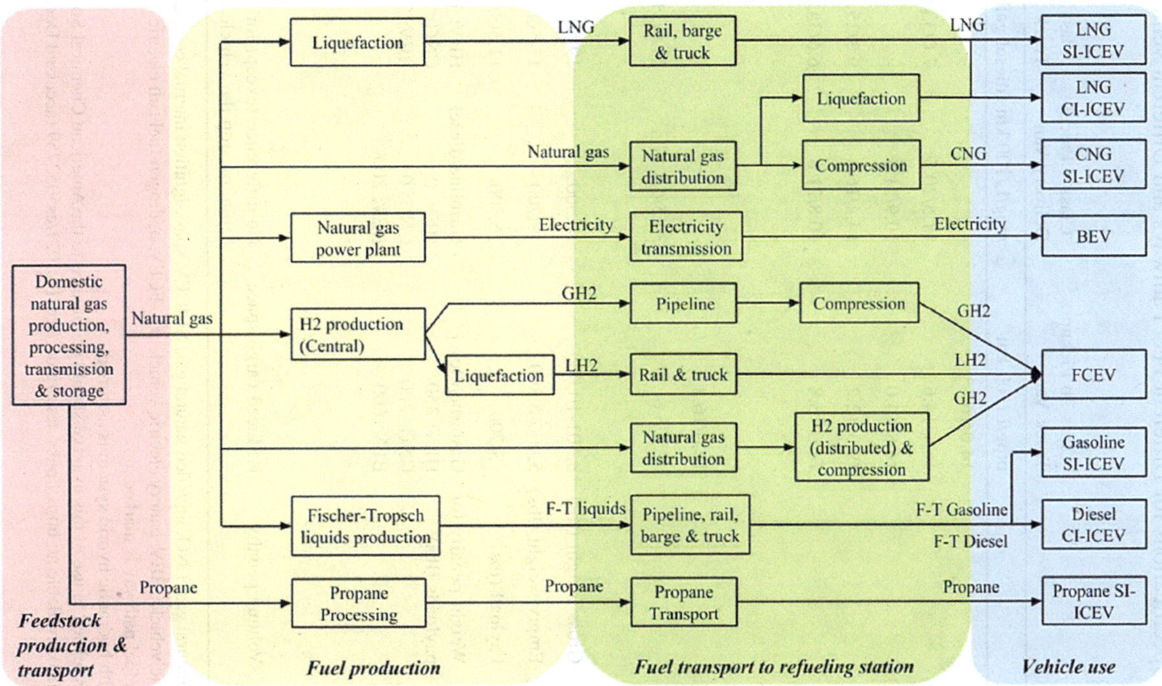

FIGURE 4-21 Pathways considered for fueling with natural gas and propane sourced fuels to a variety of different vehicle powertrains, showing feedstock production and transport, fuel production, fuel transport to refueling stations, and vehicle use.
NOTE: Further permissions related to use of this material should be directed to the American Chemical Society (ACS).
SOURCE: Tong et al. (2015), available at: https://pubs.acs.org/doi/abs/10.1021/es5052759 (accessed December 18, 2019).

[25] GWP is defined per unit mass of a gas as the ratio of its time-integrated radiative forcing to that of CO_2. The period of time integration is user defined.

TABLE 4-6 Vehicle Specifications for Different Fuel Pathways and Different Vehicle Applications

		Class 2b Pickup Truck/Van	Class 4 Parcel Delivery Van	Class 6 Box Truck	Class 8 Transit Bus	Class 8 Refuse Truck	Class 8 Tractor-Trailer Local Haul	Line Haul
Unit of fuel economy[a]		mpg/L/100 km	mpg/L/100 km (diesel gallon/liter equivalent)					
Gasoline (SI-ICEV)		14.0/16.8[a]	-	-	-	-	-	-
Diesel (CI-ICEV)		16.1/14.6	11.5/20.5[a]	7.0/33.6[a]	4.0/58.8[a]	3.3/71.3[a]	4.3/54.7[a]	6.5/36.2[a]
Gasoline-HEV (SI-ICEV)		16.8/14.0	10.9/21.5	-	-	-	-	-
Diesel-HEV (CI-ICEV)		19.3/12.2	14.4/16.4	9.3/25.3	4.8/49.0	3.6/64.8[b]	5.2/45.6	7.2/32.9
CNG (SI-ICEV)		14.0/16.8	10.8/21.8	6.6/35.7	3.6/65.3	2.9/81.0	3.9/60.8	5.9/40.2
LNG (SI-ICEV)		-	-	-	3.6/65.3	2.9/81.0	3.9/60.8	5.9/40.2
LNG (CI-ICEV)		-	-	-	-	-	4.2/55.8	6.4/36.9
Propane (SI-ICEV)		14.0/16.8	-	-	-	-	-	-
BEV		42.0/5.6	34.5/6.8	21.0/11.2	16.8/14.0	-	-	-
H_2-FCEV		-	-	-	7.6/30.9	-	-	-
Conventional Truck	Gross weight (lbs)	8,501-10,000	16,000	19,501-26,000	39,980	60,000	80,000	80,000
	Empty weight (lbs)	5,000-6,300	9,700	11,500-14,500	27,730	16,627	30,500	35,550
	Payload (lbs)	3,700	6,300	11,500	12,150	43,373	49,500	44,450
Alternative fuels	Weight penalty for payloads (lbs)	Gasoline/diesel HEV: 350 CNG: 200 BEV: 600	Gasoline/diesel HEV: 0 CNG: 0 BEV: 200	HEV: 1200 CNG: 515 BEV: 200	Gasoline/diesel HEV: 0/750 CNG: 900 LNG: 1150 H_2-FCEV: 5,400 BEV: 4,800	Hybrid: 400 CNG: 915 LNG (SI): 265	HEV: 880 CNG: 502 LNG (SI): 252 LNG (CI): 1249	HEV: 880 CNG: 2042 LNG (SI): 1142 LNG (CI): 2541
	Volume penalty	Reduced cargo space	No difference (except for transit buses) because fuel tanks are mounted on the chassis, behind the cabin, or atop the vehicle. For transit buses, volume penalty has been factored into weight penalty.					

NOTE: CNG, compressed natural gas; LNG, liquefied natural gas; SI-ICEV, spark-ignition internal combustion engine vehicle; CI-ICEV, compression-ignition internal combustion engine vehicle; HEV, hybrid electric vehicle; BEV, battery electric vehicle; H_2-FCEV, hydrogen fuel cell electric vehicle; mpg, miles per gallon.

[a]The baseline petroleum fuel pathway is marked.

[b]A diesel refuse truck with hydraulic hybrid system is assumed.

NOTE: Further permissions related to use of this material should be directed to the American Chemical Society (ACS).

SOURCES: Tong et al. (2015), available at: https://pubs.acs.org/doi/abs/10.1021/es5052759 (accessed December 18, 2019). Vehicle figures: NREL (2013).

The CMU study included

- Life-cycle assessment (LCA) of the multiple pathways using a best estimate for methane leakage and
- Analysis to determine a break-even level of methane leakage that would be needed to ensure a net GHG benefit with natural gas.

Based on extensive review of the literature on methane leakage, which is a topic of continuing study, CMU determined to use a leakage level of 1.0 to 2-2 percent for the initial LCA. We show three examples from the CMU paper (see Figures 4-22, 4-23, and 4-24) that are similar vehicles to those emphasized elsewhere in our report: Class 2b pickup, Class 6 box truck, and Class 8 line haul.

CMU estimated that, where the technology is available (pickups, buses), the path of natural gas to electricity in battery electric vehicles (BEVs) is advantageous for GHG emission reduction. However, comparing CNG to conventional diesel vehicles in delivery trucks and Class 8 line hauls suggests that natural gas vehicles would not offer a net GHG benefit for any of the pathways considered.

The second phase of the CMU study examined the question of what leakage level would be necessary to more likely ensure GHG benefits from natural gas. As described in the previous section on the ICCT study, the efficiency gap between natural gas and diesel fuel end-use devices is a key parameter as well.

Figure 4-25 is reproduced from Tong et al. (2015) and shows the break-even life-cycle methane leakage rates for CNG and LNG vehicles. The methane leakage rate is defined as the volumetric percentage of natural gas produced that is lost through venting or fugitive leaks. It includes fugitive methane emissions from the natural gas system, boil-off leaks from production and storage of LNG, and crankcase and tailpipe methane leaks from

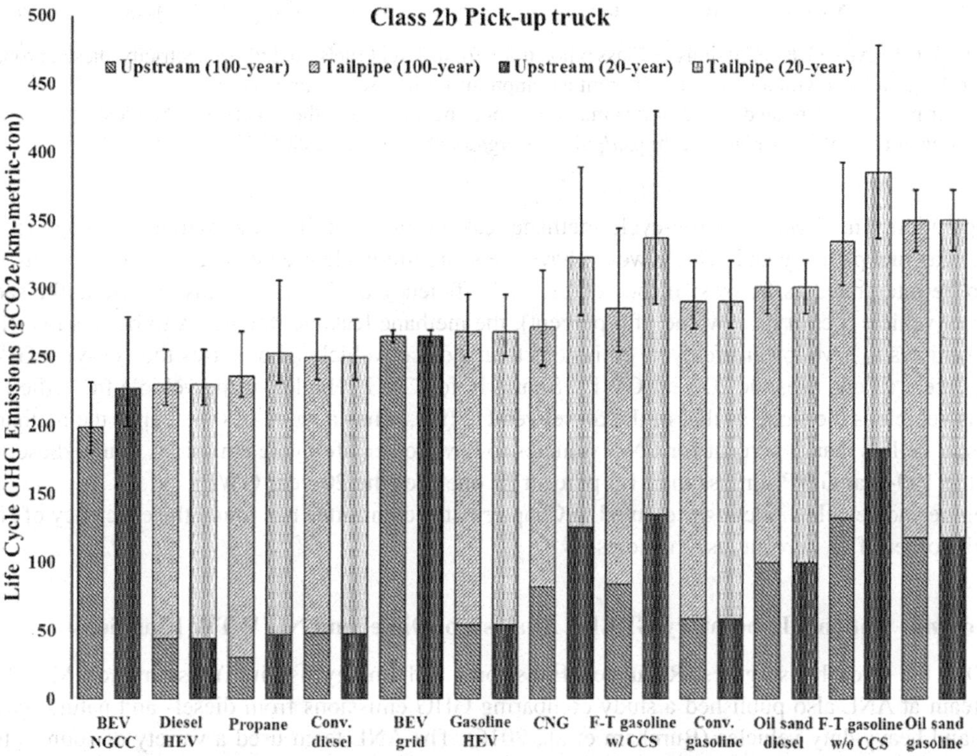

FIGURE 4-22 Life-cycle GHG emissions of Class 2b pickup truck for different fuels, including electricity, diesel, gasoline, propane, natural gas, Fischer-Tropsch fuels with a variety of different assumptions on fuel source and end use.
NOTE: Further permissions related to use of this material should be directed to the American Chemical Society (ACS).
SOURCE: Tong et al. (2015), available at: https://pubs.acs.org/doi/abs/10.1021/es5052759 (accessed December 18, 2019).

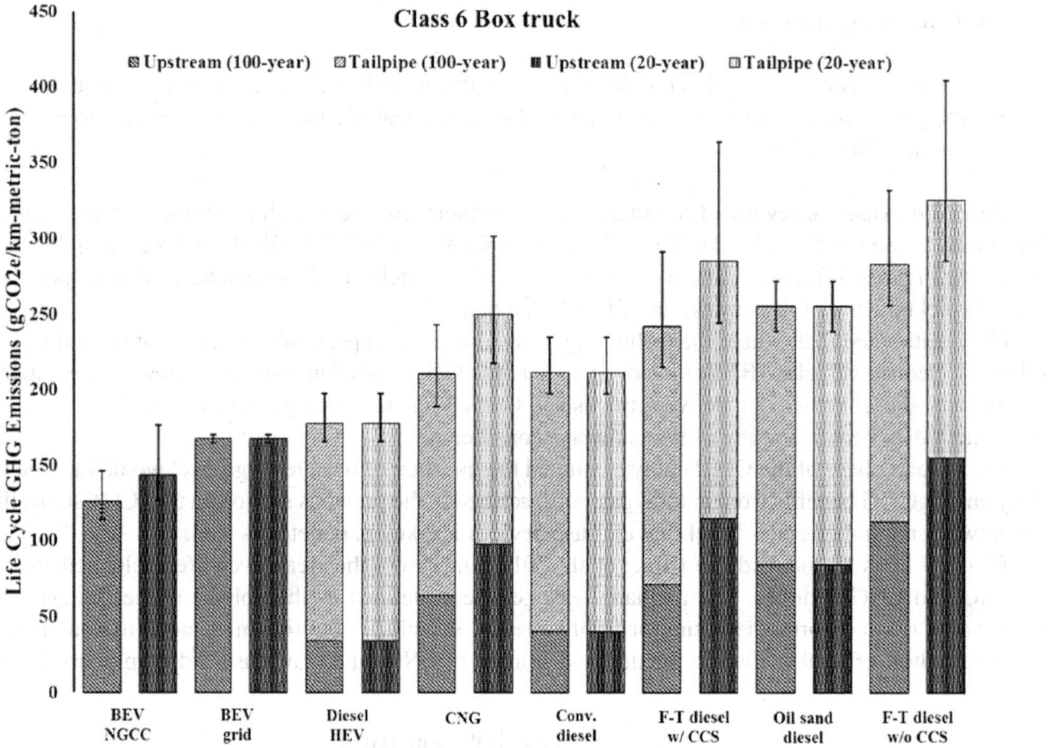

FIGURE 4-23 Life-cycle GHG emissions of Class 6 box truck for different fuels, including electricity, diesel, propane, natural gas, Fischer-Tropsch fuels with a variety of different assumptions on fuel source and end use.
NOTE: Further permissions related to use of this material should be directed to the American Chemical Society (ACS).
SOURCE: Tong et al. (2015), available at: https://pubs.acs.org/doi/abs/10.1021/es5052759 (accessed December 18, 2019).

vehicle operation. The *break-even* life-cycle methane leakage rate identifies the methane leakage rate at which the natural gas fuel pathway and vehicle would have the same life-cycle greenhouse gas emissions as diesel. The figure can be interpreted as follows: If the vehicle fuel efficiency of the natural gas vehicle is the same as that of the diesel vehicle (i.e., it has a value of 0 percent), the methane leakage rate needs to be less than 1.2 percent for LNG vehicles to have lower life-cycle emissions than a diesel vehicle (if one uses the 100-year GWP) or less than 0.5 percent (if one uses the 20-year GWP). Similarly for CNG vehicles, if the vehicle fuel efficiency of the natural gas vehicle is the same as that of the diesel vehicle (i.e., it has a value of 0 percent), the methane leakage rate needs to be less than 3 percent for LNG vehicles to have lower life-cycle emissions than a diesel vehicle (if one uses the 100-year GWP) or less than 1.2 percent (if one uses the 20-year GWP).

In engine and vehicle modeling presented in Chapter 6, the committee has found the efficiency of NG engines to be on the order of 15 percent less than diesel.

4.9.5.4 Argonne National Laboratory GREET Analysis of Diesel and NG WTW Emissions

In 2016 the Greenhouse gases, Regulated Emissions, and Energy use in Transportation Model (GREET) analysis team at ANL also published a study comparing GHG emissions from diesel- and natural gas–powered medium- and heavy-duty vehicles (Burnham et al., 2016). The ANL team used a variety of sources to arrive at inputs to GREET for diesel- and NG-fueled vehicles, resulting in an overall picture of the efficiency advantage of diesels shown in Table 4-7 below. The study scope generally does not look as far ahead as the committee's current study or the Phase II rule.

These values for projected fuel efficiency advantage for diesels are in the same parameter range as other studies, except the committee's analysis and models show the NG disadvantage much higher for combination trucks. The

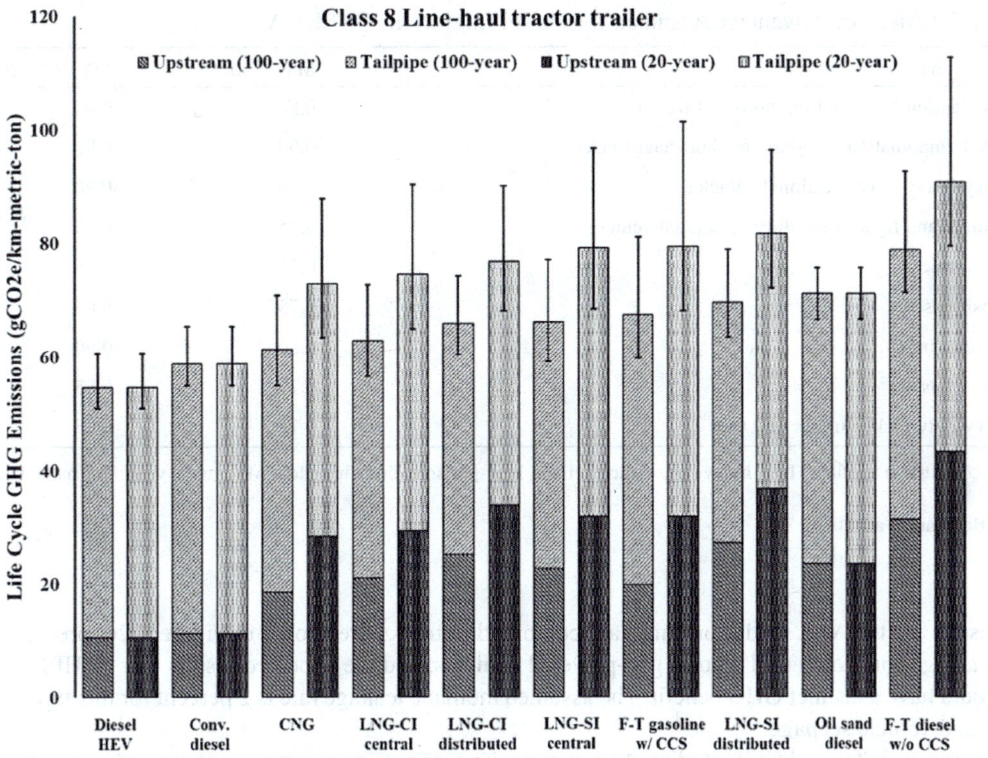

FIGURE 4-24 Life-cycle GHG emissions of Class 8 truck for different fuels, including electricity, diesel, propane, natural gas, Fischer-Tropsch fuels with a variety of different assumptions on fuel source and end use.
NOTE: Further permissions related to use of this material should be directed to the American Chemical Society (ACS).
SOURCE: Tong et al. (2015), available at: https://pubs.acs.org/doi/abs/10.1021/es5052759 (accessed December 18, 2019).

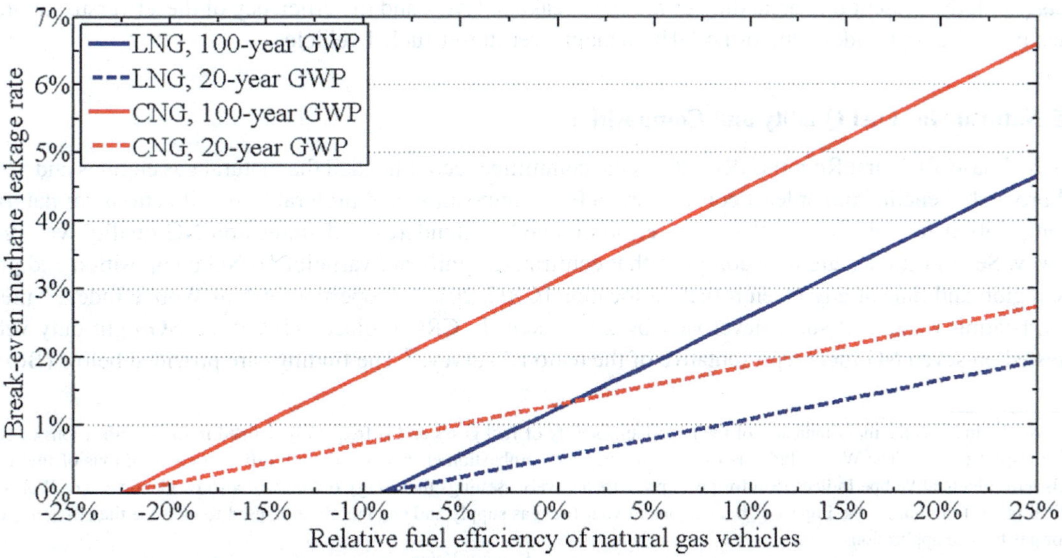

FIGURE 4-25 Break-even life-cycle methane leakage rate (the volumetric percentage of natural gas produced that is lost through venting or fugitive leaks) that enables a NG vehicle to be lower emitting than a comparable diesel vehicle, as a function of relative fuel economy of natural gas vehicles (the percentage difference between the fuel efficiency of a natural gas–fueled vehicle and that of a diesel vehicle) and GWP.
NOTE: Further permissions related to use of this material should be directed to the American Chemical Society (ACS).
SOURCE: Tong et al. (2015), available at: https://pubs.acs.org/doi/abs/10.1021/es5052759 (accessed December 18, 2019).

TABLE 4-7 Efficiency Advantage Assumed for Diesel Engines in GREET Analysis

HDV Subcategory	MY 1990-2005	MY 2010-2020
SI NG combination long-haul or short-haul trucks	0.80	0.90
CI HPDI NG combination long-haul or short-haul trucks	0.90	0.95
SI NG heavy heavy-duty vocational vehicles	0.80	0.90
SI NG medium and light heavy-duty vocational vehicles	0.75	0.85
SI NG refuse trucks	0.75	0.85
SI NG transit buses	0.75	0.85
SI NG intercity buses	0.80	0.90
SI NG school buses	0.75	0.85
SI NG heavy-duty pickup trucks and vans	0.75	0.85

NOTE: CI, combustion ignition; HDV, heavy-duty vehicle; HPDI, high-pressure direct injection; MY, model year; NG, natural gas; SI, spark ignition.
SOURCE: Burnham et al. (2016).

overall results for the ANL study for long-haul combination trucks are shown in Figure 4-26, predicting a slight GHG advantage for even the SI natural gas–powered vehicles, and the expected results that an HPDI natural gas engine would have a distinct GHG benefit. The assumed methane leakage rate is 2 percent for this figure, including upstream and vehicle seepage.

Similar to the ICCT and CMU studies, ANL prepared a parametric presentation of the trade-off between engine efficiency difference (diesel versus natural gas) and methane leakage in Figure 4-27. In summary they found that if parity were achieved in engine efficiencies, leakage could be 4 percent for a break-even GHG output for diesel or natural gas vehicles. On the other hand, if the SI natural gas engine had a 25 percent efficiency disadvantage, a leakage rate of zero would be needed for natural gas to have any GHG advantage.

Finding: With the present understanding of methane leakage levels and the efficiency of diesel versus natural gas engines, natural gas provides little or no GHG benefit over diesel-fueled vehicles.

4.9.5.5 Natural Gas Fuel Quality and Composition

In the Phase Two First Report (NRC, 2014) the committee recommended that natural gas engines and vehicles would probably benefit from at least consistency in fuel composition and preferably specifications for natural gas fuel composition for vehicles. At that time, efforts related to standards and studies on NG quality were getting under way. Several studies are now complete that confirmed significant variation in NG composition and properties by region and time at any given refueling location (CRC, 2014). Properties such as Wobbe Index,[26] methane number, heating value, and sulfur level vary by 20 percent. In CRC project E-109 three NG light-duty vehicles were tested on seven NG fuels representative of the national survey.[27] The findings are provided below (Box 4-1).

[26] The Wobbe Index is the main indicator of the interchangeability of fuel gases and is frequently defined in the specifications of gas supply and transport utilities. The Wobbe Index is used to compare the combustion energy output with different compositions of fuel gases. If two fuels have identical Wobbe Indices then for given pressure and valve settings the energy output will also be identical. The Wobbe Index is a critical factor to minimize the impact of fluctuations in your fuel gas supply and can therefore be used to increase the efficiency of your burner or gas turbine applications.

$$\textit{Wobbe Index} = \frac{\textit{Heating Value}}{\sqrt{\textit{Specific gravity}}}$$

[27] Effect of Fuel Composition on the Emissions and Performance of Modern, Light-Duty Natural Gas Vehicles – CRC Project No. E-109, April 2015.

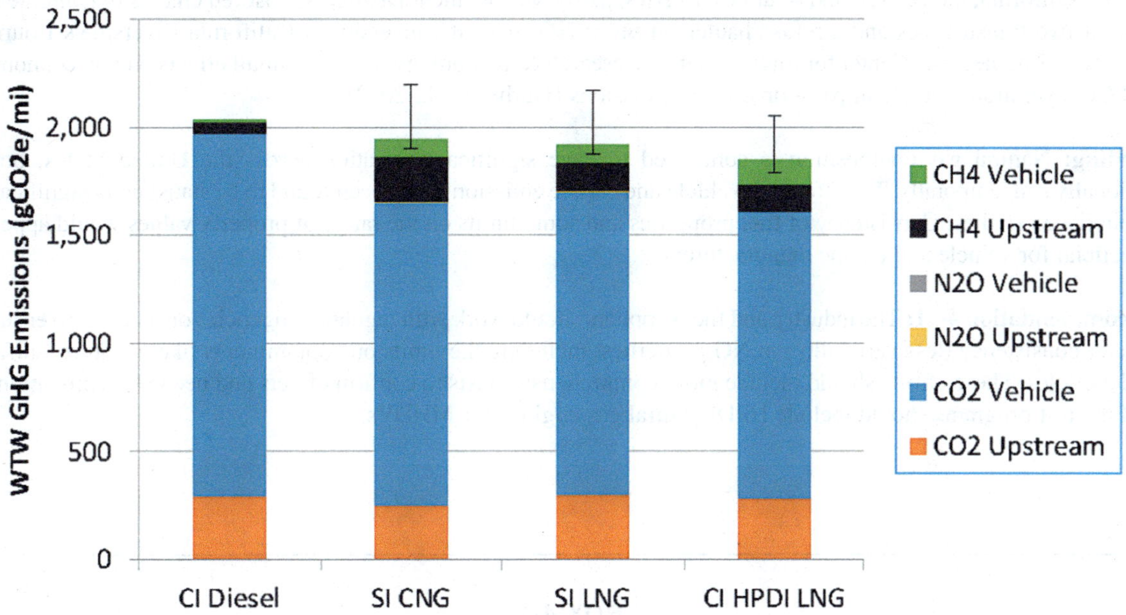

FIGURE 4-26 Comparison of well-to-wheels greenhouse gas emissions of long-haul trucks fueled by diesel and by natural gas fuels, broken down by the upstream, downstream and vehicle emissions category, and by type of gas.
SOURCE: Burnham et al. (2016).

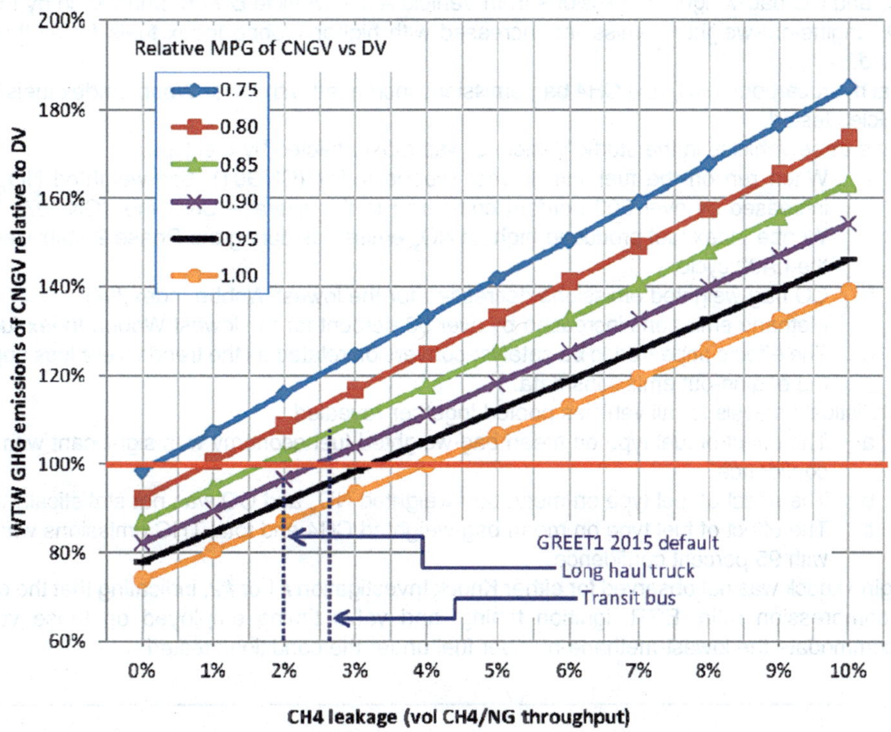

FIGURE 4-27 Well-to-wheels analysis of GHG emissions of CNG vehicles and diesel-fueled vehicles.
SOURCE: Burnham et al. (2016).

In California, the CARB- and South Coast Air Quality Management District–sponsored chassis dynamometer tests of two transit buses and a refuse hauler on seven NG fuels at University of California Riverside's Bourns College of Engineering, Center for Environmental Research & Technology observed small effects on fuel economy and CO_2, yet more notable impacts on criteria emissions (Durbin et al., 2012).

Finding: Natural gas composition is confirmed to have significant variation across the United States, both regionally and temporally. The effects on vehicle and engine emissions have been found to be statistically significant in published studies. Consistency of fuel properties and some limits on the range of property values would appear beneficial for vehicle and engine manufacturers.

Recommendation 4-11: The industry and fuel suppliers should work with regulatory agencies on a means to ensure greater consistency (less variability) in NG properties, including the limits on contaminants like moisture, sulfur, and particles. These efforts should include more comprehensive tests to confirm effects and needed specifications, and the test programs should include HPDI natural gas engines for MHDVs.

BOX 4-1

1. The bag-weighted fuel economy, in miles per gasoline gallon equivalent, varied in direct proportion to the Wobbe Index for all vehicles in the study. This effect was expected due to different energy content of the test fuels.
2. NO_x and CO bag-weighted emissions from Vehicle A and Vehicle B were unaffected by the fuel type.
3. NO_x engine-out-weighted emissions increased with higher Wobbe Index fuels for all three vehicles tested.
4. Total hydrocarbon (THC) and CH4 bag emissions increased with lower Wobbe Index fuels for all three vehicles tested.
5. Of the three vehicles in the study, Vehicle C was most affected by fuel type.
    a. When run on the fuel with lowest Wobbe Index (CNG01), bag-weighted NO_x emissions increased by over 300 percent compared to the average CNG fuel (CNG07). The lowest Wobbe Index fuel produced highest NO_x emissions during the Phase 2 stabilized portion of the LA92 cycle.
    b. CO bag-weighted emissions decreased for the lowest Wobbe Index fuel.
    c. Methane emissions increased by over 50 percent for the lowest Wobbe Index fuel.
    d. The effects appeared to be catalyst-conversion related as the trends were less apparent from the engine-out emissions data.
6. A statistical analysis for all vehicles pooled together revealed
    a. The effect of fuel type on mean bag-weighted fuel economy was significant with 95 percent confidence.
    b. The effect of fuel type on mean bag-weighted NO_x and CO was not statistically significant.
    c. The effect of fuel type on mean bag-weighted CH4 and total THC emissions was significant with 95 percent confidence.
7. Engine knock was not observed for either Knock Investigation #1 or #2, indicating that the combination of compression ratio, EGR, ignition timing, and valve timing employed on these vehicles can accommodate the lowest-methane-number fuel under the conditions tested.

4.10 TRANSMISSION AND DRIVELINES

4.10.1 Introduction

The primary job of the transmission is to match the wheel speed to the optimum engine speed for torque, horsepower, and fuel usage. Over the past two decades, the improvements in engines have necessitated transmission changes to achieve the optimum match. The number of gears in the transmission has increased to allow the engine to operate in a narrower band of speeds as dictated by the fuel map. Engine torque curves have flattened and extended to lower engine speeds. Engine horsepower and torque have continued to increase, but not as rapidly as in prior decades. There has been a continued migration from mechanical transmissions shifted by the driver to electronically controlled transmissions that automatically shift based on the speed of the vehicle, road grade, and knowledge of the engine operation, including the fuel map.

Table 4-8 shows the typical types of transmissions used in each weight class of vehicle. The lower classes generally use transmission technology and products derived from the light-duty vehicles. By example, Class 3 and 4 walk-in vans have moved toward automatic transmissions over the past decade. Automatic transmission usage in Classes 5 and 6 has also increased. Automated mechanical transmissions with dry clutches, wet clutches, or torque converters have increased penetration in the past 5 years so that OEMs are reporting penetration rates in excess of 50 percent (NACFE, 2014). Dual clutch transmissions have recently been introduced by Volvo, Mitsubishi Fuso, ZF, Allison, and Eaton. Continuously variable transmissions are not a factor in these weight classes. The remainder of this chapter will primarily focus on Class 8 vehicle transmissions, since the majority of fuel for commercial vehicles is used by this classification of vehicle.

For the purposes of this report, transmissions are categorized as follows:

1. Mechanical transmissions,
2. Automated mechanical transmissions with dry or wet clutches,
3. Automated mechanical transmissions with torque converters,
4. Automatic transmissions,
5. Continuously variable transmissions,
6. Dual clutch transmissions, and
7. Hybrid transmissions, electric and hydraulic (see Chapter 7 for more information).

Transmissions are efficient in translating rotational torque at the input to torque at the output. Efficiencies up to 98 percent are typical for steady-state operation. The fewer the number of gears that are meshed, the more efficient the transmission. The better the lubricating oil is dispersed, the more efficient. Compared to the opportunity for fuel savings through the engine, with its 42 to 48 percent efficiency today, the transmission would appear to

TABLE 4-8 Typical Transmission Types by Vehicle Weight Class

Vehicle Weight Class	Transmission Types
2b	4, 5
3	4
4	4
5	1, 2, 3, 4, 6
6	1, 2, 3, 4, 6
7	1, 2, 3, 4
8	1, 2, 3, 4

NOTE: 1, mechanical transmissions; 2, automated mechanical transmissions with dry or wet clutches; 3, automated mechanical transmissions with torque converters; 4, automatic transmissions; 5, continuously variable transmissions; 6, dual clutch transmissions; 7, hybrid transmissions, electric and hydraulic.

be a much lower priority. However, better coordination of the shifting of the transmission can create significant savings for even long-distance, line-haul applications. The 2014 North American Council for Freight Efficiency (NACFE) Confidence Report: Electronically Controlled Transmissions provides a description of the basics of vehicle/engine/transmission driveline design and recent changes that have created technology solutions to improve fuel consumption. As discussed in the first report and elsewhere, the importance of real-world fuel savings has led to improved understanding of the duty cycles of commercial vehicles and the impact of freight efficiency in ton-miles per gallon. The simplified diagram of power demand, at constant speed on level ground, used for more than a decade is being replaced by diagrams such as these in Figure 4-29 from the 21st Century Truck Partnership RoadMap. This increased understanding of real-world impacts, combined with improvements in component and system design, has led to new insights and innovations for synergistic operation of the engine and transmission.

Key points are that

- Engines produce torque at lower speeds,
- The sweet spot of the engine fuel map has become smaller,
- Smaller engines weigh less and produce sufficient horsepower/torque,
- Shift density is increasing,
- Startability and gradeability remain key design parameters, and
- Engine and transmission parameters must be optimized.

These are the factors driving increases in the number of gears for the transmission, development of sensors and software to detect grade and mass of the vehicle, lock-up of torque converters as early as possible, development of dual clutch transmissions, predictive cruise control, and deeper integration of the engine and transmission controls.

4.10.2 New Product Developments

4.10.2.1 Mechanical Transmissions

Mechanical transmission development is focused on weight, lubrication, cooling, and gearing. Weight savings come from redesign of the housing to reduce its size, often eliminating wasted space inside the housing. It also comes from greater use of aluminum where possible. Lubrication improvements come from changes in the lubricant and changes in the internal transmission design to better ensure fluid is moved to the gear interfaces effectively. Gearing changes are regularly improved to adapt to the preferred engine speeds for fuel consumption and the axle ratios available. Axle ratios continue to get faster (lower number). Continued, but small, improvements in these areas are expected in the next decade.

4.10.2.2 Automated Mechanical Transmissions with Dry or Wet Clutches

This type of transmission has seen increased penetration on new vehicle orders for over-the-road trucks and tractors. Volvo publicly stated in March 2015 at the MidAmerica Truck Show that they are seeing in excess of 75 percent use on new vehicles. One year later, they are reporting 83 percent.[28] Daimler Trucks North America indicates penetration rates of 42 percent in its high-volume Cascadia product.[29] By August of 2016, the penetration had grown to over 60 percent.[30] Both automated mechanical and automatic transmissions are an enabler for more advanced driver assistance systems and, eventually, more autonomous operation of vehicles.

---

[28] See http://www.volvotrucks.us/about-volvo/news-and-events/new-volvo-i-shift-with-crawler-gears-offers-slow-speed-maneuverability-extreme-startability/ (accessed December 18, 2019).

[29] See http://www.hdma.org/Main-Menu/HDMA-Publications/Diesel-Download/September-17-2015/DT12-Detroit-Assurance-Helping-DTNA-Defend-Market-Share.html (accessed September 17, 2019).

[30] See http://www.ttnews.com/articles/basetemplate.aspx?storyid=42839&page=2 (accessed December 18, 2019).

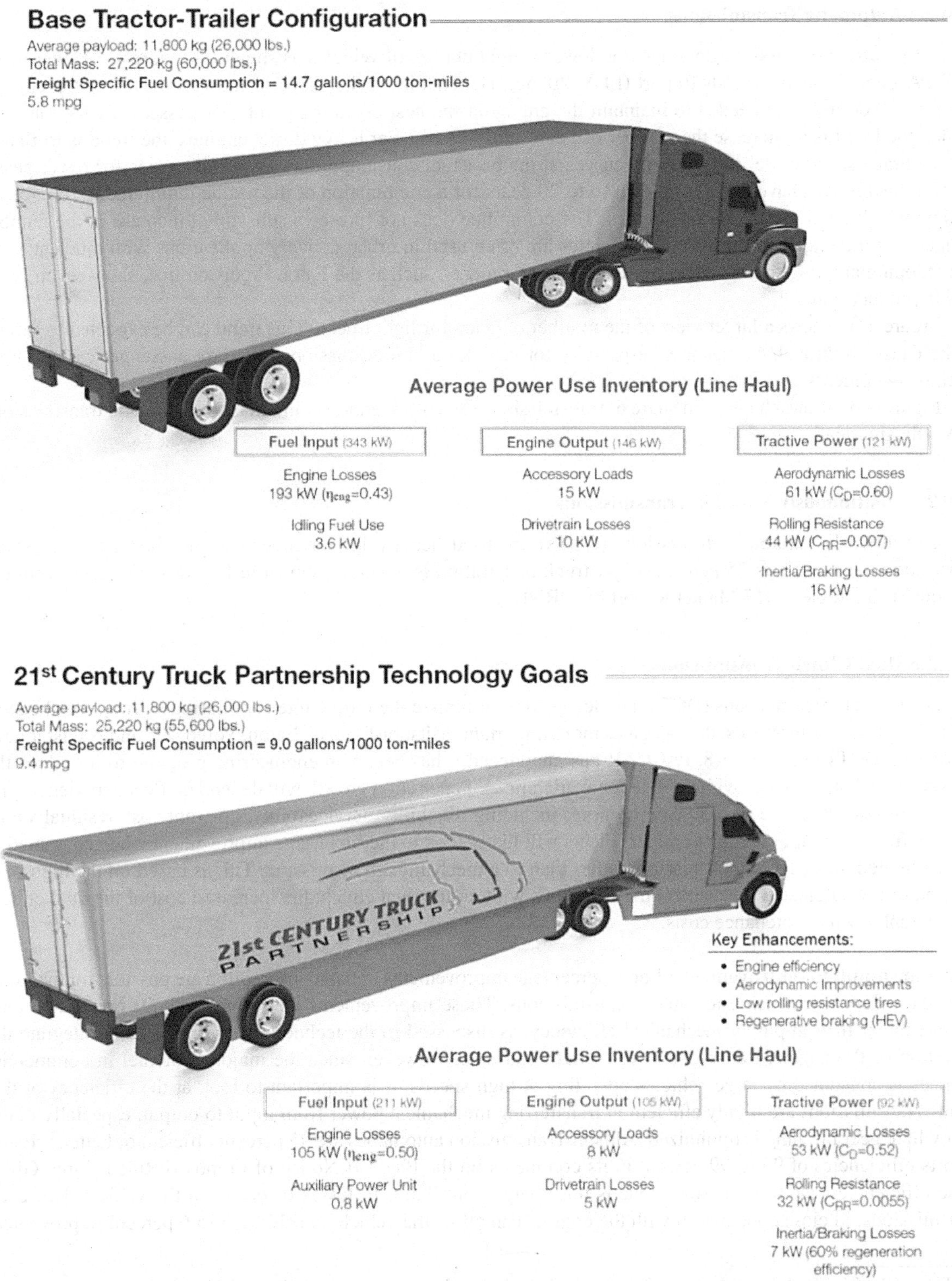

FIGURE 4-29 Power use inventory for Class 8 heavy truck: base (upper), and with 21st Century Truck Partnership goals (lower).
SOURCE: DOE (2013, p. 25).

4.10.2.3 Automatic Transmissions

Automatic transmissions dominate the lower weight classes of vehicles as shown in Figure 4-30 taken from the EPA Fuel Economy Trends Report (EPA, 2015c). Depending on the shape of the fuel curves of the engine, more or fewer gears are needed to maintain the engine at the best operating point. For passenger cars and light trucks, the trend is to increase the number of gears substantially. For heavy diesel engines, the trend is to flatten out the brake specific fuel consumption curves at the best fuel consumption points. This argues for fewer gears. Truck transmissions have had as many as 16 to 20 gears for a combination of the torque requirements, combined vehicle weight, and the engine fuel curves. The committee does not foresee a substantial increase in the number of gears for these applications. Class 6 vehicles are often used in urban delivery applications with much starting and stopping and low-speed operation. Recent developments, such as the Eaton Procision line, show seven gears in this product range.[31]

Figure 4-31 shows a larger view of the number of gears for light trucks. This trend can be expected to extend to the Class 2b through 6 categories, especially for vehicles and transmissions based on passenger car and light truck developments.

Figure 4-32 shows the Allison share of transmissions for several end-user markets for automatic transmissions in North America.

4.10.2.4 Continuously Variable Transmissions

Continuously variable transmissions (CVTs) are used heavily in passenger cars as shown above, where penetration is as much as 25 percent. Light truck penetration is lower, as shown in Figure 4-33, taken from the Vehicle Technologies 2017 Market Report by ORNL.

4.10.2.5 Dual Clutch Transmissions

Dual clutch transmissions (DCTs) are designed to minimize the torque interrupt during shifting. As the shift density for vehicles increases, this becomes more important. Mitsubishi Fuso, Eaton, Volvo, ZF, and Navistar have introduced DCTs in classes 3-8. PACCAR has announced it has begun an engineering program to introduce the transmission, but has not provided a date for availability.[32] Penetration growth will depend on fleet experience with the transmission over the next 5 years or more, including reliability, serviceability, up-front cost, residual value, and any fuel consumption improvements. Fleets will likely look to the fuel usage improvement when compared to the automated mechanical transmission, rather than to a mechanical transmission. This is based on the high penetration of the automated mechanical transmissions without the dual clutch, the increased cost of the dual clutch, and the unknown maintenance costs.

Fuel Consumption Improvements: Large percentage improvements in fuel consumption are possible for medium- and heavy-duty trucks using automatic transmissions. These improvements are related to shift points and time to shift rather than to purely mechanical efficiency, as discussed in the technologies to more fully integrate the operation of the engine, transmission, driveshaft, and axle. However, since the majority of fuel in commercial vehicles is consumed by large vehicles operating at high speeds, it is important to look at the efficiency at this point. Transmissions are highly efficient in transferring mechanical power from input to output, especially in top gear where gear meshing is minimized. Allison Transmission automatics are 95 percent efficient or better.[33] Eaton reports efficiencies of 96 to 99 percent in its comments on the Phase II Notice of Proposed Rulemaking. Given these efficiencies, the transmission alone is not a large contributor to losses of power in the vehicle. However, transmissions, in close cooperation with the engine, can allow the vehicle to achieve 3 to 6 percent improvement

[31] Eaton, "Procision Dual-Clutch Automatic Transmission," Available at http://www.eaton.com/Eaton/ProductsServices/Vehicle/Transmissions/procision/index.htm (accessed December 18, 2019).

[32] See http://www.peterbilt.com/about/media/2016/501/ (accessed December 18, 2019).

[33] Personal communication to Paul Menig from Michael J. Howenstein, Allison Transmission, December 11, 2015.

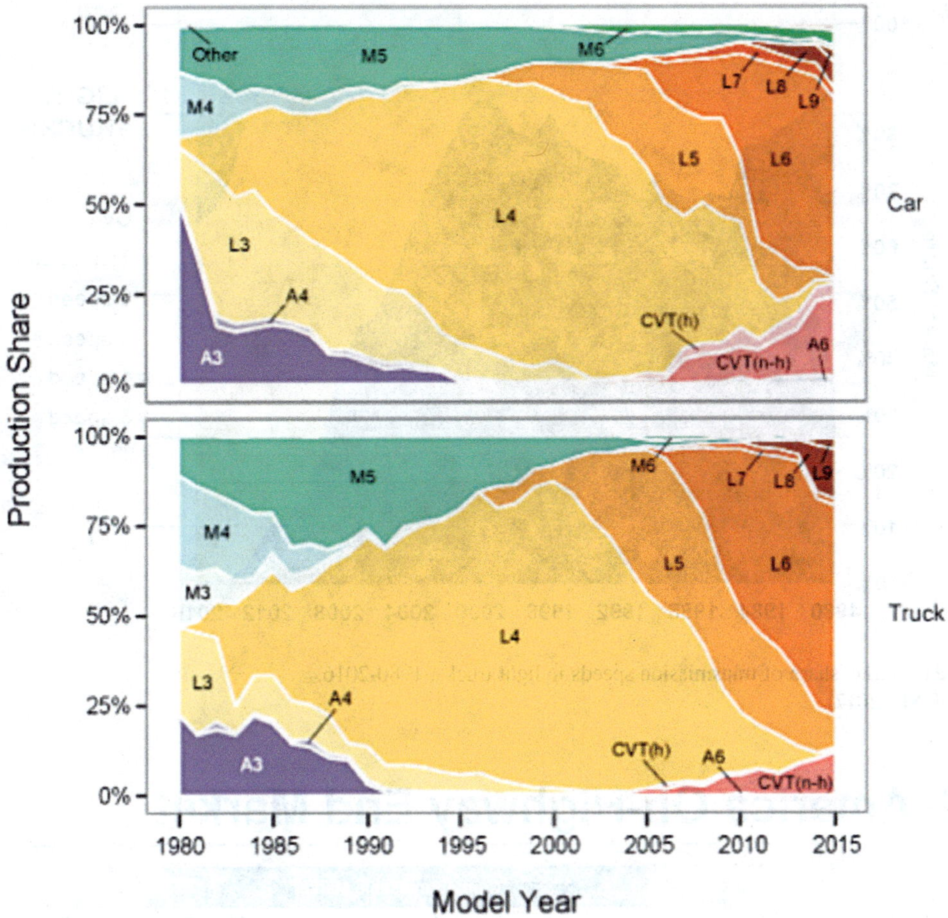

Transmission	Lockup?	Number of Gears	Key
Automatic Semi-Automatic Automated Manual	No	3	A3
		4	A4
		5	A5
		6	A6
		7	A7
	Yes	3	L3
		4	L4
		5	L5
		6	L6
		7	L7
		8	L8
		9	L9
Manual	–	3	M3
		4	M4
		5	M5
		6	M6
Continuously Variable (non-hybrid)	–	–	CVT(n-h)
Continuously Variable (hybrid)	–	–	CVT(h)
Other	–	–	Other

FIGURE 4-30 Transmission production share in passenger cars and light trucks.
SOURCE: EPA (2015c).

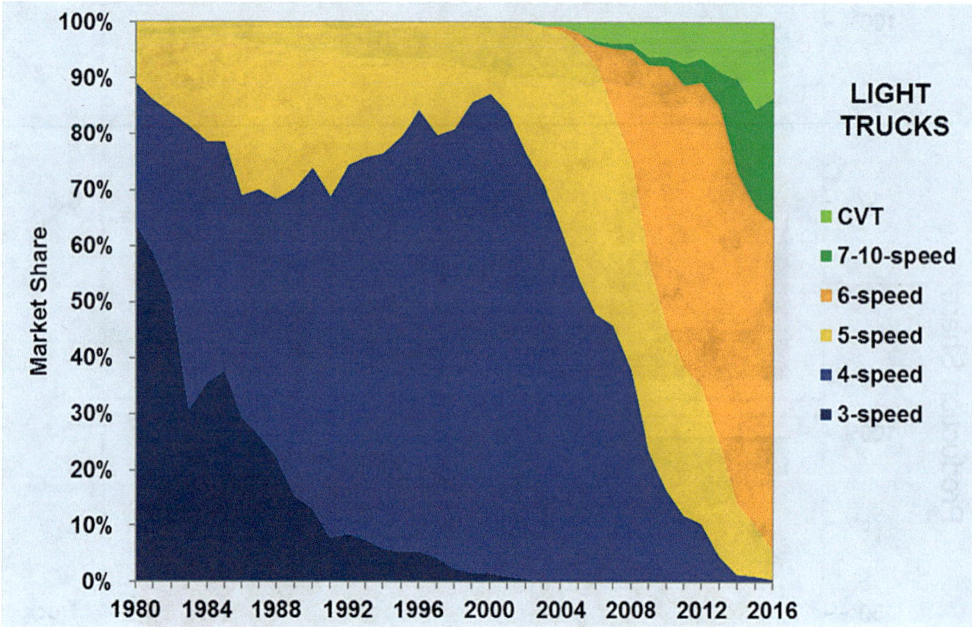

FIGURE 4-31 Market share of transmission speeds in light trucks, 1980-2016.
SOURCE: ORNL (2017).

North America On-Highway End Market

		Underserved	Core Addressable Market				Underserved	
	Class 1-3	Class 4-5	Motor Home	School Bus	Class 6-7	Class 8 Straight	Class 8 Metro	Class 8 Tractor
Vehicles								
Weight (000s of lbs)	< 14 lbs	14 – 19 lbs	16 – 33 lbs	16 – 33 lbs	19 – 33 lbs	33 lbs+	33 lbs+	33 lbs+
2015 Industry Units Produced	6,974,454	86,774	18,270	31,689	95,684	79,241	76,629	167,412
2015 Allison Share	0%	5%	40%	98%	77%	61%	7%	0%

- ~ 30-40% of Allison's North America On-Highway market volume is driven by municipal spending, reducing end-market volatility
- Significant opportunity to further penetrate the Class 8 Metro market
  - "Metro" is a term for tractors that are used primarily in urban environments, which represent ~30% of the Class 8 tractor market

Note: Analysis excludes Allison's Transit/Coach Bus and Hybrid Transit Bus volume.
Source: Class 1-3 from Wards 2015 Factory Sales North America; Core Addressable Market and Class 8 Tractor from Allison and ACT Research.

FIGURE 4-32 Allison transmission share by vehicle weight class and vehicle type.
SOURCE: Allison Transmission Inc. (2016).

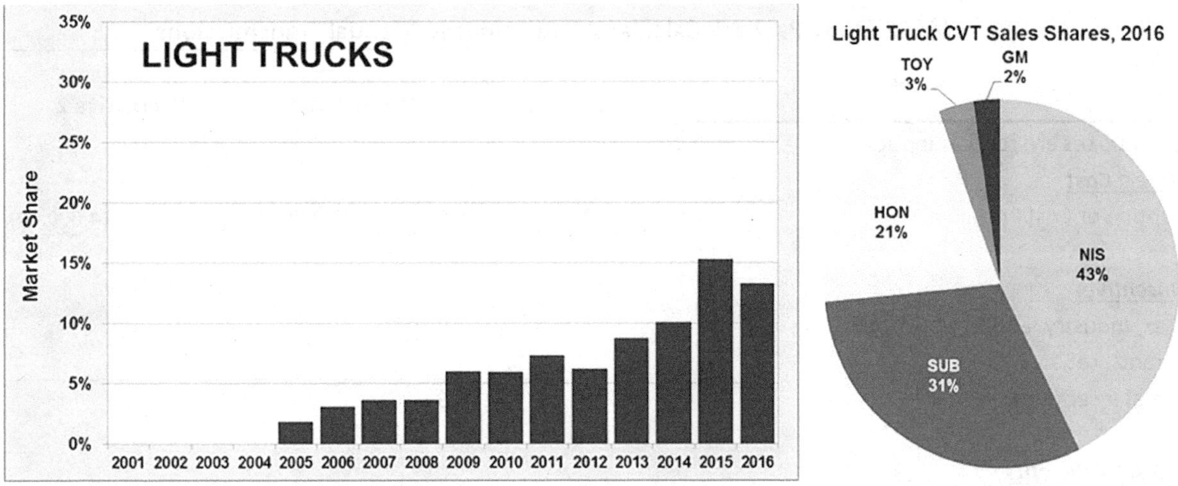

FIGURE 4-33 CVT market share in light trucks, 2001-2016, and CVT light truck sales by manufacturer, 2016.
NOTE: CHR, Chrysler; FOR, Ford; GM, General Motors; NIS, Nissan; SUB, Subaru; TOY, Toyota.
SOURCE: ORNL (2017).

in fuel consumption over defined routes. This is accomplished through efficient shifting of the transmission to avoid loss during the shift and to place the engine in the best position on its fuel map.

Costs: Purchase price is often a small part of the cost of technology in the trucking industry. Additional considerations include resale value, maintenance requirements, training, and government incentives. Figure 4-34 is taken from the Automated Transmission Study (NACFE, 2014).[34]

Initial purchase price for transmissions, as paid by fleets, is difficult to determine publicly. In general, fleets report prices of \$3,000 to \$5,000 more for an automated transmission than for a mechanical transmission (NACFE, 2014). As Class 8 vehicle OEMs continue to improve fuel consumption to meet the Phase I regulations, and the Phase II regulations, the cost of the transmission may be dwarfed by the cost of the vehicle. This will make it even more difficult to determine the added cost of the new technology. Automatic transmissions for Class 8 vehicles are likely to cost more. The fleet cost typically ranges from \$3,000 added cost to \$11,000 or more. Figure 4-35 is taken from the Q2 2016 investor relations presentation of Allison Transmission. It provides a reasonable example of the added cost of technology.

Increased Shifting Density: As engine, transmission, and vehicle OEMs work to improve fuel consumption and reduce greenhouse gas emissions, engine speeds are decreasing. In some cases, there is a shift to using smaller displacement engines as well. Alone or in combination, these trends require more manipulation of the transmission gearing to keep the engine in its most efficient status under normal driving conditions. This is leading to an increased number of shifts per mile necessary for Class 8 vehicles to achieve fuel economy goals.

Finding: Automated transmission penetration in the commercial vehicle industry is growing significantly as engine, transmission, and vehicle OEMs work to improve fuel consumption for the benefit of the customer and to meet government regulations.

Recommendation 4-13: Future regulations for fuel consumption, criteria emissions, and greenhouse gases must take account of the close relationship of the engine, transmission, and axle ratios and the effect of lubricants on

[34] The calculator itself can be accessed at http://www.nacfe.org/powertrain/automated-manual-transmissions (accessed December 18, 2019).

NACFE Study Payback Calculator: Automated Manual Transmissions

	Baseline	Alternative 1	Alternative 2
Gray boxes are for user inputs			
Direct Cost			
Option or Cost Premium		$ (3,000.00)	$ (4,000.00)
Incentives			
Tax, Industry, and/or other incentives		$ -	
Grants, rebates			$ -
Total Incentives		$ -	$ -
Societal Benefits			
Improved Emissions - Air, Noise etc.		yes	no
Usage			
Miles per vehicle per year	125,000	125,000	125,000
Years of anticipated Use	4	4	4
Operating Costs			
Fuel			
Fuel Economy Improvement Percentage		3%	3%
Fuel Expense per mile	$ 0.65	$ 0.63	$ 0.63
Fuel Expense per year per truck	$ 81,250	$ 78,813	$ 78,813
Fuel savings per year per truck		$ 2,438	$ 2,438
Driver/Operator Impact			
Driver Turnover per Year in percent	100%	75%	90%
Number of Drivers	100	100	100
Number of Trucks in Fleet	100	100	100
Number of Drivers per Truck per Year	1	0.75	0.9
Driver Recruiting Cost per Driver	$ 1,000		
Driver Training Cost per Driver	$ 5,000		
Driver Recruiting Cost Savings per Driver		$ 250	$ 100
Driver Training Cost Savings per Driver	$ 5,000	$ 1,250	$ 500
Total Recruiting & Training Savings	$ 6,000.00	$ 1,500	$ 600
Safety Cost Improvement (Cost/Year)	$ -	$ -	$ -
Total Driver/Operator Impact		$ 1,500	$ 600

FIGURE 4-34 Automated mechanical transmissions payback calculator.
SOURCE: NACFE (2014).

Maintenance Improvement/Deficit		
Additional Maintenance Cost/Year	$ (10)	$ (10)
Additional Specialized Tooling Cost/Year	$ (100)	$ (100)
Additional Mechanic Training Cost/Year	$ (10)	$ (10)
Reduced Maintenance Cost/Mile	$ -	$ -
Additional Consumable or Fluid Cost/Year	$ (100)	$ (100)
Other Vehicle Related Cost Improvements/Liabilities Cost per Year	$ -	$ -
Total Maintenance Improvement/Deficit	$ (220)	$ (220)
Other Operating Cost Change/Year	$ -	$ -
Resale Value or Impact		
Penalty / Benefit at end of first owner life	$ -	$ -
Payback in months (costs upfront)	**15**	**20**
Payback in months (costs amortized monthly)	**10**	**18**
ROI (end of year 4)	**3.96**	**1.82**

How to use:
This Payback calculator is principally an incremental analysis tool designed to compare two or more
Note: All costs should be entered as a negative numbers which usually means the number is entered in parentheses i.e.. A $4000. cost is entered as (4000). Conversely a $4000. savings is noted as 4000.

FIGURE 4-34 Continued.

these. As mentioned elsewhere, NHTSA should immediately be collecting information on the details of vehicle production.

4.11 AXLES AND DRIVELINES

For fuel consumption purposes, axle developments related to gearing, lubrication, weight, and axle ratio are the most important. Several of these are similar to the developments for transmissions. As the engines have become more fuel efficient at lower speeds and able to handle the torque, this has necessitated faster (lower numerical) axle ratios to maintain the speed of the vehicle. Axle ratios of 2.26 are readily available. A 2.16 ratio has been announced for availability in 2017 by Freightliner. Research programs are using axle ratios lower than 2.00 (Sturgess, 2016).

The slower engine speed at cruise, combined with the faster axle ratio, requires the driveline to handle more torque, especially at startup. Dana has published a white paper providing information on the issue. Meritor has also published a study.[35] For any given startability, as the axle ratio becomes faster, the driveline torque increases. Torque transients, especially during aggressive starting, create significantly more stress in the driveline. Component fractures can occur; thus, axle and driveline manufacturers have been strengthening the components to eliminate over-stress fractures.

Since there is significant fuel savings from down-speeding the engine, there will continue to be development in the axles and drivelines to match the speed and torque characteristics needed.

[35] See https://www.meritor.com/ourcompany/news/whitepapers/pdfs/Meritor-WhitePaper-EngineDownspeeding.pdf.

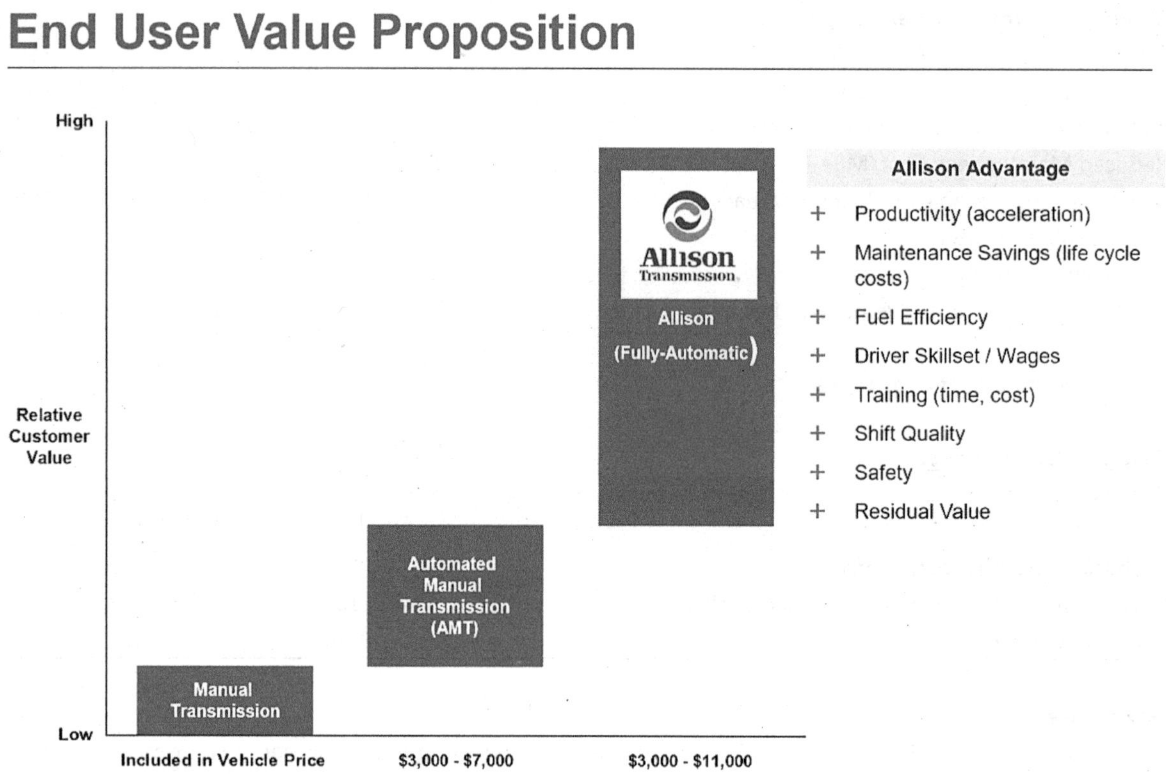

FIGURE 4-35 Typical end-user cost of automated transmissions, including advantages of an automatic transmission. SOURCE: Allison Transmission Inc. (2016).

As with the transmission, lubrication of the axle gearing is an important factor in achieving long life and high efficiency. Various types of axle lubrication have been field tested. These include systems by Meritor, Dana, and Freightliner. These systems often reduce the lubrication splash during cruise control by reducing the lubricant in the reservoir. Electronic controls and sensors are required to effect this reduction. Such systems are available at additional cost and are reported to provide improvements of less than 1 percent in fuel economy.

The 6×2 axle configurations are another development with axles. The Confidence Report on 6×2 axles from NACFE (2014) is a good summary of the potential for 2 percent improvement in fuel consumption.

Motorized axles have been researched in the past and are available on some buses. Mercedes-Benz' recent announcement of an urban electric truck uses such a motorized axle from ZF. A number of developments and early product offerings exist for trailer axles with motors.

Driveline developments are focused on reducing weight, including hollow-tube drivelines and carbon fiber materials. These developments appear to still be in the early concept and research phases. Such developments will provide small improvements in overall vehicle fuel consumption.

4.12 CLUTCHES

Another side effect of the reduced engine speeds is vibration in the driveline that is translated to the driver. Development to reduce this vibration is in process. Valeo announced the development of a clutch combining high torque, low stiffness, and a pendulum approach with floating masses. The goal is a further reduction of up to 200 RPM in engine speed.

4.13 REFERENCES

Abani, N., M. Chiang, I. Thomas, N. Nagar, and R. Zermeno. 2016. *Developing a 55+ BTE Commercial Heavy-Duty Opposed-Piston Engine without a Waste Heat Recovery System*. San Diego, CA: Achates Power, Inc.

ACEA. 2013. Worldwide Fuel Charter. Fifth Edition. Available at http://www.acea.be/uploads/publications/Worldwide_Fuel_Charter_5ed_2013.pdf (accessed December 18, 2019).

ACT (Americas Commercial Transportation Research Company, LLC). 2015. ACT Research, Natural Gas Quarterly: U.S. Class 8 NG Trends, Forecast, Analysis & Insight Q3 2015. Available at http://www.actresearch.net/wp-content/uploads/Q3-15-NGQ.pdf (accessed September 15, 2016.

Aikawa, K., and J.J. Jetter. 2013. Impact of gasoline composition on particulate matter emissions from a direct-injection gasoline engine: Applicability of the particulate matter index. *International Journal of Engine Research* 15(3):298-306.

Aikawa, K., T. Sakurai, and J. Jetter. 2010. Development of a predictive model for gasoline vehicle particulate matter emissions. *SAE International Journal of Fuels and Lubricants* 3(2):610-622.

Alleman, T. 2013. US Biofuels Quality Update. NIST Biocorrosion Workshop, Boulder, CO, July 22.

Allison Transmission, Inc. 2016. Q2 2016- Investor Relations Presentation of Allison Transmission, August 2.

Alvarez, R.A., S.W. Pacala, J.J. Winebrake, C.W. Chameides, and S.P. Hamburg. 2012. Greater focus needed on methane leakage from natural gas infrastructure. *Proceedings of the National Academy of Sciences of the United States of America* 109(17):6435-6440.

Arimilli, R.V., K. Erickson, F.T. Mottley, and J.C. Conklin. 2003. Performance evaluation of a prototype TurbX™ engine. In *Proceedings of the ASME Turbo Expo: Power for Land, Sea, and Air*, Vol. 3. International Gas Turbine Institute, pp. 295-304.

Atkinson, C., S. Petreanu, N.N. Clark, R.J. Atkinson, T.I. McDaniel, S. Nandkumar, and P. Famouri. 1999. Numerical Simulation of a Two-Stroke Linear Engine-Alternator Combination. SAE Technical Paper 1999-01-0921, doi: 10.4271/1999-01-0921.

Bhatnagar, V., and P. Owende. 2015. Energy harvesting for assistive and mobile applications. *Energy Science & Engineering* 3(3):153-173.

Blain, L. 2015. Honda files patents for brand new, fuel injected two-stroke engine. New Atlas. http://www.gizmag.com/honda-two-stroke-with-fuel-injection-patent-filing/38529/ (accessed April 13, 2016).

Burnham, A., H. Cai, and M. Wang. 2016. Critical Factors in the Development of Well-to-Wheel Analyses of Alternative Fuel and Advanced Powertrain Heavy-Duty Vehicles. SAE Technical Paper 2016-01-1284, doi: 10.4271/2016-01-1284.

Butler, A., R. Sobotowski, G. Hoffman, and P. Machiele. 2015. Influence of Fuel PM Index and Ethanol Content on Particulate Emissions from Light-Duty Gasoline Vehicles. SAE Technical Paper 2015-01-1072, doi: 10.4271/2015-01-1072.

Chadwell, C., T. Alger, J. Zuehl, and R. Gukelberger. 2014. A demonstration of dedicated EGR on a 2.0 L GDI engine. *SAE International Journal of Engines* 7(1):434-447.

Chang, J., G. Kalghatgi, A. Amer, and Y. Viollet. 2012. Enabling High Efficiency Direct Injection Engine with Naphtha Fuel through Partially Premixed Charge Compression Ignition Combustion. SAE Technical Paper 2012-01-0677, doi: 10.4271/2012-01-0677.

Clark, N.N., D.R. Johnson, D.L. McKain, W.S. Wayne, H. Li, J. Rudek, R.A. Mongold, C. Sandoval, A.N. Covington, and J.T. Hailer. 2017a. Future methane emissions from the heavy-duty natural gas transportation sector for stasis, high, medium, and low scenarios in 2035. *Journal of the Air & Waste Management Association* 67(12):1328-1341.

Clark, N.N., D.L. McKain, D.R. Johnson, W.S. Wayne, H. Li, V. Akkerman, C. Sandoval, A.N. Covington, R.A. Mongold, J.T. Hailer, and O.J. Ugarte. 2017b. Pump-to-wheels methane emissions from the heavy-duty transportation sector. *Environmental Science & Technology* 51(2):968-976.

Clarke, J., and E. O'Malley. 2013. Analytical Comparison of a Turbocharged Conventional Diesel and a Naturally Aspirated Compact Compression Ignition Engine both Sized for a Highway Truck. SAE Technical Paper 2013-01-1736, doi: 10.4271/2013-01-1736.

Cobb, J. 2014. Wrightspeed Combines Gas Turbine and Batteries for Big Fuel Savings. Hybrid Cars, February 18. http://www.hybridcars.com/wrightspeed-combines-gas-turbine-and-batteries-for-big-fuel-savings/ (accessed April 13, 2016).

Confer, K., and Delphi Advanced Powertrain. 2015. Ultra Efficient Light Duty Powertrain with Gasoline Low Temperature Combustion. Presentation at the 2015 DOE Vehicle Technologies Review, June 11.

Conklin, J.C., and J.P. Szybist. 2010. A highly efficient six-stroke internal combustion engine cycle with water injection for in-cylinder exhaust heat recovery. *Energy* 35(4):1658-1664.

CRC (Coordinating Research Council, Inc.). 2014. Biodiesel and Renewable Diesel Characterization and Testing in Modern LD Diesel Passenger Cars and Trucks, CRC Report No. AVFL-17b.

Cummins Filtration. 2009. Diesel Exhaust Fluid (DEF) Q&A. Available at https://www.cumminsfiltration.com/sites/default/files/MB10033.pdf (accessed December 18, 2019).

Cummins Westport. 2016. Introducing ISL G Near Zero. Available at https://mart.cummins.com/imagelibrary/data/assetfiles/0042998.pdf (accessed December 18, 2019).

Curran, S., Z. Gao, J. Szybist, and R.Wagner. 2014. Fuel Effects on RCCI Combustion: Performance and Drive Cycle Considerations. Presentation at the 2014 CRC Workshop on Advanced Fuels and Engine Efficiency, February 26. Baltimore, MD. Oak Ridge, TN: Oak Ridge National Laboratory.

Curran, S., Z. Gao, and R. Wagner. 2015. Reactivity-controlled compression ignition drive cycle emissions and fuel economy estimations using vehicle system simulations. *International Journal of Engine Research* 16(8):1014-1024.

Daw, C.S., et al. 2010. Report on the Transportation Combustion Engine Efficiency Colloquium Held at USCAR, March 3-4. Oak Ridge National Laboratory Report No. ORNL/TM-2010/265, October.

Delgado, O., and R. Muncrief. 2015. Assessment of Heavy-Duty Natural Gas Vehicle Emissions: Implications and Policy Recommendations. Washington, DC: ICCT. July.

DeLizo, S. 2015. Kenworth Class 8 AFV NG Micro-Turbine Hybrid Electric Vehicle, CEC-ARV-009-012, April.

Dempsey, A., S. Curran, and R. Wagner. 2016. A perspective on the range of gasoline compression ignition combustion strategies for high engine efficiency and low NO_x and soot emissions: Effects of in-cylinder fuel stratification. *International Journal of Engine Research* 17(8):897-917.

DOE (Department of Energy). 2012. Overview of the DOE Advanced Combustion Engine R&D. Presentation at the 2012 Annual Merit Review and Peer Evaluation Meeting in Washington, DC, May 14-18.

DOE. 2013. The Roadmap and Technical White Papers of the 21st Century Truck Partnership. February 27.

Durbin, T., G. Karavalakis, K. Johnson, M. Hajbabaei, and W. Miller. 2012. Effect of Natural Gas Fuel Composition on Vehicle Emissions and Performance. Presentation to the Program to Evaluate the Implications of Gas Variability in California, February 15. University of California, Riverside Bourns College of Engineering.

Eckerle, W. (Cummins Inc.). 2013. Roadmap to Energy Conversion Efficiency at Cummins. Presentation to the 2013 ERC Symposium in University of Wisconsin, June 5.

Eckerle, W. (Cummins Inc.). 2015. Engine Technologies for GHG and Low NO_x: ARB Symposium on Phase 2 GHG. Presentation to the CA Phase 2 GHG April 2015 Symposium in Diamond Bar, CA.

Edwards, K.D., R. Wagner, T.E. Briggs, and T. Theiss. 2011. Defining Engine Efficiency Limits. Presentation at the 17th DEER Conference in Detroit, MI, October 5. Oak Ridge, TN: Oak Ridge National Laboratory.

EIA. 2014. Annual Energy Outlook 2014 with Projections. Washington, DC: Environmental Protection Agency.

EIA. 2016. Annual Energy Outlook 2016 with Projections. Washington, DC: Environmental Protection Agency.

Elson, A., et al. 2015. ICF International Waste Heat to Power Market Assessment. ORNL/TM-2014/620, March.

EPA (Environmental Protection Agency). 2002. A Comprehensive Analysis of Biodiesel Impacts on Exhaust Emissions: Draft Technical Report. EPA420-P-02-001.

EPA. 2009. EPA Lifecycle Analysis of Greenhouse Gas Emissions from Renewable Fuels. EPA-420-F-09-024.

EPA. 2010. EPA Lifecycle Analysis of Greenhouse Gas Emissions from Renewable Fuels. Office of Transportation and Air Quality, EPA-420-F-10-006.

EPA. 2015a. Proposed Rulemaking for Greenhouse Gas Emissions and Fuel Efficiency Standards for Medium- and Heavy-Duty Engines and Vehicles–Phase 2 Draft Regulatory Impact Analysis. EPA-420-D-15-900. Washington, DC: EPA.

EPA. 2015b. Oil and Natural Gas Sector: Emission Standards for New and Modified Sources. *Federal Register* 80:56593-56698.

EPA. 2015c. Light Duty Automotive Technology, Carbon Dioxide Emissions, and Fuel Economy Trends. EPA 420-R-15-016. Washington DC: EPA.

EPA and NHTSA (National Highway Traffic Safety Administration). 2015. Greenhouse gas emissions and fuel efficiency standards for medium- and heavy-duty engines and vehicles—Phase 2. *Federal Register* 80:40137-40765 (July 13).

EPA and NHTSA. 2016. Greenhouse gas emissions and fuel efficiency standards for medium- and heavy-duty engines and vehicles—Phase 2. *Federal Register* 81:73478-74274 (October 25).

ET Auto (Economic Times Auto). 2016. Continental to Present Connected Technologies for CVs. Available at http://auto.economictimes.indiatimes.com/news/auto-components/continental-to-present-connected-technologies-for-cvs/53886925 (accessed December 19, 2019).

Farrell, J. (National Renewable Energy Laboratory). 2015. Optima Program Overview. Presented at Optima Stakeholder Listening Day, June 15.

Fraas, L., J. Avery, E. Malfa, J.G. Wuenning, G. Kovacik, and C. Astle. 2003. Thermophotovoltaics for combined heat and power using low NO_x gas fired radiant tube burners. *AIP Conference Proceedings* 653:61. doi: 10.1063/1.1539364.

Gao, F., M. Kollár, Y. Wang, N.M. Washton, J. Szanyi, and C. Peden. Institute for Integrated Catalysis, Pacific Northwest National Laboratory. 2015. Effects of Alkali and Alkaline Earth Cocations on the Activity and Hydrothermal Stability of Cu/SSZ-13 NH3-SCR Catalysts. Presentation at the 2015 DOE CLEERS Workshop. University of Michigan, Dearborn, MI. April 29.

General Motors. 2012. Split Cycle Internal Combustion Engine. U.S Patent 8,267,056 B2. September 18.

Government Fleet News. 2014. Half of Ford's Class 3-5 Chassis Cab Buyers Choose Gasoline. Available at http://www.government-fleet.com/news/story/2014 (accessed February 18, 2017).

Greszler, A. (Volvo Group Truck Technology). 2012. View from the Bridge—Commercial Vehicle Perspective. Presentation at the DEER Conference, Department of Energy.

Greszler, A. (Volvo Group Truck Technology). 2014. Impact of NO_x Control on Heavy Duty Diesel Vehicle Efficiency. Presentation to the National Academies of Sciences Committee on the Assessment of Technologies and Approaches for Reducing the Fuel Consumption of Medium- and Heavy-Duty Vehicles, Phase Two, Ypsilanti, MI, July 28.

Hanson, R., S. Spannbauer, C. Gross, R. Reitz, S. Curran, J. Storey, and S. Huff. 2015. Highway Fuel Economy Testing of an RCCI Series Hybrid Vehicle. SAE Technical Paper 2015-01-0837. doi: 10.4271/2015-01-0837.

Hunter, S., et al. 2011. Development of MEMS based pyroelectric thermal energy harvesters. In *Energy Harvesting and Storage: Materials, Devices, and Applications II*, edited by N.K. Dhar, P.S. Wijewarnasuriya, and A.K. Dutta. *Proceedings of SPIE* 8035. doi: 10.1117/12.882125.

Huo, H., M. Wang, C. Bloyd, and V. Putsche. 2009. Life-cycle assessment of energy use and greenhouse gas emissions of soybean-derived biodiesel and renewable fuels. *Environmental Science & Technology* 43:750-756.

Huo, M., Y. Huang, and P. Hofbauer. 2015. Piston Design Impact on the Scavenging and Combustion in an Opposed-Piston, Opposed-Cylinder (OPOC) Two-Stroke Engine. SAE Technical Paper 2015-01-1269. doi:10.4271/2015-01-1269.

ICCT (International Council on Clean Transportation). 2016. Costs of Emission Reduction Technologies for Heavy-Duty Diesel Vehicles. Washington, DC: ICCT. February. Available at https://www.theicct.org/sites/default/files/publications/ICCT_costs-emission-reduction-tech-HDV_20160229.pdf (accessed December 19, 2019).

Ikeya, K., M. Takazawa, T. Yamada, S. Park, et al. 2015. Thermal efficiency enhancement of a gasoline engine. *SAE International Journal of Engines* 8(4):1579-1586. doi: 10.4271/2015-01-1263.

Infineum. 2014. Worldwide Winter Diesel Fuel Quality Survey 2014. Available at https://www.infineum.com/media/80722/wdfs-2014-full-screen.pdf (accessed December 19, 2019).

Johansson, B. 2015. High efficiency combustion engines, what is the limit?, Society of Automotive Engineers High Efficiency Engine Symposium, Detroit, MI.

Johnson, T. 2016. Vehicular emissions in review. *SAE International Journal of Engines* 9(2):1258-1275. doi: 10.4271/2016-01-0919.

Johnson, T., and Corning Incorporated. 2014. Review of Heavy-Duty Truck Emissions and Efficiency Trends. Presentation to the National Academies of Sciences Committee on the Assessment of Technologies and Approaches for Reducing the Fuel Consumption of Medium- and Heavy-Duty Vehicles, Phase Two, Ypsilanti, MI, July 29.

Karagozian, M. 2013. A look back at Ford's Big Red turbine powered concept truck. *Examiner*, February 9. http://www.examiner.com/article/a-look-back-at-ford-s-big-red-turbine-powered-concept-truck (accessed April 13, 2016).

Keolian, R. 2011. Final Report: DOE Project DE-FC26-04NT42113 CPR/IW Project 17057 Truck Thermoacoustic Generator and Chiller.

Kierstein, A. 2014. The humble two-stroke might be the engine of the future. *Road and Truck*, December 12. http://www.roadandtrack.com/new-cars/car-technology/news/a24496/renaults-unveils-tiny-2-stroke-2-cylinder-turbo-diesel/ (accessed April 13, 2016).

Kim, J., C. Sluder, and R. Wagner. 2012. Combustion studies with FACE diesel fuels: A literature review. *SAE International Journal of Engines* 5(4):1648-1660. doi: 10.4271/2012-01-1688.

Knothe, G. 2010. Biodiesel and renewable diesel: A comparison. *Progress in Energy and Combustion Science* 36(3):364-373.

Kocher, L. 2016. Cummins 55% BTE Project. Presentation at the SuperTruck Annual Merit Review, Arlington, VA, June 16.

Koeberlein, D., and Cummins Inc. 2013. Cummins SuperTruck Program Technology and System Level Demonstration of Highly Efficient and Clean, Diesel Powered Class 8 Trucks. Presentation at the 2013 Department of Energy Annual Merit Review, May 16.

Koeberlein, D., and Cummins Inc. 2014. Cummins SuperTruck Program: Technology and System Level Demonstration of Highly Efficient and Clean, Diesel Powered Class 8 Trucks. Presentation at the 2014 DOE Hydrogen and Fuel Cells Program and Vehicle Technologies Office Annual Merit Review and Peer Evaluation Meeting, Washington, DC, June 20.

Kothari, T. U., D. Kumar, and K.D. Tandel. 2014. Design and analysis of six-stroke internal combustion engine. *International Journal of Engineering Development and Research* 2(2):2409-2416.

Kumar, S., S.D. Heister, X. Xu, J.R. Salvador, and G.P. Meisner. 2013. Thermoelectric generators for automotive waste heat recovery systems part I: Numerical modeling and baseline model analysis. *Journal of Electronic Materials* 42(4):665-674.

Lammert, M., R. McCormick, P. Sindler, and A. Williams. 2012. Effect of B20 and low aromatic diesel on transit bus NO_x emissions over driving cycles with a range of kinetic intensity. *SAE International Journal of Fuels and Lubricants* 5(3):1345-1359. doi: 10.4271/2012-01-1984.

Leone, T. (Ford Motor Company). 2015. The Importance of Fuel Octane. Presentation at the 2015 Fuels Institute Annual Meeting, Indianapolis, IN, September 23.

Leone, T., J. Anderson, R. Davis, A. Iqbal, R.A. Reesell, M.H. Shelby, and W.H. Studzinski. 2015. The effect of compression ratio, fuel octane rating, and ethanol content on spark-ignition engine efficiency. *Environmental Science & Technology* 49(18):10778-10789.

Liang, X., X. Wang, G. Shu, H. Wei, H. Tian, and X. Wang. 2015. A review and selection of engine waste heat recovery technologies using analytic hierarchy process and grey relational analysis. *International Journal of Energy Research* 39:453-471. doi: 10.1002/er.3242.

Lopes, S., P. Geng, and A. Konzack. 2014. Review of 2013 U.S. Retail Biodiesel Blends Quality Survey. SAE Technical Paper 2014-01-1379. doi: 10.4271/2014-01-1379.

Martec Group. 2016. Diesel Engine Technology and the Midterm Evaluation: An Analysis of Compliance Costs and Benefits. Available at http://www.martecgroup.com/wp-content/uploads/2016/05/The-Martec-Group-White-Paper-Diesel-Engine-Technology-and-the-Midterm-Evaluation-Summer-2016.pdf (accessed December 19, 2019).

McCormick, R.L., et al. 2006. Effects of Biodiesel Blends on Vehicle Emissions Fiscal Year 2006 Annual Operating Plan Milestone 10.4. Milestone Report NREL/MP-540-40554, October.

MECA (Manufacturers of Emission Controls Association). 2015. Statement of the Manufacturers of Emission Controls Association on the U.S. Environmental Protection Agency's Proposed Rulemaking on Greenhouse Gas Emissions Standards and Fuel Efficiency Standards for Medium- and Heavy-Duty Engines and Vehicles—Phase 2. Docket ID No. EPA-HQ-OAR-2014-0827.

Moriarty, K., M. Kass, T. Theiss. 2014. Increasing Biofuel Deployment and Utilization through Development of Renewable Super Premium: Infrastructure Assessment. NREL/TP-5400-61684. doi: 10.2172/1164886.

NACFE (North American Council for Freight Efficiency). 2014. Confidence Report: Electronically Controlled Transmissions. Available at http://nacfe.org/wp-content/uploads/2014/12/Trucking_Efficiency_Transmissions_Confidence_Report.pdf (accessed December 19, 2019).

Nadaf, S.L., and P.B. Gangavati. 2014. A review on waste heat recovery and utilization from diesel engines. *International Journal of Advanced Engineering Technology* 5(4):31-39.

Naik, S., F. Redon, G. Regner, and J. Koszewnik. 2015. Opposed-Piston 2-Stroke Multi-Cylinder Engine Dynamometer Demonstration. SAE Technical Paper 2015-26-0038. doi: 10.4271/2015-26-0038.

National Renewable Energy Laboratory (NREL). 2013. *Clean Cities Guide to Alternative Fuel and Advanced Medium- and Heavy-Duty Vehicles.* http://www.afdc.energy.gov/uploads/publication/ medium_heavy_duty_guide.pdf.

NGVAmerica. 2014. 2014 NGV Production and Sales Report. Available at https://www.ngvamerica.org/pdfs/2014%20NGV%20 Production%20and%20Sales%20Report.pdf (accessed December 19, 2019).

NRC (National Research Council). 2008. *Review of the 21st Century Truck Partnership.* Washington, DC: The National Academies Press.

NRC. 2010. *Technologies and Approaches to Reducing the Fuel Consumption of Medium- and Heavy-Duty Vehicles.* Washington, DC: The National Academy Press.

NRC. 2012. *Review of the 21st Century Truck Partnership, Second Report.* Washington, DC: The National Academies Press.

NRC. 2014. *Reducing the Fuel Consumption and Greenhouse Gas Emissions of Medium- and Heavy-Duty Vehicles, Phase Two: First Report.* Washington, DC: The National Academies Press.

NRC. 2015a. *Cost, Effectiveness, and Deployment of Fuel Economy Technologies for Light-Duty Vehicles.* Washington, DC: The National Academies Press.

NRC. 2015b. *Review of the 21st Century Truck Partnership, Third Report.* Washington, DC: The National Academies Press.

Olsson, A., B. Jönsson, and L. Sundin. 2005. The Volvo Heavy Truck Gas Turbine VT300. SAE Technical Paper 2005-01-3504. doi: 10.4271/2015-01-3054.

ORNL (Oak Ridge National Laboratory). 2016. 2015 Vehicle Technologies Market Report. Available at http://cta.ornl.gov/vtmarketreport/pdf/2015_vtmarketreport_full_doc.pdf.

ORNL (Oak Ridge National Laboratory). 2017. 2016 Vehicle Technologies Market Report. Available at https://tedb.ornl.gov/wp-content/uploads/2019/04/2016_Vehicle_Technologies_Market_Report.pdf.

Orr, B., A. Akbarzadeh, M. Mochizuki, and R. Singh. 2016. A review of car waste heat recovery systems utilising thermoelectric generators and heat pipes. *Applied Thermal Engineering* 101:490-495. doi: 10.1016/j.applthermaleng.2015.10.081.

Patel, P.S., and E.F. Doyle. 1976. Compounding the Truck Diesel Engine with an Organic Rankine-Cycle System. SAE Technical Paper 760343. doi: 10.4271/760343.

Perfetto, A., and S. Geckler. 2014. Ultra-Low Carbon Powertrain. Final Report for California Energy Commission, ARV-10-044, August.

Pihl, J., S. Daw, and Oak Ridge National Laboratory. 2015. Measuring the Impacts of Catalyst State on NH_3 Adsorption in Copper Zeolite SCR Catalysts. Presentation at the 2015 DOE CLEERS Workshop, University of Michigan, Dearborn, MI, April 27.

Pirault, J.-P., and M. Flint. 2009. *Opposed Piston Engines: Evolution, Use, and Future Applications*. Warrendale, PA: SAE International.

Posada, F., C. Bedick, N.N. Clark, A. Koslov, M. Linck, D. Boulanov, and J. Pratapas. 2007. Low Temperature Combustion with Thermal Recuperation. Presented at SAE Commercial Vehicle Engineering Conference, Rosemont, IL, October-November. SAE Paper 2007-01-4074.

Pradhan, A., D.S. Shrestha, J. Van Gerpen, A. McAloon, W. Yee, M. Haas, and J.A. Duffield. 2012. Reassessment of life cycle greenhouse gas emissions for soybean biodiesel. *Transactions of the American Society of Agricultural and Biological Engineers* 55(6):2257-2264.

Prevost/Volvo. 2010. The Volvo D13: Powertrain Product Tour. Available at https://www.prevostcar.com/sites/default/files/attachments/27volvod13engine_1_en.pdf (accessed September 30, 2019).

Propel Fuels. 2015. Propel Launches North America's First Retail Rollout of High Performance Renewable Diesel Fuels. March 18. Available at https://propelfuels.com/news_and_media/press_releases/propel_launches_diesel_hpr (accessed December 19, 2019).

PSI (Power Solutions International). 2016. Navistar & Power Solutions International Launch Gasoline-Fueled IC Bus at STN EXPO. July 26. Available at http://investors.psiengines.com/releasedetail.cfm?ReleaseID=981194 (accessed February 18, 2017).

Rappé, K., M. Stewart, C. DiMaggio, J. Theis, W.L. Se-Oh, J. Parks, J. Pihl, G. Fisher, and K. Howden (USDRIVE Low Temperature Aftertreatment Team). 2016. USDRIVE and Advanced Engine Crosscut Teams: Roadmaps and Protocols. Presentation at the 2016 DOE CLEERS Workshop, Ann Arbor, MI, April 6.

Redon, F. 2011. The Opposed-Piston Two-Stroke Engine Alternative: Performance and Emissions Results in a Medium-Duty Application. Directions in Engine-Efficiency and Emissions Research Conference, Detroit, MI, October 3-6.

Reinhart, T. 2015. Commercial Medium- and Heavy-Duty Truck Fuel Efficiency Technology Study—Report #1. Report No. DOT HS 812 146. Washington, DC: National Highway Traffic Safety Administration, June.

Reinhart, T., and M. Megel. 2016. An efficient, durable vocational truck gasoline engine. *SAE International Journal of Engines* 9(3):1437-1448. doi: 10.4271/2016-01-0660.

Reitz, R.C., and G. Duraisamy. 2015. Review of high efficiency and clean reactivity controlled compression ignition (RCCI) combustion in internal combustion engines. *Progress in Energy and Combustion Science* 46:12-71.

Renewable Fuels Association. 2016. Press Release. https://ethanolrfa.org/2016/08/rfa-applauds-wayne-fueling-systems-move-to-e25-in-standard-dispensers/ (accessed December 19, 2019).

Ricardo. 2013. Game Changing Ricardo Concept for High Efficiency Heavy Duty Diesel Moves Forward. Press Release, September 4. http://www.ricardo.com/en-GB/News--Media/Press-releases/News-releases1/2013/Game-changing-Ricardo-concept-for-high-efficiency-heavy-duty-diesel-moves-forward/ (accessed April 13, 2016).

Robbins, C., et al. 2011. Effects of Biodiesel Fuels Upon Criteria Emissions. Society of Automotive Engineers Paper No. 2011-01-1943.

Robinson, M.C. 2015. Analysis and Optimization of a Dual Free Piston, Spring Assisted, Linear Engine Generator. West Virginia University, ProQuest Dissertation Publishing.

Sax, T., S. Yoon, C. Misra, J. Collins, J. Herner, D. Hawelti, S. Lemieux, et al. California Environmental Protection Agency Air Resources Board. 2014. An Integrated Approach to Heavy-Duty GHG and NO_x Control. Presentation to the National Academies of Sciences Committee on the Assessment of Technologies and Approaches for Reducing the Fuel Consumption of Medium- and Heavy-Duty Vehicles, Phase Two, Ypsilanti, MI, July 29.

Sellnau, M., M. Foster, K. Hoyer, W. Moore, J. Sinnamon, and H. Husted. 2014. Development of a gasoline direct injection compression ignition (GDCI) engine. *SAE International Journal of Engines* 7(2):835-851. doi: 10.4271/2014-01-1300.

Sellnau, M., W. Moore, J. Sinnamon, K. Hoyer, M. Foster, and H. Husted. 2015. GDCI multi-cylinder engine for high fuel efficiency and low emissions. *SAE International Journal of Engines* 8(2):775-790. doi: 10.4271/2015-01-0834.

Senefsky, B. 2003. Chevrolet Turbo-Titan III Concept Vehicle - Space Truckin'. Truck Trend Network, September 1. http://www.trucktrend.com/cool-trucks/0309st-chevrolet-turbo-titan-3-concept-vehicle/ (accessed April 13, 2016).

Sharma, A., and F. Redon. 2016. Multi-Cylinder Opposed-Piston Engine Results on Transient Test Cycle. SAE Technical Paper 2016-01-1019. doi: 10.4271/2016-01-1019.

Sharp, C. (Southwest Research Institute). 2016. CARB Low NO_x Demonstration Program—Update. Presentation at the SAE 2016 Heavy-Duty Diesel Emissions Control Symposium (HDDEC), Gothenburg, Sweden, September 21.

Shenk, M. 2014. U.S. oil output surges to highest since 1986 on shale. *Bloomberg*, September 17. https://www.bloomberg.com/news/articles/2014-09-17/u-s-oil-output-surges-to-highest-since-1986-on-shale (accessed December 19, 2019).

Singh, S. 2011. "Exhaust Heat Driven Rankine Cycle for a Heavy Duty Diesel Engine" presentation at 2011 DEER Conference. October 5. Detroit, MI, USA. https://www.energy.gov/sites/prod/files/2014/03/f8/deer11_singh.pdf.

Singh, S. (Detroit Diesel Corporation). 2015. SuperTruck Program: Engine Project Review, Class 8 Truck Freight Efficiency Improvement Project, Project ACE 058. Presentation at the DOE Annual Merit Review, June 20.

Sluder, S., J. Szybist, B. Kaul. 2015. Preliminary Investigation of Heat of Vaporization Effects in High-Octane Fuel Experiments at ORNL. Presentation to USDRIVE Combustion and Emission Control Tech Team, July.

Splitter, D., A. Pawlowski, and R. Wagner. 2016. A historical analysis of the co-evolution of gasoline octane number and spark-ignition engines. *Frontiers in Mechanical Engineering*. January 6. doi: 10.3389/fmech.2015.00016.

Stanton, D. (Cummins Inc.). 2014. Natural Gas Engine Technologies to Meet Future U.S. GHG Regulations for Commercial Vehicles. Presented to the 5th International CTI Conference, September.

Stein, R., D. Polovina, K. Roth, M. Foster, et al. 2012. Effect of heat of vaporization, chemical octane, and sensitivity on knock limit for ethanol-gasoline blends. *SAE International Journal of Fuels and Lubricants* 5(2):823-843. doi: 10.4271/2012-01-1277.

Sturgess, S. 2016. Engine downspeeding dictates drive-axle design. *Transport Topics*, October 24. http://www.ttnews.com/articles/petemplate.aspx?storyid=43628 (accessed December 19, 2019).

Subramanian, S. (Eaton). 2014. Heavy Duty Roots Expander Heat Energy Recovery. Presentation at the 2014 DOE Annual Merit Review, June.

Szybist, J.P., R. Steeper, D. Splitter, V.B. Kalaskar, J. Pihl, and C. Daw. 2014. "Negative Valve Overlap Reforming Chemistry in Low-Oxygen Environments." SAE International Journal of Engines 7(1): 418-33. www.jstor.org/stable/26277775 (accessed December 19, 2019).

Tartakovsky Leonid et al. (2014): Table 4.1, p 90: Tartakovsky, L., V. Baibikov, M. Gutman, A. Poran, M. Veinblat. 2014. "Thermo-Chemical Recuperation as an Efficient Way of Engine's Waste Heat Recovery." *Applied Mechanics and Materials* 659:256–261. https://doi.org/10.4028/www.scientific.net/amm.659.256 (accessed December 19, 2019).

Theiss, T., T. Alleman, A. Brooker, A. Elgowainy, G. Fioroni, J. Han, S. Huff, et al. 2016. Summary of High-Octane Mid-Level Ethanol Blends Study. ORNL/TM-2016/42. Available at http://info.ornl.gov/sites/publications/files/Pub61169.pdf (accessed December 19, 2019).

Tong, F., P. Jaramillo, and I. Azevedo. 2015. Comparison of Life Cycle Greenhouse Gas from Natural Gas Pathways for Medium and Heavy-Duty Vehicles (Supporting Information). Pittsburgh, PA: Carnegie Mellon University. May.

Venkatesh, A., P. Jamarillo, W.M. Griffin, and H.S. Matthews. 2012. Uncertainty in life cycle greenhouse gas emissions from United States natural gas end-uses and its effects on policy. *Environmental Science & Technology* 45(19):8182-8189. doi: 10.1021/es200930h.

Wagner, R. 2015. "Future Engines: Incremental Advances Driving Disruptive Opportunities" presented at 2015 SAE International High Efficiency Internal Combustion Engines Symposium. April. Detroit, MI.

Warey, A., J. Hardy, M. Hennequin, M. Tatur, et al. 2010. Fuel Effects on Low Temperature Combustion in a Light-Duty Diesel Engine. SAE Technical Paper 2010-01-1122. doi: 10.4271/2010-01-1122.

Wekin, A. 2008. Characterization and Comparison of Piezoelectric Materials for Transducing Power from a Thermoacoustic Engine. Master's thesis, Washington State University.

Westport. 2015a. Westport Completes HPDI Gen 2.0 Technology Program with Daimler AG, Westport Innovations, Inc., Vancouver, BC, July.

Westport. 2015b. HPDI 2.0 - New Diesel-Gas System by Westport. http://gazeo.com/automotive/technology/HPDI-2.0-new-diesel-gas-system-by-Westport,article,7689.html (accessed September 19, 2015).

Westport. 2015c. Westport HPDI 2.0 on Track for Commercial Production, November. http://www.westport.com/news/2015/westport-hpdi-2.0-on-track-for-commercial-production?utm_source=Westport+Communications&utm_campaign=c1a37b053d-Westport+WatCH4+News&utm_medium=email&utm_term=0_f905da2430-c1a37b053d-84079581 (accessed November 5, 2015).

Wissink, M., and R. Reitz. 2015. Direct dual fuel stratification, a path to combine the benefits of RCCI and PPC. *SAE International Journal of Engines* 8(2):878-889. doi: 10.4271/2015-01-0856.

Yu, Z., A.J. Jaworski, and S. Backhaus. 2012. Travelling-wave thermoacoustic electricity generator using an ultra-compliant alternator for utilization of low-grade thermal energy. *Applied Energy* 99:135-145. doi: 10.1016/j.apenergy.2012.04.046.

Zigler, B., W. Pitz, N. Killingsworth, D. Flowers, S. Aceves, M. Ratcliff, R. Wagner, et al. (National Renewable Energy Laboratory). 2012. Fuels for Advanced Combustion Engines, Project FT002. Presentation at the DOE Annual Merit Review, May 15.

Zukouski, R. 2014. "SuperTruck–Development and Demonstration of a Fuel-Efficient Class 8 Tractor and Trailer," presented to NRC Committee on Review of the 21st Century Truck Partnership, Phase 3, Washington, D.C., November 18.

5

Technologies for Reducing the Power Demand of MHDVs

5.1 INTRODUCTION

This chapter describes the disposition of the fuel energy processed by a heavy-duty engine and particularly that converted into mechanical work and delivered to the vehicle. It reviews the findings of studies on truck fuel consumption, focusing on the largest vehicle segment, Class 8. The net power output of a truck engine is consumed by numerous components. The power consumption of *all* such components is proportional to truck speed. In addition there are several other factors responsible for power consumption of a tractor-trailer, the most significant of which are aerodynamic loads, which depend on the cube of truck speed, truck frontal area, and friction of the air drag; and tire rolling resistance, due to flexing within the tires and road friction. Depending on truck speed, these two components consume 60 to 85 percent of this power. The power inventory includes numerous truck auxiliary devices plus major driveline components, typically those related to engine speed. When encountering positive grade or accelerating, power demand is proportional to vehicle weight and grade.

5.1.1 Base Case for Over-the-Road Tractor-Trailer, Circa 2013 Specifications

The nature of heavy truck power use can be better appreciated if it is summarized in a use inventory. This class of truck has a regulated gross vehicle weight (GVW) of 80,000 pounds on interstate highways. Table 5-1 shows an inventory developed by the U.S. Department of Energy's 21st Century Truck Partnership (21CTP) for a typical Class 8 high-roof sleeper cab tractor-van trailer combination carrying a payload of 26,000 pounds and operated on an unspecified over-the-road drive cycle. (Duty cycles have a highly variable characterization throughout the varying elevation contours, speed limits, and road congestion in the United States.) The chosen payload was 57 percent of the allowable interstate capacity for this truck. The 21CTP has designed this inventory which appears more actual-duty representative compared to a traditional, useful, and long-used inventory described as a fully loaded, steady speed, and level road characterization (see NRC, 2010).

Notice this inventory reported a fuel economy figure, and a load-specific fuel consumption (LSFC) result. The LSFC is perhaps the most useful metric to evaluate heavy-truck consumption reduction because it embodies use of the freight payload weight. Efficient payload transport is the objective of the tractor-trailer industry.

Last, the 21CTP inventory developer coined a useful phrase, tractive power (TP), to describe the three vehicle power demand (VPD) energy consumers: aerodynamic drag, rolling resistance, and inertia/braking loss.

TABLE 5-1 Vehicle Power Demand Inventory for Over-the-Road Tractor-Van Trailer with Average Payload

Operating vehicle weight: 60,000 lbs (27,220 kg)	Vehicle power demand (VPD): 146 kW
Average payload: 26,000 lbs (11,800 kg); short tons: 13	Aerodynamic drag: 61 kW (C_d: 0.60)
Miles per gallon: 5.8	Rolling resistance: 44 kW (C_{rr} 0.007)
Load[a] specific fuel consumption: 13.3 gallons/1000 ton-miles	Inertia/braking: 16 kW
Fuel input: 343 kW	Auxiliary load: 15 kW
Engine loss (EL): 193 kW (BTE: 0.43)	Drivetrain loss: 10 kW
Idling fuel use (IF): 3.6 kW	Total fuel consumed (EL+IF+VPD): 343 kW
Engine output: 146 kW	

[a] For payload only, calculations were done using "long tons" (long ton = 2,200 lbs).
SOURCE: 21CTP (2013).

5.1.2 Multiplicative Effect of Vehicle Power Demand Reduction on Engine Fuel Consumption

In conventional vehicles the road load power is supplied by an engine via a transmission and one or more drive axles characterized by an effective efficiency (ή). A corresponding simple engine power demand (P_{eng}) model is given by the following (in this equation, the intended VPD components are those described TP consumers):

$$P_{eng} = (\text{VPD} / ή) + (\text{auxiliary load}) / \text{BTE}$$

The significance of this relationship can be illustrated by considering the reduction in any of the VPD energy consumers. For example, consider the accomplishment of a 20 percent aerodynamics drag reduction (producing about a 10 percent fuel consumption reduction) in the case illustrated in Table 5-1. That reduction would be 10 percent × 61 kW = 6.1 kW. Applying the above equation, P_{eng} (reduction) = 6.1 kW / 0.43 = 14.2 kW. So, VPD reduction has a powerful leverage in reducing actual engine fuel energy (power). Indeed, we have observed that two of the SuperTruck (ST) programs have reduced their engine size and maximum power rating.

While this relationship is valid, its numerical application requires the user to apply the achieved VPD reductions over the engine operating cycle speed-load points to achieve the cycle-average efficiency (ή-avg) in the equation.

5.1.3 Vehicle Power Demand Reduction Goals Established by 21CTP

Many trucking fleets have not only routinely evaluated and adopted truck components that were demonstrated to reduce fuel consumption within their particular return-on-investment (ROI) hurdle, but they have also paid increasing attention to operational fleet management system tools, which, they found, would better manage typical driving cycles, as well as monitor and control driver behavior (TMC, 2014). The 21CTP is a government-industry partnership with the goals of research and development, demonstrations, validation, and deployment of cost-effective, reliable, and durable technologies that reduce vehicle power demand with the goal of improving overall efficiency of heavy vehicles. The program represents some of the highest achieved efficiencies in research vehicles and in deployed technologies, and provides a guide to future achievable power demand reductions. The 21CTP Roadmap and Technical White Papers (2013) offer specific technology goals for vehicle power demands, which might be effectively achieved within a 10-year window, as follows.

Technology Goal 1: Develop and demonstrate advanced technology concepts that reduce the aerodynamic drag of a Class 8 highway tractor-trailer combination by 20 percent (from a drag coefficient [C_d] of 0.69 to 0.55). Evaluate a stretch goal of 30 percent reduction in aerodynamic drag (from C_d=0.69 to C_d=0.48). The baseline for this goal is the proposed Environmental Protection Agency (EPA)/National Highway Traffic Safety Administration (NHTSA) baseline of C_d=0.69 with 9.2 m^2 frontal area for a conventional Class 8 tractor with high-roof sleeper.

Technology Goal 2: Develop and demonstrate low-rolling-resistance tires that can reduce vehicle rolling resistance and wheel weight for a Class 8 tractor-trailer. Demonstrate 35 percent reduction in rolling resistance from C_{rr}=8.2 kg/metric ton for drive wheels to a goal of C_{rr}=5.33 kg/metric ton. The baseline for this goal is the EPA/NHTSA proposed baseline for a Class 8 tractor-trailer equipped with low-rolling-resistance dual tire drive wheel configurations having C_{rr}=8.2 kg/metric ton.

Technology Goal 3: Develop and demonstrate technologies that reduce essential auxiliary loads by 50 percent (from current 20 horsepower to 10 horsepower) for Class 8 tractor-trailers. The baseline for this goal is a Class 8 highway tractor-trailer with sleeper operating 5-day over-the-highway operations at 36,000 kg (80,000 pounds) combined gross vehicle weight (CGVW).

Technology Goal 4: Develop and demonstrate lightweight material and manufacturing processes that lead to a 10 percent reduction in tare weight for a 15,500 kg (34,000 pounds) tractor-trailer combination. Establish a long-term stretch goal of reducing combined vehicle weight by 20 percent. The baseline for this goal is a Class 8 highway tractor-trailer with high-roof sleeper and dry van trailer capable of 36,000 kg CGVW.

Technology Goal 5: Increase heat load rejected by thermal management systems by 20 percent without increasing radiator size. Develop and demonstrate parasitic friction reduction technologies that reduce driveline losses by 50 percent, thereby improving Class 8 fuel efficiencies by 3 percent. The baseline for this goal is a Class 8 highway tractor-trailer with sleeper operating at steady-state 65 mph at 36,000 kg CGVW.

5.1.4 Technology Goal for Operational Efficiency and Intelligent Transportation Technologies Development and Deployment

21CTP has not devised a single specific goal in this multifaceted technology field. The following remarks were drawn from this same 21CTP report (21CTP, 2013, p. 124): "With [the 21CTP's] analysis-backed sense of the magnitude of potential energy savings available from ITS technologies and systems, it is important that DOE and DOT work very closely in this domain so that the magnitude of energy savings attainable with Connected Vehicle systems does not go untapped. Research and development of these applications must be sufficiently funded to assure attainment of the maximum benefits in fuel savings and emissions reductions, in addition to safety and mobility. DOE (as well as EPA) sponsorship of activities in support of other Connected Vehicle research is strongly recommended, and regular communications between the DOE and DOT regarding ITS [Intelligent Transportation System] advancement should be maintained."

5.1.5 21CTP Goal's Relation to U.S. Phase II Rule on GHGs and Fuel Consumption

These 21CTP goals all identify viable technologies and potentially achievable reductions. The reductions and timetable requirements differ in the proposed rule but pose no significant contradictions. The proposed Phase II rule states, "The proposed vehicle and engine performance standards would cover model years 2021-2027, and apply to semi-trucks, large pickup trucks and vans, and all types and sizes of buses and work trucks. They would achieve up to 24 percent lower CO_2 emissions and fuel consumption than an equivalent tractor in 2018, based on the fully phased-in standards for the tractor alone in a tractor-trailer vehicle" (EPA and NHTSA, 2015) Note the reported 24 percent reduction is for Class 8 tractor high-roof sleeper cabs and relative to the Phase I regulation. Taken together, Phase I and the proposed Phase II amount to a 42 percent reduction in fuel use and greenhouse gas (GHG) emissions from the described model year 2010 baseline tractor.

The International Council on Clean Transportation (ICCT) made an analysis of the anticipated technologies and their estimated production costs. Figure 5-1 reports that analysis for a composite 2027 tractor-trailer. It shows how a Class 8 high-roof sleeper-cab tractor-trailer combination could reduce its overall fuel consumption by approximately 33 percent under Phase II (ICCT's analysis).[1] This is an alternate and slightly more recent view of achievable consumption reductions compared to the 21CTP technology goals forecasts. The status of the research, development, and deployment achieved for each of the six vehicle power demand fuel consumption domains is found in the following sections.

[1] Sharpe, B., ICCT. "5 numbers you need to know about the proposed U.S. truck efficiency rule." Available at http://www.theicct.org/blogs/staff/5-numbers-you-need-know-about-proposed-us-truck-efficiency-rule (accessed December 19, 2019).

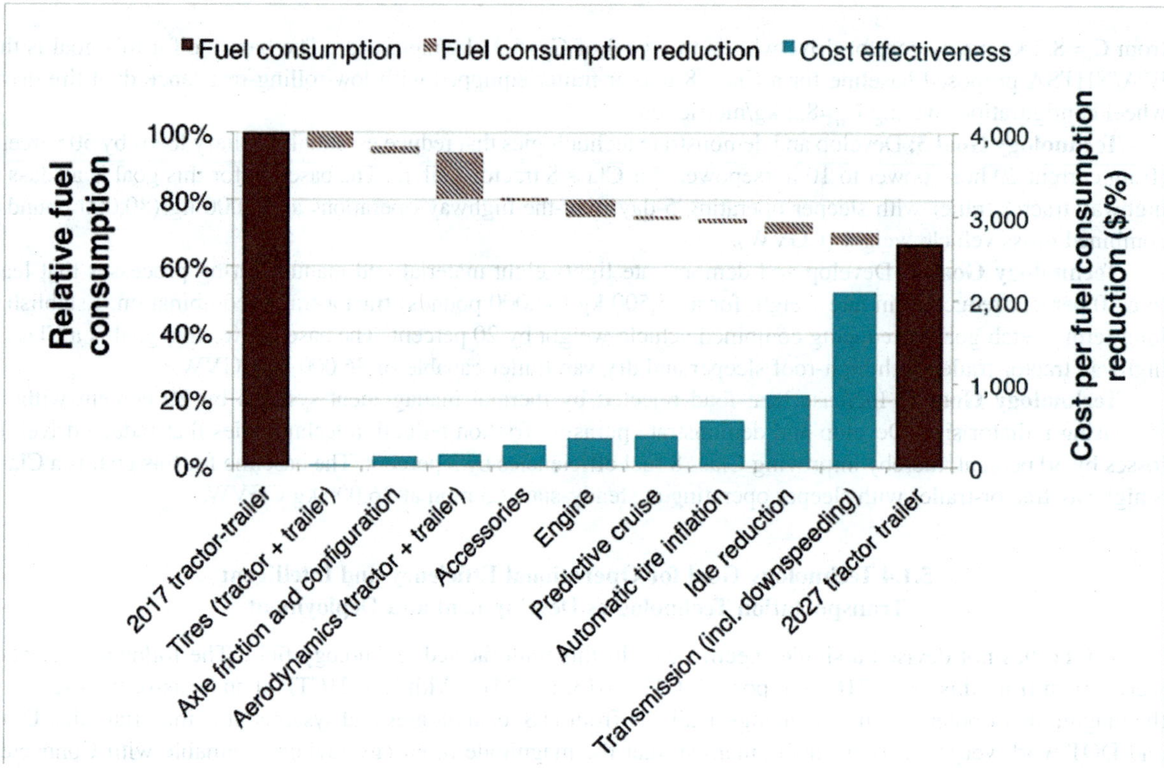

FIGURE 5-1 Estimated fuel consumption technology improvements and costs to reach Phase II compliant fuel consumption in 2027 from a Phase I (2017) Class 8 high-roof sleeper tractor-trailer.
SOURCE: ICCT (2015), © ICCT, reproduced with permission.

Finding: The 21st Century Truck Partnership has taken a step forward in redefining a new baseline with the proposed average power use inventory in the 2013 roadmap and technical white papers. The continuing improvement in reducing fuel consumption from vehicle operations will benefit from evaluating with new baselines over a drive cycle(s) based on real-world operation. The new baselines would take into account such factors as load, grade, speed, torque, distance-based target schedules, and drive cycles over an extended period of time as well as rest periods required by law.

Recommendation 5-1: NHTSA and EPA should coordinate with the Department of Energy (DOE), the California Air Resources Board (CARB), and the trucking industry to develop an inclusive new baseline for vehicle power demands as described in the above finding, for a circa 2025 vehicle that includes an extended period of operation.

5.2 AERODYNAMIC DRAG REDUCTION

This section addresses aerodynamic development only for Class 8b combination tractor-van trailers. This class of trucks consumes 60 percent of medium- and heavy-duty vehicle fuel (see Table 5-2). Notice the application descriptions for the balance of the classes are predominantly city or regional, with substantially smaller annual consumption shares. However, in Chapter 6 of this report, the effects of reducing aerodynamic drag on Class 2b and 6 trucks are shown in simulation along with Class 8 vehicles.

To clarify the emphasis on Class 8b tractors (and trailers) as being the most pertinent trucks for aerodynamic development, refer to the perspective from a recent California Hybrid, Efficient and Advanced Truck Research Center study. That study "classifies CL7-8 over the road tractors as younger trucks, high annual VMT, mostly

TABLE 5-2 Medium- and Heavy-Duty Truck Classes, Applications, and Percent of Fuel Use

Class	2b	3	4	5	6	7	8a	8b
Application	Large Pickup Utility Van, Multipurpose, Minibus, Step Van	Utility Van, Multipurpose, Mini-Bus, Step Van	City Delivery, Parcel Delivery, Large Walk-in, Bucket, Landscaping	City Delivery, Parcel Delivery, Large Walk-in, Bucket	City Delivery, School Bus, Large Walk-in, Bucket	City Bus, Furniture, Refrigerated, Refuse, Fuel Tanker, Dump, Tow, Concrete, Fire Engine, Tractor-Trailer	Dump, Refuse, Concrete, Furniture, City Bus, Tow, Fire Engine (straight trucks)	Tractor-Trailer: Van, Refrigerated, Bulk Tanker, Flat Bed (combination trucks)
Fuel %	11.6	3.1	1.1	0.5	12.7	4.1	7.4	59.4

SOURCE: NRC (2015).

higher average speed, and highway driving, typically with sleeper cabs … Over the Road trucks represent the majority of emissions and vehicle miles traveled … while only accounting for 10 percent of the population of these vehicles in the state" (CARB, 2014).

Nevertheless, since Class 2b and 3 vehicles are tested on a chassis dynamometer, a C_d must be measured to complete the loading demand. Class 2b and 3 aerodynamic designs are borrowed principally from light-duty vehicle practice (and substantially, that tooling).

The NHTSA and EPA Phase I fuel consumption and GHG rules (EPA and NHTSA, 2011), promulgated September 11, 2011, placed aerodynamic measurement requirements on combination tractor-trailer vehicles. Standards were set with metrics for measurement in grams/ton-mile for CO_2 results, and gallons/1,000 ton-mile for fuel consumption. The fuel gallons/1,000 freight-ton-miles metric is also referred to as "freight efficiency." Certification was achieved via exercise of the EPA's Greenhouse Gas Emissions Model (GEM) simulation utilizing supplier data for the incumbent combination vehicle.

Aerodynamic performance data for the combination were required to be verified by coast-down testing to support the vehicle performance in the GEM. If enhanced aerodynamic performance was needed to achieve the Phase I standards, they would be incumbent on tractor design only. The EPA standards became effective in January 2014. The NHTSA standards became effective in January 2017.

5.2.1 Baseline Aerodynamic Drag (C_dA)

In the Phase I rule, and continued in Phase II, NHTSA and EPA defined bins of aerodynamic performance with specific values of C_dA to use in the GEM simulation. Each original equipment manufacturer (OEM) vehicle is assigned to a bin based on test and simulation results. Each bin has a range of C_dA values to accommodate the specific vehicle aerodynamic performance, but then in the GEM simulation a single C_dA is used for each bin. Since the discussion in industry and the regulations frequently refer to these bins, we include Table 5-3, which shows the example C_dA range in each bin as well as the specific GEM value. Note that day cab tractors have a different and distinct set of bins than sleeper cabs.

Additional details on this feature of the rules are in the Regulatory Impact Analysis (RIA) or final rules for Phase I and Phase II. The Phase rule also requires special aerodynamic provisions for the prescribed closed van trailers. (In NHTSA's technical support document [Reinhart, 2015], the baseline vehicle—a Class 8 tractor-trailer with high-roof sleeper cab—corresponds to bin II.[2]) The EPA trailer standards would begin to take effect in model year 2018 for certain trailers, while NHTSA's standards would be in effect as of 2021. The baseline aerodynamic C_dA values published in the Phase II standards are shown in Table 5-4 for reference.

[2] The value for C_d is 0.639 and that of the effective frontal area is 10.14 m^2, yielding a C_dA of 6.48 m^2.

TABLE 5-3 Aerodynamic Bins Used in GEM

	Class 7	Class 8	
	Day Cab High Roof	Day Cab High Roof	Sleeper Cab High Roof
		Aerodynamic Test Results (C_dA_{wad} in m^2)	
Bin I	≥7.2	≥7.2	≥6.9
Bin II	6.6-7.1	6.6-7.1	6.3-6.8
Bin III	6.0-6.5	6.0-6.5	5.7-6.2
Bin IV	5.5-5.9	5.5-5.9	5.2-5.6
Bin V	5.0-5.4	5.0-5.4	4.7-5.1
Bin VI	4.5-4.9	4.5-4.9	4.2-4.6
Bin VII	≤4.4	≤4.4	≤4.1
		Aerodynamic Input to GEM (C_dA_{wad} in m^2)	
Bin I	7.45	7.45	7.15
Bin II	6.85	6.85	6.55
Bin III	6.25	6.25	5.95
Bin IV	5.70	5.70	5.40
Bin V	5.20	5.20	4.90
Bin VI	4.70	4.70	4.40
Bin VII	4.20	4.20	3.90

SOURCE: EPA and NHTSA (2016b).

TABLE 5-4 GEM Inputs for the Baseline Class 7 and 8 Tractor

Class 7			Class 8					
Day Cab			Day Cab			Sleeper Cab		
Low Roof	Mid Roof	High Roof	Low Roof	Mid Roof	High Roof	Low Roof	Mid Roof	High Roof
Engine								
2017 MY 11L Engine 350 HP	2017 MY 11L Engine 350 HP	2017 MY 11L Engine 350 HP	2017 MY 15L Engine 455 HP	2017 MY 15L Engine 455 HP	2017 MY 15L Engine 455 HP	2017 MY 15L Engine 455 HP	2017 MY 15L Engine 455 HP	2017 MY 15L Engine 455 HP
Aerodynamics (C_dA in m^2)								
5.41	6.48	3.68	5.41	6.48	6.38	5.41	6.48	5.90
Steer tires (C_{rr} in kg/metric ton)								
6.99	6.99	6.87	6.99	6.99	6.87	6.87	6.87	6.54
Drive tires (C_{rr} in kg/metric ton)								
7.38	7.38	7.26	7.38	7.38	7.26	7.26	7.26	6.92

SOURCE: EPA and NHTSA (2016b, p. 73588).

Using the previously mentioned typical frontal area of 9.2 square meters (m^2) and other data showing up to 10.5 m^2 (Bhatnagar et al., 2011), the C_d range for Class 8 is 0.51-0.69 which, as expected, is roughly consistent with the literature, as in following examples:

- The Southwest Research Institute (SwRI) simulation work for the committee used a baseline C_dA for the Kenworth T700 of 6.84, which was grounded in prior work for NHTSA (Reinhart, 2015, 2016). For NHTSA, SwRI used C_dA of 6.26-6.48 for the baseline.
- In the recent North American Council for Freight Efficiency (NACFE) Confidence Report (NACFE, 2016a), "OEM representatives told the study team that current market penetration estimates from EPA suggest that nearly 70 percent of 2015 high-roof sleeper tractor production would fall into the proposed Bin 2, with a C_dA between 7.2 and 6.6" (Sproull, 2015).
- In NHTSA and EPA's RIA (EPA and NHTSA, 2016a, Figure 3-17), results of vehicle tests to support the rulemaking showed wind-averaged C_dA of 5.4-6.0 for high-roof sleeper-cab trucks, indicating a moving baseline. Further accounting of the EPA work has been published by the Society of Automotive Engineers (SAE; Gururaja, 2016), which includes the effects of trailer skirts now in use.
- In comments to the proposed Phase II rule, PACCAR showed C_dA data for high-roof sleeper tractors concentrated between 5.5 and 7.0.[3]
- Work conducted and published by National Research Council Canada with a 2014 model sleeper vehicle also showed a baseline C_dA of about 6.2, reduced to about 5.0 with skirts and boat tail on the trailer (McAuliffe and Chuang, 2016).

The committee examined recent work such as in SuperTruck to assess the potential reductions in C_d, as well as considering NHTSA and EPA's estimates for 2027 regulations. Fortunately, aerodynamic enhancements for both tractors and trailers had been pursued by the industry for years. Arguably, tractor enhancements on engine-forward models were first offered by Kenworth in 1985, as described in the Phase I report (NRC, 2010). Trailer aerodynamics achieved via add-on fairings to the leading top edge of vans had been offered for two decades, but with limited carrier uptake. Then beginning in 2004, EPA's SmartWay initiative provided increased encouragement for carriers to achieve voluntary fuel savings (and associated CO_2 reductions) via addition of aerodynamic (and other) features on combination tractor-trailers. This initiative particularly recognized the potential benefits of emerging trailer aerodynamic devices, developed principally by third-party manufacturers. A key to successful deployment of increasingly sophisticated aerodynamic features is that they should not interfere with normal carrier and shipper processes to any substantial amount, or could even improve processes.

The universe of trailers pulled principally by Class 8 tractors is dramatically diverse. Yet dry and refrigerated van trailers constitute just over 70 percent of all trailers (using 2013 trailer sales distribution as a surrogate for the total in-use trailers). The preponderance of fuel to be saved by aerodynamic improvements is in the van pulling combination tractor-trailer, as it is these trucks that have consistently above average road speed and mileage accumulation. This includes all lengths of van trailers.

For this committee's work to estimate achievable aerodynamic efficiency and overall reduction in truck fuel consumption, it was necessary to glean the projected C_dA of future vehicles from the literature. Some of those valuable estimates can be found in the 21st Century Truck Partnership activities and the SuperTruck projects. Also the examples in SmartWay give insights on the drag reduction of particular devices. The projected drag coefficients were then used in the committee's simulation work (via SwRI) as described in Chapter 6. The validity of measured or modeled C_d is a highly important topic, as we want to ensure that reducing fuel consumption is actually realized in real-world freight movement in contrast to just being seen in a vehicle simulation. Progress is being made here, but a few shortcomings are noted in the following sections.

[3] "Comments from PACCAR." NHTSA-2014-0132-0223. Available at https://www.regulations.gov/document?D=NHTSA-2014-0132-0223 (accessed April 6, 2017).

5.2.2 SmartWay

EPA's SmartWay initiative now involves over 3,000 U.S. carriers and other partners in the program. A program enhancement introduced in March 2014 is the EPA-designated "SmartWay *Elite*" trailer, which designates traditional 53-foot box dry van or refrigerated trailers equipped with a combination of SmartWay-verified low-rolling-resistance tires[4] and SmartWay-verified aerodynamic devices (assuming a constant truck speed of 65 mph) to obtain a total of 10 percent or more fuel savings over traditional trailers (EPA, 2016b). EPA requires a carrier to adopt the 10 percent fuel savings combination to be eligible to receive their Elite designation. According to EPA, at least 9 percent of these fuel savings come from a combination of two or more SmartWay-verified aerodynamic devices (EPA, 2015a). The devices can be installed on the front, on the side, under, or on the rear of the trailer. Further detail on available packages is included at the end of this chapter in the Annex, "Description of Available SmartWay-Certified Trailer Packages."

EPA SmartWay engineering has expended considerable effort to improve both the fidelity and user flexibility of the requisite verification procedures. EPA and the Department of Transportation (DOT) have summarized the test procedures and unit specifications, including revisions to the previous rule, in their Phase II Proposed GHG and Fuel Efficiency Rulemaking, Draft Regulatory Impact Analysis, Section 3.2 (EPA and NHTSA, 2015).

Finding. EPA through SmartWay and regulation development has significantly improved the fidelity of the required aerodynamics verification procedures. For example, their requisite C_dA determination has displayed a standard error in the range of 0.5 to 2.1 percent. They also encourage manufacturers to utilize multiple testing processes to increase confidence of observed improvements.

Concern over the fidelity of standard test procedures relative to real-world conditions remains a constant in the industry. With increasing congestion on even interstate highways, the ambient turbulence intensity on roadways is significant and not accounted for in current tests. Additionally, the effects of other vehicles traveling in the same or opposite direction are not well represented in coast-down or track tests. It is likely that computational fluid dynamics (CFD) modeling could help determine the magnitude of these effects and whether they must somehow be accounted for in certifying devices or full vehicles. There are published works on modeling of platooning vehicles, meaning vehicles running in tandem in close proximity (Hammache et al., 2002; Mihelic et al., 2015), but this needs to be extended to unstructured groups of vehicles moving at different speeds as in the real world. Peterbilt reported (Smith et al., 2014) that on-highway testing of its SuperTruck in Texas, depending on the route, showed that other vehicles aerodynamically affected the SuperTruck between 37 and 53 percent of the miles traveled (NACFE, 2016b).

Finding: The issue of ambient turbulence and close-proximity vehicles likely creating new aerodynamic loading not managed by current standard testing seems to represent opportunity currently unmanaged, possibly resulting in unaccounted losses.

Recommendation 5-2: The magnitude of the ambient turbulence intensity and close-proximity vehicle passing turbulence scenario should be evaluated by EPA and NHTSA to quantify its significance.

[4] The rolling resistance coefficient (kg force per metric ton) is calculated at a load of either 4,250 pounds and an inflation pressure of 100 pounds per square inch (psi) (drive or trailer tires) or 6,000 pounds and an inflation pressure of 110 psi (steer tires). See EPA (2016a).

5.2.3 Aerodynamics Information from SuperTruck

The DOE-developed SuperTruck projects[5] have generated significant aerodynamic innovations and are analyzed in a recent National Research Council report (NRC, 2015). In addition, some expected results are discussed from yet-to-be-completed projects.[6] Four project teams variously identifying similar aerodynamic component modifications, which are described in detail in the Annex to this chapter entitled "Summary of SuperTruck Projects."

In most cases, the ST project work will first seek optimum performance of potential aerodynamic features. While not a requirement of the SuperTruck project dossiers, critical to successful production deployment will be compatible fit-up of ST trailers with "today's standard" legacy trailers and more importantly, legacy tractors. These issues become necessary questions to answer as ST-like aerodynamic tractors and trailers enter the universe of about 1.2 million existing tractors and nearly 4 million van trailers. This "excess" of trailers is purposeful and necessary to facilitate trailer staging to on- and off-load trailers, while others are on road in transit. For the most part, the SmartWay configurations have been substantially nonintegrated and therefore have not posed the fit-up question raised here. But, the performance-aggressive SuperTruck goals have changed this dynamic.

The SuperTruck project teams variously identified and implemented similar aerodynamic component modifications: to the tractor (bumper, radiator louvers, chassis extenders, hood, roof, and wheel covers) and to the trailer (skirts, tail, bogie cover, wheel covers, and front gap fairing) (see Figure 5-2). Yet, most have shape uniqueness as illustrated in Figures 5-3 and 5-4. Apparent in both figures are large trailer tails and very close road clearance skirts and tractor fairings.

The Cummins-Peterbilt project team indicated that aerodynamics-specific upgrades contributed 28.7 percent to the freight efficiency (ton-miles/gallon) increment and corresponding 1.6 mpg increase on the long-haul drive cycle (Koeberlein and Cummins Inc., 2015). The C_d reduction for the system was reported as 46 percent, which would put this vehicle's C_d at about 0.32-0.35 or C_dA at approximately 3.6 (depending of course on the baseline selected).

Daimler included a number of aerodynamic features in their final SuperTruck tractor's cab, such as widening it to match the sleeper width and raking back both the grill and windshield. The shape of the hood has been extensively modified to optimize aerodynamic efficiency with the smaller engine (10.7 liter versus 14.6 liter) and angled cooling system.

The wind-averaged C_d of the final demonstrator tractor and trailer was reported as 54 percent lower than the baseline, as determined in a scale-model wind tunnel (this would mean a C_dA of about 3.1). As an indication of the relative design maturities of tractor aerodynamics versus trailer aerodynamics, at the start of the project, trailer improvements delivered 72 percent of the total vehicle aerodynamic improvement while requiring only one-third of the aerodynamic engineering effort.

Another unique feature included on the tractor was a louvered radiator grill that closes when cooling demand is not detected, thus not interrupting the air flow over the tractor nose and contributing 6 percent to the C_d reduction (Rotz and Ziegler, 2014).

Navistar utilized an active ride height system to lower the front of the tractor as well as the rear of the trailer to achieve an overall profile closer to that of an airfoil (Zukouski, 2015). Navistar reported to the committee that they achieved a 50 percent reduction in C_d at 55 mph in their SuperTruck vehicle. Volvo reported a 30 percent drag reduction from their baseline.

---

[5] SuperTruck is a major initiative of DOE's 21st Century Truck Partnership and is supported by 3 other federal agencies in cooperation with 15 industrial partners. The latter include all six U.S. heavy-duty truck manufacturers and many heavy-duty engine and powertrain system manufacturers in order to accelerate technology development and provide focus for research and development (R&D) efforts. The four industry-led projects, spanning 2010 through 2016, were established to fund R&D and demonstration of full vehicle systems integrating a number of technologies into Class 8 heavy-duty, long-haul trucks. There are three major goals: (1) demonstrate a 50 percent increase in freight efficiency (measured in ton-miles per gallon) on a defined drive cycle with a 20 percent engine contribution; (2) demonstrate engine 50 percent brake thermal efficiency (BTE) at a level road load of 65,000 pounds at 65 mph; and (3) conduct technology development and scoping for a path to 55 percent BTE. SOURCE: Ken Howden, DOE, "21st Century Truck Partnership and SuperTruck Initiative." Presentation to the committee, March 29, 2013.

[6] The Volvo group and Navistar truck projects will not complete until June and October 2016, respectively, when final results are expected to be reported (Amar, 2015; Zukouski, 2015).

FIGURE 5-2 Features identified to reduce aerodynamic drag of Cummins-Peterbilt SuperTruck.
SOURCE: Peterbilt Motors Co. (2014).

FIGURE 5-3 Cummins-Peterbilt SuperTruck in show paint.
SOURCE: Koeberlein and Cummins Inc. (2014).

FIGURE 5-4 Daimler Freightliner SuperTruck.
SOURCE: Daimler.

The American Council for an Energy-Efficient Economy reported that the SuperTruck aerodynamic performance was similar to bin VI in the Phase II Rule, meaning a C_dA of 4.5.

Other than SuperTruck, for completeness we note there are a number of additional advanced heavy-duty vehicle demonstrations that are described in the RIA (EPA and NHTSA, 2016) and in the NACFE Confidence Report on aerodynamic devices (NACFE, 2016c).

5.2.4 Expect Continuing Advancements in Aerodynamics and Computational Fluid Dynamics Methods

Discovery of new approaches to reducing aerodynamic drag continues as well as the development of more powerful CFD tools for analysis. As an example, a sail- or wing-like tractor-trailer shape was recently modeled and tested in a Department of Energy project that drastically changed the understanding of the impact of wind direction (Figure 5-5) (Salari, 2016). Whereas typical vehicle shapes have minimum drag at zero yaw angle, this particular shape generates forward force in side winds much as a sail at a proper angle (Figure 5-6). Now, there would be additional side forces on the tires (much as would be seen on a boat keel), but the potential for overall drag reduction seems clear.

Mihelic et al. (2015) have projected the minimum C_d for heavy trucks being about 0.3 (Mihelic, 2015), and Greenberg projected C_d could be taken below 0.3 (Greenberg, 2014). Note in the previous sections the SuperTruck C_d values were estimated to be in this same range. At present, these sources provided the most aggressive suggestions for reduction of drag coefficient. Future aerodynamic features are thoroughly, yet qualitatively, discussed by NACFE (2016a,b,c), with examples of

- Active aerodynamic flow surfaces,
- On-board aero sensing, and
- Reconfigurable truck-trailer shape ("geometry morphing").

Continued progress and innovation are evident in medium- and heavy-duty vehicle (MHDV) aerodynamics, such as seen in the SuperTruck projects and other demonstrations in the private sector.

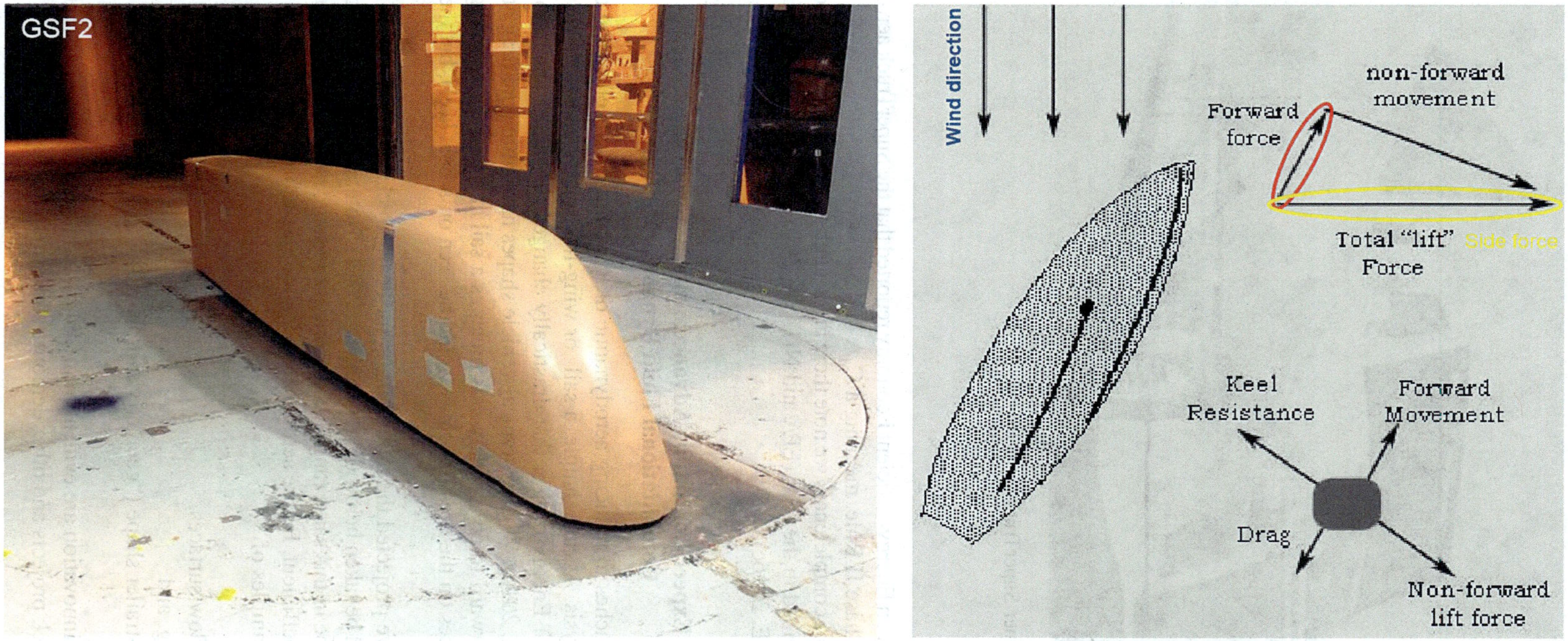

FIGURE 5-5 Scale-model demonstration of sail effect, as well as the physical forces involved.
SOURCE: Salari (2016).

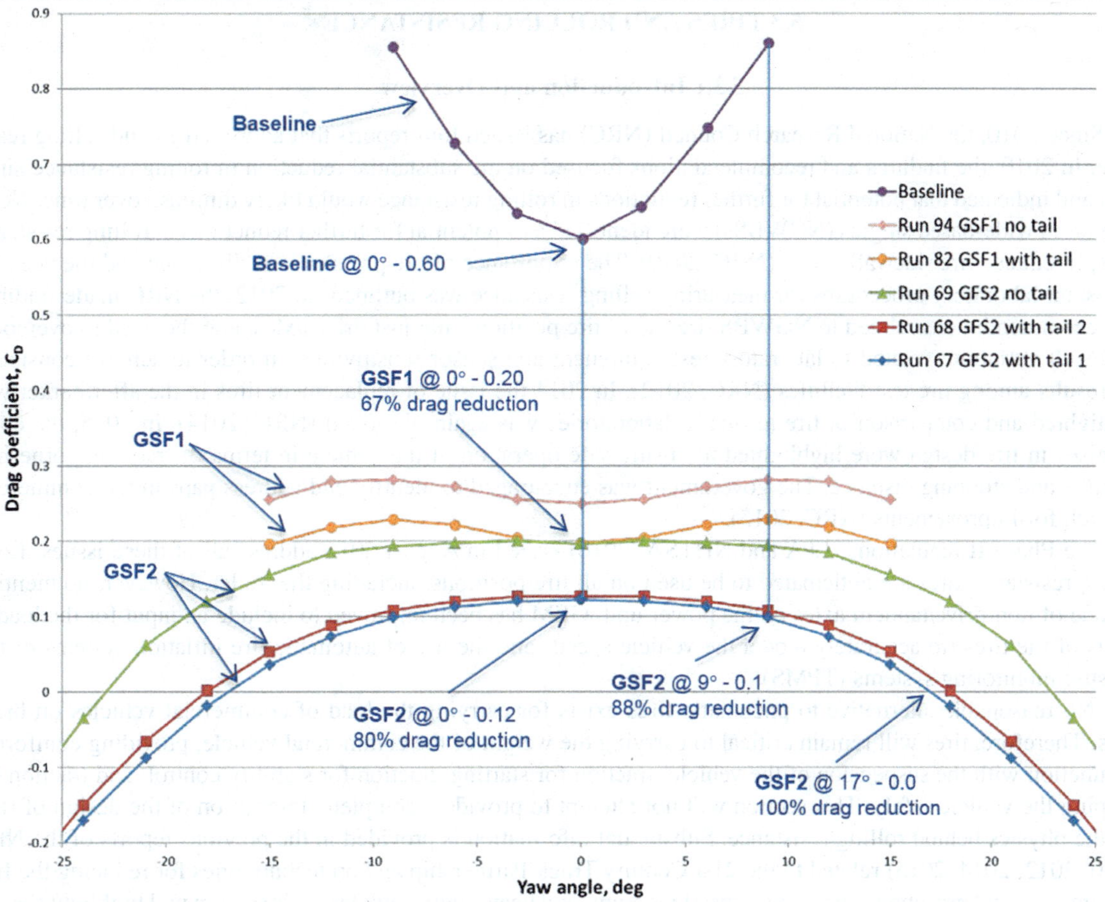

FIGURE 5-6 Plot of drag coefficient vs yaw angle, showing radically different response of drag to yaw angle with GFS2 shape.
SOURCE: Salari (2016).

Finding: Integration conceptually results in better aerodynamic performance. This situation is especially observed and documented with the Cummins-Peterbilt and Daimler aerodynamics SuperTruck configurations. Visually, the aerodynamic-appearance differences may not seem so significant. The Cummins-Peterbilt project illustrated the fuel consumption effect of fitting a legacy trailer to their ST tractor, increasing fuel consumption by 14 percent. The Volvo project advised their intent to conclude with both an integrated ST and a purposeful nonintegrated ST, permitting legacy fit-up.

Finding: Integrated aerodynamics development has been shown to be an appropriate process to achieve optimum performance. However, to achieve relatively seamless adoption of higher-performing aero tractors and higher-performing trailers when fit up to their legacy mates, manufacturers will need to find performance-efficient nonintegrated designs to broadly achieve real-world fuel consumption gains.

Recommendation 5-3: Given the incentive for aerodynamic drag reduction, the rapid progress in CFD, and emerging innovations, the post-2027 potential for reducing C_d should be reevaluated in an interim review.

5.3 TIRES AND ROLLING RESISTANCE

5.3.1 Introduction and Overview

Since 2010, the National Research Council (NRC) has issued four reports that address tires and rolling resistance. In 2010, the findings and recommendations focused on the substantial reduction in rolling resistance since 1990 and indicated that potential for further reductions in rolling resistance would likely diminish over time. Next-generation wide-based singles (NGWBSs) were identified as a potential for further reductions in rolling resistance through reduced tire sidewall losses (NRC, 2010). The importance of tire pressure was discussed and the need for consistent laboratory procedures for measuring rolling resistance was outlined. In 2012, the NRC made findings and recommendations related to NGWBS tires at all tire positions, not just drive axles, and the need to overcome test result variations related to laboratory test equipment and sensor sensitivities, in order to achieve consistent test results among tire test facilities (NRC, 2012). In 2014 the issue of replacement tires in the aftermarket was highlighted and comparison of tire testing at laboratories was again discussed (NRC, 2014). In 2015, the compromises in tire design were highlighted to ensure safe operation of the vehicle in terms of traction, cornering, stability, and stopping distance. The government was encouraged to identify and address gaps in the commercial research for improvements (NRC, 2015).

The Phase II regulations (EPA and NHTSA, 2016) issued in August 2016 address all of these issues. Low-rolling-resistance tires are anticipated to be used on all tire positions, including the trailer. However, no mention is made of non-drive/tandem axles on the power unit. GEM has been improved to include an input for the loaded radius of the tires (to accurately model the vehicle speed) and the use of automatic tire inflation systems or tire pressure monitoring systems (TPMS).

No reasonable alternative to pneumatic tires exists for carrying the load of commercial vehicles on highways. Therefore, tires will remain critical to carrying the weight of the commercial vehicle, providing comfort in conjunction with the suspension of the vehicle, traction for starting, traction for stability control, and traction for stopping the vehicle safely. This section will not attempt to provide a complete description of the design of tires and the physics behind rolling resistance. Substantial information is provided in the previous reports of the NRC (2010, 2012, 2014, 2015) related to the 21st Century Truck Partnership and on technologies for reducing the fuel consumption and greenhouse gas emissions of medium- and heavy-duty vehicles. This section will highlight several areas of concern and areas for potential development of tires and other methods for reducing the rolling resistance of commercial vehicles today and into the future. Box 5-1 lists a number of parameters that must be considered in the design of a tire. Progress to reduce the rolling resistance of tires has been steady since the introduction of internal combustion engines. Figure 5-7 shows that progress (Lew, 2013).

If one were to take a straight line from the value of 30 kilograms per ton (kg/t) in 1890 to the value of roughly 8 kg/t in 2010, and then extend it to 2030, the result would be an unrealistic value of about 1-2 kg/t. Therefore, it is reasonable to expect that we are approaching an asymptotic portion of the development curve. This inflection is shown in another curve, Figure 5-8.

More current developments in the past 10 years are shown in Figure 5-9 (Lew, 2013).

The notice of proposed rulemaking (NPRM) for the Phase II regulations (EPA and NHTSA, 2015) was released in July 2015 and made projections for rolling resistance. These are detailed in Table 5-5, which appears in Trucking Efficiency Confidence Report: Low Rolling Resistance Tires, published by NACFE/Carbon War Room in 2015 (NACFE, 2015b). The EPA SmartWay has also published values using the International Organization for Standardization (ISO) 28250 method of measurement.[7]

The final Phase II regulation (EPA and NHTSA, 2016) expects all trailer tires to reach a coefficient of rolling resistance or C_{rr} of 4.7. Twenty-five to 35 percent of steer axles are expected to reach 4.9. Up to 35 percent of drive tires are expected to reach 5.0.[8] The values provided by OEMs for emissions certification were used to inform NHTSA and EPA and give them confidence in somewhat more aggressive numbers.

[7] See http://www3.epa.gov/smartway/forpartners/documents/verified/420f12024.pdf. Accessed December 12, 2015.

[8] See Table III-30 in Phase 2 Regulatory Impact Analysis and 2-30, 2-34 through 2-37 in Phase 2 Final Rule.

BOX 5-1
Factors in the Design of a Tire

- Construction
  - Casings
  - Treads
  - Sidewalls
- Key Parameters of Tires
  - Aspect ratio
  - Radius
  - Rolling radius
  - Tread width
  - Vertical stiffness
  - Cornering stiffness
  - Adhesion
  - Type, dual or single
  - Tread life or wear
  - Tread pattern
  - Tread depth
  - Structural durability
  - Driver comfort
  - Rain/snow traction
  - Stopping distance
  - Rolling resistance—C_{rr}
  - Retread rolling resistance—C_{rr}
  - Tire construction type
  - Weight
  - Load carrying capacity/capability
  - Tire pressure
  - Speed rating

At the SAE Commercial Vehicle Congress in October 2015, both PACCAR and Daimler Trucks North America presented concerns with the proposed numbers as shown in Figure 5-10. PACCAR is making the point that vehicles in the North American Free Trade Agreement region must traverse roads safely across the North American continent. Requirements for safe operation in Canada are different than in the United States, with higher axle weights that may limit the use of next-generation wide-based single tires in that country (Figure 5-11).

There must be some limit of rolling resistance that is needed to support the weight of the vehicle and safe operation under all driving situations. NHTSA studied this issue in 2009, primarily for passenger vehicles. Figure 5-12 highlights the trade-offs among rolling resistance, tread wear, and traction.

Figure 5-13 graphs the results of tests done on wet surfaces, showing a correlation between reduced rolling resistance and reduced traction. The report states, “The sliding values all have a strong and significant relationship between better rolling resistance and poorer wet traction.”

For the Phase II regulations, NHTSA (2015) ran a small test to determine the correlation between the coefficient of rolling resistance and the stopping distance of a combination vehicle. The conclusion from that report is quoted in the RIA: “The results of this research suggest that tire rolling resistance is not a reliable indicator of Class 8 tractor-trailer stopping distance” (EPA and NHTSA, 2016a, p. 9-8). Those tests were run with vehicles loaded between 57,700 and 80,000 pounds. The two studies suggest that heavier vehicles are less impacted by rolling resistance when the weights are much higher. An empty combination vehicle might weigh one-third to one-half the test loads used.

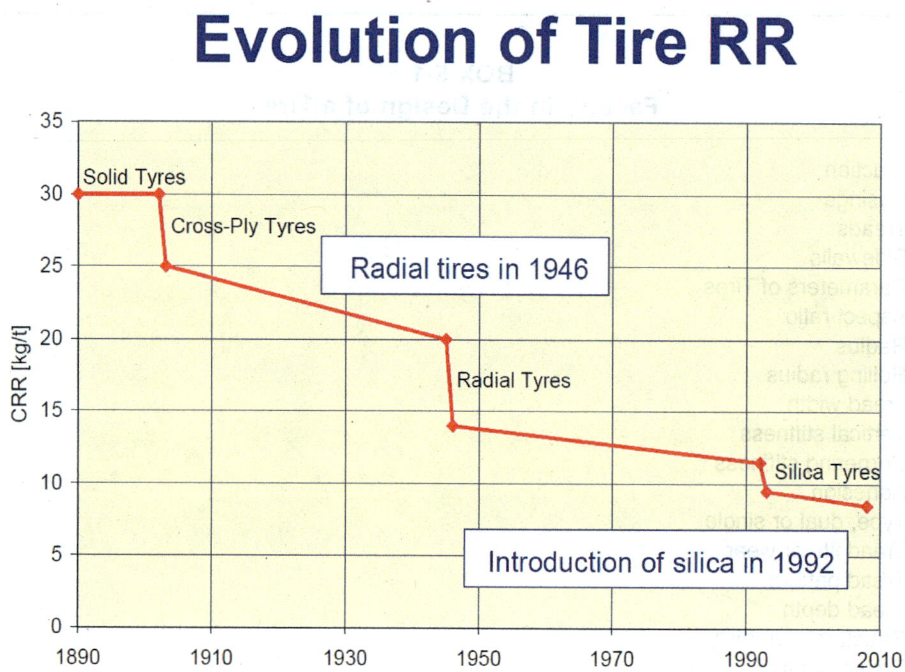

FIGURE 5-7 Evolution of tire rolling resistance from 1890 to 2010, showing different tire technologies and their impact on the coefficient of rolling resistance.
SOURCE: Lew (2013), Michelin North America, Inc.

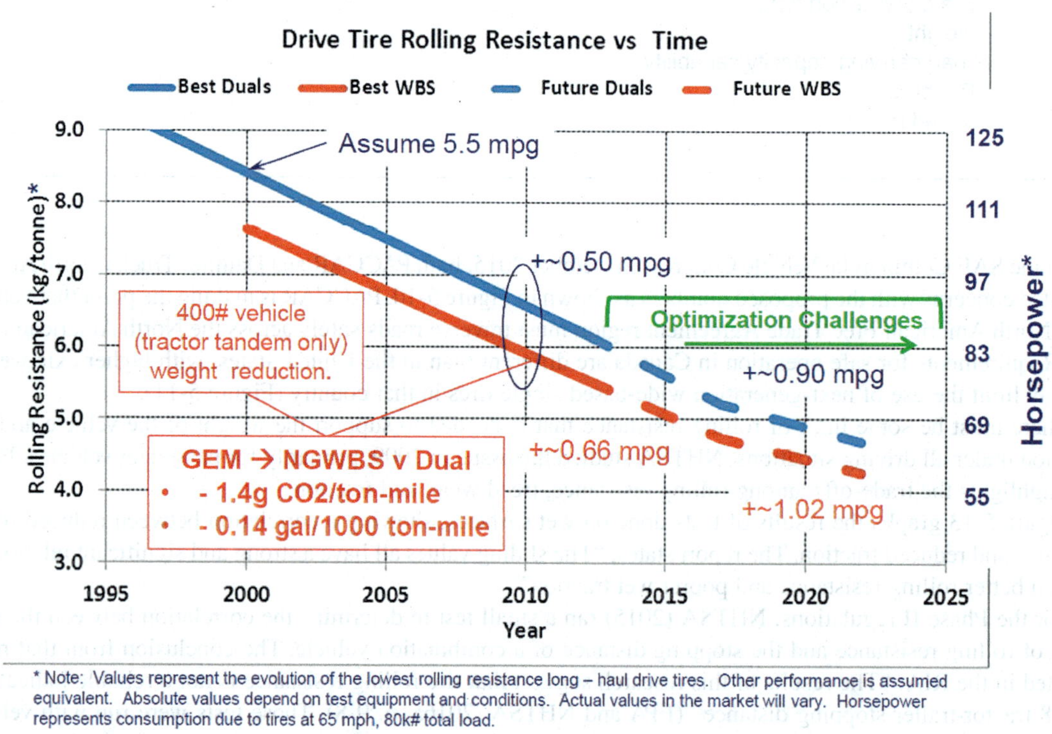

FIGURE 5-8 Tire rolling resistance progress estimated from 1995 to 2013, and projected to 2022, showing dual tires (blue) and wide-base singles (red).
SOURCE: Lew (2013), Michelin North America, Inc.

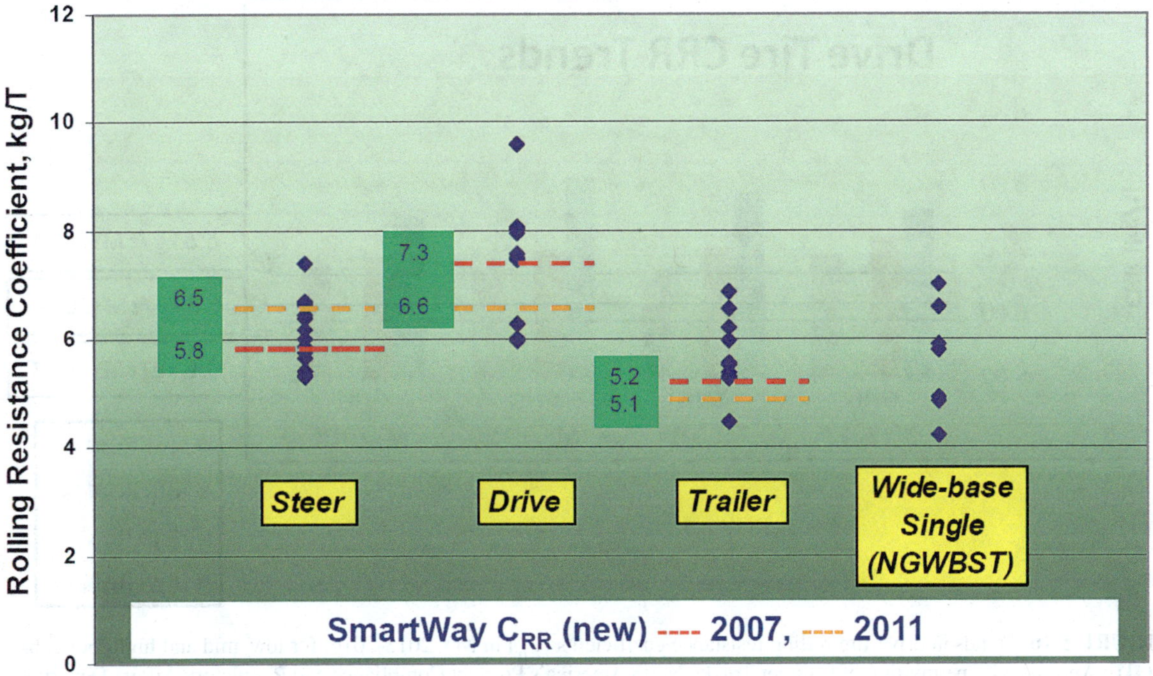

FIGURE 5-9 Example range of rolling resistance coefficients for steering, driving, trailer and wide-base single heavy truck tires, with the 2007 and 2011 SmartWay rolling resistance coefficient goals.
SOURCE: Lew (2013), Michelin North America, Inc.

TABLE 5-5 Projections for Rolling Resistance in Notice of Proposed Rulemaking for the Phase II Regulations.

Tire Location	EPA SmartWay	GHG1'14	GHG1'17	GHG2'18	GHG2'21	GHG2'24	GHG2'27
Steer	6.5	7.7	7.7		6.2	5.9	5.6
Drive	6.6	7.7	7.7		6.6	6.2	5.9
Trailer	5.1			5.1	5.1	4.7	4.7

SOURCE: NACFE (2015a).

Finding: The minimum safe value of C_{rr} for a commercial vehicle with a defined weight, aerodynamic value of C_d, a defined number of wheels with tire patches of specific size, and traveling at a specific speed (such as the requirements of NHTSA regulations for stopping distance) is not currently available.

Recommendation 5-4: NHTSA with the help of the national laboratories and academics should determine the minimum value of the coefficient of rolling resistance or C_{rr} needed to safely operate a vehicle in terms of stability, stopping distance, and traction, taking into account all variations of vehicles from straight trucks to long combination vehicles with additional weight. Both modeling and physical tests should be completed.

Finding: Limited testing on tires has been done to date to establish performance parameters for traction and safe stopping distance.

Recommendation 5-5: Prior to the promulgation of the next phase of regulations, additional work should be done by NHTSA with current commercial vehicle tires and loadings to determine the balance needed among tread wear, rolling resistance, and traction. Vehicles should be outfitted with all required safety systems, such as stability control. Typical tread depths and retreads should be investigated.

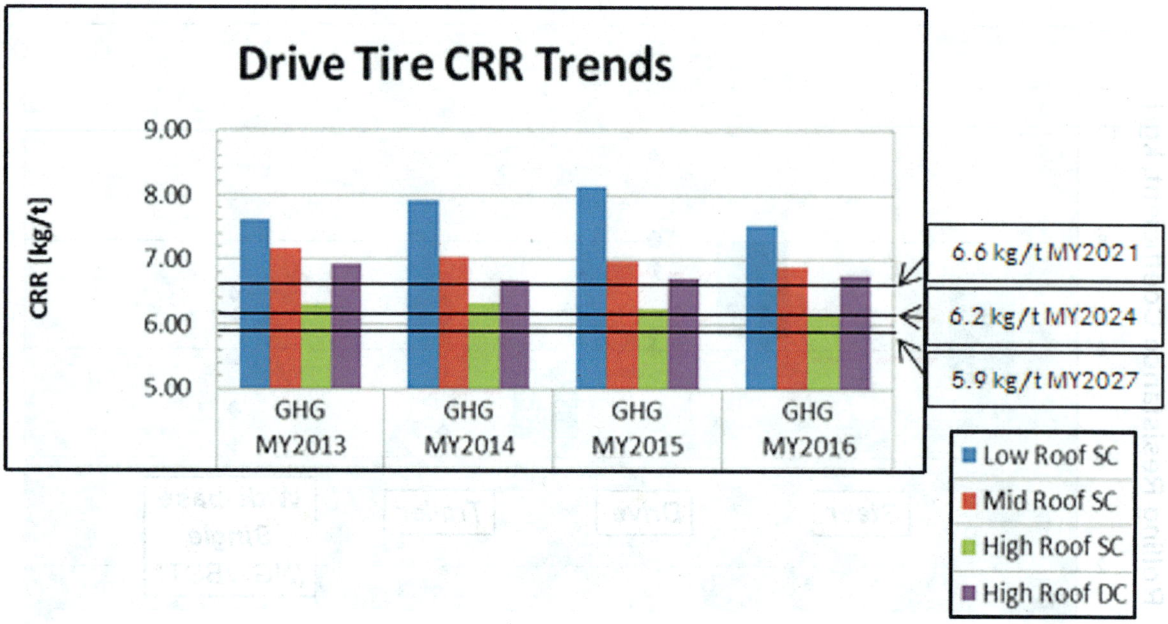

FIGURE 5-10 Trends in drive-tire rolling resistance coefficients (C_{rr}) in MY 2013-2016, for low, mid and high roof cabs.
NOTE: Appears with permission of Daimler Trucks North America's Product Compliance and Regulatory Affairs Department.
SOURCE: Kayes (2015), Daimler Trucks North America's Product Compliance and Regulatory Affairs Department.

Canadian Tire Example

	Current		Phase II Targets		
	US	Canada	2021	2024	2027
High Roof Sleeper Tractors	6.2	6.4	6.2	5.9	5.6
	6.2	6.8	6.6	6.2	5.9
Other Tractors	6.4	6.4	6.2	5.9	5.6
	6.9	7.9	6.6	6.2	5.9
Vocational	6.4	7.5	7.1	6.8	6.4
	6.3	8.1	7.5	7.3	7.0

FIGURE 5-11 Differences in drive-tire C_{rr} in the United States and Canada.
SOURCE: Sproull (2015), courtesy of PACCAR, all rights reserved.

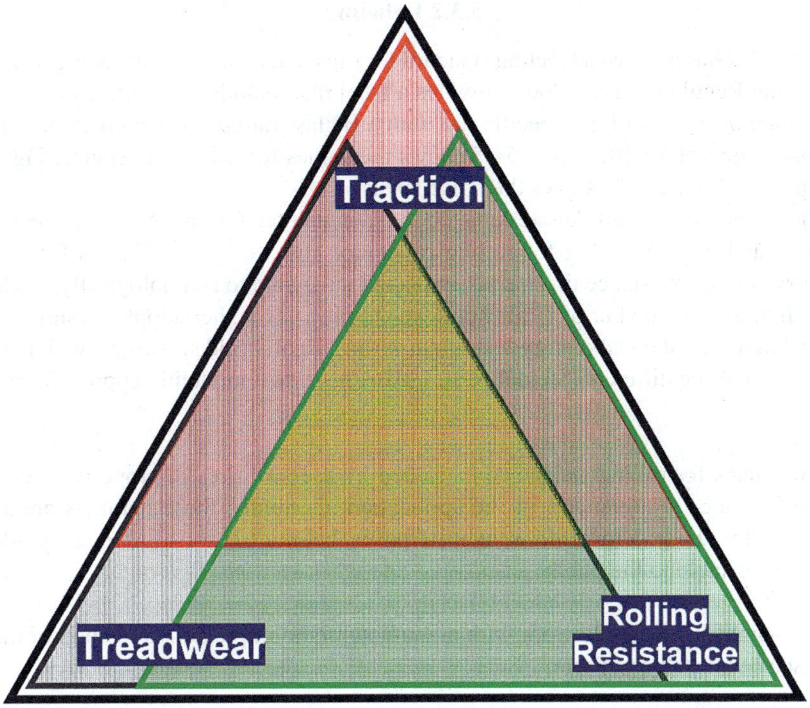

FIGURE 5-12 Traction, tread wear and rolling resistance—the "magic triangle" of tire performance characteristics.
SOURCE: NHTSA (2009).

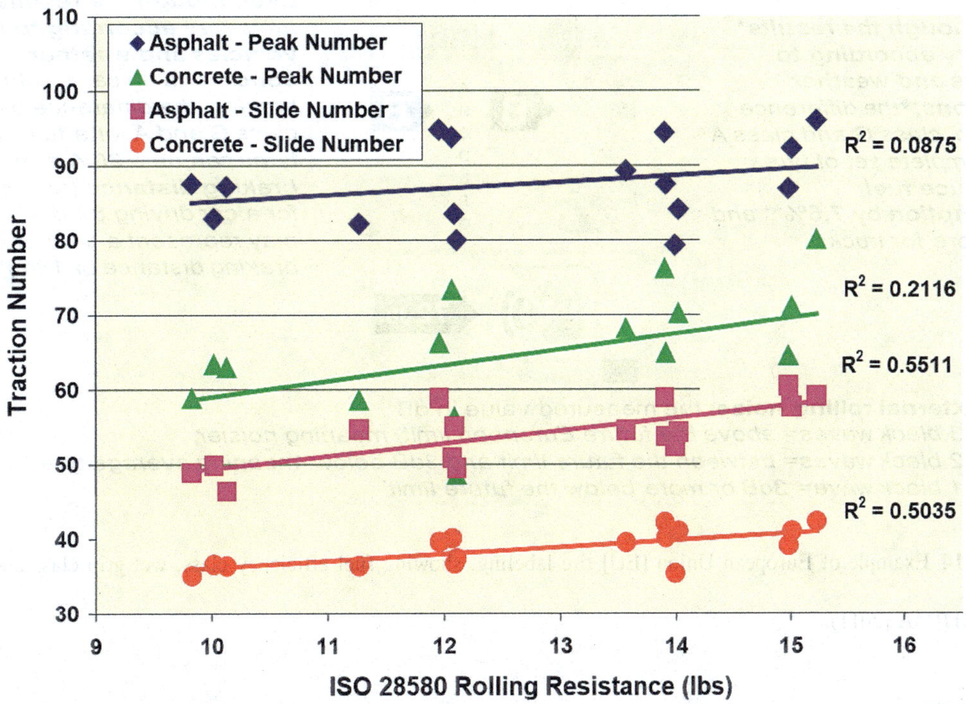

FIGURE 5-13 Wet traction peak and slide numbers vs ISO 28580 rolling resistance for asphalt and concrete.
SOURCE: NHTSA (2009).

5.3.2 Labeling

The European market has adopted a labeling standard for tires that includes both rolling resistance and traction. The EU Tyre Labelling Regulation 1222/2009 provides a label that includes fuel efficiency, wet grip, and noise.[9] Figure 5-14 is a picture of the label for a specific tire with a B class rating for fuel efficiency, a B class rating for wet grip, and a noise rating of 72 dB. Figure 5-15 shows the values for rolling resistance. Figure 5-16 shows the values for wet grip. Last, Figure 5-17 shows the noise levels.

At the Internationale Automobil-Ausstellung 2016, Continental introduced a tire with a triple A rating providing low noise, and the best rating for rolling resistance and wet grip.[10] This is for both drive and steer axles. Therefore, low-rolling-resistance tires can be expected to be able to technologically reach lower levels than currently forecast. It becomes a matter of field experience to push for higher adoption rates.

The above labeling is based on laboratory tests on new tires. Real-world situations, with reduced tread and the possibility of retreads, will be different. Steidel et al. (2016) provide a modeling approach for estimating rolling resistance.

Finding: Access to values for rolling resistance is needed by users of tires to properly assess the ability of the tire to meet the performance requirements of the application. Currently, these numbers are available to OEMs for the GEM inputs and to the government for audit purposes, but not to users. European regulations require this information be made available to the public.

Recommendation 5-6: NHTSA should work with manufacturers to implement a method of making information on tire rolling resistance available to users.

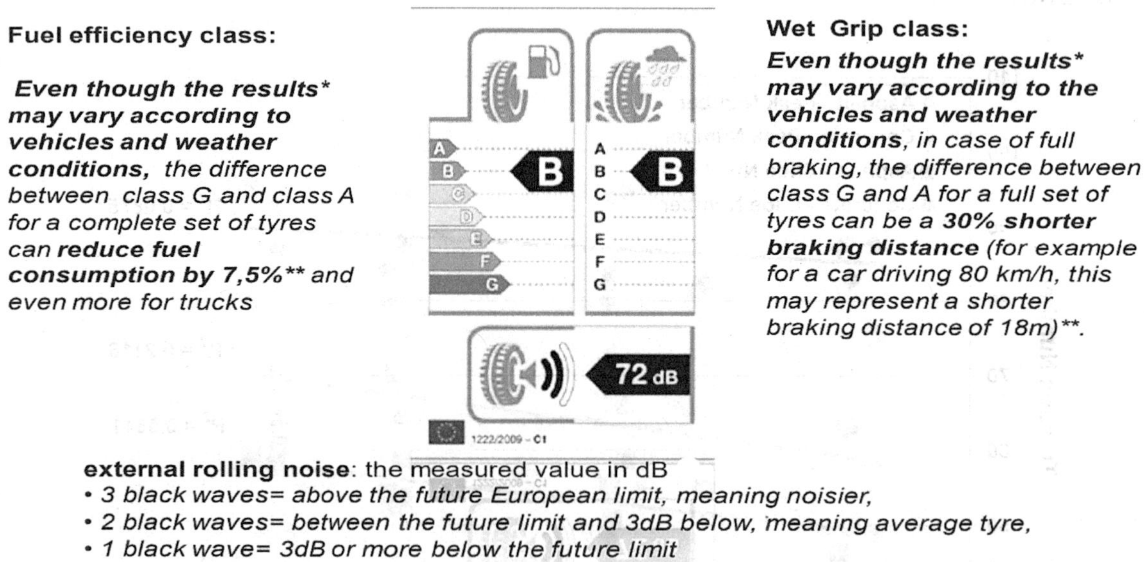

FIGURE 5-14 Example of European Union [EU] tire labeling, showing fuel efficiency class, wet grip class and external rolling noise.
SOURCE: ETRMA (2011).

[9] See http://www.etrma.org/uploads/Modules/Documentsmanager/2011-11-30--industry-guideline-on-tyre-labelling-(vers3).pdf.
[10] See http://www.continental-tires.com/transport/media-services/newsroom/20160921-efficient-tires.

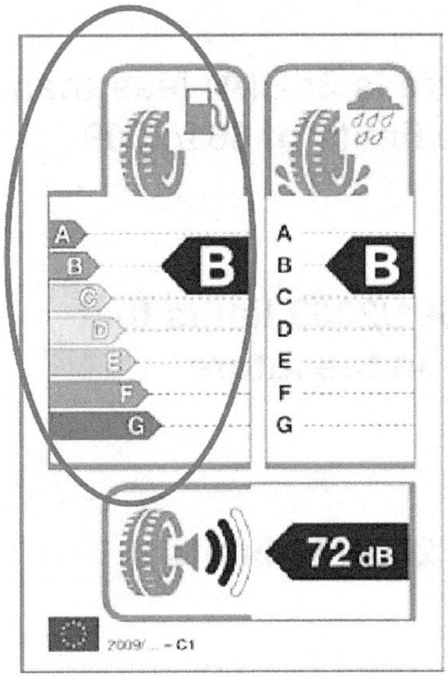

Passenger car C1 Tyres		Light Truck C2 Tyres		Truck & Bus C3 Tyres	
RR C in kg/t	Energy Efficiency class	RR C in kg/t	Energy Efficiency class	RR C in kg/t	Energy Efficiency class
RRC≤6,5	A	RRC≤5,5	A	RRC≤4,0	A
6,6≤RRC≤7.7	B	5,6≤RRC≤6.7	B	4,1 ≤RRC≤ 5,0	B
7,8≤RRC≤9,0	C	6,8≤RRC≤8,0	C	5,1≤RRC≤6,0	C
Empty	D	Empty	D	6,1≤RRC≤7.0	D
9,1≤RRC≤10,5	E	8,1≤RRC≤9,2	E	7,1≤RRC≤8,0	E
10,6≤RRC≤12,0	F	9,3≤RRC≤ 10,5	F	RRC≥8,1	F
RRC≥12,1	G	RRC≥10,6	G	Empty	G

FIGURE 5-15 EU labeling for tire rolling resistance, showing parameters for fuel efficiency by vehicle class.
SOURCE: ETRMA (2011).

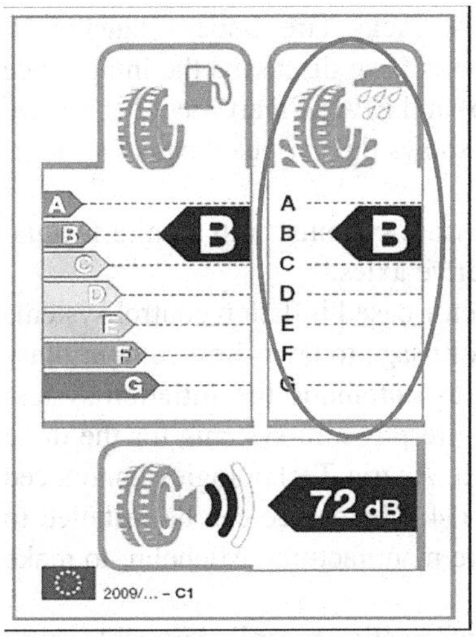

Passenger car C1 tyres		Light Truck C2 tyres		Truck & Bus C3 tyres	
G	Wet grip class	*G*	Wet grip class	*G*	Wet grip class
1,55≤*G*	A	1,40≤*G*	A	1,25≤*G*	A
1,40≤*G*≤1,54	B	1,25≤*G*≤1,39	B	1,10≤*G*≤1,24	B
1,25≤*G*≤1,39	C	1,10≤*G*≤1,24	C	0,95≤*G*≤1,09	C
Empty	D	Empty	D	0,80≤*G*≤0,94	D
1,10≤*G*≤1,24	E	0,95≤*G*≤1,09	E	0,65≤*G*≤0,79	E
G≤1,09	F	*G*≤0,94	F	*G*≤0,64	F
Empty	G	Empty	G	Empty	G

FIGURE 5-16 EU tire labeling for wet grip, showing parameters by vehicle class.
SOURCE: ETRMA (2011).

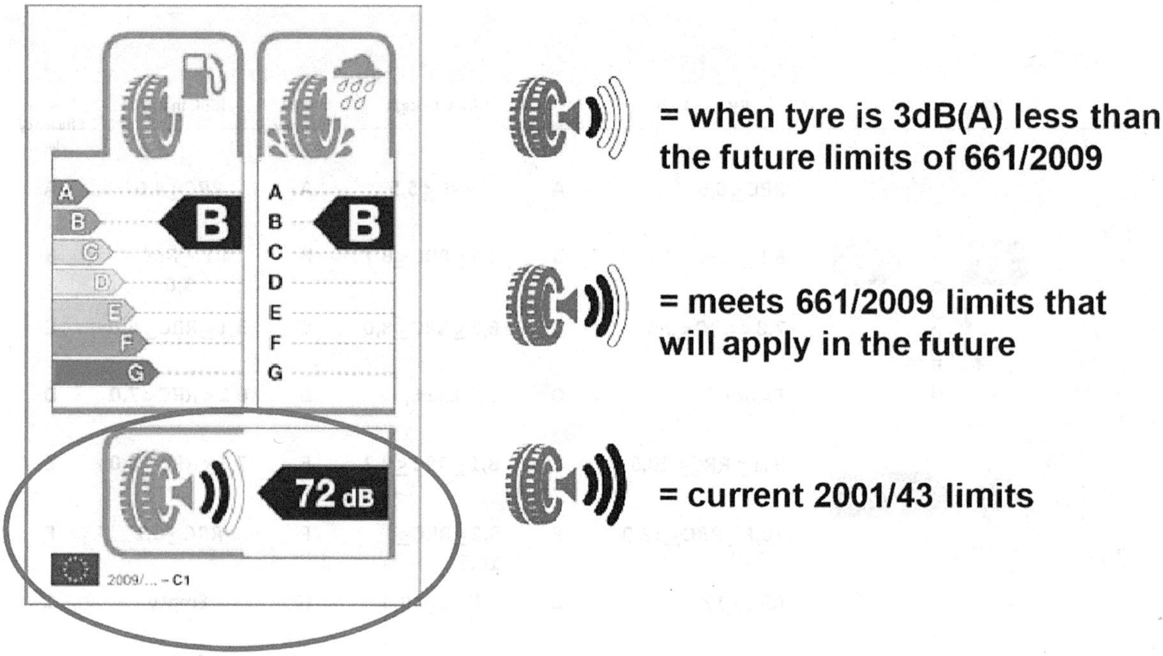

FIGURE 5-17 EU tire labeling for noise.
SOURCE: ETRMA (2011).

5.3.3 Tire Pressure

Tire pressure is well known to be an important factor in fuel consumption of trucks. Tires underinflated by 10 psi can decrease fuel economy by 0.5 to 1 percent (Sidders, 2017). Other reports have discussed the importance of tire pressure to optimize tire rolling resistance and traction. NACFE (2013b) and EPA's SmartWay (2019) have issued reports on the subject. Figure 5-18 is taken from the NACFE report. It shows the various types of systems that are available.

The data in Exhibit 1.1 of the report (NACFE, 2013b) suggest that tire pressure systems, both monitoring and inflation, are growing in acceptance for trailer applications, which have non-drive axles.

Issues with sealing on drive axle tires have delayed the introduction of tractor-based inflation control systems that meet the industry requirements for safety, durability, and reliability. Tractor usage today is limited to military and heavy haul applications, often using the Central Tire Inflation System (an automatic tire inflation system [ATIS]) of Dana.[11] Other companies are continuing to work on the issue of tire pressure systems for the drive axles of tractors, including Airgo[12] and Aperia Technologies. In March 2016, Aperia Technologies announced they were installed on the vehicles of 100 fleets following introduction in 2014. The device can be installed to existing vehicles in 10 minutes per wheel end. They have partnered with a tire manufacturer, Michelin, to make the system available.[13]

The Phase II proposed rule (EPA and NHTSA, 2015) from 2015 asked for information regarding possible credit for tire pressure monitoring systems. The concern from EPA's perspective was that there is no guarantee a driver or fleet will ensure the tire is inflated properly after a monitoring system finds an issue. Considering the Federal Motor Carrier Safety Administration has various means for auditing the operations of fleets, and the Federal Highway Administration can monitor weights, it would seem possible for the nation to find a way to ensure that issues with

---

[11] See http://www.dana.com/commercial-vehicle/products/tire-pressure-management#searchRootId=e2131c3b-468f-443d-87ae-0bca7e2ef08e/facets= (accessed December 19, 2016).

[12] See http://www.airgo.ca/tire_inflation.html (accessed December 19, 2016).

[13] See http://www.michelintruck.com/services-and-programs/auto-inflate/ (accessed December 19, 2016).

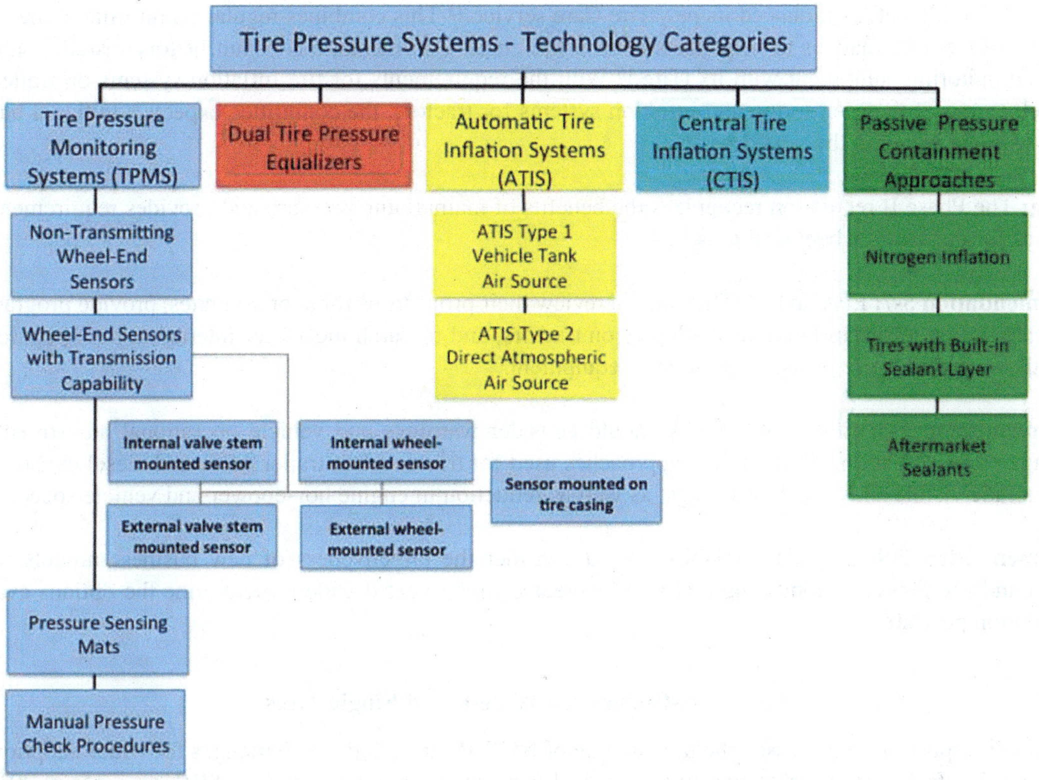

FIGURE 5-18 Tire pressure systems listed by technology category.
SOURCE: NACFE (2013b).

tire pressure that are highlighted by a monitoring system are addressed in a reasonable time frame. The industry has other safety and emissions devices that rely on the driver to take action. These include the trailer anti-lock braking system (ABS) malfunction indicator, the tractor ABS malfunction indicator, and the malfunction indicator lamp for the engine emissions system. In addition, methods of reducing vehicle speed and engine horsepower are used to encourage driver action. Transmission controls also have methods of controlling the vehicle to encourage driver intervention. Such techniques could be used to ensure action is taken to properly inflate tires when using a tire pressure monitoring system. Numerous people responded to the request. Based on the comments received, NHTSA and EPA were persuaded to make changes in the final rule, allowing for a credit in GEM of 1.0 percent for TPMS and 1.2 percent for ATIS. The RIA for the Final Rule (EPA and NHTSA, 2016a) projects adoption rates for 2021 as 20 percent TPMS and 20 percent ATIS, increasing in 2024 to 50 percent TPMS and 25 percent ATIS, and further increasing in 2027 to 70 percent TPMS and 30 percent ATIS.

Goodyear has been testing a pressure maintenance device based on a peristaltic pump. In October 2014, an 18-month field test on commercial trucks was announced with funding from the Department of Energy. The original project, System for Automatically Maintaining Pressure in a Commercial Truck Tire, was last reported on at the 2014 DOE Annual Merit Review (Benedict, 2014). The extended field test appears to be work beyond the original effort. This effort has the potential to provide a reliable, low-cost air pressure maintenance device in the time frames envisioned by the current NPRM and subsequent rulemaking.

The importance of maintaining tire pressure to achieve reductions in fuel usage has spawned new business opportunities. A major truck stop operator, Love's, has introduced a service at its fuel islands to also check and, if necessary, inflate tires while the vehicle is being fueled. TirePass is the name given to this service.[14] In

[14] Love's. "TirePass" Available at <http://www.loves.com/HomeLinks/TruckTireCare/TIREPASS.aspx> (accessed September 22, 2016).

February 2015, Michelin announced its new Tire Care service.[15] This combines regular monitoring at the vehicle with back office data analysis to help fleets control tire costs. Continental has begun factory installation of tire pressure monitoring equipment with its iTire.[16] With the requirements for tire inflation systems on trailers and the development of tire monitoring and inflation systems for tractors, the committee expects additional business models will be created in the next decade.

Finding: The Phase II regulation recognizes the benefits of maintaining pressure and provides requirements and incentives for use on all wheel positions.

Recommendation 5-7: EPA and NHTSA should review their projections for adoption rates, provide programs for large-scale testing of tire inflation technologies on tractors, and research means for interagency cooperation that would allow credit for tire pressure monitoring equipment.

Recommendation 5-8: EPA and NHTSA should consider warnings and vehicle operational adjustments that encourage proper inflation, similar to the approaches used for tractor ABS, trailer ABS, and diesel exhaust fluid. These include visible and audible warnings, as well as reduction in engine horsepower and vehicle speed.

Recommendation 5-9: EPA and NHTSA should consider the effectiveness of new business models for tire inflation and tire pressure monitoring that will be created in the next decade to determine the options available for regulation post-2027.

5.3.4 Next-Generation Wide-Based Single Tires

Previous reports have discussed the take-up rate of NGWBS tires and the challenges for wider adoption. The cost of the tire itself, compared to two tires in a dual configuration, was greater in 2010, on a par in 2015 and projected to be several percentage points less in the future.[17] Thus, cost of the rubber is not the issue for a higher adoption rate for these tires. The complete cost of the axle and wheel end, ancillary equipment (such as tire pressure monitoring to detect deflation), number of allowed retreads, quality of cores for retreading, and logistics issues continue to slow the penetration. The chart in Figure 5-19 shows that fleets are adopting low-rolling-resistance tires, but NGWBS tire penetration is declining.

Chapter 7 of the NACFE report details the study group's information on wide-based tires (Figure 5-19). Little factual information is provided as fleets and tire manufacturers continue to carefully control the dissemination of such information as costs, replacement rates, number of times a tire can be and is retread, and percentage of available tires that can be retread.

Finding: Little factual and statistical information is available regarding wide-based single tires and wheel ends for performance or cost.

Recommendation 5-10: EPA and NHTSA should work with European counterparts and complete reliable tests in the United States to validate the costs and benefits of wide-based single tires.

5.3.5 Alternative Methods to Reduce Rolling Resistance

Tire Configuration. Rolling resistance can be reduced by minimizing the amount of rubber on the road other than by using next-generation wide-based single tires. Alternative configurations of axles that can carry the

[15] Michelin® Tire Care™ Added to Michelin® Commercial Service Network Offer, February 15, 2015, and http://www.michelintruck.com/services-and-programs/tire-care/ (accessed December 15, 2015).

[16] See http://www.continental-tires.com/transport/media-services/newsroom/20160921-itire (accessed December 19, 2019).

[17] Stan Lew, personal communication to Paul Menig, October 9, 2015.

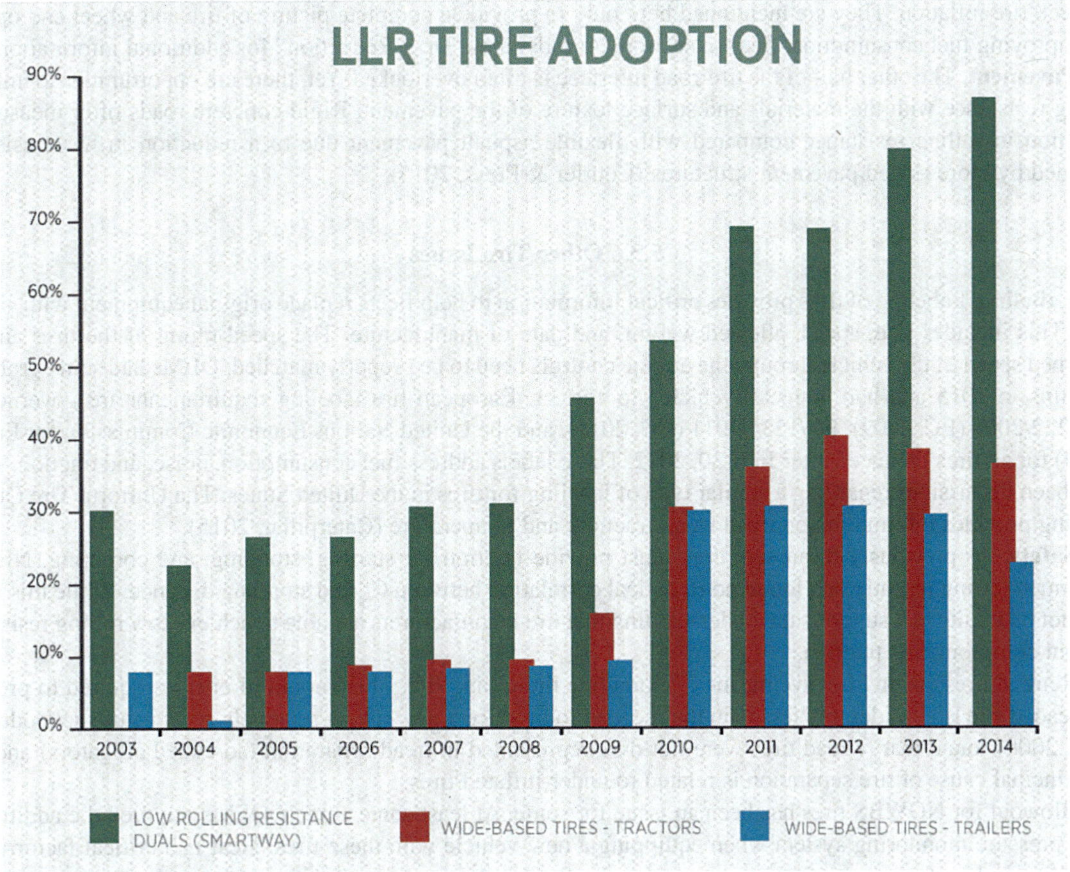

FIGURE 5-19 Percent adoption of low-rolling-resistance tire duals (green), wide-based singles for tractors (red) and wide-based singles for trailers (blue) in 2003-2014.
SOURCE: NACFE (2015b).

legal weight, and lift axles that are raised when the weight is less than the maximum are two such ways. Other sections of this report cover the issue of longer combination vehicles which have the potential to reduce the rolling resistance by eliminating several axles and wheel ends, as well as a power unit. NACFE published a report on 6×2 axle configurations in 2014. The NPRM addresses this technology, expecting an adoption rate that was commented upon in the record. Steerable axles will reduce the scrubbing of tires in turns, effectively reducing the rolling resistance. While this has generally been applied to low-speed maneuvering of specialty vehicles, it can be applied to vehicles at speeds to reduce rolling (scuffing) resistance in highway turns. Knorr-Bremse and Kässbohrer have taken a step to reduce the cost of steerable axles by using the brakes to accomplish the steering. Self-steering axles are used in many jurisdictions.

Additional Axles to Allow More Weight per Vehicle Adds to Rolling Resistance. Other sections of this report discuss how different tire and axle configurations can improve the productivity of commercial vehicles. Some configurations suggest adding an additional axle to a trailer to allow greater loads. This will increase the vehicle load which will increase the rolling resistance of the combination vehicle. However, the improvement in load-specific fuel consumption will outweigh the increase in rolling resistance (Woodrooffe, 2014).

Spats for Aerodynamics. These devices are primarily meant to improve the air flow around the wheel end, covering up the wheel. These have been shown to improve the aerodynamics of the vehicle. SAE testing done by suppliers suggests a fuel economy improvement of 1 to 2 percent when used on tractor rear wheels and trailer wheels combined. Design of such devices needs to account for the need to check the wheels for loose lug nuts and to check tire inflation. They are mentioned here only to provide a complete picture of tire and wheel end options for improving fuel consumption. See the section "Aerodynamic Drag Reduction" for additional information.

Pavement. The other half of the tire-road interface is often overlooked. Yet, there are opportunities to improve rolling resistance with the materials and surface texture of the pavement. Rigid concrete roads offer measurable reduction in rolling resistance compared with flexible asphalt pavement due to a reduction in hysteresis loss provided by more ridged pavement structures (Courier & Press, 2015).

5.3.6 Other Tire Issues

Labeling. Labeling of tires provides critical information to help fleets replace original equipment with similar tires. This includes size, speed, allowed weight, and date of manufacture. The speed rating of the tires and the governed speed of the vehicle through the engine controls need to be properly matched. OEMs had some problems with this in 2015 and had to recall vehicles to correct. European tire labeling requirements are governed by EC/1222/2009 (EC, 2009), EC/1235/2011 (EC, 2011), and the United Nations Economic Commission for Europe (2014) for all tires produced after June 30, 2012. These labels address fuel consumption, noise, and traction. There have been discussions regarding a similar type of labeling for tires in the United States. The Uniform Tire Quality Grading provides information on tread wear, traction, and temperature (Caterpillar, 2015).

Safety. As previously discussed, tires must provide traction for starting, stopping, and cornering. NHTSA performed testing recently and found no statistical correlation between C_{rr} and stopping distance. While this result may not be intuitive, it supports the understanding that tire manufacturers are able to achieve low rolling resistance without compromising traction.

Retread tires are an effective means of extending tire casing life. The amount of energy required to produce a retreaded tire is considerably less than that needed to produce a new tire. A relatively recent study (Woodrooffe et al., 2008) found that retread tires were not over-represented in tread separation (so-called alligators) and that the principal cause of tire separation is related to under-inflated tires.

Blowout for NGWBS tires has been an issue for some. At least some vehicle OEMs required the addition of a tire pressure monitoring system when equipping a new vehicle with these tires. Heat is a critical factor in the durability and performance of a tire. As previously discussed, aerodynamic improvements have the potential to increase the heat at the wheel ends, creating issues for durability and safety.

Tire Damage. Tires can only maintain their rolling resistance when undamaged. The Technology and Maintenance Council of the American Trucking Associations (TMC, 1994) has an extensive guide on the maintenance of tires and damage, entitled "Radial Tire Conditions Analysis Guide: A Comprehensive Review of Tread Wear and Tire Conditions."

Aftermarket Tires. New tire production for new commercial vehicles produced by OEMs is on the order of 5 million units. More than 15 million units are for replacement of tires on existing vehicles; 15.7 million units a year are retread for use on existing vehicles.[18] Therefore, new production for new vehicles represents only 25 percent of new tire production and 14 percent of all tire placements including new, aftermarket, and retread. Penetration of low-rolling-resistance tires into the existing fleet of commercial vehicles could be increased with incentives or regulations (see Chapter 2).

5.3.7 Future Developments

A limited number of items are known to the committee to be in development related to both tires and alternative methods for reducing the rolling resistance of commercial vehicles.

Airless Tires. Several manufacturers have begun to develop and test airless tires, including for commercial vehicles.[19] Bridgestone, Michelin, and Hankook have all reported activities publicly.[20] Applications for passenger

[18] Modern Tire Dealer, January 2014, "Facts Issue 2014." Available at http://www.moderntiredealer.com/uploads/stats/2014-facts-sans-ads.pdf (accessed June 12, 2017).

[19] "Airless Tires." http://www.bridgestonetire.com/tread-and-trend/tire-talk/airless-concept-tires (accessed September 6, 2015).

[20] Michelin Tweel Technologies, http://www.michelintweel.com (accessed September 6, 2015).

cars and small, off-road vehicles are currently being sold and prototyped (Collins, 2017). Indications suggest that these tires will have benefits but not reduce the rolling resistance overall (Collins, 2017). Other benefits will be reduction in heat generation in the tire, as there is less flexure in the sidewalls (Collins, 2017). As vehicles get more aerodynamic and employ fairings that reduce air cooling to the tire and wheel, this could become an important benefit for future tires.

Alternative Materials. Rubber (from the *Hevea brasiliensis* tree) and synthetic rubber remain a major material in the construction of tires, along with steel belts and glue. Supply of rubber has been an issue in the past. Therefore, research continues into alternative sources of rubber for tires. This includes rubber from the guayule plant and a special species of dandelions.[21] Continental is in field tests with the Taraxagum tire.[22] Major research is being carried out by the manufacturers and by academic institutions, such as the Ohio State University PENRA program,[23] which was created to integrate and accelerate the incubation, demonstration, market entry, and growth of a domestic natural rubber industry. PENRA focuses on the creation of the science and technology and the private partnerships needed to support the introduction and scale-up of natural rubber alternatives. Europe has a program called DRIVE4EU that is similarly engaged.[24] DRIVE4EU—Dandelion Rubber and Inulin Valorization and Exploitation for Europe—is a demonstration project, aimed at the development of the production chain of natural rubber and inulin from Russian dandelions. The objective of the project is to set up a new European chain for the production and processing of natural rubber. Cooper Tires is involved in a $6.9 million research program by the U.S. Department of Agriculture and DOE to develop a domestic natural rubber source. These developments are aimed at securing the supply of rubber and, potentially, controlling the cost of supply. Bridgestone announced on February 24, 2017, that it has succeeded in creating synthetic isoprene rubber. There is a potential for this new material to surpass the performance of natural rubber. The press release mentions evaluations suggesting it has better fuel efficiency. More research is needed to establish developments that will lead to tires with reduced rolling resistance or better construction to prevent losses from flexure, either in the sidewalls or in the treads.[25]

Retreading of Next-Generation Wide-Based Singles. NGWBS have a small advantage in rolling resistance compared to dual tires. Dual tires continue to improve but are not expected to ever equal NGWBS tires due to the latter's lower wall flexure and smaller tread patch making contact with the ground. Dual tires on commercial vehicles are routinely retreaded three or more times. NGWBS tires have only been able to achieve one or two retreads. Research continues to allow these tires to be retread as many times as duals. This is needed to overcome a barrier to their adoption in the industry (Erb, 2015; Britt, 2015).

Self-healing Rubber. Recently, researchers at the Leibniz Institute for Polymer Research in Germany have found a way to have rubber heal itself while the vehicle is not being driven (Bohme, no date). This involves new materials and new processing. Instead of the well-known vulcanization process, a different process is used to add carbon and nitrogen. Given the long process of taking something from the research laboratory to production in the commercial vehicle world, the committee does not expect this to be in production in the time frames considered. Furthermore, while this addresses an important issue for tires, it is not expected to provide benefit in terms of low rolling resistance. It may help with issues related to failure of next-generation wide-based singles and improve the acceptance of them.

Aerodynamic Improvements Causing Additional Heat. Heat is the biggest concern for tire durability. Many of the aerodynamic improvements already in production and expected in the future will reduce the cooling air flow around the tire and wheel end. Cooling of the engine takes precedence in design over aerodynamic improvements

---

[21] "Cooper Tire Begins Testing Tires Made with Guayule Component." http://coopertire.com/News/Corporate-News-Releases/Cooper-Tire-Begins-Testing-Tires-Made-with-Guayule.aspx (accessed October 14, 2015); Natural Rubber from Dandelions. Available at http://www.fraunhofer.de/en/press/research-news/2015/June/natural-rubber-from-dandelions.html (accessed September 6, 2015).

[22] "Continental Trials First Car Tires with Taraxagum Dandelion-Rubber Tread." http://www.continentaltire.com/content/continental-trials-first-car-tires-taraxagum-dandelion-rubber-tread (accessed September 6, 2015).

[23] PENRA, the Program of Excellence in Natural Rubber Alternatives. See https://u.osu.edu/penra/ (accessed October 14, 2015).

[24] See http://www.drive4eu.eu (accessed December 19, 2019).

[25] Automotive World. February 24, 2017. "Bridgestone succeeds in creating synthetic isoprene rubber through precise molecular structure control – development of next-generation rubber with performance surpassing that of natural rubber." Available at http://www.automotive-world.com/news-releases/bridgestone-succeeds-creating-synthetic-isoprene-rubber-precise-molecular-structure-control-development-next-generation-rubber-performance-surpassing-natural-rubber/ (accessed March 7, 2017).

at the front of the vehicle. Care must be taken to ensure that aerodynamic improvements provide adequate cooling to the tires and brakes.

5.4 MASS REDUCTION

5.4.1 General Weight Characteristics of Medium and Heavy Vehicles

The typical Class 8 sleeper tractor weighs about 20,000 pounds and the popular 53-foot dry van trailer weighs roughly 14,000 pounds. The overall weight distribution is shown in Figure 5-20 below, and further details were presented in the Phase I report (NRC, 2010) and other recent reports (NACFE, 2015a).

With a maximum allowed weight in most states of 80,000 pounds, the maximum payload is about 46,000 pounds. The majority of trucks on the road do not operate at the weight limit, but the trend is toward denser freight loads, which increases overall freight efficiency.

5.4.2 Impacts of Vehicle Mass on Freight Efficiency

The effects of mass and mass reduction on fuel consumption and load-specific fuel consumption were explained in the NRC Phase One report (NRC, 2010) and recently reviewed in the NRC report on the 21st Century Truck Partnership (NRC, 2015). Truck weight impacts fuel consumption (e.g., gallons per mile) primarily through tire rolling resistance and unrecovered energy used when accelerating or grade climbing. The energy needed to overcome these resistances is essentially linearly dependent on truck weight. Truck weight effects are more conspicuous and dominate fuel consumption over duty cycles with frequent stops and accelerations, or on roads with notable grade. Revisiting the example from the Phase I report, in powering an 80,000-pound vehicle, maintaining 50 mph on a mere 3 percent grade triples the engine power demand compared to a level road (e.g., 150 hp on level road to 450 hp on the 3 percent grade) (Caterpillar, 2006). In the National Academy of Sciences (NAS) Phase I

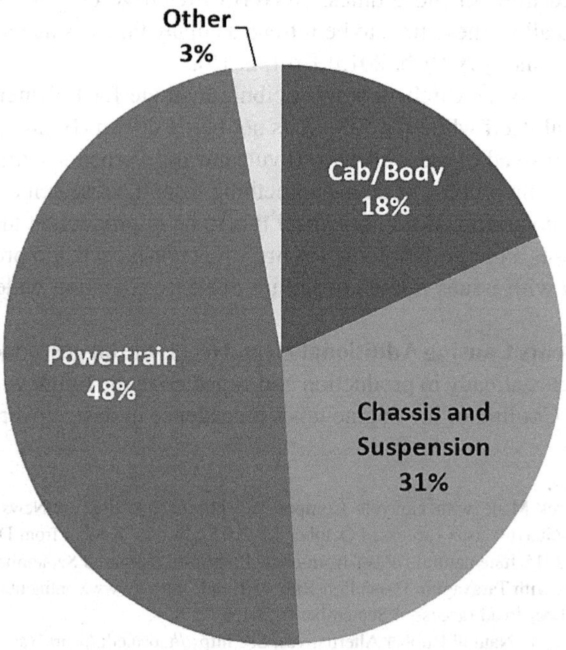

FIGURE 5-20 Weight distribution of typical Class 8 tractor (DOE EE-0867) in its powertrain, chassis and suspension, cab and body, and other components.
SOURCE: DOE (2013).

study, based on reviewed test data and models (ANL, 2009; NRC, 2010), the impact of weight reduction was found to range from approximately 0.5 percent per 1,000 pounds reduction, to approximately 2 percent per 1,000 pounds reduction. Shortly after the NAS Phase I study concluded, results of on-road monitoring of a small fleet of Class 8 trucks were published in final form by Oak Ridge National Laboratory (ORNL, 2011) that provided a look at payload effects in the real world. An example is shown in Figure 5-21, where it is seen that on average for an upslope terrain the effect of mass reduction on fuel consumption (gallons per mile) would be only about 1 percent per 1,000 pounds, and the impact on mass-specific fuel consumption even less.

A study by Ricardo Inc. (2009) for the Aluminum Association produced a similar generalization for Class 8 trucks: 0.7 to 0.9 percent reduction in fuel consumption per 1,000 pounds weight removed.

Reducing the unloaded (tare) weight of the truck and trailer always benefits the metric of gallons per ton-mile, and usually directly reduces fuel consumption in gallons per mile. In cases where the vehicle is weight limited (meaning at its maximum allowable weight), the mass reduction can be translated to additional payload for higher freight efficiency (fewer gallons per ton-mile). Vehicles with maximum GVW loads may still operate at reduced loads on the same trip after off-loading at stops, to the extreme example of tankers running empty on a return trip. During reduced-load vehicle miles traveled, the fuel consumption (gallons per mile) will conspicuously improve with reduced weight. For vehicles that are at maximum volume capacity—"cubed out"—weight savings will always provide the more conspicuous reduction in fuel use. As the density of freight shipments, meaning the mass carried by each truck, has been increasing over the past decade, the importance of reducing the vehicle tare

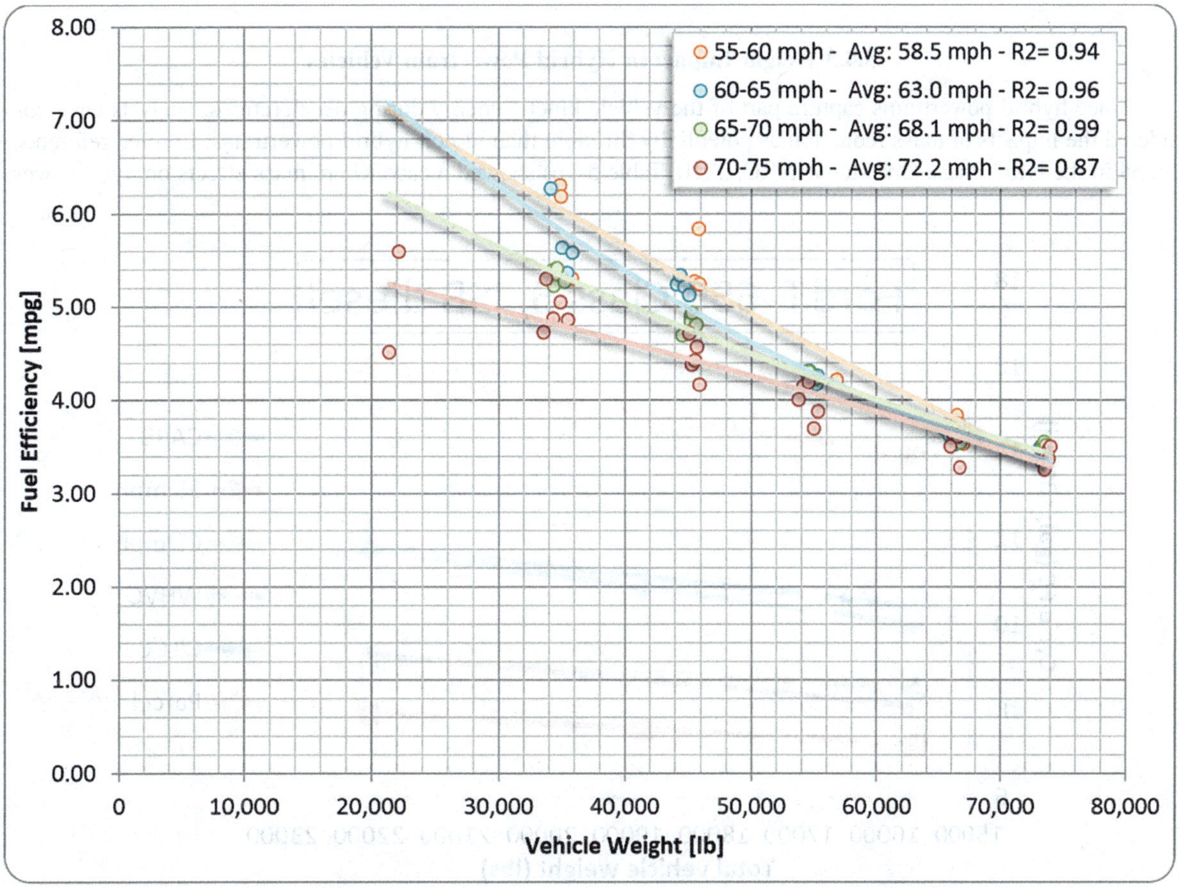

FIGURE 5-21 Effect of varying vehicle weight on fuel economy under a range of vehicle speeds.
SOURCE: ORNL (2011).

weight is more for allowing additional payload for higher freight efficiency (reduced gallons per ton-mile). At the fleet level, this can lead to fewer trucks and truck trips needed to deliver the same overall freight mass, a huge incentive for lightweighting (NACFE, 2015a).

Weight reduction technologies are also needed to maintain current freight efficiency levels because of legal requirements, notably the selective catalytic reduction (SCR) exhaust aftertreatment systems, which added approximately 400 pounds in 2010 (21CTP, 2013). There are similar trade-offs to be considered with weight added by aerodynamic devices, hybrid powertrain components, driver amenities, and auxiliary power units. Overall, Class 8 trucks have become about 1,000 pounds heavier in the past decade (NACFE, 2015a). Weight increases in tractor and trailer caused loss of payload of about 2,000 pounds between 2005 and 2011.

Will higher-efficiency powertrains and lower aerodynamic drag, expected for MHDVs of 2025 and beyond, alter the impacts of weight reduction? The committee contracted with SwRI to examine vehicle fuel consumption with advanced powertrains through simulation, and this allows assessing the effects of mass for a future vehicle. Multiple vehicle classes and drive cycles were examined as further described in Chapter 6.

The results from the vehicle simulations are summarized in Table 5-6 (EDF, no date). The previously identified trends appear to hold true, that a 0.5 to 1.0 percent decrease in fuel consumption is likely for every 1,000 pounds removed from Class 8 vehicles, with a greater effect seen in a few cases. For diesel-powered Class 6 box trucks, there is a greater impact of weight reduction, generally exceeding 1.0 percent per 1,000 pounds reduced, and for the simulations of Class 5 diesel trucks the weight reduction effect was found to be over 2.0 percent per 1,000 pounds. The Class 2b and 3 pickup exhibited a weight sensitivity of about 2.5 percent per 1,000 pounds. Representative results from simulations showing weight effects are shown in Figures 5-22, 5-23, and 5-24 below.

5.4.3 Weight Impact in Hybrid Powertrain Vehicles

Since hybrid powertrains capture part of the vehicle kinetic energy during decelerations, analysts have considered the impacts of mass reduction as potentially different than in non-hybrid powertrains. Limited references found in the NRC Phase One study (NRC, 2010, Table 5-16) showed a case where mass effects on hybrids were

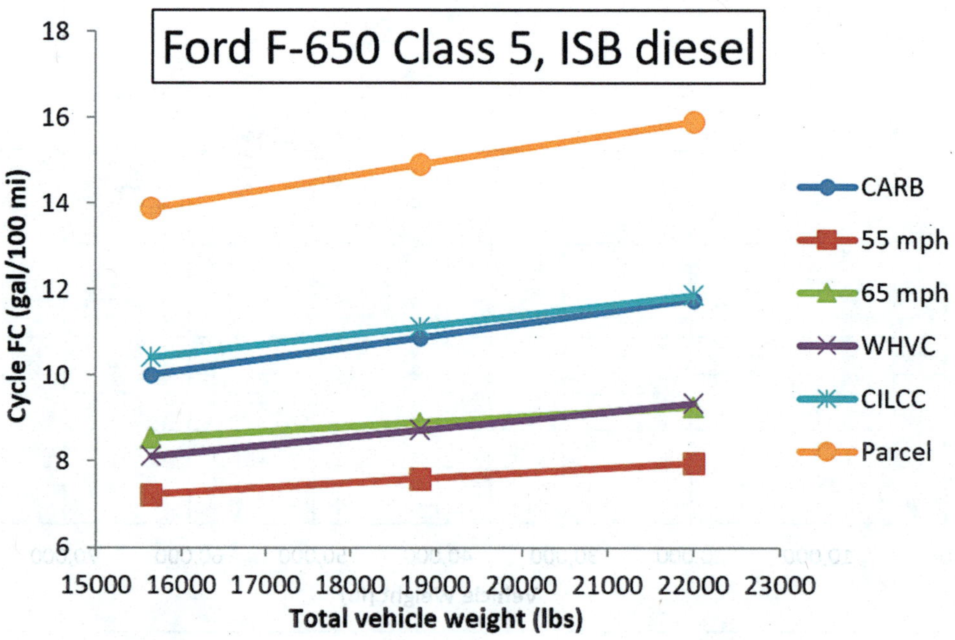

FIGURE 5-22 Sensitivity of fuel consumption to vehicle weight of Class 5 vehicle simulated on 6 test cycles. Further details about the SwRI vehicle modeling are provided in Chapter 6.

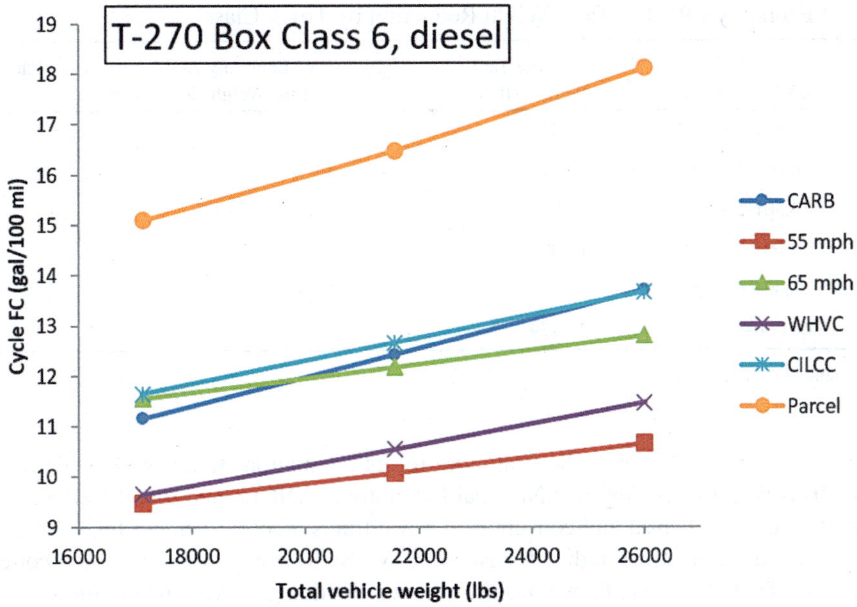

FIGURE 5-23 Sensitivity of fuel consumption to vehicle weight of Class 6 vehicle simulated on 6 test cycles. Further details about the SwRI vehicle modeling are provided in Chapter 6.

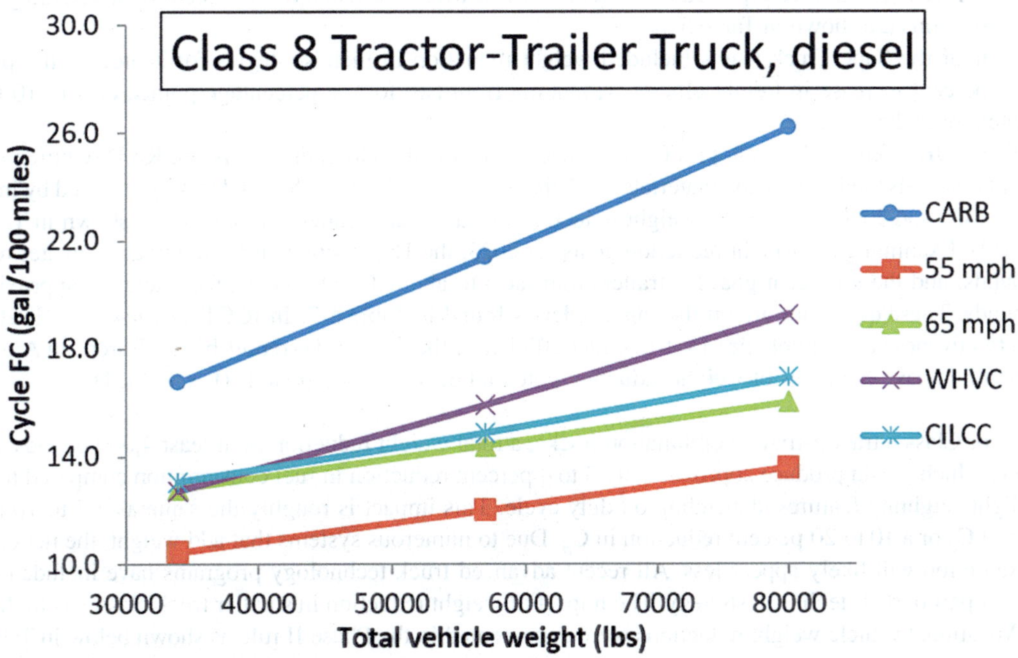

FIGURE 5-24 Sensitivity of fuel consumption to vehicle weight of Class 8 vehicle simulated on 5 test cycles. Further details about the SwRI vehicle modeling are provided in Chapter 6.

TABLE 5-6 Fuel Economy Effect of Tare Weight Reduction by Truck Class

Truck Class	GVWR Range	Average Tare Weight (lbs)	Fuel Economy Improvement for Each 1,000 lbs of Tare Weight Reduction
4	14,001-16,000	10,343	5.6%
5	16,001-19,500	10,413	4.7%
6	19,501-26,000	13,942	3.9%
7	26,001-33,000	18,094	2.8%
8A	33,001-60,000	23,525	1.9%
8B	60,000+	28,979	1.1%

SOURCE: Environmental Defense Fund.

greater than 2 percent per 1,000 pounds. Several more-recent simulation studies were reviewed for the Phase II study. For Class 2B pickup trucks, Argonne National Laboratory (ANL) found that the effects of mass reduction were essentially the same for hybrid and conventional powertrains, about 2 percent for a 200 kg mass reduction (ANL, 2009). In a set of simulation studies for NESCAUM, SwRI compared hybrid and conventional Class 8 trucks and found that, for both types of powertrains, weight reduction gave very nearly the same reduction in fuel consumption. Heavy-duty (HD) vehicle simulations performed by the ORNL analyzed the benefits of hybrid Class 8 powertrains over a range of payloads, showing a slight nonlinearity (optimum payload) for fuel consumption benefits (Gao et al., 2015).

The NRC Phase One study reported a potential 1.25 percent fuel consumption reduction through the 2020 time frame (NRC, 2010, Table 5-18) for Class 8 trucks, which translates to approximately a net 2,000 pound weight reduction. Higher impacts were expected in lighter classes of MHDVs. Examples of recent lightweighting studies and demonstrations are shown in Table 5-7.

All four of the SuperTruck teams included weight reduction in their strategies for achieving the program goal of 50 percent increase in freight efficiency, ranging from two to five percentage points, or 4 to 10 percent improvement over the baseline.

In 2011 DOE's Vehicle Technologies Office convened a workshop to define goals and R&D requirements for lightweight materials (and propulsion materials) with the inclusion of heavy vehicles. Heavily attended by industry, targets for 2025 and 2050 were set for weight reduction of tractor and trailer components as shown in Table 5-8 (DOE, 2013). Examining the weight reduction goals for 2025, the 16 percent goal for the tractor equates to about 3,200 pounds, and the 9 percent goal for trailers translates to about 1,200 pounds, for a total of approximately 4,400 pounds. These quantities are on the same order as found in Table 5-7. In ICCT's review (ICCT, 2015) of the cost effectiveness of technologies for truck fuel efficiency, they noted the report by the European Aluminum Association that gave estimates of weight-reduction potential of over 6,000 pounds (EAA, 2014).

Finding: For Class 8 tractor-trailer combination trucks, a total weight reduction of at least 4,000 pounds appears achievable, which would produce approximately 3 to 4 percent reduction in fuel consumption compared to trucks without lightweighting features, depending on duty cycle. This impact is roughly the same as a 5 to 10 percent reduction in C_d or a 10 to 20 percent reduction in C_{rr}. Due to numerous systems that add weight, the net effect on fuel consumption will likely appear less. All recent advanced truck technology programs have included weight reduction as part of their technical strategy. The impact of weight reduction in smaller truck classes is higher than Class 8. Vocational vehicle weight reductions were documented in the Phase II rule as shown below in Table 5-9.

TABLE 5-7 Summary of Recent Lightweighting Studies Including Approach and Estimated Resulting Weight Reduction

Weight Reduction Approach	Estimated Weight Reduction	Comments	Source
Cummins Peterbilt SuperTruck, multiple approaches: Aluminum fifth wheel and driveshaft, lighter axles and wheels, silicon carbide infused aluminum brake drums, magnesium cross members, lightweight air suspension, no lead-acid batteries	Net 1,305 lbs, total 5,500 lbs	Weight increases associated with waster-heat reduction on engine, aerodynamic aids, etc.	NAS review of 21CTP
Daimler SuperTruck, multiple methods: Aluminum frame and cross members, lightweight air suspension	Net 2,800 lbs	Initial weight reduction on prototype vehicle was 1,550 lbs	NAS review of 21CTP
Volvo SuperTruck, selected examples	Downsized engine, 11L instead of 13L, 400 lbs Aluminum frame, 800 lbs		NAS review of 21CTP
Navistar SuperTruck, multiple, see next column	Total tractor weight reduction: 3,250 lbs Trailer weight reduction: 3,700 lbs -Composite cab (240 lbs), -Metal matrix composite brake brakes (tractor and trailer, 600 lbs), -Light driveshaft and axles (190 lbs), -Aluminum wheel hubs and bearings (150 lbs), -Aluminum engine and transmission mounts (50 lbs), -Aluminum intensive rear suspension (370 lbs), -Thin wall, advanced material, and downsized fuel tanks (500 lbs), -Aluminum wide-based single wheels (350 lbs), and -Michelin wide-based single tires (350 lbs).		NAS review of 21CTP R. Zukouski presentation November 2014, Washington, DC
Walmart report on Freightliner 125BBC[a] and 113BBC Engine downsize, wide single tires/wheels, details next column.	Total weight reduction 2010 (125BBC) to 2013 (113BBC)=1,396 lbs Includes engine downsize, DD15 vs DD13, dual vs wide base, CastLite drums vs steellite, steel rear axle carrier vs aluminum, and lighter frame rails, additional 485 lbs reduction identified in other components		E. Fretheim presentation to NAS committee, April 28, 2014

continued

TABLE 5-7 Continued

Weight Reduction Approach	Estimated Weight Reduction	Comments	Source
Walmart WAVE Truck Trailer, carbon fiber composite materials and lightweight assembly methods	Weight reduction of 4,000 lbs on 53 ft trailer (10,000 lbs instead of 14,000 lbs)		http://corporate.walmart.com/_news_/news-archive/2014/03/26/walmart-debuts-futuristic-truck (accessed December 19, 2019) E. Fretheim, presentation to NAS committee, April 28, 2014. http://www.latimes.com/business/la-fi-walmart-truck-20150920-story.html (accessed December 19, 2019)
Multiple methods for tractor and trailer	Mild lightweighting: 2,000 lbs Aggressive lightweighting: 4,000 lbs	See reference for detailed component weight savings	North American Council on Freight Efficiency Confidence Report on Lightweighting, August 25, 2015. http://www.nacfe.org/downloads/confidence-report-lightweighting/ (accessed December 19, 2019)

[a]Bumper to back of cab.

5.4.4 Cost Effectiveness of Weight Reduction

Although there is a tendency to think of materials substitution as the primary weight reduction path, there are a number of operational, or "right-sizing," approaches that are sometimes lower cost than the baseline vehicle:

- Smaller truck cab,
- Downsized engine,
- Reduced fuel tank capacity,
- 6×2 tandem axle, and
- Innovative payload packaging.

Studies of cost effectiveness of weight reduction have often found lightweighting to be one of the less attractive options, perhaps because the materials substitution options are the most considered. Meszler and colleagues; 2015 review of technology cost effectiveness reviewed the prior studies and produced a range of costs for lightweighting shown in Figure 5-25.

In the NACFE study of weight reduction, they found that, per pound of weight reduction, "On average, bulk carriers will pay $6.00–$11.00, reefer or certain dry van dedicated routes $2.00–$5.00, and general dry van freight $0–$2.00." Interestingly in the EPA/NHTSA RIA, weight reduction is generally dismissed for future standard setting.

TABLE 5-8 Weight Reduction Targets from DOE Workshop 2011

	2020	2025	2030	2040	2050
Class 8 tractor Component group					
Wheels and tires	10%	20%	20%	25%	25%
Chassis/frame	3%	10%	10%	20%	20%
Drivetrain & suspension	3%	5%	10%	15%	20%
Misc. accessories/systems	5%	15%	25%	30%	35%
Truck body structure	15%	35%	45%	55%	60%
Powertrain	5%	10%	15%	15%	20%
Total Class 8 HDV	7%	16%	22%	27%	31%
Trailer (53 ft) Component group					
Wheels and tires	10%	20%	20%	25%	25%
Chassis/frame	3%	10%	10%	20%	20%
Suspension	3%	5%	10%	15%	20%
Box/other	5%	10%	15%	20%	25%
Total trailer	3%	9%	13%	19%	23%
Truck and trailer combined totals	5.0%	13.2%	18.0%	23.6%	27.4%

SOURCE: DOE (2013).

The agencies propose setting the 2021 through 2027 model year tractor standards without using weight reduction as a technology to demonstrate the feasibility. The agencies view weight reduction as a technology with a high cost that offers a small benefit in the tractor sector. For example, our estimate of a 400-pound weight reduction would cost $2,050 (2012$) in MY2021, but offer a 0.3 percent reduction in fuel consumption and CO_2 emissions (EPA and NHTSA, 2016a, p. 2-88).

Finding: The cost of weight reduction is highly variable and dependent on the technical approach, and can range from a negative cost to over $10 per pound.

Recommendation 5-11: Future MHDV fuel consumption standards should include expectations for weight reduction adoption with likely potential of 3 to 4 percent reduction in fuel consumption at market-acceptable ROI.

5.5 AXLE AND DRIVETRAIN LOSSES

As with transmissions, there continue to be minor improvements in the efficiency, lubrication, and weight of axles. The primary developments are in the areas of faster gearing to match the lower engine speeds, increased torque capability as a result of the lower engine speeds and flatter torque curves, lubrication improvements to minimize losses, and innovative ways to eliminate gearing when not needed. NACFE (2013a) completed a study of 6×2 axle configurations in 2013. Table 5-10 summarizes the fuel savings possible, with some caveats. For instance, the report indicates the savings achieved by Con-way (NACFE, 2013a, p. 27) is an estimate that removes the effect of different configurations of transmissions and tires used in the test vehicles. It appears reasonable to project a fuel savings of 2 percent for this axle configuration.

Driveline weight continues to be improved slightly with reports of savings of 30 pounds in some newer designs. Carbon fiber prototypes of drivelines have been created. Aluminum, tubular drivelines have also been introduced with weight savings of 40 percent compared to typical two-piece steel driveshafts.

TABLE 5-9 Weight Reduction Technologies Evaluated for the Phase II Rule

		Vocational Vehicle Class		
Component	Material	Class 2b-5	Class 6-7	Class 8
Axle Hubs – Non-Drive	Aluminum	40	40	
Axle Hubs – Non-Drive	High Strength Steel	5	5	
Axle – Non-Drive	Aluminum	60	60	
Axle – Non-Drive	High Strength Steel	15	15	
Brake Drums – Non-Drive	Aluminum	60	60	
Brake Drums – Non-Drive	High Strength Steel	42	42	
Axle Hubs – Drive	Aluminum	40	80	
Axle Hubs – Drive	High Strength Steel	10	20	
Brake Drums – Drive	Aluminum	70	140	
Brake Drums – Drive	High Strength Steel	37	74	
Suspension Brackets, Hangers	Aluminum	67	100	
Suspension Brackets, Hangers	High Strength Steel	20	30	
Crossmember – Cab	Aluminum	15	15	15
Crossmember – Cab	High Strength Steel	5	5	5
Crossmember – Non-Suspension	Aluminum	15	15	15
Crossmember – Non-Suspension	High Strength Steel	5	5	5
Crossmember – Suspension	Aluminum	15	25	25
Crossmember – Suspension	High Strength Steel	6	6	6
Driveshaft	Aluminum	12	40	50
Driveshaft	High Strength Steel	5	10	12
Frame Rails	Aluminum	120	300	440
Frame Rails	High Strength Steel	40	40	87
Wheels – Dual	Aluminum	150	150	250
Wheels – Dual	High Strength Steel	48	48	80
Wheels – Wide Base Single	Aluminum	294	294	588
Wheels – Wide Base Single	High Strength Steel	168	168	336
Permanent 6x2 Axle Configuration	Multi	N/A	N/A	300

SOURCE: EPA and NHTSA (2016).

Driveline angles are an important factor in the overall efficiency of the driveline. Vehicle OEMs routinely install the engine at a slight angle in order to minimize the losses in the universal joints of the driveline that take up the difference in angle from the engine flywheel to the input to the drive axle.

5.6 AUXILIARY LOADS

Both auxiliary and accessory loads are consumers of fuel energy delivered by engine power, and through various paths. We consider here only diesel engines, the favored selection for Class 6 through 8 trucks.

In the simplest description, accessory loads are those specific to engine subsystems performing certain functions required to produce power at the engine's crankshaft/flywheel. Consequently, they are not debited to "engine brake horsepower." Needed are a fuel pump supplying fuel to the cylinder, an oil pump supplying lubricating

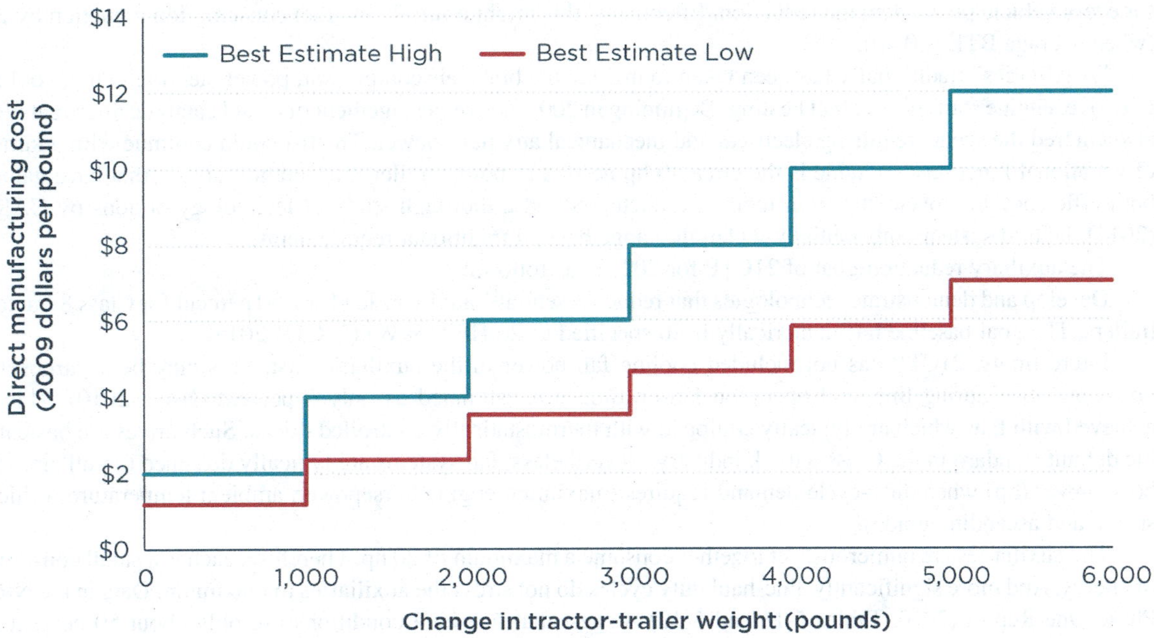

FIGURE 5-25 Cost curve for tractor-trailer weight reduction.
SOURCE: Meszler et al. (2015).

TABLE 5-10 Cost and Weight Reduction of Various Axle and Drivetrain Changes

Item	Cost	Weight Impact
125BBC vs 113BBC	0	280 lbs
DD15 vs DD13	-$1,610	-390 lbs
Duals vs wide base	-$349	-376 lbs
CastLite drums vs steellite	+$136	-40 lbs
Steel rear axle carrier vs aluminum	+$255	-130 lbs
11/32 frame vs 9/32	-$106	-180 lbs
Total savings	-$1,674	-1,396 lbs

SOURCE: NACFE (2013a).

oil to various sliding engine surfaces, and a water pump to help coolant regulate engine temperature, and more recently exhaust-gas recirculation temperature. Notice that accessory power reduction directly increases engine brake thermal efficiency (BTE), as does internal engine friction reduction.

An alternator might be considered both an accessory and auxiliary. As an accessory, electricity is provided to the engine electronics system managing fuel and other engine sensor/actuator control and perhaps an electric fuel pump. The alternator, as auxiliary, provides the vehicle electricity for lighting, heating/ventilating, powertrain electronics, battery charge, and myriad other functions.

Auxiliaries satisfy vehicle functions, so are included as "vehicle power demands." They regularly evolve, often to satisfy occupant comfort-hotel loads (heating, ventilation, and air conditioning [HVAC], computers, entertainment systems, and on-board appliances like refrigerators, microwaves, coffee pots), but increasingly communications. Because auxiliaries consume energy from the engine's "brake horsepower," when auxiliary load is reduced, the fuel consumption is reduced by "auxiliary consumption ÷ duty-cycle-average BTE." This is true for every vehicle power demand reduction. Effectively, this might multiply the fuel consumption reduction by 2.5 (when average BTE is 0.40).

”Auxiliaries” traditionally has been taken to include the brake air compressor, power-steering pump, cooling fan drive, engine starting, and fuel heating. Beginning in 2007, active management of diesel catalytic aftertreatment also entered the scene, requiring electrical and mechanical auxiliary power. The list could continue with addition of vocational functions—notable is the circa 40 hp needed to power trailer refrigeration units—that have gained both efficiency improvements from their manufacturers and a thorough study of technology options by CARB (2014). Hybrid systems substantially widen the range beyond traditional requirements.

The auxiliary reduction goal of 21CTP for 2021 is as follows:

Develop and demonstrate technologies that reduce essential[26] auxiliary loads by 50 percent for Class 8 tractor-trailers. The goal baseline has numerically been specified as 20 HP (15kW) (21CTP, 2013).

Interestingly, 21CTP has not included cooling fan power in the auxiliaries list. This may be a purposeful omission since during line-haul operation fan-on time was estimated as only 5 percent (NRC, 2010). This is achieved with fans which are typically equipped with thermostatically controlled drives. Such drives are basically the default standard in the Class 8 truck industry. Nevertheless, fan systems are typically designed to pull circa 50 horsepower (hp) when duty-cycle demand requires (maximum engine horsepower, ambient temperature, vehicle speed, and ascending grade).

The auxiliaries are numerous, yet together consume a maximum of 20 hp. Therefore, each is a small consumer of energy. And more significantly, line-haul duty cycles do not stress the auxiliaries to maximum. Data in the NRC Phase One Report (2010, Tables 5-11 and 5-12) suggest that line-haul conditions use only about 50 percent of the maximum auxiliary power. Reducing small auxiliary loads still requires significant design, development, and quality assurance effort. That effort has not been absent in engine manufacturing companies because engine fuel consumption has always been a competitive factor to attract truck customers. And prior to SCR, more stringent NO_x regulation was routinely accompanied by a substantial challenge to maintaining good fuel consumption. But now priority for that effort has been increasing as emphasis moves from regulatory attention to GHG reductions.

5.6.1 More Electric Truck

DOE (2006) and industry collaborated in an effort to reduce certain auxiliary and accessory loads in about 2000. The More Electric Truck[27] uses electrically driven auxiliaries powered by a 30 kilowatt (kW) generator located inside the flywheel housing, which also serves as the starter motor. The auxiliaries converted to electrical power on the research demonstration vehicle were the HVAC, brake air compressor, and accessories: oil and water pumps. This system was prototyped and demonstrated on a Class 8 truck. Evidently, the prototype was never moved in entirety into a production application. A description of the project status following completion of the tractor-trailer demonstration can be found in DOE (2006).

5.6.2 Cummins Medium- and Heavy-Duty Accessory Hybridization Cooperative Research and Development Agreement

This joint DOE project, also known as VSS133, was initiated in 2013 (Deter, 2014). A cooperative between Cummins and ORNL, the project seeks solutions to these two issues:

- Accessories (and auxiliaries) have optimal running conditions based on load requirement. The ability to separate accessories from the engine speed allows the opportunity to run at ideal conditions without engine speed dependencies.

---

[26] The Partnership has not clearly defined “essential.”

[27] The More Electric Truck program was a collaboration between industry partners and the U.S. Department of Energy’s Office of Vehicle Technology’s FreedomCAR program. See http://www1.eere.energy.gov/vehiclesandfuels/pdfs/success/more_electric_truck.pdf (accessed December 19, 2019).

TABLE 5-11 Auxiliaries and Accessories Addressed by Energy-Efficient Truck Initiatives of Cummins, Daimler, Volvo, and Navistar

Auxiliary or Accessory	Description of Component or System			
	Cummins	Daimler	Volvo	Navistar
Auxiliary	Cab & electric sleeper HVAC	Variable-speed fan	Unidentified	3-speed fan clutch
		Clutched compressor		48 V AC compressor
		Power steering pump		Alternator/generator
Accessory	Lube pump	Variable water pump	Variable coolant pump	Variable water pump
	Fuel pump		Variable oil pump	

SOURCE: June 2015 engine DOE Annual Merit Reviews from each project.

- HD line-haul trucks often idle for long periods in traffic and overnight for hotel loads. This typically requires the engine to idle for little reason other than driver comfort and system readiness, which uses excessive fuel and causes more emissions.

A thorough description of this project's status can be found in the third review of the 21CTP program by NRC (2015).

5.6.3 SuperTruck Projects' Development Work on Auxiliaries and Accessories

The four projects identified auxiliary systems or components where development could reduce consumption. These are identified in Table 5-11. As typical for equally funded, joint agency-industry R&D projects, little detail is required to be reported for smaller projects within the complete project. Yet it is valuable to the intended project objectives to have had close engineering scrutiny of these components.

Finding: There has been a modicum of agency funding to improving the vehicle power demand of auxiliaries. Yet, even where DOE-sponsored/shared projects have been pursued, reduction successes have not been well quantified.

Recommendation 5-12: When there are auxiliary power reductions achieved from agency-funded projects, NHTSA and DOE should appropriately document and publicly distribute these accomplishments.

5.7 ANNEX: DESCRIPTION OF AVAILABLE SMARTWAY-CERTIFIED TRAILER PACKAGES

There are 10 aerodynamic trailer packages, as of October 2015, produced by five companies that are Elite verified. Only one trailer manufacturer produces such a combination. All other devices and packages are produced by trailer component manufacturers. SmartWay has introduced a "basket" of three test procedures providing manufacturers choice for aerodynamics verification beginning in February 2015.[28] The protocols are a full truck track test, a wind-tunnel test, and a coast-down test. All of these protocols have been built on relatively long-standing SAE procedures.

[28] Dennis Johnson and Sam Walter, U.S. Environmental Protection Agency, June 25, 2015, "SmartWay Technology Programs, USEPA Technology Assessment Center," personal communication with committee member Charles Salter.

- **Track Test.** An EPA-approved full truck track test remains the default procedure for SmartWay approvals and general industry verifications (EPA, 2016b). EPA's "2014 SmartWay Track Test" test procedure provides data on the incremental fuel consumption of the test vehicle. The 2014 SmartWay track test includes a number of updates. Prior to 2014, SmartWay verification track testing was based on industry methods commonly referred to as SAE J1321 procedures adopted in 1986. For the new 2014 SmartWay track test, EPA updated the protocol to incorporate the more recent (2012) version of the SAE J1321 procedure, overlaid with new SmartWay testing enhancements, such as tractor-trailer set-up specifications. These new enhancements will improve truck comparisons by limiting some sources of variability. One main addition is to require more than one complete track test in order to achieve SmartWay verification. This is an industry-recommended practice and will give fleets additional confidence that the aerodynamic technologies verified with this procedure will perform comparably in similar conditions on the road.
- **Wind Tunnel.** Wind-tunnel testing provides data on the incremental coefficient of drag (C_d) and incremental wind-averaged drag coefficient (C_{dwa}).[29] Interestingly, the 10 Elite packages have all been verified by EPA-accepted wind-tunnel testing. EPA requires the use of the C_{dwa} results, as they contain a better real-world drag estimate by providing effects of crosswinds similar to real truck environments. Most wind-tunnel facilities have correlation factors to forecast fuel consumption from the C_d results. EPA provides guidelines for wind-tunnel selection and operation (EPA, 2015b).
- **Coast-down.** EPA also approves a full truck coast-down procedure for trailer aerodynamics performance verification, having collaborated in test procedure updates with the SAE Task Force on the SAE J1263 (with J2263) procedures. Coast-down, too, provides C_d figures after replicate tests and specified data analysis, including extraction of the aerodynamic (C_d) load from the total of the unpowered road loads imposed by the procedure. Incremental C_d for the experimental component (trailer, for example) is acquired by comparing the results to a baseline vehicle. Furthermore, these data are required for the truck performance verification analysis in the GEM protocol. EPA and NHTSA have enhanced this procedure in their proposed Phase II GHG and fuel consumption rules. It now must include a high-low iteration method, described in the proposal. The process determines mechanical losses in a low-speed portion of the coast-down, then subtracts those from a high-speed portion which will finally result in determining the C_dA (product of the C_d and frontal area A). The C_dA becomes a GEM input parameter. EPA has shown the C_dA standard error to be in a range of about 0.5 to 2.1 percent, in small test samples (EPA and NHTSA, 2015).

Finally, SmartWay encourages component design variable analysis using CFD. However, this method alone cannot be used for the component/system validation.

5.8 ANNEX: SUMMARY OF SUPERTRUCK PROJECTS

The cooperative R&D agreements established four project teams variously identifying similar aerodynamic component modifications: to the tractor (bumper, radiator louvers, chassis extenders, hood, roof, and wheel covers) and to the trailer (skirts, tail, bogie cover, wheel covers, and front gap fairing) (see Figure 5-26). Yet, most have shape uniqueness as illustrated in Figures 5-27 and 5-28. Apparent in both figures are large trailer tails and very close road clearance skirts and tractor fairings. It is not clear if complete results dossiers will be publicly available.

5.8.1 Cummins-Peterbilt Project

The baseline tractor-trailer at the target 65,000 pounds GVW, performed in freight efficiency terms at 101 ton-miles per gallon, corresponding with the more useful LSFC of 9.9 gallons per 1,000 ton-miles, and the more familiar fuel economy figure of 6.45 mpg, over the long-haul duty cycle. This project has reported final performance results of a 76 percent increase in freight efficiency (50 percent goal), arriving at 178 ton-miles per gallon

[29] Wind-averaged C_d: a weighted calculation of drag based on a range of yaw angle (typically ± 12 degrees, and wind speed). See https://www.sae.org/publications/technical-papers/content/870001/ (accessed December 19, 2019).

FIGURE 5-26 Cummins-Peterbilt SuperTruck tractor-trailer configuration-aero enhancements.
SOURCE: Peterbilt Motors Inc. (2014).

FIGURE 5-27 Cummins-Peterbilt final SuperTruck in show paint.
SOURCE: Koeberlein (2015).

on their selected long-haul drive cycle. The corresponding LSFC was improved by 43 percent to 5.64 gallons per 1,000 ton-miles, and a fuel economy figure of 10.7 mpg,[30] over the long-haul duty cycle.

Aerodynamics-specific performance data indicated a contribution of 28.7 percent to the freight efficiency (ton-miles per gallon) increment, and corresponding 1.6 mpg increase on the long-haul drive cycle (Koeberlein, 2015). The C_d reduction for the system was reported as 46 percent, which would put this vehicle's C_d at about 0.32-0.35.

Additional final combination SuperTruck results over the long-haul cycle reported the full vehicle load-specific fuel consumption (gallons per 1,000 ton-miles) at 5.72, but increasing to 6.52 when the standard (baseline) trailer with skirt was substituted (Peterbilt, 2014). This 14 percent fuel consumption increase with a legacy-aero trailer is significant. The Cummins-Peterbilt project self-imposed the ability for their SuperTruck tractor to achieve compatibility with standard, legacy trailers. Figures 5-26 and 5-27 display the Cummins-Peterbilt final Supertruck and its various aerodynamic improvements.

Two design accommodations to manage real-world issues encountered by drivers was to make the trailer skirt pneumatically retractable for access to loading docks and for crossing crowned areas such as railroad tracks. An access panel covering the trailer tandem was modified so that it could open 180 degrees for tire inspection.

5.8.2 Daimler Project

The "signature" freight efficiency increase achieved over the chosen Texas long-haul cycle route was 119.5 percent (to 206 ton-miles per gallon), with a corresponding 54.5 percent reduction in LSFC (to 4.85 gallons per 1,000 ton-miles). These figures plus the corresponding remarkable fuel economy was 12.2 mpg, all tested at 65,000 pounds GVW in their Texas long-haul cycle.[31]

Daimler included a number of aerodynamic features in their final SuperTruck tractor's cab such as widening it to match the sleeper width, and raking back both the grill and windshield (see Figure 5-28). The shape of the hood has been extensively modified to optimize aerodynamic efficiency with the smaller engine (10.7L versus 14.6L) and angled cooling system.

FIGURE 5-28 Daimler Final SuperTruck Tractor-Trailer Configuration.
SOURCE: Rotz (2015).

[30] The headline trip result of 10.7 mpg was achieved during a later test, where the average wind was 6 mph, with no gusts. See NRC (2015).
[31] Rotz (2015).

The wind-averaged C_d of the final demonstrator tractor and trailer is 54 percent lower than the baseline, as determined in a scale model wind tunnel (this would mean a C_d of about 0.3). Telling of the relative tractor versus trailer aerodynamic design maturities, at the start of the project, trailer improvements delivered 72 percent of the total vehicle aerodynamic improvement, but requiring only one-third of the aerodynamic engineering effort. Another unique feature included on the tractor was a louvered radiator grill that closes when cooling demand is not detected, thus not interrupting the air flow over the tractor nose and contributing 6 percent to the C_d reduction (Rotz and Zeigler, 2014).

5.8.3 Volvo Group Project

Volvo Trucks North America's SuperTruck project successfully demonstrated fuel efficiency of greater than 12 miles per gallon, 88 percent greater freight efficiency, and 70 percent better fuel efficiency than the baseline 2009 Volvo model VNL 670.[32] Lightweighting was a key approach for Volvo. They used a down-sized D11 model engine, which also helped with improving hood slope for aerodynamics (see Figure 5-29). Carbon fiber was used for the roof and side fairings (Park, 2016). Recycled carbon fiber was used in places to save cost. An aluminum frame was used to cut weight by nearly 50 percent (Park, 2016). The supertruck is about 3,200 pounds lighter (Park, 2016).

The Volvo SuperTruck program enabled several technologies that were put into production in its latest product offering. This includes turbocompounding to recover wasted energy, a wave piston design for improved fuel efficiency of the engine of 2.2 percent. The common-rail fuel injection system was optimized, providing an injector with enhanced flow rate, spray angle, and pressure characteristics, higher injection pressures, and finer control of the injection event. The D11 engine has a flatter torque curve from 900 to 1,300 RPM.

Another key to the Volvo truck was taking a complete systems approach to the combination vehicle that optimized the tractor, trailer, and the powertrain to work together. An example of the integration is the Volvo iSee 12-speed transmission, which learns the road's topography to enable better shifting and intelligent cruise control including coasting in neutral with the engine at idle speed. Cameras with less aerodynamic drag replaced the conventional mirrors. Solar panels are built into the roof to aid battery charging and HVAC operation. A 10 kW alternator is capable of using wasted kinetic energy during braking (Marsh, 2016). This effort will expedite commercial product availability.

FIGURE 5-29 Volvo SuperTruck aerodynamics view.
SOURCE: Amar and Gibble (2016).

[32] See https://www.volvotrucks.us/innovation/supertruck/ (accessed December 19, 2019).

5.8.4 Navistar Inc. Project

Navistar completed their SuperTruck project in 2016, achieving

- 104 percent overall improvement in freight efficiency compared to 2009 baseline,
- 50.3 percent brake thermal efficiency for the engine, and
- 13 miles per gallon.

According to Navistar,[33] the key features of their CataLIST truck are

- Advanced integration of the Navistar® N13 Engine utilizing proprietary intelligent controls and high-efficiency combustion;
- Reduction in aerodynamic drag through replacement of cab- and hood-mounted mirrors with a series of cameras and interior-mounted monitors, which also yield equal or better indirect vision for the driver;
- A new light-emitting diode (LED) headlamp system that reduces lamp size for a more aerodynamic shape and cuts electrical power requirements by greater than 80 percent, while improving luminous output and light color for improved night-time direct driver vision and reduced driver fatigue;
- An all-new shape with a sloped windshield and wedged cab for improved aerodynamics, and innovative use of lighter-weight carbon-fiber panels in the upper body, roof headers, back panel, and dash panel;
- A hybrid front suspension and lightwei.ght rear suspension that leverages lightweight alloys with composite materials, reducing weight and enabling an electronic ride height management system, which provides dynamic ride height and pitch control for improved aerodynamics;
- An electronic ride height management system, which provides dynamic ride height and pitch control for improved aerodynamics; and
- Aerodynamic improvements that reduce the trailer's drag coefficient by more than 30 percent.

Considerable wind-tunnel testing has produced both significant trailer and tractor results. The trailer in Figure 5-30 delivered up to 19 percent freight efficiency improvement. Features included a boat tail, segmented trailer skirts, bogie fairing with diffuser-extends trailer skirts, and wheel covers on trailer and tractor bogies. Tractor design features have delivered up to a 12 percent improvement in freight efficiency. The design features include treatments to the bumper, hood, cab fairing, and extended chassis fairings.

Navistar considered three methods to treat the gap between cab and trailer: bellows, inflatable, and ball/socket concepts. Navistar acknowledges that the highest-benefit concepts also carry the highest risk of failure (Zukouski, 2014). Navistar worked to improve aerodynamic performance with an electronically controlled ride height feature. At speeds above 50 mph, the front and rear suspension are dropped 1.5 to 2 inches to create a slightly more teardrop, aerodynamic profile. Figure 5-30 shows the concept as reported by Navistar investigators for the DOE SuperTruck project (Zukouski, 2015).

[33] See https://news.navistar.com/2016-09-28-Navistars-SuperTruck-Exceeds-Department-Of-Energys-Freight-Efficiency-Goal (accessed December 19, 2019).

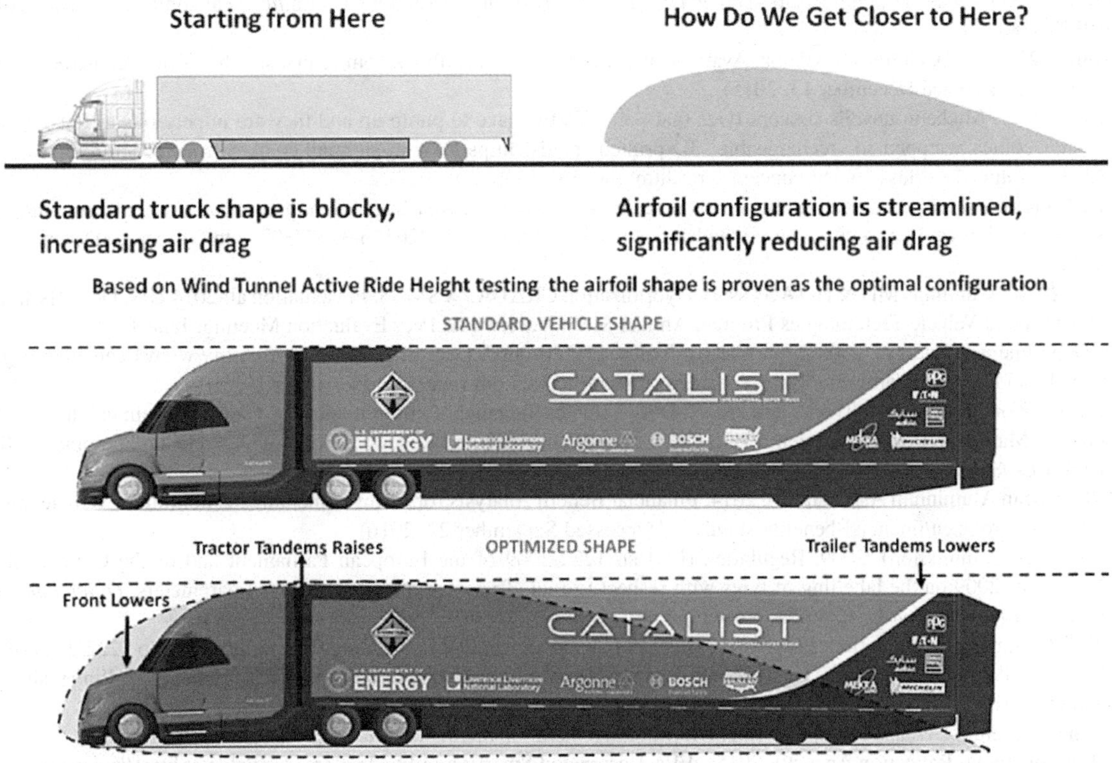

FIGURE 5-30 Navistar aerodynamics concepts showing modifications to approach airfoil shape.
SOURCE: Zukouski (2015).

5.9 REFERENCES

21CTP (21st Century Truck Partnership). 2013. 21st Century Truck Partnership, Roadmap and Technical White Papers. Department of Energy. Available at http://energy.gov/sites/prod/files/2014/02/f8/21ctp_roadmap_white_papers_2013.pdf (accessed December 19, 2019).

Amar, P. (Volvo). 2015. Volvo Technology of America SuperTruck Program, DOE Project ID VSS081. June.

Amar, P., and J. Gibble. 2016. SuperTruck Powertrain Technologies for Efficiency Improvement, DOE Contract DE-EE0004232. June.

ANL (Argonne National Laboratory). 2009. Evaluation of Fuel Consumption Potential of Medium and Heavy Duty Vehicles through Modeling and Simulation. Lemont, IL: ANL.

Benedict, R. (The Goodyear Tire & Rubber Company). 2014. A System for Automatically Maintaining Pressure in a Commercial Truck Tire: Project ID# VSS085. Presentation to the Annual Merit Review, Washington, DC, June 19.

Bhatnagar, B., M. Dinesh, and C. Mohan. 2011. A Study of Reynolds Number Effects on Production Class 8 Tractor-Trailer Combinations. Chicago: Daimler. June 28.

Bohme, F. No date. Self-healing Materials. http://www.ipfdd.de/en/departments/institute-of-macromolecular-chemistry/polymer-structures/self-healing-materials/ (accessed December 19, 2019).

Britt, M.G. 2015. SAE Government Industry Meeting 2015. Oral Presentation, Fleet Perspective on Fuel-Saving Tires. Presented at the SAE Government Industry Meeting 2015. American Trucking Associations Technical Advisory Group.

Burger, L. 2014. Tire Makers Race to Turn Dandelions into Rubber. Available at http://www.reuters.com/article/2014/08/20/us-dandelion-rubber-idUSKBN0GK0LN20140820 (accessed September 6, 2015).

CARB (California Air Resources Board). 2014. Transport Refrigerators Technology Assessment. Presentation in Sacramento, CA, September 3.

Caterpillar. 2006. Understanding Tractor-Trailer Performance. Brochure. Available at http://ohe.cat.com/cda/files/867819/7/LEGT6380.pdf.

Caterpillar. 2015. CDX Online eTextbook. Availabe at http://www.cdxetextbook.com/steersusp/wheelsTires/construct/tireratings.html (accessed December 13, 2015).

Collins, T. 2017. Michelin unveils concept tyres that you NEVER have to pump up and they are impervious to flats: Airless design comes wrapped in 'rechargeable' 3D-printed treads. https://www.dailymail.co.uk/sciencetech/article-4610728/Michelin-unveils-airless-Vision-concept-tyres.html (accessed July 30, 2017).

Courier & Press. 2015. Special coating applied to curve on Crittenden Co.'s Rosebud Hill. http://archive.courierpress.com/news/special-coating-applied-to-curve-on-crittenden-cos-rosebud-hill-ep-1253426326-324773071.html (accessed December 19, 2019).

Deter, D. 2014. "Cummins MD & HD Accessory Hybridization CRADA." VSS133. Presentation at: 2014 U.S. DOE Hydrogen Program and Vehicle Technologies Program Annual Merit Review and Peer Evaluation Meeting. June 16-20.

DOE (Department of Energy). 2006. More Electric Truck, 21CTP-0003. December. Available at http://www1.eere.energy.gov/vehiclesandfuels/pdfs/program/21ctp_roadmap_appendix_2007.pdf (accessed December 19, 2019).

DOE. 2013. Workshop Report: Trucks and Heavy-Duty Vehicles Technical Requirements and Gaps for Lightweight and Propulsion Materials. U.S. Department of Energy, Energy Efficiency and Renewable Energy, Vehicle Technologies Office. DOE/EE-0867.

EAA (European Aluminium Association). 2014. Financial Benefit Analysis for an Investment in an Aluminium Vehicle. http://www.alueurope.eu/financial-benefits-simulator/ (accessed September 22, 2016).

EC (European Commission). 2009. Regulation (EC) no 1222/2009 of the European Parliament and of the Council of 25 November 2009 on the labelling of tyres with respect to fuel efficiency and other essential parameters. *Official Journal of the European Union* L342:46-58. November 25.

EC. 2011. Commission Regulation (EU) No 1235/2011 of 29 November 2011 amending Regulation (EC) No 1222/2009 of the European Parliament and of the Council with regard to the wet grip grading of tyres, the measurement of rolling resistance and the verification procedure. *Official Journal of the European Union* L317:17-23. November 29.

EDF (Environmental Defense Fund). No date. *GHG Management for Medium Duty Truck Fleets.* New York: EDF.

EPA (Environmental Protection Agency). 2015a. EPA Designated SmartWay Elite Trailers. Available at http://www3.epa.gov/smartway/forpartners/documents/aerodynamic/420f15008.pdf (accessed October 17, 2015).

EPA. 2015b. EPA Aerodynamic Wind Tunnel Testing for SmartWay Verification. Available at http://www3.epa.gov/smartway/forpartners/documents/aerodynamic/420f15012.pdf (accessed October 19, 2015).

EPA. 2016a. SmartWay Verified Low Rolling Resistance Tires Performance Requirements. EPA 420-F-12-024. Available at https://www.epa.gov/sites/production/files/2016-02/documents/420f12024.pdf (accessed December 19, 2019).

EPA. 2016b. Track Testing for SmartWay Verification. Available at https://www.epa.gov/sites/production/files/2016-02/documents/420f15011.pdf (accessed December 19, 2019).

EPA. 2019. Maintaining Proper Tire Inflation Pressure: A Glance at Clean Freight Strategies. Available at https://www.epa.gov/sites/production/files/2019-09/documents/420f19028.pdf (accessed December 19, 2019).

EPA and DOT. 2015. Greenhouse Gas Emissions and Fuel Efficiency Standards for Medium- and Heavy-Duty Engines and Vehicles—Phase 2: Draft Regulatory Impact Analysis. EPA-420-D-15-900. Washington, DC: EPA. June.

EPA and DOT. 2016. Greenhouse Gas Emissions and Fuel Efficiency Standards for Medium- and Heavy-Duty Engines and Vehicles—Phase 2: Regulatory Impact Analysis. EPA-420-R-16-900. Washington, DC: EPA and DOT. August.

EPA and NHTSA (National Highway Traffic Safety Administration). 2011. Greenhouse Gas Emissions Standards and Fuel Efficiency Standards for Medium- and Heavy-Duty Engines and Vehicles; Final Rule. September 15, 2011. *Federal Register* 76:57106-57513.

EPA and NHTSA. 2015. Greenhouse gas emissions and fuel efficiency standards for medium- and heavy-duty engines and vehicles—Phase 2. *Federal Register* 80:40137-40765. July 13.

EPA and NHTSA. 2016a. Regulatory Impact Analysis, Final Rule for Greenhouse Gas Emissions and Fuel Efficiency Standards for Medium- and Heavy-Duty Engines and Vehicles, Phase II. Washington, DC: NHTSA.

EPA and NHTSA. 2016b. Greenhouse gas emissions and fuel efficiency standards for medium- and heavy-duty engines and vehicles—Phase 2. *Federal Register* 81:73478-74274. October 25.

Erb, A. 2015. The Place of Low Rolling Resistance Tires in EPA's Greenhouse Gas Reduction Strategy. U.S. Environmental Protection Agency. 2015 SAE Government/Industry Meeting, Washington, D.C., January 23.

ETRMA (European Tyre and Rubber Manufacturers' Association). 2011. Tyre Labelling. http://www.etrma.org/tyres/tyre-labelling (accessed July 30, 2017).

Gao, Z., D.E. Smith, C.S. Daw, K.D. Edwards, B.C. Kaul, N. Domingo, and J.E. Parks II. 2015. The evaluation of developing vehicle technologies on the fuel economy of long-haul trucks. *Energy Conversion Management* 106:766-781.

Greenberg, M. (U.S. Environmental Protection Agency). 2014. A Look at Tractor Trailer Aerodynamics and Fuel Efficiency. Presentation to the ACEEE/ICCT Workshop on Emerging Technologies for Heavy-Duty Vehicle Fuel Efficiency, Washington, DC, July 22.

Gururaja, P. 2016. Wind-Averaged Drag Determination for Heavy-Duty Vehicles Using On-Road Constant-Speed Torque Tests. SAE Technical Paper 2016-01-8153. doi: 10.4271/2016-01-8153.

Hammache, M., M. Michaelian, and F. Browand. 2002. Aerodynamic Forces on Truck Models, Including Two Trucks in Tandem: California PATH Research Report. UCB-ITS-PRR-2001-27. Berkeley, CA: University of California, Berkeley.

ICCT. 2015. Parsing Phase 2 Tractor-Trailers Proposed Regulation. Washington, DC: ICCT. July. Available at http://www.theicct.org/blogs/staff/parsing-phase-2-tractor-trailers-proposed-regulation (accessed December 19, 2019).

Kayes, D. (Daimler Trucks North America). 2015. Executive Panel: Greenhouse Gas Legislation. Presentation at the SAE 2015 Commercial Vehicle Engineering Congress in Rosemont, IL, October 7.

Koeberlein, D., and Cummins Inc. 2014. Cummins SuperTruck Program: Technology and System Level Demonstration of Highly Efficient and Clean, Diesel Powered Class 8 Trucks. Presentation at the 2014 DOE Hydrogen and Fuel Cells Program and Vehicle Technologies Office Annual Merit Review and Peer Evaluation Meeting, Washington, DC, June 20.

Koeberlein, D. (Cummins). 2015. Cummins Engine Co. SuperTruck Program. DOE Project ID ACE057. June.

Lew, S. 2013. Test Methods for Truck Tire Rolling Resistance and Reducing Fuel Consumption of M-D and H-D Vehicles. Michelin North America, Inc. Presentation to NRC Committee on Assessment of Technologies and Approaches for Reducing the Fuel Consumption of Medium- and Heavy-Duty Vehicles, Phase 2, Irvine, California, November 21.

Marsh, A. 2016. For Volvo, SuperTruck Project was 'Knowledge Accelerator.' https://www.fleetowner.com/technology/volvo-supertruck-project-was-knowledge-accelerator (accessed July 30, 2017).

McAuliffe, B., and D. Chuang. 2016. Track-based aerodynamic testing of a heavy-duty vehicle: Coast-down measurements. *SAE International Journal of Commercial Vehicles* 9(2):381-396. doi: 10.4271/2016-01-8152.

Meszler, D., N. Lutsey, and O. Delgado. 2015. Cost Effectiveness of Advanced Efficiency Technologies for Long-Haul Tractor-Trailers in the 2020-2030 Time Frame: White Paper. Washington, DC: ICCT. April.

Mihelic, R. (Mihelic Vehicle Consulting LLC). 2015. Heavy Truck Aerodynamics Beyond 2025. Presentation at the SAE 2015 Commercial Vehicle Engineering Congress, Rosemont, IL, October 6.

Mihelic, R., J. Smith, and M. Ellis. 2015. Aerodynamic comparison of tractor-trailer platooning and A-train configuration. *SAE International Journal of Commercial Vehicles* 8(2):740-746. doi: 10.4271/2015-01-2897.

NACFE (North American Council for Freight Efficiency). 2013a. Confidence Findings on the Potential of 6x2 Axles. Available at https://nacfe.org/report-library/confidence-reports/ (accessed March 8, 2016).

NACFE. 2013b. Confidence Report: Tire Pressure Systems. Available at http://truckingefficiency.org/tires/tire-pressure-inflation-trailer (accessed September 6, 2015).

NACFE. 2015a. Confidence Report on Lightweighting, Available at http://www.truckingefficiency.org/tractor-aerodynamics/weight-reduction-tractors (accessed September 6, 2015).

NACFE. 2015b. Confidence Report: Low Rolling Resistance Tires. Available at http://truckingefficiency.org/tires/low-rolling-resistance-duals (accessed September 6, 2015).

NACFE. 2016a. Confidence Report: Tractor Aerodynamics. Washington, DC: NACFE. May 4.

NACFE. 2016b. Confidence Report: Two-Truck Platooning. Washington, DC: NACFE.

NACFE. 2016c. Confidence Report: Trailer Aerodynamic Devices. Washington, DC: NACFE. February.

NHTSA (National Highway Traffic Safety Administration). 2009. NHTSA Tire Fuel Efficiency Consumer Information Program Development: Phase 2—Effects of Tire Rolling Resistance Levels on Traction, Treadwear, and Vehicle Fuel Economy. DOT HS 811 154. Washington, DC: DOT. August.

NHTSA. 2015. Effects of Tire Rolling Resistance on Class 8 Tractor-Trailer Stopping Distance. DOT HS 812 176. Washington, DC: NHTSA. June.

NRC (National Research Council). 2010. *Technologies and Approaches to Reducing the Fuel Consumption of Medium- and Heavy-Duty Vehicles.* Washington, DC: The National Academies Press.

NRC. 2012. *Review of the 21st Century Truck Partnership, Second Report.* Washington, DC: The National Academies Press.

NRC. 2014. *Reducing the Fuel Consumption and Greenhouse Gas Emissions of Medium- and Heavy-Duty Vehicles, Phase Two First Report.* Washington, DC: The National Academies Press.

NRC. 2015. *Review of the 21st Century Truck Partnership, Third Report.* Washington, DC: The National Academies Press.

ORNL (Oak Ridge National Laboratory). 2011. Effect of Weight and Roadway Grade on the Fuel Economy of Class-8 Freight Trucks. ORNL/TM-2011/471. October.

ORNL (Oak Ridge National Laboratory). 2017. 2016 Vehicle Technologies Market Report. Available at https://tedb.ornl.gov/wp-content/uploads/2019/04/2016_Vehicle_Technologies_Market_Report.pdf.

Park, J. 2016. Volvo's SuperTruck Demonstrates the Art of the Possible. Available at https://www.truckinginfo.com/157083/volvos-supertruck-demonstrates-the-art-of-the-possible (accessed July 30, 2017).

Peterbilt Motors Co. 2014. DOE SuperTruck Program, Technology and System Level Demonstration of Highly Efficient and Clean, Diesel Powered Class 8 Trucks, Project ID AARAVT081, August.

Reinhart, T.E. 2015. Commercial Medium- and Heavy-Duty Truck Fuel Efficiency Technology Study—Report #1. DOT HS 812 146. Washington, DC: NHTSA. June.

Reinhart, T.E. 2016. Commercial Medium- and Heavy-Duty Truck Fuel Efficiency Technology Study—Report #2. DOT HS 812 194. Washington, DC: NHTSA. February.

Ricardo Inc.. 2009. Impact of Vehicle Weight Reduction on Fuel Economy for Various Vehicle Architectures. Arlington, VA: The Aluminum Association.

Rotz, D. (Daimler Truck NA). 2015. SuperTruck Program: Vehicle Program Review, DOE Project ID ARRVAT080. June.

Rotz, D., and M. Ziegler (Daimler Truck North America). 2014. SuperTruck Program: Vehicle Program Review. DOE Project ID ARRVAT080. Presentation to the Annual Merit Review, Washington, DC, June 19.

Salari, K. (Lawrence Livermore National Laboratory). 2016. DOE's Effort to Improve Heavy Vehicle Fuel Efficiency through Improved Aerodynamics. Presentation at DOE Annual Merit Review, Washington, DC, June 6-10.

Sidders, J. (Dana Commercial Vehicles). 2017. "Optimized Tire-Pressure Management." http://danacv.com/downloads/TirePressureManagement_WhitePaper.pdf (accessed August 28, 2019).

Smith, J., R. Mihelic, B. Gifford, and M. Ellis. 2014. Aerodynamic impact of tractor-trailer in drafting configuration. *SAE International Journal of Commercial Vehicles* 7(2):619-625.

Sproull, L. (PACCAR Inc.). 2015. Executive Panel: Greenhouse Gas Legislation. Presentation at the SAE 2015 Commercial Vehicle Engineering Congress in Rosemont, IL, October 7.

Steidel, S., T. Halfmann, M. Baecker, and A. Gallrein. 2016. Prediction of Rolling Resistance and Tread Wear of Tires in Realistic Commercial Vehicle Application Scenarios. SAE Paper 2016-01-8027.

TMC (The Maintenance Council). 1994. *Radial Tire Conditions Analysis Guide: A Comprehensive Review of Tread Wear and Tire Conditions*. Alexandria, VA: TMC.

TMC. 2014. Fuel Efficiency and Vehicle Emissions Regulations: Now and for the Foreseeable Future. S11 Tech Session, March 11. Available at http://www.tech-i-m.com/insight/s11-tech-session-march-11-2014-fuel-efficiency-and-vehicle-emissions-regulations-now-and-for-the-foreseeable-future/ (accessed March 27, 2015).

United Nations Economic Commission for Europe. 2014. Addendum 116: Regulation No. 117, Revision 3, Uniform Provisions Concerning the Approval of Tyres with Regard to Rolling Sound Emissions and/or to Adhesion on Wet Surfaces and/or to Rolling Resistance. New York: United Nations. February.

Woodrooffe, J.F., O. Page, D. Blower, and P.E. Green. 2008. Commercial Medium Tire Debris Study. Report No. UMTRI-2008-34. Washington, DC: NHTSA.

Woodrooffe, J. 2014. "Reducing Truck Fuel Use and Emissions: Tires, Aerodynamics, Engine Efficiency, and Size and Weight Regulations." Ann Arbor: University of Michigan Transportation Research Institute. http://hdl.handle.net/2027.42/109749 (accessed December 19, 2019).

Zukouski, R. (Navistar). 2015. Navistar SuperTruck—Development and Demonstration of a Fuel-Efficient Class 8 Tractor & Trailer Vehicle. DOE Contract DE-EE0003303. June.

6

Projected Benefits of Technologies on Fuel Consumption

The projected CO_2 and fuel consumption benefits of technologies were analyzed in two steps:

- The benefits of engine (and natural gas) technology were studied via simulation in the absence of vehicle technology improvements, using a baseline approximately 2016 vehicle.
- Next the advanced engines were coupled with advanced vehicle characteristics (aero, mass, and tires) for simulation to estimate a best achievable CO_2 and fuel consumption.

6.1 IMPACT OF ENGINE TECHNOLOGIES ON VEHICLE FUEL CONSUMPTION

Engine technologies and fuels are described in some detail in other sections of this report. The committee made projections of the likely achievable efficiencies of several classes of engine (additional discussion in Chapter 4) and contracted with Southwest Research Institute (SwRI) to construct performance maps of these plausible engines. The maps were then used in vehicle simulations to estimate the whole vehicle fuel consumption and CO_2 emissions compared to baseline engines for each vehicle category. The analysis process and the complete results, including sample engine maps, are provided in Appendix D.

The simulations included the following vehicle-engine combinations which are detailed in Table 6-1. Multiple drive cycles were used in the simulations, including several similar to the Environmental Protection Agency (EPA) and National Highway Traffic Safety Administration (NHTSA) cycles used in the Phase I rule. For the Phase II rule, EPA and NHTSA modified their 55 and 65 mph steady cruise cycles to include a variable grade ± 2 percent.

In the vehicle simulation analyses conducted by SwRI for this committee, only the Northeast States Center for a Clean Air Future cycle includes grade. Descriptions of the drive cycles can be found in the NHTSA report DOT HS 812 146 (Reinhart, 2015) and in Appendix D of the present report. As an example, the so-called California Air Resources Board (CARB) cycle is shown in Figure 6-1, which is one of the Greenhouse Gas Emissions Model (GEM) cycles used in certification. The CARB cycle was used for all vehicles except the pickup truck.

Fuel energy consumption and CO_2 emissions are shown for the selected classes of vehicles in the various figures below. Each vehicle-engine combination was simulated at three weights, including two typical payloads

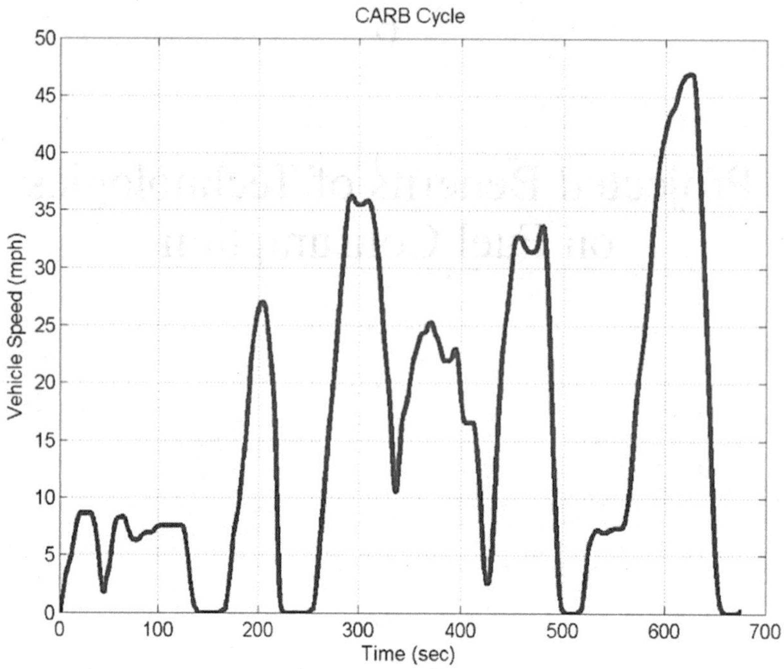

FIGURE 6-1 CARB transient cycle, which is used in GEM certification simulation.
SOURCE: Reinhart (2015).

and a zero-payload case. Fuel energy consumption and CO_2 emissions are calculated for the selected classes of vehicles in Table 6-1 as follows.

6.1.1 References to Vehicle and Engine Fuel Consumption and CO_2 Targets from the EPA-NHTSA Phase I Rules and Phase II Rules for 2027

The analytical results in this section can be placed in context by approximate comparison to the fuel consumption and CO_2 standards of the Phase I Rule (EPA and NHTSA, 2011) and the Phase II Rule (EPA and NHTSA, 2016c), released in August 2016. More complete discussion of the Phase II rules can be found in the EPA and NHTSA report (2016a). The change between 2017 and 2027 is of interest to compare to the findings of the SwRI analysis for the committee.

For combination trucks, the current standards are shown below in Table 6-2 and the Phase II standards are copied here from the Final Rule document as Table 6-3.

As a reference point, the difference between 2027 and 2017 CO_2 standards for the Class 8 sleeper cab (high roof) is a 10.7 percent reduction in CO_2. The 2027 standard reflects a 27 percent reduction compared to the baseline vehicle (87.8 g CO_2/mile) in the GEM simulation (Table 6-4).

The 2027 standards for the vocational vehicles[1] are provided in Table 6-5 for CO_2, followed by the NHTSA fuel consumption standards in Table 6-6.

The baseline vocational vehicle performance defined in the Phase II rule is provided below in Table 6-7 for compression-ignition engines and in Table 6-8 for spark-ignition engines.

[1] The term *vocational vehicle* "includes any vehicle that is equipped for a particular industry, trade or occupation such as construction, heavy hauling, mining, logging, oil fields, refuse and includes vehicles such as school buses, motorcoaches and RVs." See Code of Federal Regulations, Title 49, Section 523.2.

TABLE 6-1 Vehicle Engine Combinations, Class, and Simulation Specifications for Committee-Commissioned SwRI Study

Vehicle-Engine Combinations and Class	Simulation Specifications
T700 Tractor-Trailer (Class 8) 	• 2019 baseline DD15 diesel (derived from a 2013 engine as described in Appendix D -SwRI) • Westport high-pressure direct-injection (HPDI) natural gas engine (from SwRI experimental data) • Simulated 15 liter stoichiometric natural gas spark-ignited engine with exhaust-gas recirculation (EGR) • Simulated 15 liter 55% brake thermal efficiency (BTE) diesel, based on 2019 DD15 engine plus waste-heat recovery (WHR) and further refinements • A 50% peak BTE diesel, a more conservative version of the above • Gas turbine based on efficiency map from Brayton Energy
Kenworth T270 Box Truck (Class 6) 	• 2019 baseline 6.7 liter 300 horsepower (hp) interact system B (ISB) diesel (Reinhart, 2015) • Simulated 6.7 liter stoichiometric spark-ignited natural gas engine with EGR • 48% BTE 6.7 liter 300 hp diesel • 42% peak BTE 3.5 liter turbocharged EGR spark-ignited gasoline engine
Ford F-650 Tow Truck (Class 5) 	• 2019 baseline 6.7 liter 300 hp ISB diesel (Reinhart, 2015) • Simulated 6.7 liter stoichiometric spark-ignited engine with EGR • 48% BTE 6.7 liter 300 hp diesel • 42% BTE 3.5 liter turbocharged EGR spark-ignited gasoline engine
Ram 2500 Pickup Truck (Class 2b) 	• 2019 baseline 6.7 liter 385 hp ISB diesel (Reinhart, 2015) • Simulated 6.7 liter stoichiometric spark-ignited natural gas engine with EGR • 48% BTE 6.7 liter 385 hp diesel • 42% BTE 3.5 liter turbocharged EGR spark-ignited gasoline engine

TABLE 6-2 Phase I Standards for Combination Trucks

	EPA CO_2 Emissions g/ton-mile			NHTSA Fuel Consumption gal/1,000 ton-mile		
Category	Low Roof	Mid Roof	High Roof	Low Roof	Mid Roof	High Roof
Day cab Class 7	104	115	120	10.2	11.3	11.8
Day cab Class 8	80	86	89	7.8	8.4	8.7
Sleeper cab Class 8	66	73	72	6.5	7.2	7.1

SOURCE: EPA and NHTSA (2011).

TABLE 6-3 Phase II Standards through 2027

	Day Cab		Sleeper Cab	Heavy-Haul
	Class 7	Class 8	Class 8	Class 8
2021 Model Year CO_2 Grams per Ton-Mile				
Low roof	105.5	80.5	72.3	52.4
Mid roof	113.2	85.4	78	
High roof	113.5	85.6	75.7	
2021 Model Year Gallons of Fuel per 1,000 Ton-Mile				
	Day Cab		Sleeper Cab	Heavy-Haul
	Class 7	Class 8	Class 8	Class 8
Low roof	10.36346	7.90766	7.10216	5.14735
Mid roof	11.11984	8.389	7.66208	
High roof	11.14931	8.40864	7.43615	
2024 Model Year CO_2 Grams per Ton-Mile				
	Day Cab		Sleeper Cab	Heavy-Haul
	Class 7	Class 8	Class 8	Class 8
Low roof	99.8	76.2	68	50.2
Mid roof	107.1	80.9	73.5	
High roof	106.6	80.4	70.7	
2024 Model Year and Later Gallons of Fuel per 1,000 Ton-Mile				
	Day Cab		Sleeper Cab	Heavy-Haul
	Class 7	Class 8	Class 8	Class 8
Low roof	9.80354	7.48527	6.67976	4.93124
Mid roof	10.52063	7.94695	7.22004	
High roof	10.47151	7.89784	6.94499	
2027 Model Year CO_2 Grams per Ton-Mile				
	Day Cab		Sleeper Cab	Heavy-Haul
	Class 7	Class 8	Class 8	Class 8
Low roof	96.2	73.4	64.1	48.3
Mid roof	103.4	78	69.6	
High roof	100	75.7	64.3	
2027 Model Year and Later Gallons of Fuel per 1,000 Ton-Mile				
	Day Cab		Sleeper Cab	Heavy-Haul
	Class 7	Class 8	Class 8	Class 8
Low roof	9.4499	7.21022	6.29666	4.7446
Mid roof	10.15717	7.66208	6.83694	
High roof	9.82318	7.43615	6.31631	

SOURCE: EPA and NHTSA (2016c).

TABLE 6-4 Class 7 and 8 Tractor 2017 Baseline Vehicle CO_2 Emissions and Fuel Consumption in the Phase II Standards

	Class 7			Class 8					
	Day Cab			Day Cab			Sleeper		
	Low Roof	Mid Roof	High Roof	Low Roof	Mid Roof	High Roof	Low Roof	Mid Roof	High Roof
CO_2 (g/ton-mile)	119.1	127.2	129.7	91.3	96.6	98.2	84	90.2	87.8
Fuel Consumption (gal/1000 ton-mile)	11.69941	12.49509	12.74067	9.96587	9.48919	9.64637	8.25147	8.86051	8.62475

SOURCE: EPA and NHTSA (2016a).

TABLE 6-5 EPA CO_2 Standards for Model Year (MY) 2027 Class 2b through 8 Vocational Vehicles

EPA Standard for Vehicle with Compression-Ignition (CI) Engine Effective MY 2027 (g CO_2/ton-mile)			
Duty cycle	Light Heavy-Duty Class 2b-5	Medium Heavy-Duty Class 6-7	Heavy Heavy-Duty Class 8
Urban	367	258	269
Multipurpose	330	235	230
Regional	291	218	189
EPA standard for Vehicle with Spark-Ignition (SI) Engine Effective MY 2027 (g CO_2/ton-mile)			
Duty cycle	Light Heavy-Duty Class 2b-5	Medium Heavy-Duty Class 6-7	
Urban	413	297	
Multipurpose	372	268	
Regional	319	247	

SOURCE: EPA and NHTSA (2016a).

TABLE 6-6 NHTSA Fuel Consumption Standards for MY 2027 Class 2b through 8 Vocational Vehicles

NHTSA Standard for Vehicle with CI Engine Effective MY 2027 (fuel consumption gallon per 1,000 ton-mile)			
Duty cycle	Light Heavy-Duty Class 2b-5	Medium Heavy-Duty Class 6-7	Heavy Heavy-Duty Class 8
Urban	36.0511	25.3438	26.4244
Multipurpose	32.4165	23.0845	22.5933
Regional	28.5855	21.4145	18.5658
NHTSA Standard for Vehicle with SI Engine Effective MY 2027 (fuel consumption gallon per 1,000 ton-mile)			
Duty cycle	Light Heavy-Duty Class 2b-5	Medium Heavy-Duty Class 6-7 (and Class 8 Gasoline)	
Urban	46.4724	33.4196	
Multipurpose	41.8589	30.1564	
Regional	35.8951	27.7934	

SOURCE: EPA and NHTSA (2016a).

TABLE 6-7 Baseline Vocational Vehicle Performance with CI Engines in the Phase II Standards

Baseline Emissions Performance in g CO_2/ton-mile			
Duty cycle	Light Heavy-Duty Class 2b-5	Medium Heavy-Duty Class 6-7	Heavy Heavy-Duty Class 8
Urban	482	332	338
Multipurpose	420	294	287
Regional	334	249	220
Baseline Fuel Efficiency Performance in gallon per 1,000 ton-mile			
Duty cycle	Light Heavy-Duty Class 2b-5	Medium Heavy-Duty Class 6-7	Heavy Heavy-Duty Class 8
Urban	47.3477	32.6130	33.2024
Multipurpose	41.2574	28.8802	28.1925
Regional	32.8094	24.4597	21.6110

SOURCE: EPA and NHTSA (2016c, p. 73,706).

TABLE 6-8 Baseline Vocational Vehicle Performance with SI Engines in the Phase II Standards

Baseline Emissions Performance in g CO_2/ton-mile		
Duty cycle	Light Heavy-Duty Class 2b-5	Medium Heavy-Duty Class 6-7 (and Class 8 Gasoline)[a]
Urban	502	354
Multipurpose	441	314
Regional	357	275
Baseline Fuel Efficiency Performance in gallon per 1,000 ton-mile		
Duty cycle	Light Heavy-Duty Class 2b-5	Medium Heavy-Duty Class 6-7
Urban	56.4870	39.8335
Multipurpose	49.6230	35.3325
Regional	40.1710	30.9441

[a]Vocational vehicles with gross vehicle weight rating (GVWR) over 33,000 pounds powered by alternate fueled engines must certify to the vehicle standard corresponding with the applicable engine standard.
SOURCE: EPA and NHTSA (2016c, p. 73,707).

6.1.2 Description of Vehicles Simulated and Results

The committee's work included simulations of four vehicle types from the lightest vehicles covered in the Phase II rule, Class 2b, up to the largest, Class 8. Specifically, these vehicles included a Dodge Ram 2500/3500 pickup truck (Class 2b/3), a Ford F-650 tow truck (Class 5), a Kenworth T270 box delivery truck (Class 6), and a Kenworth T700 long-haul tractor-trailer truck (Class 8). The results of the simulation provide estimates both of vehicle fuel consumption—gallons consumed and fuel mass consumed—and of greenhouse gas (GHG) emissions. (See Appendix D for additional description.) The results also included GHG emissions. The following four subsections discuss the results of these simulations in different drive cycles (described in Appendix D) for the four vehicles noted above, each with a variety of engines.

6.1.2.1 Class 8 Tractor Trailer

The vehicle characteristics for the simulations performed by SwRI for the committee are summarized below in Tables 6-9 and 6-10, and also see the NHTSA study report (Reinhart, 2015) for additional details.

The derivation of advanced engine features and maps is summarized in Table 6-11.

The vehicle simulation results for the baseline CO_2 and fuel consumption performance for the T700, over the CARB cycle, are shown below in Table 6-12.

For the Class 8 T700 vehicle, compared to a 2019 baseline diesel engine, the 55 percent peak diesel and natural gas engines all show a reduction in CO_2 emissions compared to a diesel baseline (Figure 6-2), but the relative inefficiency of both natural gas options compared to the simulated advanced 55 percent diesel is evident in Figure 6-3. The gas turbine engine, which generally exhibits inherently low NO_x emissions, was included here in consideration of a future reduction in NO_x emissions standards. As explained in Chapter 4, in the section "Alternative Configuration Engines," this type of engine has a relatively high sensitivity to ambient temperature. The projected peak efficiency was just under 50 percent at 59 degrees Fahrenheit (°F). Results are shown at 50 percent payload for this vehicle. The sensitivity of fuel consumption to weight is provided later in this section.

For completeness, the committee refers to a study published by Carder (2014)—issued after its own first report of NRC Phase Two (NRC, 2014)—where comparisons were made in full vehicle dynamometer tests between diesel (2011 model), HPDI, and SI natural gas (NG) engines in several types of vehicles. For goods movement vehicles, they found the CO_2 emissions and global warming potential emissions for the natural gas engines to be about 10 percent lower than the 2011 diesel engine with selective catalytic reduction.

For this vehicle, the drive cycle does not significantly impact the overall trends. The benefits of a low-carbon fuel are very evident for CO_2 emissions, with the caveat that well-to-tank emissions of methane are not found to offset this tailpipe benefit (see Section 4.9.5.1 of this report). At low ambient temperatures, the simulated gas turbine is projected to have advantage over the baseline diesel, but not so much compared to other advanced engines. For full consistent comparison, one would need to have a diesel engine map for an engine with emission controls for much lower NO_x output.

TABLE 6-9 Kenworth T700 Vehicle Characteristics Used for This Simulation

Parameter	Value
Aerodynamic drag (C_d) with trailer	0.639
C_dA	6.841 m^2
Tire rolling resistance (C_{rr})	0.005608 (average all axles)
Transmission	10 speed AMT
Base accessory power demand	6650 W
Mass	see Table 6-10 below

NOTE: A, effective frontal area of vehicle; AMT, automated mechanical transmission; W, watts.

TABLE 6-10 Mass and Payload Characteristics of Class 8 Vehicle Used in the Simulation

	T700 Diesel		
Weights in Pounds	Tare	Payload	Total
0% payload	33,960	0	33,960
50% payload	33,960	23,020	56,980
100% payload	33,960	46,039	79,999

NOTE: In comparing the simulation results in this report to the Phase I or II standards and GEM simulation, note that the payload used by NHTSA and EPA is higher than 50 percent used here.
SOURCE: Reinhart (2015).

TABLE 6-11 Technologies Used for the Class 8 Tractor-Trailer Simulation

Technology	Hardware Content	Comments
2019 baseline DD15	Based on the 2013 production DD15 without turbocompound and with an asymmetric turbo. Includes reduced combustion duration.	Complies with the 2017-2019 GHG requirement, but with no margin
15 liter SI natural gas	A stoichiometric spark-ignited natural gas variant of a 15 liter diesel, with performance and efficiency characteristics derived from experimental data in the smaller ISL-9G engine. Data from the 12 liter ISX-12G were not used, because of its lower BTE.	Represents the expected performance of a SI natural gas engine with EGR. Cummins was developing a version of their 15 liter engine along these lines, but it was put on hold when oil prices dropped.
HPDI diesel pilot ignited natural gas engine	Experimentally derived fuel map from the 2010 Westport HPDI engine, which is based on the Cummins ISX 15 liter diesel. This is a lean-burn, diesel pilot ignited natural gas engine.	Reflects 2010 level performance. Could be improved with additional development, but Westport has dropped the HPDI product line because of limited sales volume.
55% BTE diesel	Based on technology combination 3f from Reinhart (2015). Includes down-speeding, a partial friction mean effective pressure (FMEP) reduction, and waste-heat recovery. This map was then scaled to achieve a peak BTE of 55%.	Represents a very high efficiency potential future diesel powertrain. The shape of the fuel map is speculative, because the technologies required to achieve it are not defined.
Gas turbine	Based on an efficiency map provided by Brayton Energy.	Represents a very high efficiency potential future gas turbine powertrain. Requires a 50:1 speed reduction geartrain ahead of the conventional truck transmission.

SOURCE: Reinhart (2015).

TABLE 6-12 T700 Baseline CARB Drive Cycle CO_2 and Fuel Consumption

Payload	g CO_2/ton-mile	Fuel Consumption gallon/1,000 ton-mile
50%, 23,020 lbs	189.2	18.6
100%, 46,039 lbs	116.2	11.4

SOURCE: Reinhart (2015).

As a reference point, the difference between 2027 and 2017 CO_2 standards for the Class 8 sleeper cab (high roof) is a 10.7 percent reduction in CO_2. The 2027 standard reflects a 27 percent reduction compared to the baseline vehicle (87.8 g CO_2/mile) in the GEM simulation.

It is evident that, if realized commercially, the 55 percent diesel would produce reductions well toward this CO_2 reduction, even without improvements in other parts of the vehicle. For general comparisons, a 50 percent peak efficiency diesel engine would have proportionally lesser improvement. These results, along with the strong progress in diesel engines described in Chapter 4, warrant continued development in this class of engine.

The overall impacts of integrating an advanced engine in vehicles featuring anticipated reductions in mass, aerodynamic drag, and rolling resistance are described later in this chapter.

Finding: Highly developed advanced diesel engines appear capable of achieving 15 to 17 percent reduction in CO_2 emissions compared to the 2019 state of technology diesels in base vehicles. Achieving this level of reduction with natural gas engines would require considerable improvement in their efficiency, such as with HPDI technology. The impacts of unconventional engines such as gas turbines are inconclusive at this time.

Finding: The effect of improved diesel engines on CO_2 and fuel consumption is significant and warrants continued development of this family of power plant.

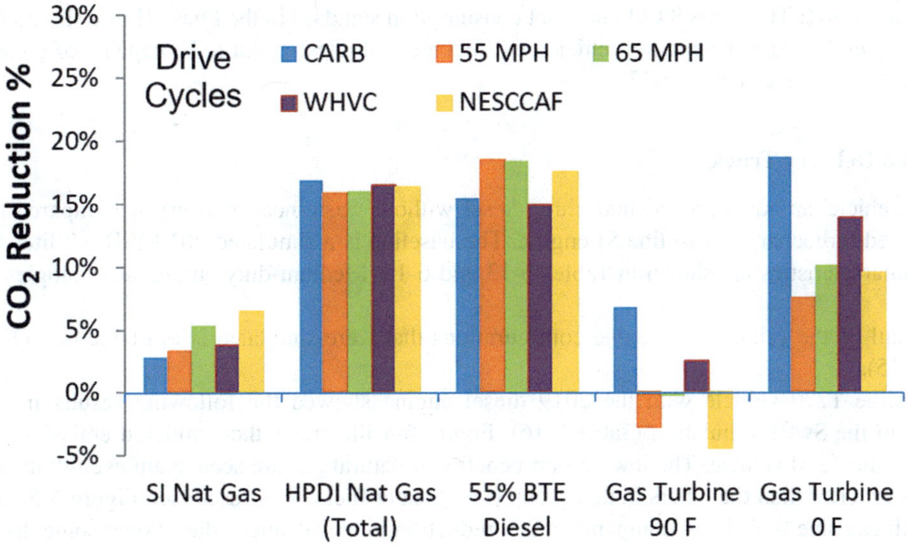

FIGURE 6-2 CO_2 reduction in T700 at 50 percent payload versus 2019 baseline diesel simulated over 5 drive cycles.

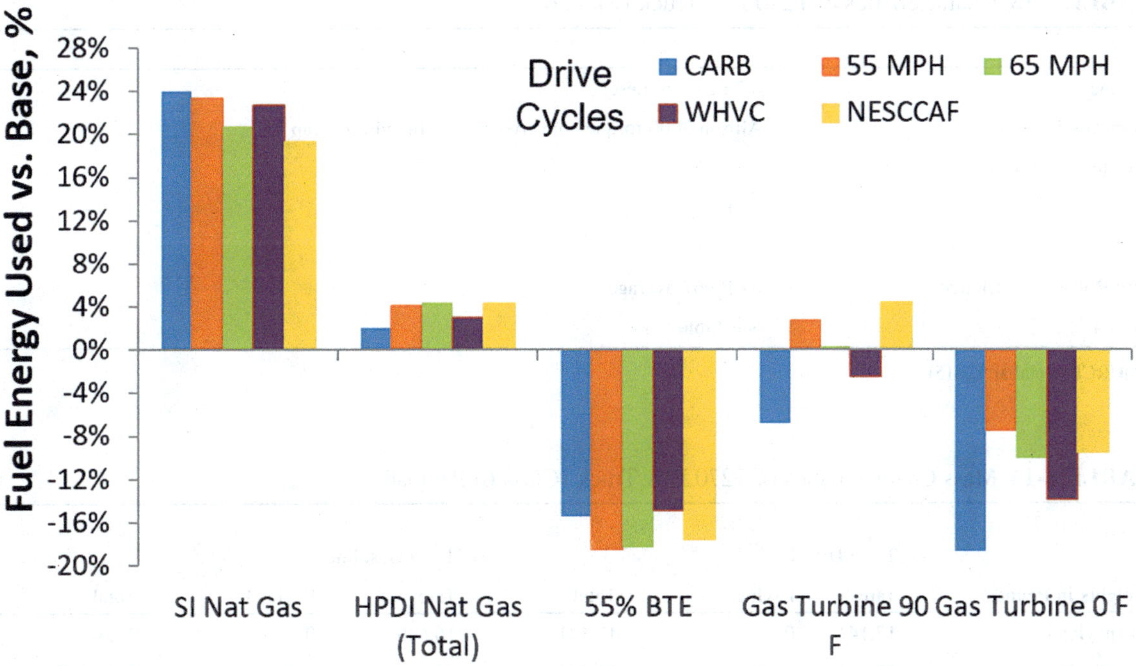

FIGURE 6-3 Fuel energy change in T700 at 50 percent payload with 5 advanced powertrains versus 2019 diesel baseline simulated over 5 drive cycles.

Recommendation 6-1: The Class 8 CO_2 and fuel consumption standard in the Phase II rule should be revisited in a mid-phase review because the engine, vehicle, and fuel technologies appear to be capable of greater reductions than the standards taking effect in 2027.

6.1.2.2 Class 6 Delivery Truck

For this vehicle, an advanced medium-duty diesel without waste-heat recovery is compared to natural gas and an advanced turbocharged gasoline SI engine. The baseline is a simulated 2019 ISB 6.7 liter diesel engine. The vehicle characteristics are shown in Tables 6-13 and 6-14. Medium-duty engine technologies are shown in Table 6-15.

More detail on the vehicle and engine configurations that were simulated is available in DOT HS 812 146 (Reinhart, 2015).

The baseline T270 vehicle with the 2019 diesel engine showed the following results in CO_2 and fuel consumption in the SwRI simulation (Table 6-16). Figure 6-4 illustrates the simulated emissions for advanced powertrains in the T270 vehicle. The low-carbon benefits of natural gas are seen again even though the modeled engine is less efficient than the others in this group (i.e., greater fuel energy used; see Figure 6-5). The advanced SI engine with gasoline fuel shows comparable CO_2 reduction to the advanced diesel over some drive cycles, and roughly 8 percent reduction compared to a 2019 baseline diesel that complies with prevailing standards with a small margin. As described in the discussion on engines (see the section "Compression-Ignition-Dominated Engines" in Chapter 4), the cost of the emissions control in an SI engine is likely to be less than the diesel, suggesting a possible shift in market preference. In the committee's simulation, this gasoline engine vehicle has a CO_2 output

TABLE 6-13 Characteristics of T270 Box Truck (Class 6)

T270	
Engine	6.7 liter I-6 diesel
Transmission	Allison 2000 torque converter automatic with lockup
Mechanical Accessory Power	5,750 W
C_d	0.514
C_dA	5.033 m^2
Tire Rolling Coefficient	0.010967 average
Mass 1	See Table 6-14

SOURCE: Reinhart (2015).

TABLE 6-14 Mass Characteristics of T270 Box Truck (Class 6) Payload

	T700 Diesel			T270 Gasoline		
Weights in Pounds	Tare	Payload	Total	Tare	Payload	Total
0% payload	17,141	0	17,141	16,640	0	16,640
50% payload	17,141	4,430	21,571	16,640	4,430	21,070
100% payload	17,141	8,860	26,001	16,640	8,860	25,500

SOURCE: Reinhart (2015).

TABLE 6-15 Engine Technologies Used in Medium-Duty Trucks

Technology	Hardware Content	Comments
2019 baseline ISB diesel	Derived from the 2013 production engine, including a reduction in FMEP and a reduction in combustion duration.	Complies with 2017-2019 GHG requirements, but with little margin.
ISB SI natural gas	A stoichiometric SI natural gas variant of the ISB diesel, with performance and efficiency characteristics derived from experimental data in the larger ISL-9G engine.	Represents the expected performance of the natural gas version of the ISB expected in 2016 or 2017.
48% BTE diesel	Created by taking the fuel map of the best medium-duty diesel from the NHTSA project, and scaling it so that the peak BTE reached 48%. This map is derived from Package 9, described in Reinhart (2015).	Represents a very high efficiency medium-duty diesel without waste-heat recovery. This engine has approximately the BTE of the SuperTruck program engines (not including WHR).
42% BTE gasoline	Created by taking the fuel map of the most efficient 3.5 liter turbocharged engine from the NHTSA project (with EGR, variable valve actuation [VVA], and down-speeding), and scaling it so that peak BTE reached 42%. This map is derived from the 3.5 V-6 Package 17, described in Reinhart (2015).	Represents a very high efficiency gasoline engine. Achieving this level of efficiency would probably require full authority VVA/variable valve timing (VVT), and several other efficiency features in reduction of friction and heat losses. See Section 4.3.

NOTE: These engines were used in the simulations of both the Class 6 delivery truck and the Class 5 tow truck.

of about 300 g/ton-mile in the 65 mph cycle at 100 percent payload (8,860 pounds).[2] Perhaps owing to a difference in vehicle tare (i.e., empty) weight—16,640 pounds in the committee's analysis versus 13,950 pounds[3] in the Regulatory Impact Analysis (RIA) (Section 3.5.1)—this value is higher than the Phase I standard of 225 g CO_2/ton-mile for SI-powered delivery trucks.

6.1.2.3 Ford F-650 Tow Truck (Class 5)

For this vehicle, an advanced medium-duty diesel without waste-heat recovery is compared to natural gas and an advanced turbocharged gasoline SI engine. The baseline is a simulated 2019 ISB 6.7 liter diesel engine. The vehicle characteristics are shown in Tables 6-17 and 6-18.

The comparison of fuel consumption at zero payload for the different engines simulated for this vehicle is shown in Figure 6-6. The most striking result again is how competitive the advanced gasoline SI engine is in emissions as compared to the baseline and advanced diesel on CO_2 (see Figure 6-7). The primary reasons for this appear to be the following:

- The SI engine has a much smaller displacement than the base diesel and, hence, has lower weight. Since its maximum torque is lower, the transmission also weighs less than the baseline. In general, SI engines have lower peak cylinder pressures than CI engines, which can also be a factor in their structural robustness and weight.
- The SI engine has a broader speed range for reasonable level of torque delivery, making the transmission-engine speed matching more optimal (the diesel has a narrow usable speed range), and the component efficiencies slightly higher.
- The idle fuel consumption characteristics with automatic transmission tend to favor the SI engine.

[2] The gross vehicle weight (GVW) is 25,500 pounds (Table 6-14).

[3] Making the tare weight lower has the effect of "returning" weight to payload. This addition of weight to the payload results in a lower load-specific fuel consumption (LSFC). (When calculating LSFC the payload appears in the denominator.)

TABLE 6-16 Baseline T270 Box Truck (Class 6) Load-Specific CO_2 and Fuel Consumption with 2019 Baseline Diesel

Payload	g CO_2/ton-mile	Fuel Consumption gallon/1,000 ton-mile
50%, 4,430 lbs	559	54.9
100%, 8,860 lbs	315	30.9

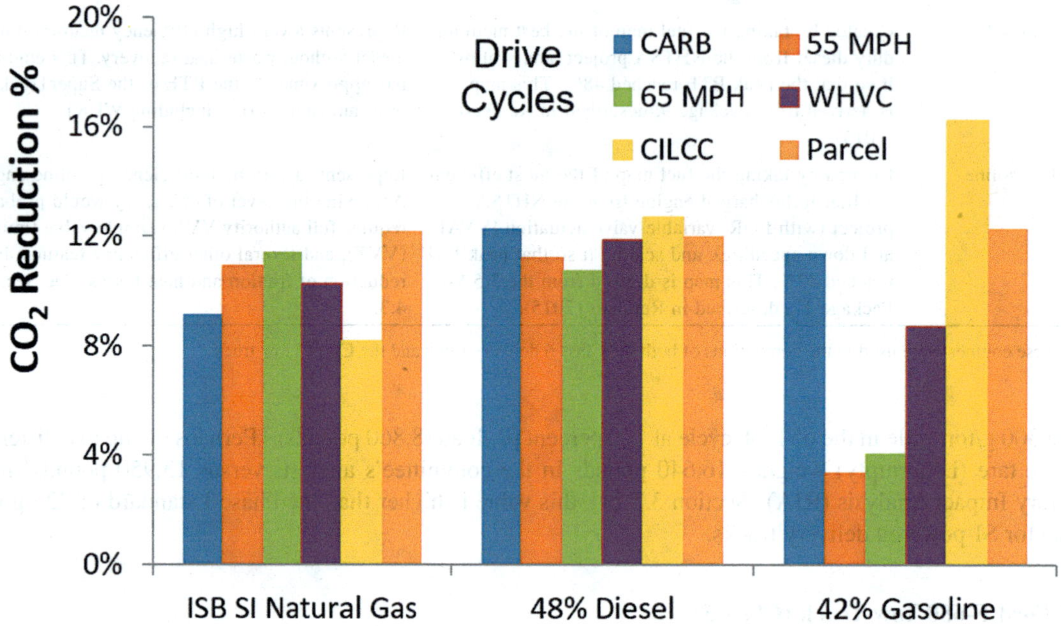

FIGURE 6-4 Relative reductions in CO_2 using advanced natural gas, diesel, and gasoline powertrains in the T270 at 50 percent payload simulated over 6 drive cycles.

In their modeling and analysis effort for NHTSA, SwRI (Reinhart, 2016) observed that a modern turbocharged SI engine approached or equaled the CO_2 reduction levels of diesel engines in some duty cycles. The report further describes an economic comparison of diesel and SI engines considering the cost premium of the diesel initial purchase (approximately $10,000 in some cases) as well as the recent trends in higher diesel fuel prices than gasoline. As described in the discussion on engines (see the section "Compression-Ignition-Dominated Engines" in Chapter 4), the overall cost of purchase and operation of the emissions control in an SI engine is likely to be less than the diesel, suggesting a possible shift in market preference. These factors favor a shift toward SI engines in vocational vehicles, potentially in larger Class 7 sizes as long as the annual mileage does not place fuel consumption as the heavily dominant criterion. Further study of this type of engine in medium- and perhaps heavy-duty vehicles appears warranted to clarify or validate these apparent advantages (see also Recommendation 6-2).

Finding: For vocational vehicles, advanced diesel engines are forecast to achieve 12 percent CO_2 reduction compared to the 2019 diesel baseline. Natural gas–powered engines in these vehicles were found to be of similar effectiveness according to the simulation results. In some duty cycles, advanced gasoline engines are on par with the simulated advanced diesels. These levels of reduction are comparable to the percent reductions for the *entire vehicle* for the proposed 2027 rules.

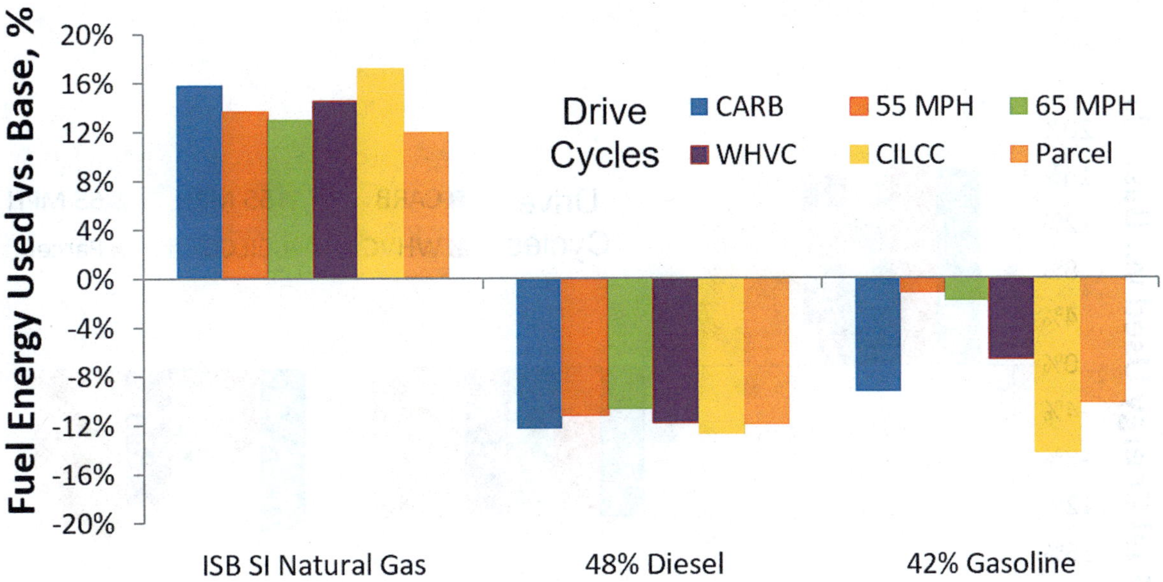

FIGURE 6-5 Fuel energy change for advanced natural gas, diesel, and gasoline powertrains versus 2019 diesel baseline in T270 at 50 percent payload simulated over 6 drive cycles.

TABLE 6-17 Characteristics of F-650 Tow Truck (Class 5)

F-650	
Engine	6.7 liter I-6 diesel
Transmission	TBD
Mechanical accessory power	5750 W
C_d	0.619
C_dA	3.151 m^2
Tire rolling coefficient	0.010068 average
Mass characteristics	See Table 6-18

SOURCE: Reinhart (2015).

TABLE 6-18 Payload Mass Characteristics of Ford F-650 Tow Truck (Class 5)

	F-650 Diesel			F-650 Gasoline		
Weights in Pounds	Tare	Payload	Total	Tare	Payload	Total
0% payload	15,640	0	15,640	15,139	0	15,139
50% payload	15,640	3,180	18,820	15,139	3,180	18,319
100% payload	15,640	6,360	21,999	15,139	6,360	21,499

SOURCE: Reinhart (2015).

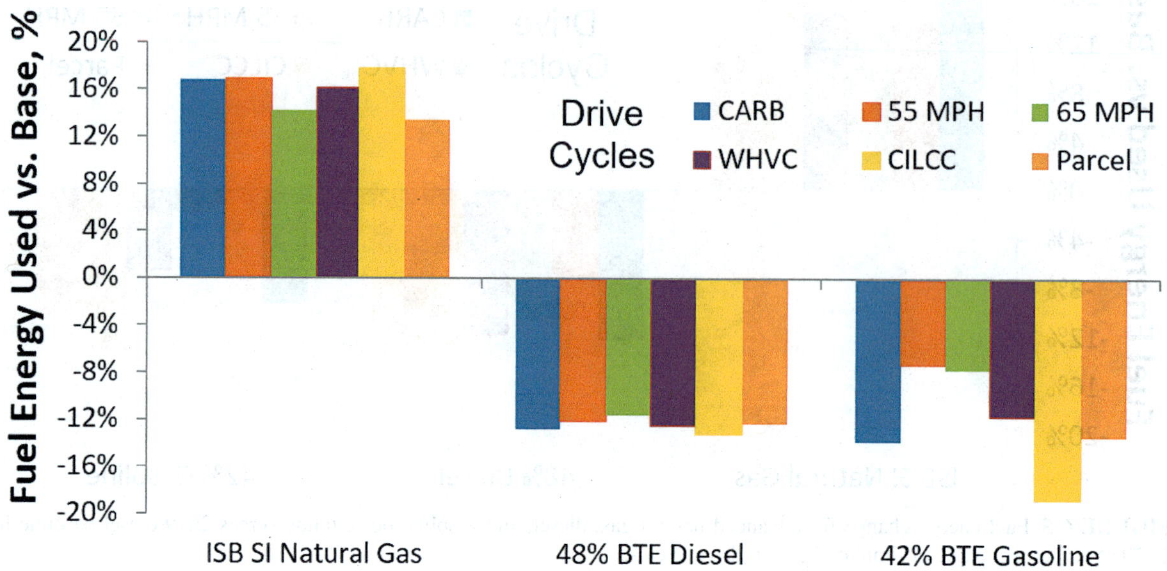

FIGURE 6-6 Fuel energy change for advanced natural gas, diesel, and gasoline powertrains versus 2019 diesel baseline in F-650 at 0 percent payload simulated over 6 drive cycles.

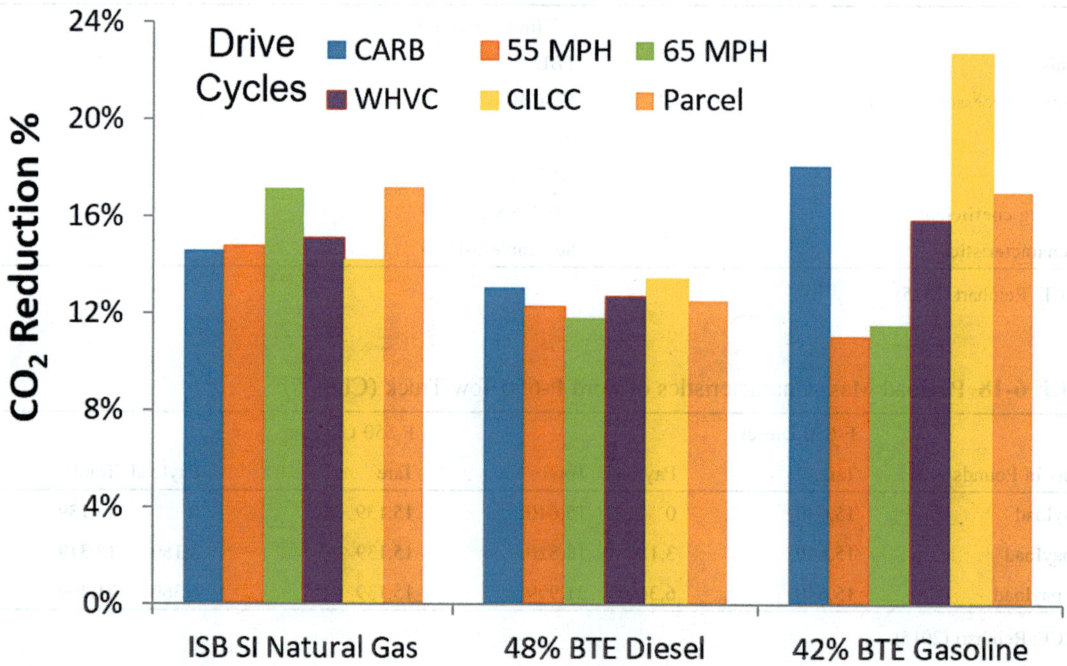

FIGURE 6-7 Reduction in CO_2 emissions of an F-650 across advanced natural gas, diesel, and gasoline powertrains at 50 percent payload simulated over 6 drive cycles relative to simulated 2019 baseline diesel engine.

Recommendation 6-2: Government and industry should continue the development of higher-efficiency SI gasoline and natural gas engines for vocational vehicles and should continue to ensure substantial CO_2 and fuel consumption reductions are realized, because (1) higher efficiency appears feasible and (2) market forces may cause a shift toward the SI engines.

6.1.2.4 Class 2b Pickup Truck

The Class 2b and 3 pickups and vans use about 15 percent of the total fuel in the medium- and heavy-duty vehicle (MHDV) sector and include those vehicles of gross vehicle weight rating from 8,501 to 14,000 pounds that are not regulated under the light-duty vehicle regulations. The Class 2b pickup truck simulation was based on a Dodge Ram 2500, with the basic characteristics shown in Table 6-19 and mass payload characteristics shown in Table 6-20.

Additional details may be found in EPA and NHTSA (2016b). The sources of the engine maps used are shown in Table 6-21.

TABLE 6-19 Characteristics of Ram 2500 (Class 2b)

T270	
Engine	6.7 liter I-6 diesel and 3.5 liter turbocharged SI gasoline
Transmission	6-speed automatic, torque converter
Mechanical accessory power	4450 W
Mass (curb weight)	6,876 lbs diesel, 6,376 lbs gasoline
Gross Combined Weight (GCW)	25,000 lbs
C_d	0.400
C_dA	1.505 m^2
Tire rolling coefficient	0.0078 average
Mass characteristics	See Table 6-20

SOURCE: Reinhart (2015).

TABLE 6-20 Dodge Ram Class 2b Pickup Truck Mass Payload Characteristics

	RAM diesel			RAM gasoline		
Weights in Pounds	Tare	Payload	Total	Tare	Payload	Total
0% payload	6,876	0	6,876	6,376	0	6,376
ALVW	6,876	1,562	8,438	6,376	1,562	7,938
100% GCW	6,876	18,124	25,000	6,376	18,124	24,500

SOURCE: Reinhart (2015).

TABLE 6-21 Engines Used in the Simulation of the Dodge Ram 2500 Pickup

Technology	Hardware Content	Comments
2019 baseline ISB Diesel	Derived from the 2013 production engine, including a reduction in FMEP and a reduction in combustion duration	Complies with 2017-2019 GHG requirements, but with little margin
ISB SI natural gas	A stoichiometric SI natural gas variant of the ISB diesel, with performance and efficiency characteristics derived from experimental data taken using the larger ISL-9G engine	Represents the expected performance of the natural gas version of the ISB that is due to launch in 2016 or 2017.
48% BTE diesel	Created by taking the fuel map of the best medium-duty diesel from the NHTSA project, and scaling it so that the peak BTE reached 48%. This map is derived from Package 14, described in Reinhart (2015)	Represents a very high efficiency future medium-duty diesel without WHR. This engine has approximately the BTE of the SuperTruck program engines (not including WHR)
42% BTE gasoline	Created by taking the fuel map of the most efficient 3.5 liter turbocharged engine from the NHTSA project (with EGR, VVA, and down-speeding), and scaling it so that peak BTE reached 42%. This map is derived from the 3.5 V-6 Package 17, described in Reinhart (2015)	Represents a very high efficiency gasoline engine in the 2025 time frame. Achieving this level of efficiency would require full authority VVA/VVT, and improvements in all aspects of friction and air handling.

SOURCE: Reinhart (2015).

NHTSA and EPA have different standards for gasoline- and diesel-powered pickups and vans versus Class 4 and above, using a "work factor"[4] instead of simple payload in standards (EPA RIA; Lutsey, 2015), as well as basing regulations on sales-weighted fuel consumption. This makes the analysis presented here, with a diesel baseline, somewhat less clear if attempting to make direct comparison to the regulations. The International Council on Clean Transportation (ICCT) added clarity to NHTSA and EPA's rules with a chart (Figure 6-8) that gives a percentage improvement in gasoline- or diesel-powered vehicles in context with the proposed regulations.

ICCT reports that NHTSA and EPA's projected compliance would result in the gasoline and diesel heavy-duty pickup and van fleets each seeing a CO_2 reduction of 24 percent from 2014 (the first year of standards) through 2027. Compared to the 2018-2020 standard, the reduction for Phase II would be 16 to 17 percent. The American Council for an Energy-Efficient Economy (Langer, 2015) prepared a current status of pickup and van CO_2 emissions compared to the standards through 2018 shown in Figure 6-9. Even for the initial 2014 standards, gasoline-engine vehicles appeared challenged, which was echoed by the original equipment manufacturers (OEMs) in meetings with the committee.[5] The debate about stringency of the pickup and van regulations was further highlighted by ICCT in their review of this specific class of vehicles (Lutsey, 2015), with questions of how much of the emerging light-duty (LD) vehicle technology for Corporate Average Fuel Economy 2025 standards would be applicable to HD pickups, and how much of that LD technology was already being used in the Class 2b and 3 vehicles.

As with the other vehicles, the following figures show fuel energy use (Figure 6-10) and CO_2 production (Figure 6-11) for the various engines compared to the baseline I-6 diesel engine.

At zero payload, all engines show a notable reduction in CO_2 (Figure 6-10) and fuel consumption (Figure 6-11), with the advanced SI engine showing the greatest reductions. This again is due to the lighter weight of the SI engine plus transmissions, as well as torque-speed characteristics. In Figure 6-12, at gross combined weight both

[4] EPA established CO_2 standards in the form of a set of target standard curves, based on a "work factor" that combines a vehicle's payload, towing capacity, and whether or not it has four-wheel drive (instead of a simple vehicle mass or gross vehicle weight rating [GVWR]):

Work Factor = [0.75 × (Payload Capacity + xwd)] + [0.25 × Towing Capacity]

Payload Capacity = GVWR (lb) – Curb Weight (lb)

Cross wheel drive (xwd) = 500 lb if the vehicle is equipped with four-wheel drive, otherwise equals 0 lb

Towing Capacity = GCWR (lb) – GVWR (lb)

[5] Mark Allen and Barbara Kiss, GM, "GM Heavy-Duty Presentation," presentation to the committee, Sacramento, California, July 31, 2013.

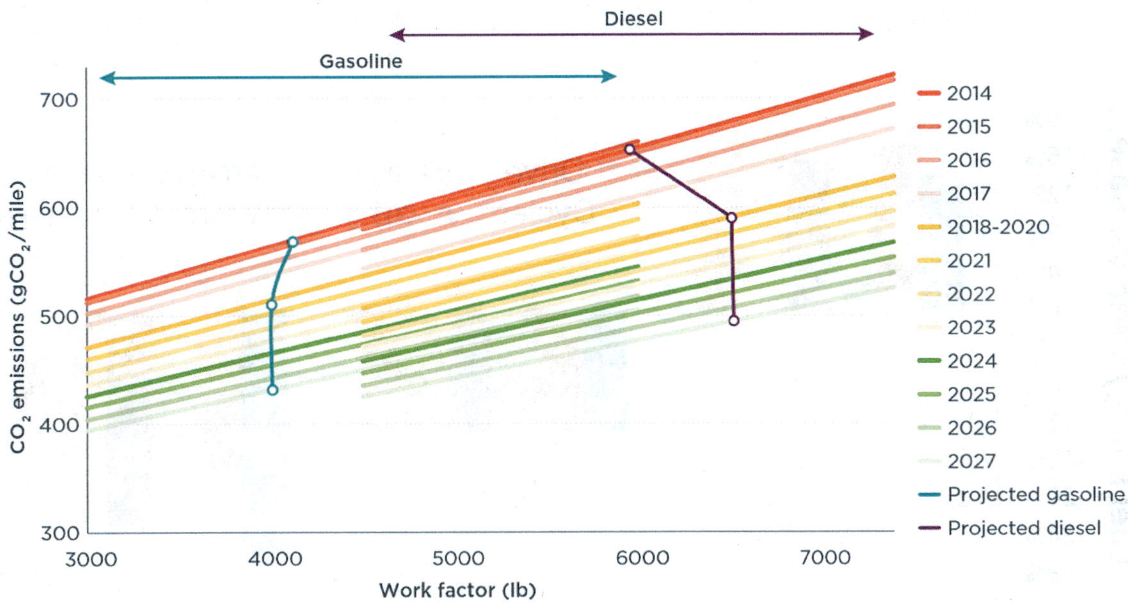

FIGURE 6-8 Phase II rule proposed CO_2 regulatory targets for heavy-duty (HD) pickups and vans based on work factor. SOURCE: Lutsey (2015).

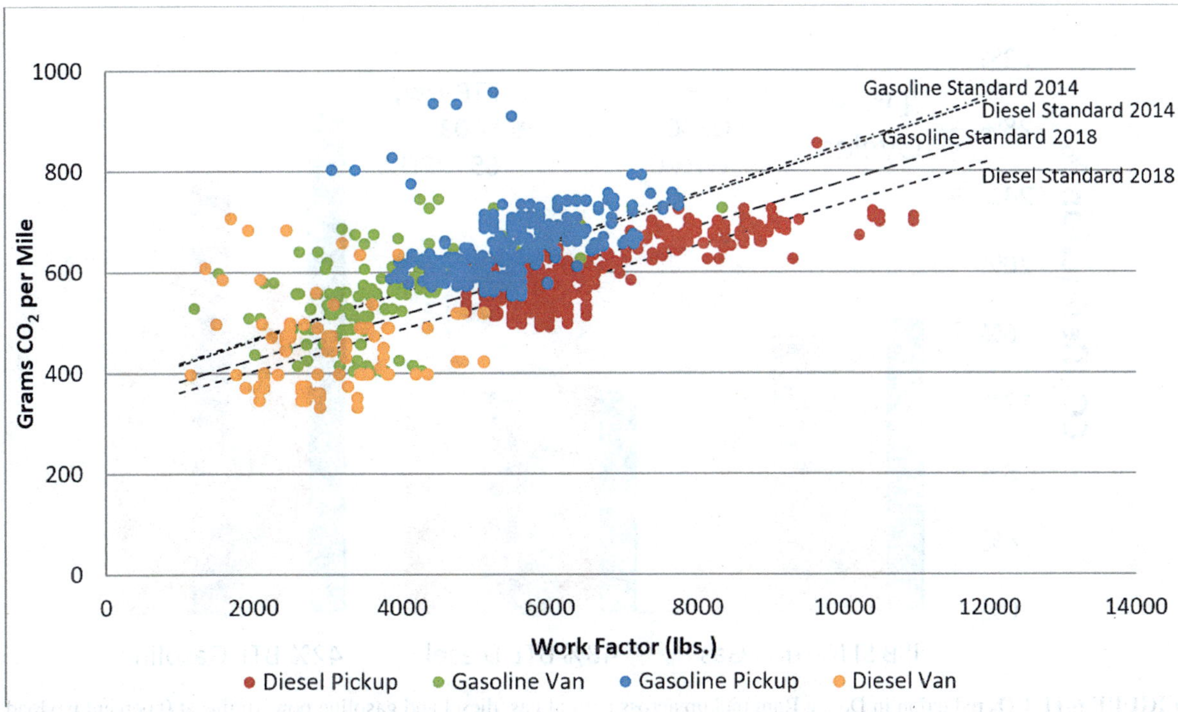

FIGURE 6-9 Current status of pickup and van CO_2 emissions compared to standards. SOURCE: Langer (2015).

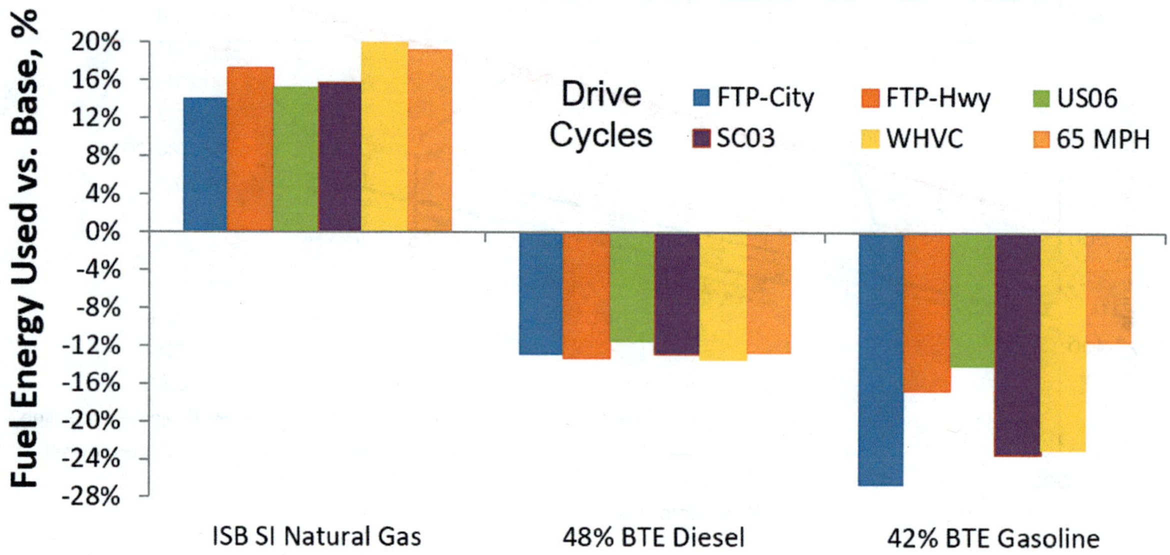

FIGURE 6-10 Fuel energy change across advanced natural gas, diesel and gasoline powertrains in Dodge Ram pickup (Class 2b) at 0 percent payload compared to 2019 baseline diesel simulated over 6 drive cycles.

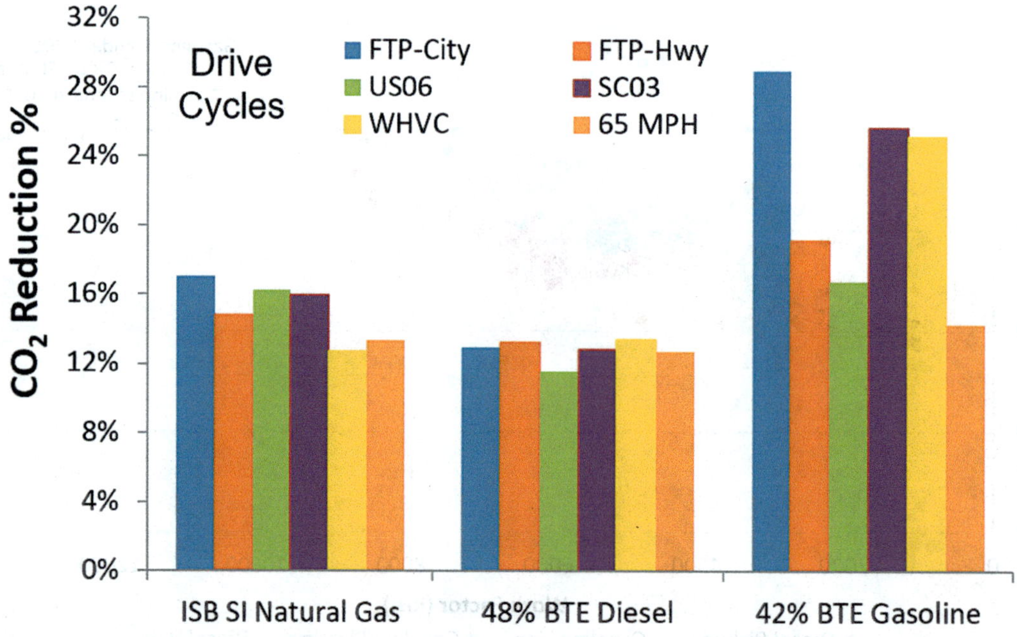

FIGURE 6-11 CO_2 reduction in Dodge Ram pickup across natural gas, diesel and gasoline powertrains at 0 percent payload compared to baseline 2019 diesel simulated over 6 drive cycles.

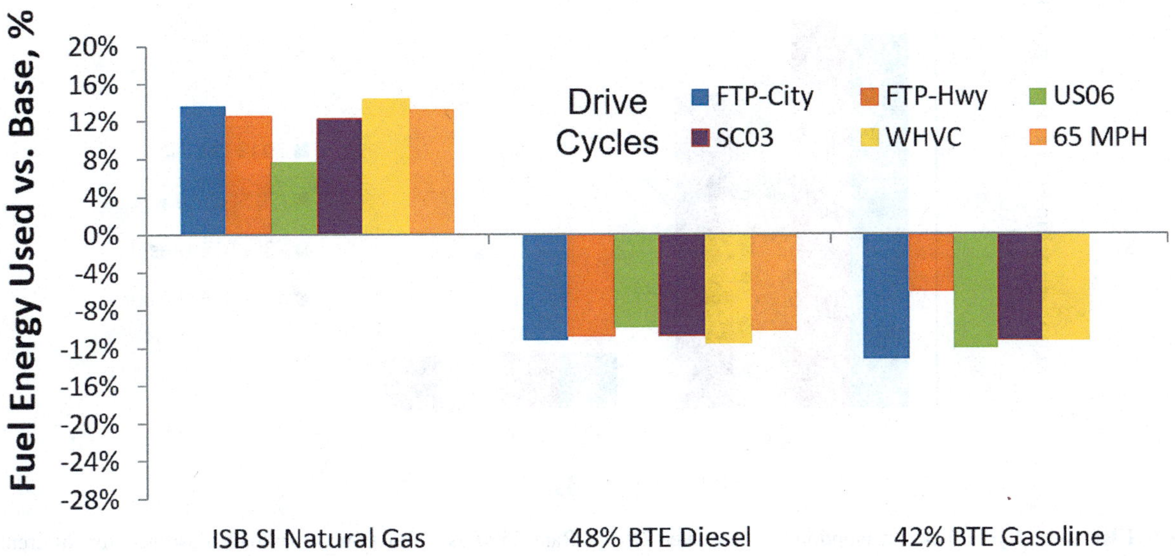

FIGURE 6-12 Fuel energy change across natural gas, diesel, and gasoline powertrains in a Dodge Ram 2500 at 25,000 pounds GCW simulated over 6 drive cycles.

the advanced diesel and the high-efficiency SI engine show greater vehicle efficiency than the baseline diesel engine. The sensitivity of fuel consumption to payload in the Dodge Ram is illustrated in Figure 6-13 for different powertrains.

Shown in Figure 6-14, the CO_2 reduction is greatest for the natural gas vehicle, although both the advanced SI and diesel engines show a 9 to 12 percent improvement relative to the baseline.

Just examining the diesel-powered Dodge Ram 2500 results, these simulations indicate that engine improvements alone would probably not achieve the approximately 17 percent CO_2 reduction in the proposed rules for 2027. Significant vehicle-level improvements would appear necessary with even the stretch diesel (48 percent) engine, with the caveat that the cycles and weighting are not an exact comparison. This is discussed further below.

To facilitate comparisons at the vehicle level on an absolute CO_2 emission basis, in Figure 6-15 the committee overlaid the simulated Dodge Ram 2500 with its advanced engines on the ICCT graph of CO_2 versus work factor. The weighted-average city-highway cycle results were used for the committee-simulated Dodge Ram. The data points are from EPA certification data using 2010 as a baseline, as reported by ICCT.

In this comparison, only the advanced gasoline engine appears close to the CO_2 levels in the Phase II proposal of 2015 (see Figure 6-8). This vehicle as simulated does not incorporate features such as stop-start. The efficiency characteristics of the advanced SI engine have been discussed also in Chapter 4 and in the section on vocational vehicles earlier in this chapter. Based on emerging experimental and modeling results, the expected future efficiency of SI engines appears well beyond the Phase II regulatory requirements depending on the extent that advanced technologies for passenger vehicle engines are adapted to MHDV gasoline engines. This adaptation is already in progress as noted in NHTSA and EPA's review and response to docket comments (EPA and NHTSA, 2016b, p. 1327).

Finding: For large pickups, advanced diesel and SI engines may achieve a moderate 12 to 15 percent CO_2 reduction compared to a baseline 2019 diesel at high payloads, with natural gas engines even more effective if their efficiency can be approved as noted. The values of CO_2 per ton-mile of the advanced SI engine were on par with the advanced diesel in this analysis. When compared at lighter payloads and when using the work factor metric,

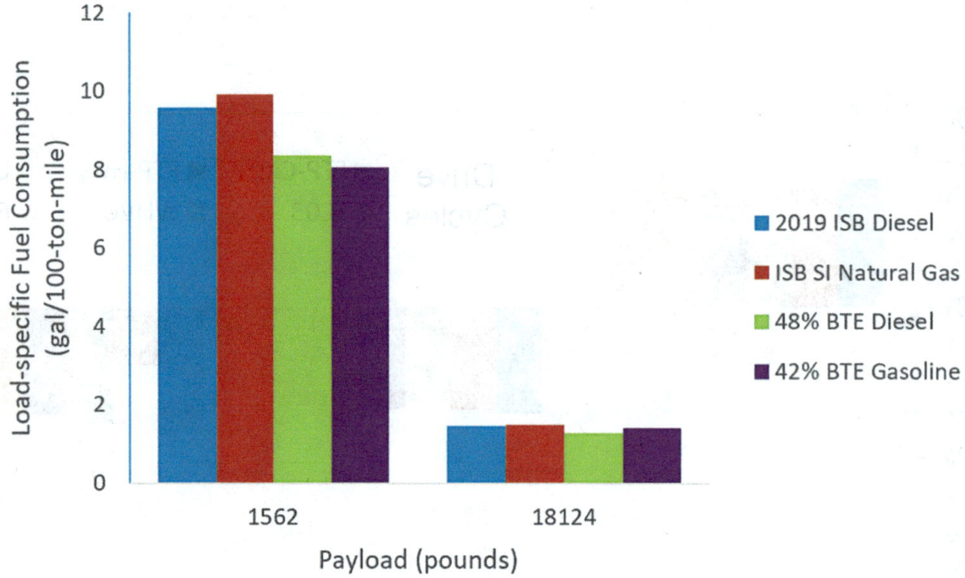

FIGURE 6-13 Cycle fuel consumption for simulated Dodge Ram 2500 as a function of payload illustrated for different powertrains. The payloads chosen represent the adjusted loaded vehicle weight (1,562 pounds) and the maximum payload (18,124 pounds) such that gross combined weight is at the maximum for the vehicle class.

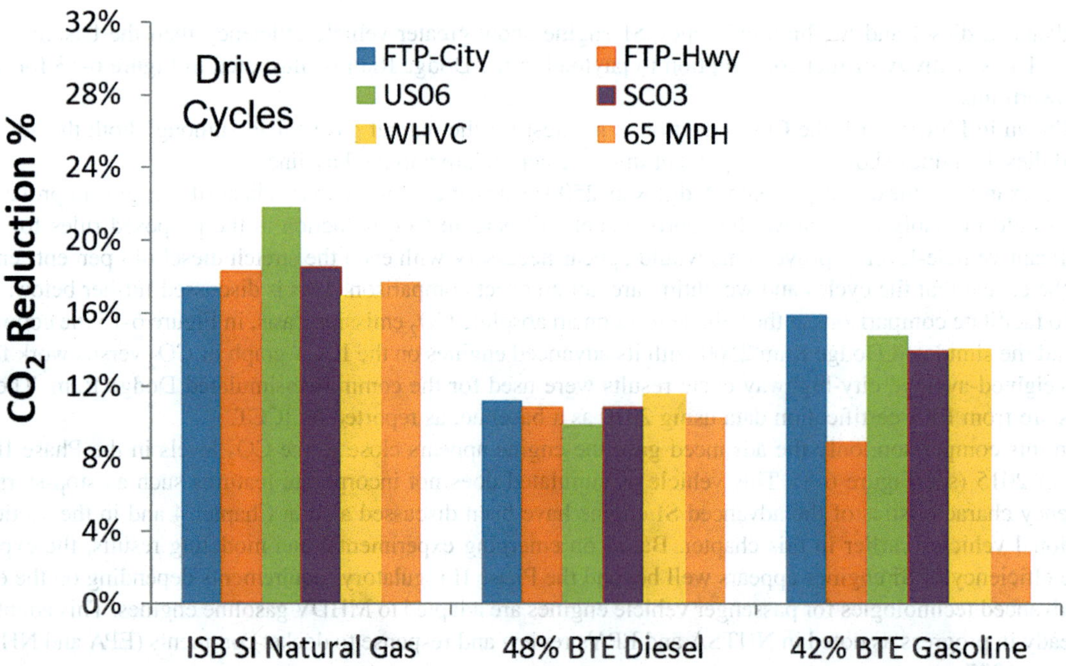

FIGURE 6-14 CO_2 reduction across natural gas, diesel, and gasoline powertrains in a Dodge Ram 2500 at full 25,000 GCW simulated over 6 drive cycles.

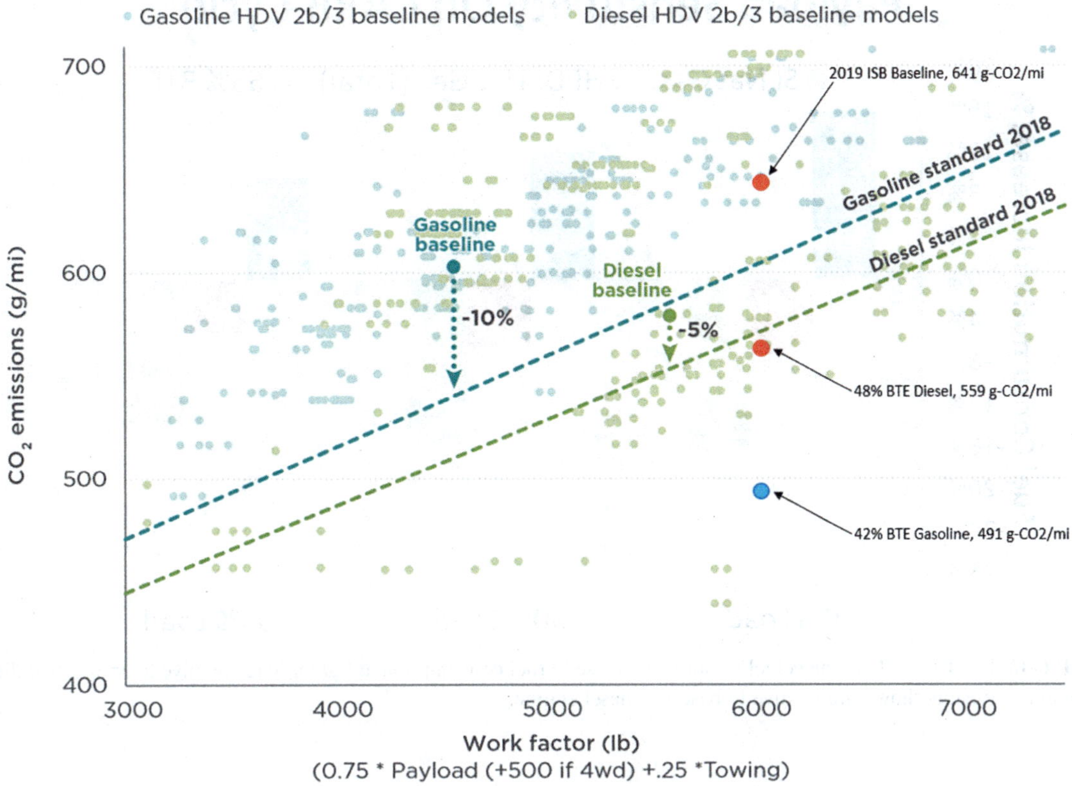

FIGURE 6-15 Comparison of Dodge Ram 2500 simulated vehicle with advanced natural gas, diesel, and gasoline engines to current vehicle data and regulated CO_2 levels.

the advanced gasoline engine stands out at approximately 24 percent GHG reduction. However, engine improvements alone do not appear adequate for Class 2b and 3 pickups to meet 2027 requirements.

6.2 IMPACT OF MASS OR PAYLOAD ON FUEL CONSUMPTION BENEFITS OF DIFFERENT ENGINES

The impact of payload on the mass-specific fuel consumption was reviewed with an expectation that the SI engine vehicles might exhibit a distinctive loss of efficiency at lighter loads and low-speed drive cycles due to more use of throttling in those conditions. The reader may wish to revisit the discussion of mass reduction in Chapter 4. In Figures 6-16 and 6-17, the comparison of diesel and NG engines in the Class 8 truck shows the payload sensitivities are somewhat as expected—that the SI natural gas engine exhibits less efficiency at lighter loads on both transient and steady-speed drive cycles. The HPDI, on the other hand, behaves more like the advanced diesel with relative insensitivity to load.

The expectation might be that this characteristic of throttled versus unthrottled engines would hold true for most vehicle classes—that the SI engines sustain an efficiency penalty at lighter loads (we also note they may suffer efficiency loss at very high loads if fuel enrichment is required). Our analyses showed similarly higher sensitivity to payload in the SI NG engine in the T700, but the trend did not hold for the T270 truck where the *advanced gasoline engine showed the least sensitivity* to payload (Figure 6-18). This places importance in the regulatory process to select an appropriate payload (or multiple payloads, or even payload sensitivity/derivative) to capture real-world effects in the certification process. If all engines had relatively flat efficiency across many payloads, this would be a moot point.

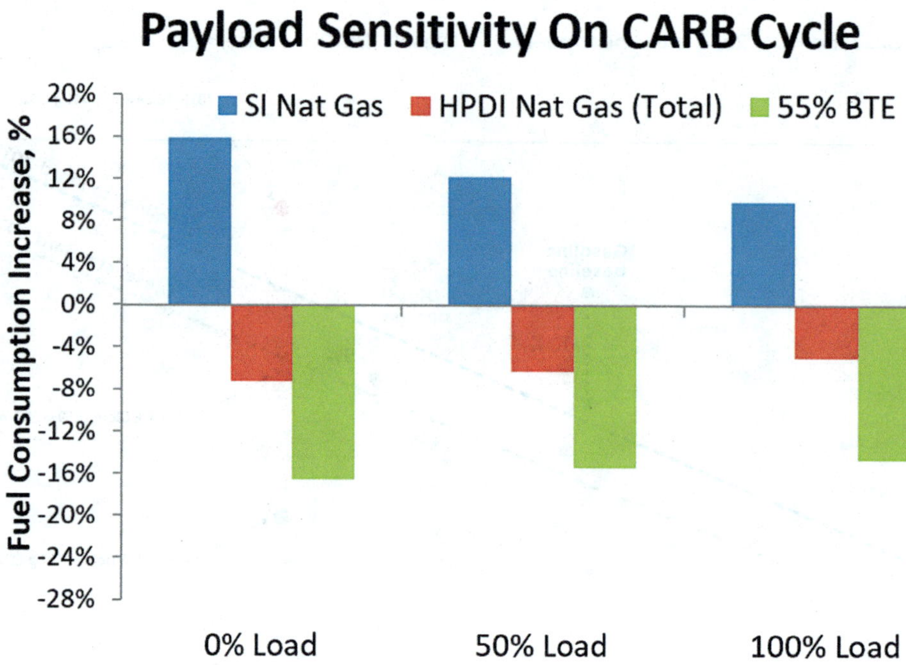

FIGURE 6-16 For T700, SI engine vehicle exhibits increase in fuel consumption at lighter loads relative to unthrottled diesel or NG engines. Results shown are relative to baseline diesel engine.

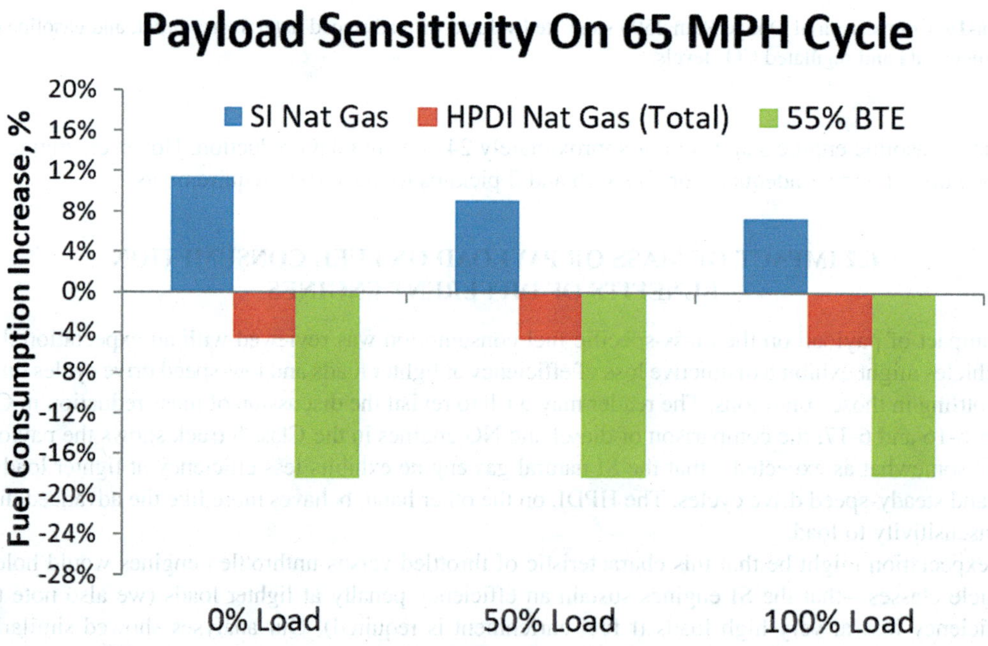

FIGURE 6-17 Payload sensitivity on 65 mph cycle for T700 (Class 8 vehicle). Sensitivity to payload is minimal for unthrottled engines relative to baseline diesel engine. Modest sensitivity seen in SI engine.

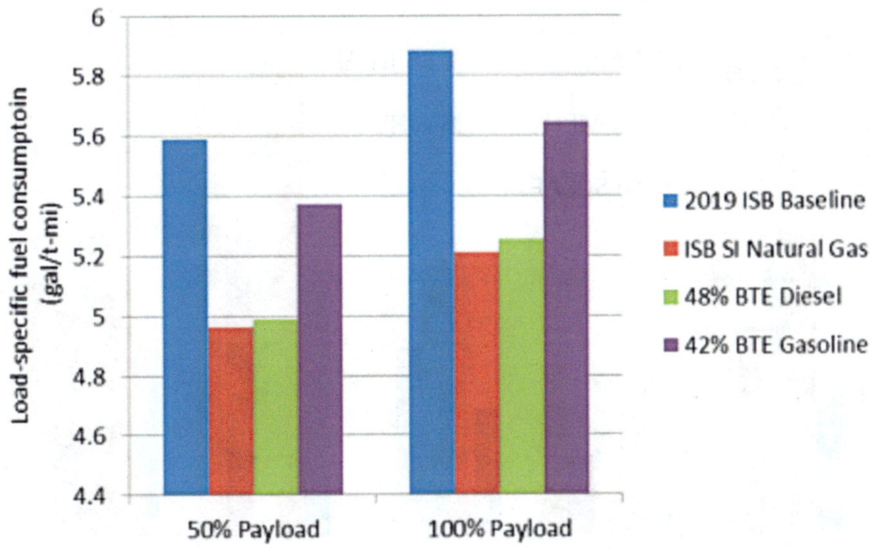

FIGURE 6-18 Sensitivity of load-specific fuel consumption to payload across four advanced powertrains for a Class 6 delivery truck (T270) on the 65 mph cycle.

6.3 DISCUSSION OF ENGINE EFFICIENCIES OVER DRIVE CYCLES

SwRI used their engine maps and vehicle simulations to calculate the cycle average engine efficiency, allowing an approximate comparison to the EPA/NHTSA engine standards for fuel consumption and GHG emissions. For general comparison, we provide here the Phase I and proposed Phase II standards for engine efficiency (Table 6-22). The comparisons are approximate, of course, for a number of reasons, such as test and drive cycles not being matched.

The engine results are presented in Figure 6-19 as BTE instead of CO_2/hp-h to provide a clearer indication of the degree of engine technology stretch in this study, not confounded by fuel carbon content or less familiar CO_2 values. This also facilitates comparison to aggressive research and development efforts such as in SuperTruck Phase II. The NHTSA/EPA proposed Phase II standards for engines called for a reduction in CO_2 of about 4 percent between 2017 and 2027, and the final rule reduction is about 6 percent. If industry were to be successful with achieving near a 55 percent peak efficiency diesel engine—roughly the goal noted above for Phase II of Super-

TABLE 6-22 Comparison of Phase I and Phase II Engine Standards

Vehicle Type		Engine Class	Standard (g CO_2/bhp-h) Baseline (2010)	Phase I (2017)	Phase II (2027)	Percent CO_2 Reduction Phase I Only	Phase II Only	Phase I+II
Spark ignition			660	627	627	-5%	0%	-5%
Compression ignition	Vocational	Light	630	576	553	-9%	-4%	-12%
		Medium	630	576	553	-9%	-4%	-12%
		Heavy	584	555	533	-5%	-4%	-9%
	Tractor	Medium	518	487	466	-6%	-4%	-10%
		Heavy	490	460	441	-6%	-4%	-10%

NOTE: Spark-ignited engines and compression-ignition vocational engines are tested under the heavy-duty Federal Test Procedure cycle while compression-ignition tractor engines are tested under the Supplemental Emissions Test.
SOURCE: Committee generated from EPA Engine Standards Data.

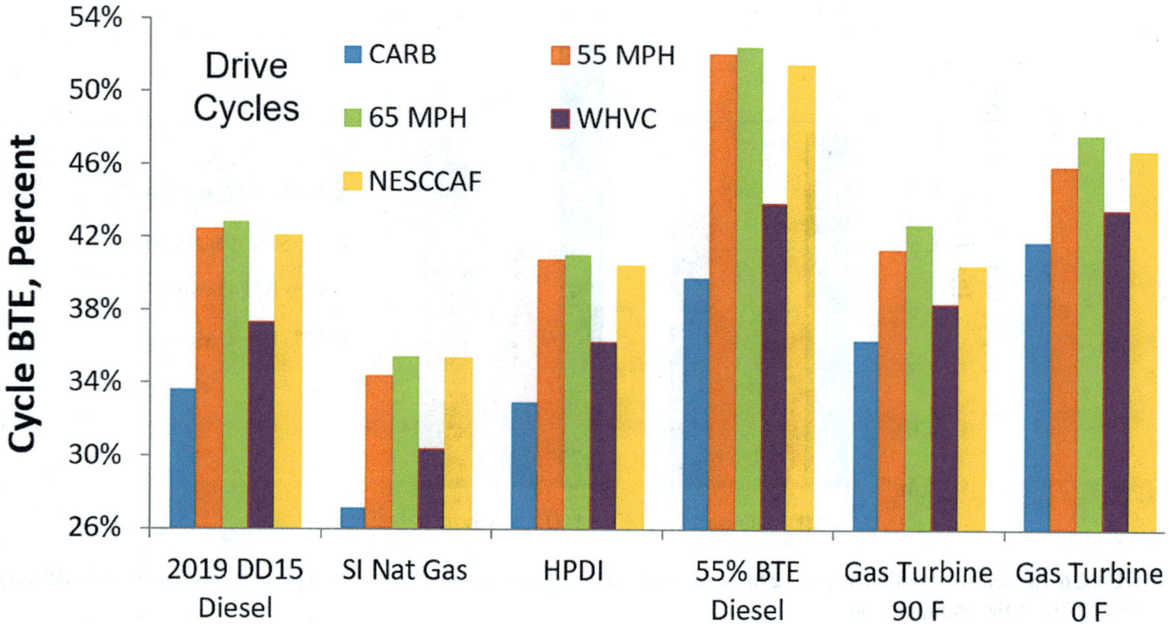

FIGURE 6-19 Cycle average brake thermal efficiencies calculated across 6 advanced powertrains for the Class 8 T700 tractor trailer on 5 simulated drive cycles. Note the 2027 engine standard for this type vehicle is approximately 43 percent, or 432 g CO_2/hp-h over the engine test cycles.

Truck—the potential improvement at the engine level appears to be roughly 17 percent beyond the 2019 baseline engine used in the present study. This finding is approximately echoed by Cummins (Eckerle, 2015), as well as ICCT (Lutsey, 2015). EPA's CO_2 standard for 2017 engines, converted to BTE, is about 41 percent and the 2027 standard (432 g/hp-h for heavy-heavy) is approximately 43.7 percent, somewhat short of the committee's expected performance of a feasible diesel engine.

For the medium-duty vocational vehicles, the committee's projected feasible efficiencies for advanced diesel as well as the gasoline-based engines appear more optimistic than that of NHTSA and EPA (Figure 6-20).

Gasoline-powered large pickups and loose engines for use in vocational vehicles are noted by the committee as somewhat of a paradox. The standards set for full Class 2b vehicles are quite challenging according to the OEMs,[6] yet the SI engine standard (627 g CO_2/bhp-h) for vocational vehicles is conservative relative to the committee's view (shown at different payloads in Figures 6-21 and 6-22) of what SI engines can achieve. The proposed standards for this size of vehicle are generally less aggressive than nearly comparable light-duty vehicles (Lutsey, 2015), but one must recognize that about half the Class 2b pickups are already diesel powered. NHTSA and EPA have generally explained that most of these engines are certified in whole vehicles and that is where the aggressive standards are being applied.

Finding: In general the Phase II engine fuel consumption standards taking effect in 2027 are higher (i.e., less efficient) than is projected to be possible for vocational and pickup engines from the committee's numerical simulations. This difference, however, may be offset by the relatively challenging whole-vehicle standards with which vocational vehicles and pickups must comply in Phase II.

Recommendation 6-3: Future standards for vocational truck engines (especially SI) and vehicle standards such as for Classes 2b and 3 should be in harmony regarding their technology stretch and both should require more ambitious progress in efficiency. This could be considered in the next phase of regulations.

[6] Mark Allen and Barbara Kiss, GM, "GM Heavy-Duty Presentation," presentation to the committee, Sacramento, CA, July 31, 2013.

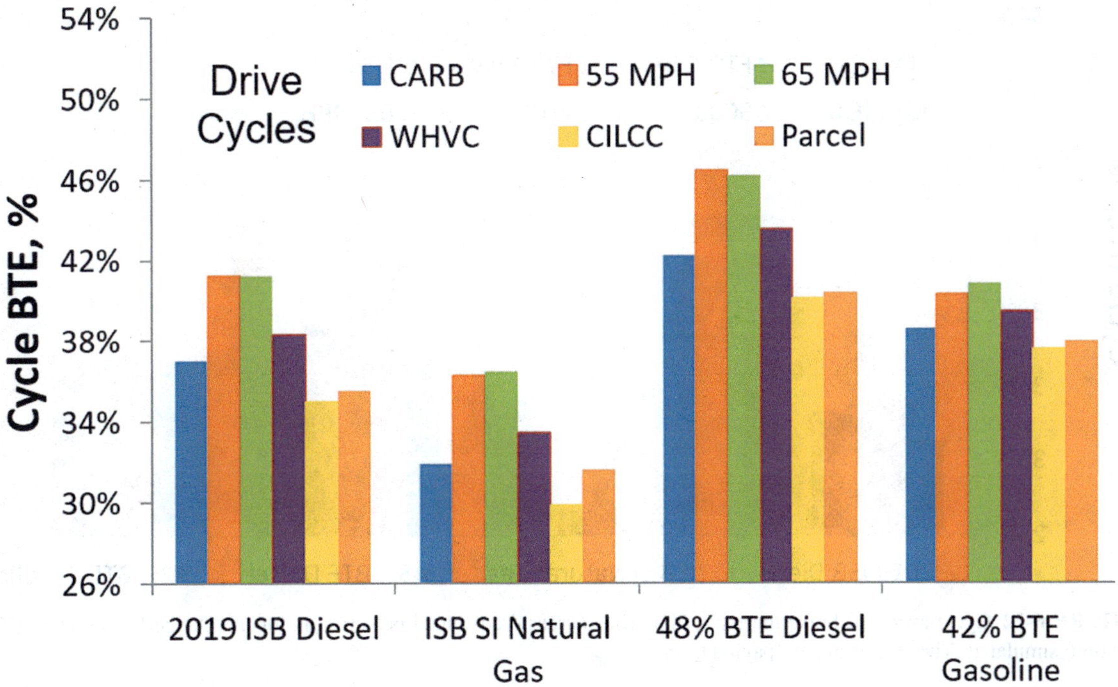

FIGURE 6-20 Cycle average brake thermal efficiency calculated across 4 advanced powertrains for the Class 6 T270 vocational vehicle on 6 simulated drive cycles at 100 percent payload.

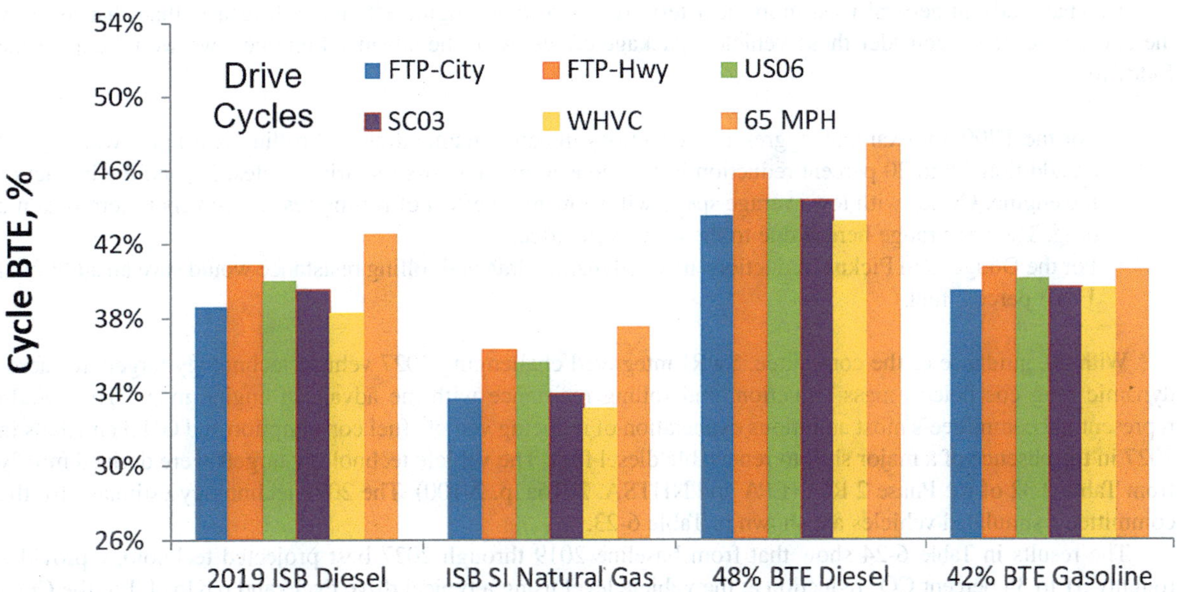

FIGURE 6-21 Cycle-averaged brake thermal efficiencies across 4 advanced powertrains for the Dodge Ram pickup (Class 2b) on 6 simulated drive cycles at full gross combined weight.

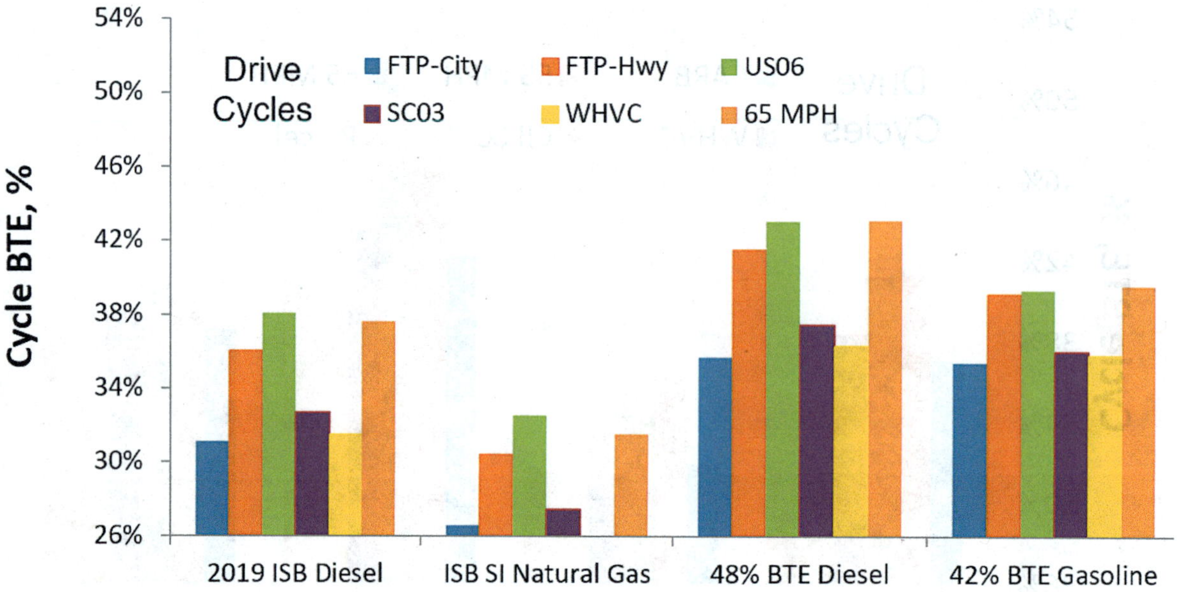

FIGURE 6-22 Cycle-averaged brake thermal efficiencies across 4 advanced powertrains for the Dodge Ram pickup (Class 2b) on 6 simulated drive cycles at zero payload.

6.4 OUTLOOK FOR COMBINATIONS OF ENGINE AND VEHICLE IMPROVEMENTS

In Reinhart (2016), SwRI and NHTSA examined many whole vehicle packages including technologies to reduce aerodynamic drag, rolling resistance, and mass, as well as reductions in auxiliary loads and transmission effects. That study in general used more near-term (conservative) engine efficiency forecasts than that used by the committee. If we consider those vehicles' package effects with the advanced engines, we would expect the following:

- For the T700, for example, aggressive reductions in aerodynamic drag and rolling resistance would yield an additional 8 to 20 percent reduction in fuel consumption across the drive cycles, beyond the impact of the engine. Cycles with low average speed will show more effect of rolling resistance versus aerodynamic drag. The wide range here is due to the duty cycle effects.
- For the Dodge Ram Pickup, reductions in aerodynamic drag and rolling resistance would save an additional 1 to 3 percent fuel.

With the guidance of the committee, SwRI integrated challenging 2027 vehicle technology targets for aerodynamic drag coefficient, mass reduction, and rolling resistance with the advanced engine maps. This would represent the committee's most ambitious expectation of reducing vehicle fuel consumption and GHG emisions in 2027 in the absence of a major shift to renewable diesel fuel. The vehicle technology targets were derived mostly from Table 2.32 of the Phase 2 RIA (EPA and NHTSA. 2016a, p. 2-100). The 2027 technology estimates for the committee's simulated vehicles are shown in Table 6-23.

The results in Table 6-24 show that from baseline 2019 through 2027 best projected technology provides roughly 31 to 35 percent CO_2 reduction at the vehicle level using a typical drive cycle and payload. For the Class 8 truck, the committee's forecast is comparable to NHTSA and EPA's Phase II regulation, and is further similar to other analyses (ICCT, 2015a; ICCT, 2015b). The committee estimates for vocational and heavy pickups are more optimistic, mostly due to the engine efficiency projections.

TABLE 6-23 Attributes for 2027 Vehicles Used in Simulation of Integrated Advanced Engines and Vehicle Technology

	T700	T270	Ram Pickup
Aero drag coefficient, C_dA	4.83 Approx C_d is 0.45[a]	4.278[b]	1.354[a]
Rolling coefficient, kg/tonne	Trailer 4.7 Steer-5.6 Tractor-5.9	7.67[b]	5.46[b]
Mass reduction, lbs	2,000	1,100[b]	500[b]
Other improvements (idling reduction, cruise control, lubes, etc.)	Use RIA Table 2-32, and NHTSA 812-146		

[a]Note that for a fully integrated tractor-trailer as in the SuperTruck projects we could expect even lower C_dA.
[b]From NHTSA 812-146, Tables C8, C9, and C11.

TABLE 6-24 CO_2 and Fuel Consumption of Simulated 2027 Vehicles with Base through Advanced Engines, Vehicle CO_2 (g/ton-mile) and Fuel Consumption (gallons per 1,000 ton-mile)

	Base Vehicle and Base Engine	Baseline 2019 Diesel, Adv Vehicle	55% HD Diesel, Adv Vehicle	50% HD Diesel, Adv Vehicle	48% MD Diesel, Adv Vehicle	Percent Change, Base to Most Efficient
T700	127.4[a]	98.9[a] g/ton-mi 9.7[a]	82.3	90.4		-35%
	12.5[a]		8.1	8.9		
T270	559[b]	442[b] g/ton-mi 43.4[b]			385	-31%
	54.9[b]				37.8	
Ram	821[c]	642 g/ton-mi 63.3[c]			558.6	-32%
	80.7[c]				54.8	

[a]At 24,020 lbs payload (50%), 65 mph cycle. If comparing to 2027 standards note that the Agencies' payload is 38,000 lbs. A rough interpolation of results using NHTSA and EPA's payload would reduce the committee advanced vehicle CO_2 from 82.3 to about 60 g CO_2/ton-mile, This is about 9% below the 2027 standard.
[b]At 4,430 lbs (50%) payload, CARB cycle.
[c]At 1,562 lbs payload, ALVW, average of 55% city and 45% highway cycles.

Finding: Combining projections for advanced engines and vehicle improvements, simulations for the committee show potentially over 30 percent reduction in CO_2 and fuel usage for all vehicles studied. The technology projections are generally aggressive (in contrast to conservative).

Recommendation 6-4: NHTSA and EPA and future committees should determine whether these ambitious simulations of vehicle-level reductions in CO_2 emissions would be consistent with national goals and international agreements to which the United States is a party and, if not, should consider alternative pathways and regulatory approaches[7] to achieve greater reductions than just at the vehicle level.

[7] Examples would be larger and/or heavier combination vehicles, low-GHG fuels, and reducing traffic congestion.

6.5 REFERENCES

Eckerle, W. 2015. Beyond GHG Phase 2 Rule, 2015 SAE ComVEC Executive Panel on GHG Legislation, Rosemont, IL, October 7.

EPA (Environmental Protection Agency) and NHTSA (National Highway Traffic Safety Administration). 2011. Greenhouse Gas Emissions Standards and Fuel Efficiency Standards for Medium- and Heavy-Duty Engines and Vehicles; Final Rule. *Federal Register* 76:57106-57513. September 15.

EPA and NHTSA. 2016a. Greenhouse Gas Emissions and Fuel Efficiency Standards for Medium- and Heavy-Duty Engines and Vehicles—Phase 2: Regulatory Impact Analysis. Publication No. EPA-420-R-16-900, August.

EPA and NHTSA. 2016b. Greenhouse Gas Emissions and Fuel Efficiency Standards for Medium- and Heavy-Duty Engines and Vehicles—Phase 2: Response to Comments for Joint Rulemaking. Publication No. EPA-420-R-16-901, August.

EPA and NHTSA. 2016c. Greenhouse Gas Emissions and Fuel Efficiency Standards for Medium- and Heavy-Duty Engines and Vehicles—Phase 2. *Federal Register* 81:73478-74274, October 25.

ICCT (International Council on Clean Transportation). 2015a. United States Efficiency and Greenhouse Gas Emission Regulations for Model Year 2018-2027 Heavy-Duty Vehicles, Engines, and Trailers: Policy Update. Washington, DC: ICCT, June.

ICCT. 2015b. Parsing Phase 2 Tractor-Trailers Proposed Regulation. Washington, DC: ICCT, July.

Langer, T. 2015. Fuel Efficiency and GHG Standards for Heavy-Duty Vehicles in the U.S. 8th Forum on Energy Efficiency in Transport, Mexico City, September 29.

Lutsey, N. 2015. Proposed U.S. Heavy-Duty Standards: Technologies, Costs, and Global Context SAE Commercial Vehicle Engineering Congress, Rosemont, IL, October 7.

NRC (National Research Council). 2014. *Reducing the Fuel Consumption and Greenhouse Gas Emissions of Medium- and Heavy-Duty Vehicles, Phase Two: First Report.* Washington, DC: The National Academies Press.

Reinhart, T.E. 2015. Commercial Medium- and Heavy-Duty Truck Fuel Efficiency Technology Study—Report #1. Report No. DOT HS 812 146. Washington, DC: NHTSA, June.

Reinhart, T.E. 2016. Commercial Medium- and Heavy-Duty Truck Fuel Efficiency Technology Study—Report #2. Report No. DOT HS 812 194. Washington, DC: NHTSA, February.

7

Hybrid and Electric Powertrain Technologies

7.1 INTRODUCTION

The quest to reduce fuel consumption and greenhouse gas emissions of vehicles has generated significant interest in electrification and hybridization of powertrains and other vehicle systems. Various government programs for vehicle efficiency have been in place during the commercialization of hybrid electric vehicles in the light-duty sector, such as the Toyota Prius. At the national level, the Corporate Average Fuel Economy program has been in place since the 1970s (NRC, 2015). The State of California has its Zero Emission Vehicle Standard.[1] The federal government supports research and development of relevant technologies. Specifically, a significant portion of the U.S. Department of Energy's (DOE's) vehicle technology budget (>$2 billion) has been devoted to hybrid, electric, and battery technology development over the past 10 years, primarily directed toward light-duty vehicles.[2] To a lesser extent, DOE has supported research and development in the application of electric powertrains and related systems through their 21st Century Truck Partnership activity under the Vehicle Technologies Office within the Office of Energy Efficiency & Renewable Energy (EERE).

The Environmental Protection Agency (EPA), National Highway Traffic Safety Administration (NHTSA), and California Air Resources Board have all published rules, proposed rules, and performed regulatory impact analyses that show a significant penetration of hybrid, plug-in hybrid, and battery electric vehicles is required in order to meet future light-duty vehicle fuel economy and greenhouse gas (GHG) emission standards, despite lackluster sales of hybrid and electric vehicles. For example, the segment-leading Toyota Prius saw its sales reach 236,000 units in 2012 only to decline to 136,000 in 2016.[3] Unfortunately, due to the difference in vehicle size, weight, duty cycle, and the energy necessary to propel the vehicles, the market application of electrification and hybridization technology to commercial vehicles significantly lags that of passenger vehicles for reasons discussed below.

7.1.1 Hybrid Powertrains and Uses

The term "hybrid" refers to the combination of two or more energy conversion sources. A typical hybrid electric system includes an internal combustion engine (ICE) and an electric motor/generator system that can

[1] See https://www.arb.ca.gov/msprog/zevprog/factsheets/zev_tutorial.pdf (accessed December 1, 2019).
[2] See https://energy.gov/eere/budget/past-and-present-eere-budget (accessed December 1, 2019).
[3] See http://carsalesbase.com/us-car-sales-data/toyota/toyota-prius/ (accessed December 1, 2019).

extract energy in the form of electricity from a battery to propel the vehicle. The battery system can also store the energy that has been recovered during braking or from the excess power generated by the ICE while running in a high-efficiency point. Other hybrid systems employ different combinations of energy conversion devices such as an ICE and a hydraulic pressure accumulator storage system or a fuel cell propulsion energy system combined with an ultra-capacitor to store high levels of electric power transferred during vehicle braking and subsequent launch.

The goal of a hybrid propulsion system is to manage energy flows most efficiently throughout the drivetrain system, thereby reducing fuel consumption and GHG emissions, by

1. Allowing the engine to be turned off during inefficient operating conditions, such as idling, and to be restarted very quickly when necessary (stop-start);
2. Operating the engine as efficiently as possible by running it near the curve of highest efficiency as much as practicable;
3. Following functional control strategies that supplement the engine power during periods of acceleration (launch assist), which can reduce enrichment or transient control approaches that increase fuel consumption, due to the excellent low-speed torque characteristics of electric motors;
4. Providing supplemental power and launch assist, which also offers the possibility to down-size the ICE while maintaining equal performance, so long as operational strategy maintains sufficient state of charge in the battery;
5. Operating the vehicle accessories, such as air conditioning compressors or power steering pumps, independent of engine speed, which allows both vehicle operation while the engine is not running and the possibility to operate the accessories at closer to peak efficiency points;
6. Using batteries and drive motors to support the potential for other forms of energy transfer, such as generation of electric current via waste-heat recovery systems (NHTSA and EPA, 2016) or the adaptation of fuel cells;
7. And, perhaps most importantly, using hybrid systems, both electrical and hydraulic (discussed below), which allow a mechanism to store energy that is normally lost to heat during braking, that can be recovered and applied during subsequent acceleration of the vehicle (regenerative braking). The significance of regeneration becomes apparent when one considers that approximately 60 percent of the total energy spent in the Federal Urban Driving Schedule (FUDS) is used to overcome the effect of inertia and that, theoretically, up to 50 percent of this energy could be recovered (NRC, 2015).

7.1.2 Hybrid Vehicle Architectures

The National Research Council (NRC, 2010a) report *Technologies and Approaches to Reducing the Fuel Consumption of Medium- and Heavy-Duty Vehicles* provides a complete and exhaustive description of the various forms of hybrid vehicle architectures which are applied to light-, medium-, and heavy-duty vehicles. Since this NRC Phase One Report was published, there have been no significant changes in overall architecture, although the costs, system efficiencies, and results of recent DOE SuperTruck research programs are included within the present report as an update and state-of-the-art statement.

7.1.2.1 Series Hybrid Electric Powertrains

In a series hybrid electric vehicle (HEV), a generator, coupled to an ICE, supplies electricity to the electric motor to propel the vehicle and/or to the energy storage system for recharging the battery. Normally, the engine/generator charges the energy storage system to between 50 and 70 percent for charge sustainability.

As shown in Figure 7-1, in this configuration only the electric motor is mechanically connected to the wheels; the engine is decoupled. Consequently, the engine can run near its best efficiency zones and it never idles, thereby reducing fuel consumption. However, the engine/generator set must be sized for sustained gradeability. Therefore, a sufficiently large energy storage system pack (needed so as not to cycle the engine on and off too frequently), an electric generator and a separate electric motor, sufficiently large to propel the vehicle, and an engine of sufficient power for performance and gradeability are required.

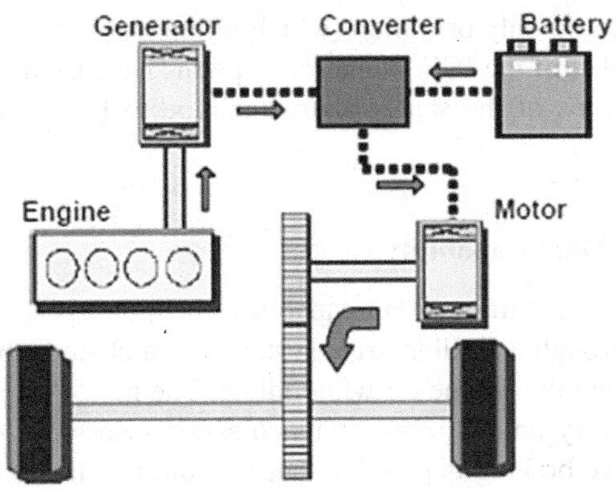

FIGURE 7-1 Series HEV configuration.
SOURCE: NRC (2010a).

For this system to be viable, it must possess an overall high efficiency in total power processing. While the added mass and the component inefficiencies make that configuration unlikely for small trucks, it is more viable for large vehicles, such as buses, which are less sensitive to increases in overall weight. The elimination of the mechanical driveshaft can be an important advantage of adopting the series configuration.

7.1.2.2 Parallel Hybrid Electric Powertrains

Parallel hybrids have mechanical connections to the wheels from both the motor/generator and the engine, as illustrated in Figure 7-2. The motor/generator can be positioned in any of multiple locations between the output shaft of the ICE and the wheels. Unlike the series HEV configuration, a single motor/generator unit can be used to drive the wheels or generate electricity during regenerative braking to recharge the battery.

In a parallel HEV, the motor/generator can supplement the ICE during start-up or vehicle acceleration because both are coupled to the wheels. Therefore, in theory, the ICE can be reduced in size and power compared to series hybrids unless the intended operational mode requires extended high-power operation, such as climbing a long hill. In some configurations the ICE has the potential to run near its optimum efficiency, with the motor/generator either supplementing the drive power or recharging the battery if all of the ICE power is not needed to drive the wheels. The type of parallel hybrid is normally classified by the size of the motor generator, the amount of time

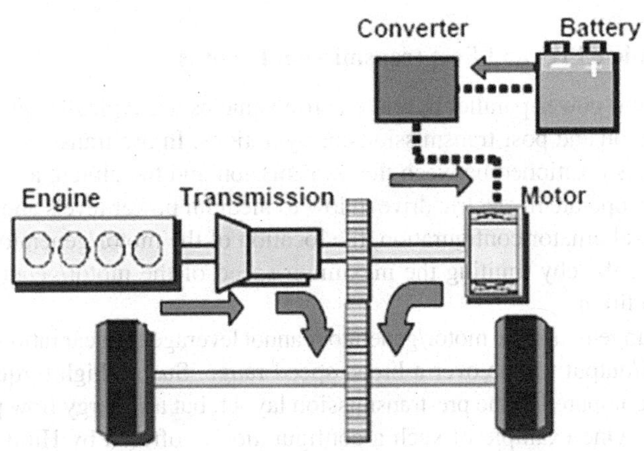

FIGURE 7-2 Parallel HEV configuration.
SOURCE: NRC (2010a).

the powertrain can function in electric-only operation (which influences the size of the battery), and the relative power of the electric machine to that of the ICE. Numerous options have been studied for parallel architectures, ranging from micro or mild versions of the starter alternator type to full-featured pre- and post-transmission hybrids, as described below.

7.1.2.3 Micro Hybrid with Stop-Start Capability

In a starter-alternator-type system with stop-start capability, the motor/generator is fixed to the engine output shaft (crankshaft) or connected through a flexible drive system like a chain or belt as shown in Figure 7-3. The main benefit of this system is to shut the engine off while idling. The motor/generator is connected to the engine with a fixed speed ratio, so some very limited level of launch assist is possible, but only up to a very low speed. Also, a certain level of regenerative braking is possible, but it is limited due to the torque transfer capability of the motor/generator drive belt and the size and voltage of the battery. Electrification of climate control accessories (e.g., the air conditioner compressor) is also possible, again depending on the size of the battery. A micro hybrid is typically a 12 to 24 volt (V) system with the maximum power limited to about 4 to 5 kilowatts (kW), although some system suppliers offer a 48 V system that provides a higher level of regenerative energy storage combined with a 7-8 kW motor/generator.[4] Energy storage capability is typically in the 1.5 to 3 kilowatt-hour (kWh) range.

7.1.2.4 Mild Hybrid with Stop-Start, Launch Assist, and Regenerative Braking

A starter-alternator-type system may also be designed with somewhat higher power, torque, and energy storage capability. This would classify the system as a mild HEV, although the largest fuel consumption benefit is still derived from stop-start functionality. The system configuration would also be represented by Figure 7-3. In a mild hybrid, the amount of power generated during regenerative braking is increased to about 7 to 12 kW depending on the size of the electric motor, the system voltage, and the battery size. System voltages for mild hybrids which offer higher levels of regenerative braking and some launch assist typically range from 24 to 48 V. The total energy storage capability varies significantly, depending on the weight classification of the vehicle, but is typically in the 3-7 kWh range. Electrification of certain accessories (e.g., the air conditioner compressor, power-steering pump) is also common to assist in driving while the engine is stopped during deceleration and low-speed launch. Many automotive suppliers, including, but not limited to, Delphi, Bosch, and Continental, are pursuing 48 V systems which can be integrated into the belt drive, thereby replacing the conventional alternator. Such systems provide stop-start, regenerative braking, and propulsion launch assist. A 48 V battery stores the recuperated braking energy, which can then be used for torque assist during launch. A direct current converter (DC/DC) steps down the voltage to 12 or 24 V depending on the application.

7.1.2.5 Hybrid Electric Vehicle Pre- and Post-transmission Layouts

Larger energy storage and power parallel hybrid electric vehicles are typically referred to as HEVs and can be designed in pre-transmission and post-transmission configurations. In pre-transmission parallel HEV configurations, the motor/generator is positioned between the transmission and the clutch, as shown in Figure 7-4. This position allows the system to operate in electric drive in low to medium power levels and support engine stop-start operation. Unlike the starter-alternator configuration, the location of the motor/generator before the transmission allows torque multiplication, thereby limiting the maximum speed of the motor/generator and allowing torque assist under high-power conditions.

In post-transmission arrangements, the motor/generator cannot leverage the gear ratios through the transmission. This necessitates high input/output torque over a broad speed range. Such a high torque requirement leads to a larger-size motor/generator compared to the pre-transmission layout, but an energy flow path that is more efficient during regenerative braking. One example of such a configuration is offered by Hino Trucks in its Model 916

[4] See http://www.johnsoncontrols.com/content/us/en/products/power_solutions/products/micro-hybrid1.html (accessed July 2, 2013).

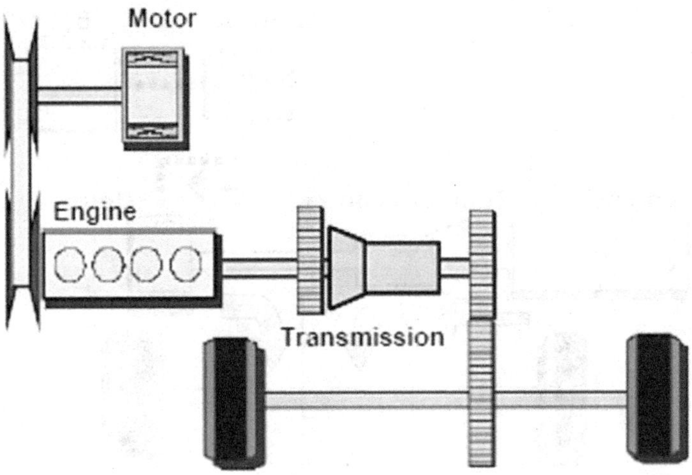

FIGURE 7-3 Micro/mild HEV configuration.
SOURCE: NRC (2010a).

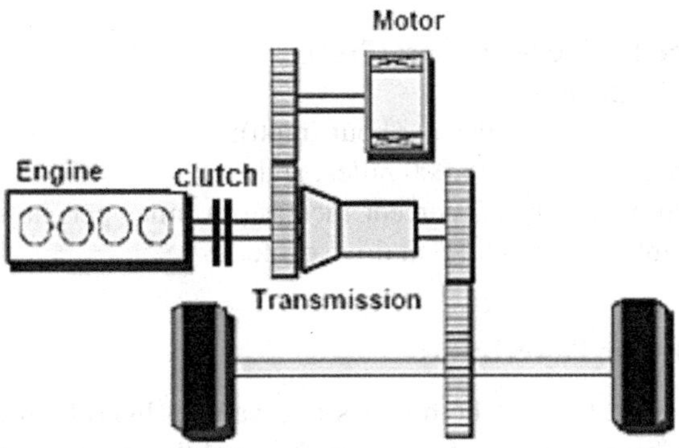

FIGURE 7-4 Pre-transmission parallel HEV configuration.
SOURCE: NRC (2010a).

diesel electric hybrid.[5] According to Hino, the hybrid system components have been redesigned to reduce weight, reduce volume, and offer improved performance.

With the redesign, Hino has implemented the use of its own "ProShift 5" automated mechanical transmission and relocated the clutch, placing it between the diesel engine and electric motor. This placement of the clutch and traction motor allows electric-only launch. Also, during deceleration, the clutch may be decoupled to maximize the amount of regenerative kinetic energy recovered via the electric generator. All Hino hybrid trucks use a nickel-metal hydride (NiMH) battery for their hybrid driveline systems, which is the same Toyota battery technology that has been used in over 3 million HEVs.

An example of a post-transmission parallel HEV configuration (Figure 7-5) is offered by XL Hybrid, Inc. (Figure 7-6). The XL system is available on a range of new commercial fleet vehicles from Chevrolet, GMC, Ford, and Isuzu. The so-called XL3® Hybrid Electric Drive System is compatible with Class 2, 3, 4, 5, and 6 vans and passenger wagons, commercial vans and shuttles, box trucks, and delivery vans. XL Hybrids reports that installation takes less than 6 hours per vehicle and the performance specifications are as follows:

- System voltage, 259 V;
- Lithium-ion battery, 1.8 kWh;

[5] See http://www.hino.com/hino-trucks-hino-195h.html (accessed December 1, 2019).

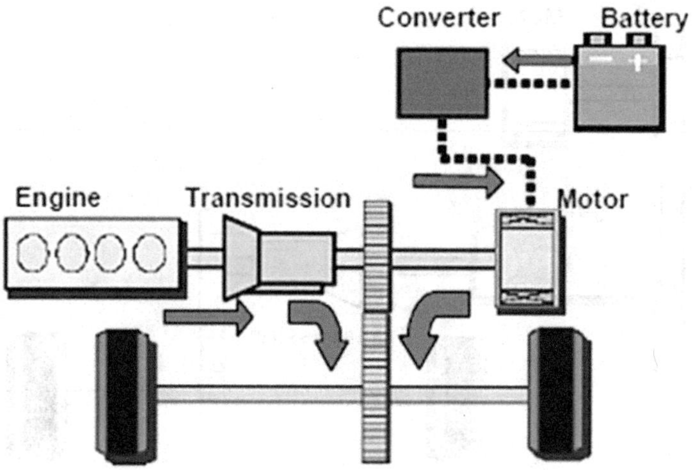

FIGURE 7-5 Post-transmission parallel HEV configuration.
SOURCE: NRC (2010a).

- Hybrid system torque (peak), 220 pound-feet (lb-ft);
- Hybrid system weight, 385 pounds;
- Electronic speed governor up to 85 miles per hour (mph);
- Hybrid component warranty, 3-year / 75,000 miles; and
- Hybrid fuel savings, approximately 25 percent increase in miles per gallon (mpg) or about 20 percent reduction in fuel consumption, depending on drive cycle.

7.1.2.6 Power-Split Hybrid Electric Powertrains

Power-split hybrids combine features of both series and parallel hybrids in an attempt to produce a more efficient drive system. The term "power-split" refers to the ability of the system to "split" the output engine power between two paths: one through a mechanical transmission and drivetrain system to the wheels and another path that drives the generator. Also, regenerative braking can generate electricity through the normal drive motor, which is then stored in the battery or other energy storage system.

An input-split system, shown in Figure 7-7, utilizes an engine, a power-split transmission, a generator, and a drive motor. Operational modes can include

- All engine power through the transmission to drive the wheels if the battery is depleted;
- Partial engine power to the wheels and the remaining engine power to the generator, which charges the battery or energy storage device;
- Engine off and electric power provided to the drive motor from the battery;
- Engine on and additional electric power provided from the drive motor for maximum acceleration; and
- During deceleration and braking, the drive motor acting as a generator and recharging the battery.

During engine starting, electricity powers the generator, thereby causing it to act temporarily as a motor and serving as the "starter motor." It is interesting to note that generator and separate drive motor, combined with the planetary gear set transmission, "behave as a continuously variable transmission (CVT) whose instantaneous gear ratio depends on the amount of power fed from the generator to the motor" (NRC, 2010a).

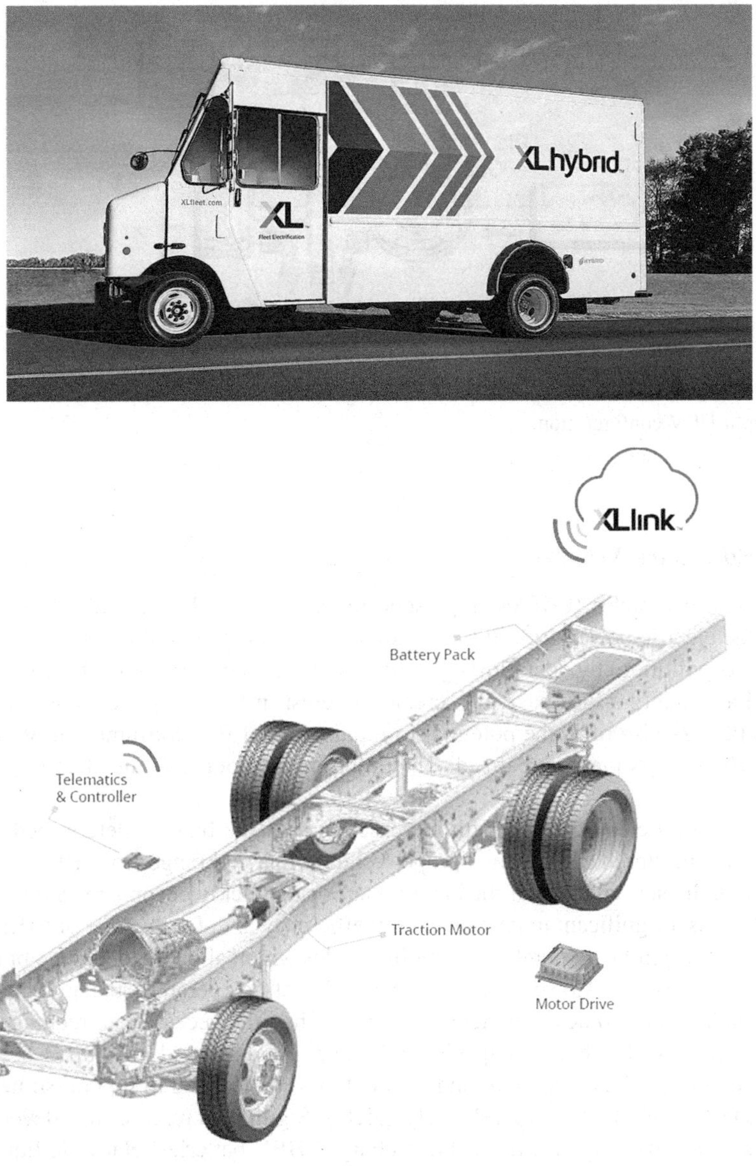

FIGURE 7-6 XL Hybrid, Inc. parallel hybrid.
SOURCE: XL Hybrids, Inc. (2018).

Variations of the power split have been implemented, each providing different advantages. These include

1. A single-mode power-split hybrid, where the first generator controls engine speed and the drive motor supplements any power needed to propel the vehicle; and
2. A dual-mode power-split hybrid with increased complexity but somewhat higher efficiency by minimizing the motor sizes and energy recirculation by using fixed gears.

The reader is again directed to the Phase I report for further details (NRC, 2010a).

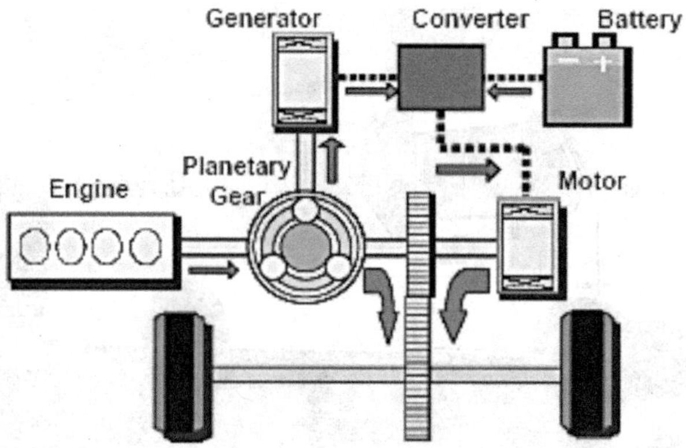

FIGURE 7-7 Power-split HEV configuration.
SOURCE: NRC (2010a).

7.1.2.7 Plug-in Hybrid Electric Vehicles

Plug-in hybrid electric vehicles (PHEVs) represent an extension of HEVs with the potential to charge the battery system from the electric power grid. Initially motivated by the potential to reduce local criteria pollutants (smog and health effects pollution) by operating light-duty vehicles over a period of time in electric mode without the ICE engine, local air quality issues have also created interest in PHEV applications for urban medium- and heavy-duty vehicles. PHEVs also offer the potential to replace liquid fuel consumption with grid electric power whereby CO_2 and GHG benefits may be realized, depending on the source of the electric power (nuclear, wind, hydro, solar, etc.).

PHEVs gain extended all-electric range with increased size of the battery, determined by goals for electric operation and cost. Since the stored energy must propel vehicles that can range from 10,000 pounds (Class 2b) to 80,000 pounds (Class 8), the size and cost of the battery packs can quickly become prohibitive. As with all hybrid vehicles, the duty cycle has a significant impact on energy efficiency and, for the case of PHEVs, the potential all-electric range. If there are a significant number of braking cycles which allow regenerative braking to recharge the battery, additional range is possible. However, during charge-depleting operation, the high power requirements for acceleration of heavy vehicles can quickly deplete the battery. The time necessary to recharge the PHEV batteries is also an important issue related to battery capacity and power. As discussed in Chapter 8, batteries designed to provide higher power output to propel medium- and heavy-duty vehicles tend to support somewhat faster recharge cycles due to their need to transfer high power relatively quickly. Significant research and development are currently ongoing to address the relatively long time needed to recharge PHEV batteries relative to liquid or gaseous fuels.[6] Vehicle manufacturers must undertake trade-off studies to determine the optimum battery size and its associated cost when developing PHEVs. Information concerning cost, size, and complexity of battery systems is presented in Chapter 8.

For medium- and heavy-duty vehicles (MHDVs), there are two general applications in which PHEVs seem most attractive: (1) relatively short-route urban delivery or service applications and (2) vocational vehicles that travel to a job site and then perform operational functions where use of electric power instead of operating the combustion engine can significantly reduce liquid fuel consumption.

An example of the latter application is in electric utility service trucks, sometimes called "bucket trucks." In such applications, travel to a job site may be followed by many hours of stationary vehicle operation of an extension arm, at the end of which is a worker elevated in a capsule or "bucket." Two commercial bucket-truck applications are offered by Odyne Hybrid Systems and Efficient Drivetrain, Inc. (Figure 7-8). The Odyne system utilizes the mechanical power-take-off (PTO) of Allison transmissions and is a parallel system configuration.

[6] See https://energy.gov/sites/prod/files/2015/06/f23/es000_faguy_2015_o.pdf (accessed December 1, 2019).

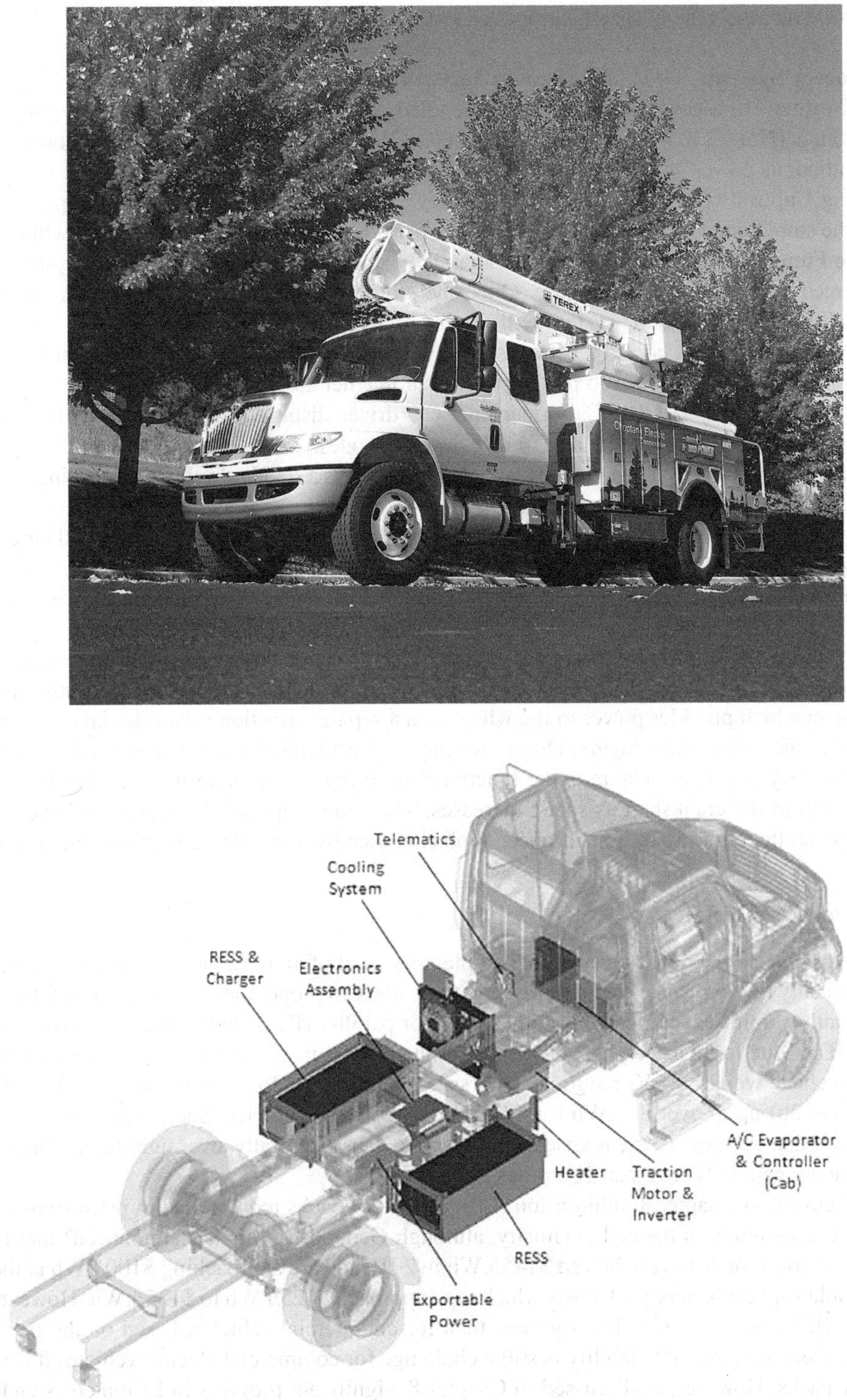

Standard hybrid system layout with Optional HVAC and exportable power

FIGURE 7-8 Odyne hybrid system PHEV electric utility service truck.
SOURCE: Odyne Hybrid Systems, LLC (2014).

According to Odyne, the vehicle specifications are as follows:

- **Stored Energy System:** 300 V lithium-ion battery pack with 28.4 kWh capacity.
- **Electric Motor:** High-power, compact liquid-cooled, permanent magnet motor providing up to 311 Newton-meters (Nm) (232 lb-ft) peak torque with 71 kW peak power and 200 Nm continuous torque and 42 kW continuous power with launch assist while driving and regenerative braking.
- **Recharging Capability:** Plug-in capability with level two 240 V single-phase III0-amp service to fully recharge the stored energy system within an 8-hour period. Engine-powered recharging while in the field.
- **Work-Site Functionality:** The engine is automatically started when the stored energy system is depleted and recharged with the vehicle's engine. Once the stored energy system is recharged, the engine will automatically shut down and the system will resume all-electric stationary mode.
- **Inverter Assembly:** Liquid-cooled inverter assembly converting the 300 V stored direct current (DC) power to alternating current (AC) power to power permanent magnet motor.
- **Liquid-Crystal Display (LCD):** Cab-mounted LCD driver display screen providing the operator with valuable plug-in hybrid system data and troubleshooting codes.
- **Override:** Hybrid override system to allow vehicle to operate as a standard vehicle for driving and stationary application if the PHEV system is not functional.
- **Auxiliary Heating and Air Conditioning System:** Auxiliary cab heating, ventilation, and air conditioning system for use during all-electric mode while the engine is not running.
- **Exportable Power:** Frame-mounted inverter assembly providing 6 kW of pure sine wave power.

Efficient Drivetrains, Inc. (EDI) offers a similar product application; however, there are significant differences in the layout and construction of the hybrid powertrain.[7] EDI replaces the traditional transmission with a parallel/series system which provides power to the wheels via a separate traction motor driven by the battery or a generator attached to the combustion engine. During regenerative braking, the traction motor acts as a generator and recharges the battery. A unique feature is the functionality of the hybrid drive transmission to mechanically connect the driveshaft to the crankshaft as speed increases. This feature allows the system to have no electrical losses when driving on the highway to, for instance, a job site, thereby maximizing highway fuel economy.

7.1.2.8 Battery Electric Vehicles

Battery electric vehicles (BEVs) represent the complete removal of an ICE, fuel cell, or other energy conversion device, instead only relying on the electric charge in the battery to propel the vehicle. Although the propulsion system is significantly simplified, compared to either series or parallel HEVs, large battery systems are required, especially for MHDVs due to their weights and power/torque requirements for acceptable vehicle performance. Depending on the vehicle weight, BEVs require battery systems that range from about 30 to 40 kWh for a 10,000-pound vehicle (Class 2b) up to over 80 kWh for Class 4 to 6 delivery vehicles. Such applications tend to favor lightweight payloads that cube out (such as snack food delivery vans); centrally, overnight fueled fleets; and local delivery routes that are generally less than 100 miles per day.

As will be discussed in Chapter 8, lithium-ion (Li-ion) battery packs today range in price from about $400/kWh to $600/kWh, depending on the cell chemistry, although General Motors has announced[8] that battery cell costs for the Bolt electric vehicle have achieved $145/kWh in 2016 and are approaching $100/kWh at the cell level by 2022, which could suggest battery pack costs which may approach $125/kWh to $135/kWh. However, application for light-duty BEVs are somewhat less rigorous than for commercial vehicles, based on the power transfer requirements. Therefore, commercial viability is still a challenge for commercial electric vehicles due to the large size of the battery packs. However, as discussed in Chapter 8, significant progress in Li-ion costs and reliability offer greater confidence in the commercial viability for certain commercial vehicle applications.

---

[7] See https://efficientdrivetrains.com/edi-powerdrive/ (accessed May 21, 2019).

[8] See http://insideevs.com/gm-chevrolet-bolt-for-2016-145kwh-cell-cost-volt-margin-improves-3500/ (accessed December 1, 2019).

7.1.2.9 Fuel Cell Electric Vehicles

Fuel cells for use as an electric power source in transportation have been under development for over 30 years (DOE, 2015a). Fuel cell buses include fuel cell and fuel cell hybrid systems. In fuel cell systems, the electrical power produced by the fuel cell goes directly to the traction motors which propel the vehicle. In hybrid hydrogen fuel cell configurations, the electric energy is transferred to an energy storage device, such as a battery or capacitor. The electrical power is then transferred to the traction motors. The energy storage device allows brake energy to be recovered as the traction motor(s) operates as a generator during braking. This regenerative braking feature can provide significant improvements in fuel consumption during operating cycles which reflect a high number of acceleration and subsequent braking modes. The federal government has supported hydrogen fuel cell research and development. In fiscal year 2015, Congress appropriated approximately $117 million for the DOE Hydrogen and Fuel Cells Program. In addition, $30 million was appropriated for activities related to solid oxide fuel cells.[9] According to DOE, the year 2015 was a landmark year for hydrogen and fuel cells. After decades of research and development (R&D) and the demonstration of various generations of prototype fuel cell electric vehicles (FCEVs) by a number of global automakers, 2015 saw the first commercial FCEVs offered for sale to the general public.

One of the most important fuel cell metrics is the projected high-volume manufacturing cost for automotive fuel cells, which DOE tracks on an annual basis. In 2015, the cost of an 80 kWnet automotive polymer exchange membrane (PEM) fuel cell (FC) system based on next-generation laboratory technology and operating on direct hydrogen was projected to be $53/kWnet when manufactured at a volume of 500,000 units/year (Figure 7-9) and $60/kWnet when manufactured at 100,000 units per year. (For comparison, the expected cost of automotive PEM

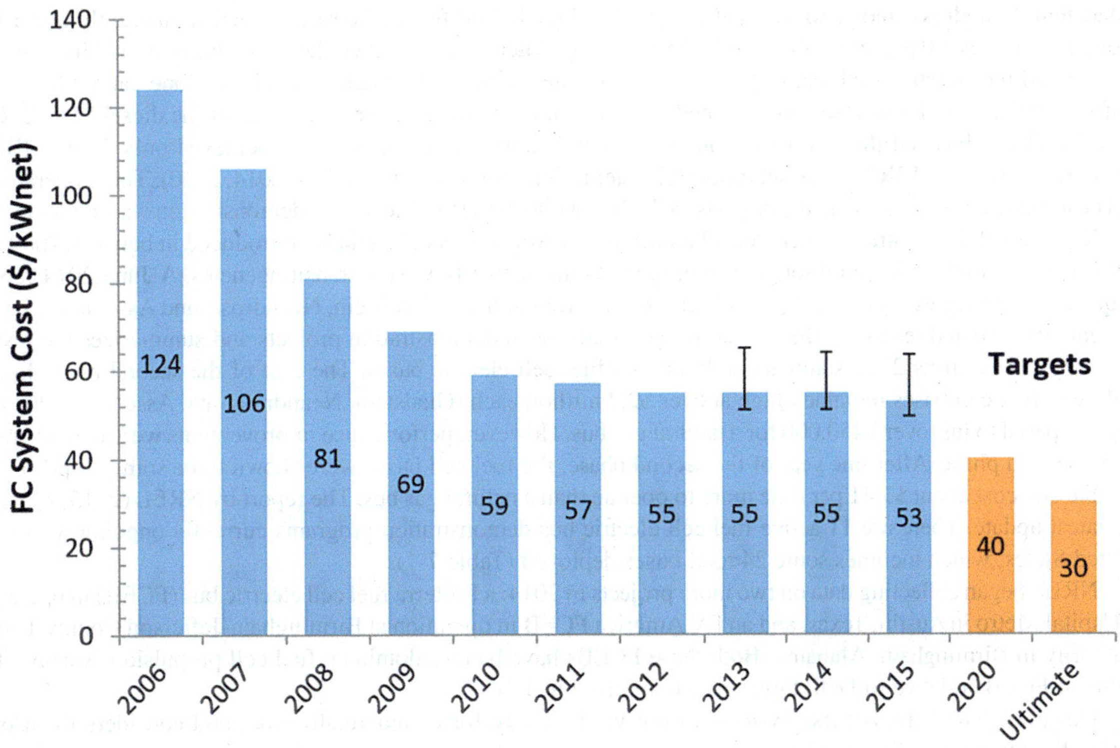

FIGURE 7-9 Modeled cost of an 80 kWnet PEM fuel cell system based on projection to high-volume manufacturing (500,000 units per year).
SOURCE: DOE (2015a).

[9] See https://energy.gov/eere/budget/past-and-present-eere-budget (accessed December 1, 2019).

fuel cell systems that are based on current technology is approximately $280/kW when manufactured at a volume of 20,000 units per year.) To enable commercially competitive fuel cell vehicles, DOE and its contractors are targeting a cost reduction to $40/kW by 2020. Long-term competitiveness with alternative powertrains is expected to require further cost reduction to $30/kW, which represents DOE's ultimate cost target.

7.1.2.10 Commercial Vehicle Experience with Fuel Cells

Although there have been recent introductions of light-duty fuel cell electric vehicles for sale by Toyota, Honda, and Hyundai, significant challenges remain in the commercialization of medium- and heavy-duty fuel cell vehicles. As a point of reference, the California Air Resources Board initiated a Transit Agency Fleet Rule in 2000 which requires large transit agencies to purchase buses with zero exhaust emissions. This requirement was originally planned to begin in 2008, after a demonstration phase wherein the transit agencies purchased and operated qualifying buses. A second demonstration phase was later added to the plan.

The first phase of demonstrations involved seven hydrogen fuel cell buses across three transit agencies. The demonstration featured two fuel cell bus configurations: hydrogen fuel cell buses and hybrid hydrogen fuel cell buses. As reported by the National Renewable Energy Laboratory (NREL, 2015), in "Fuel Cell Buses in U.S. Transit Fleets: Current Status 2015," after the initial demonstration projects, fuel cell buses were some 4.5 times as expensive, were only 15 percent as reliable, and cost $4.60 per mile more to operate than a natural gas bus, which is the de facto standard vehicle in the South Coast Air Basin where the demonstrations took place.

The initial demonstration fuel cell buses cost more than $3 million each (Gladstein, Neandross, and Associates, 2014, p. 10). The three hydrogen fuel cell buses did not contain energy storage, demonstrated fuel economy worse than diesel buses, and exhibited miles between road calls (MBRC) which were only 10 percent of the diesel buses (Gladstein, Neandross, and Associates, 2014, p. 10). Three hybrid fuel cell buses showed significantly better fuel economy—almost 70 percent better—than their diesel counterparts. However, the diesel buses do not have energy storage and regenerative braking, so the benefit associated with the fuel cell technology alone cannot be stated. Unfortunately, these buses also demonstrated poor reliability, with just over 10 percent of the diesel MBRC. One hybrid fuel cell bus had the best fuel economy of the demonstrations; however, it achieved only 2,200 MBRC, compared to 14,000 MBRC for diesel buses (Gladstein, Neandross, and Associates, 2014, p. 10). These operational and cost issues caused a delay in the purchase schedule while the second advanced demonstration was implemented.

The second demonstration included 12 additional hydrogen fuel cell vehicles introduced in between 2009 and 2011 and is being conducted through a joint partnership of five Bay Area transit agencies. A June 2014 report, "Equivalent Strategies for the ARB Zero Emission Bus Regulation" (Gladstein, Neandross, and Associates, 2014), presents the updated results of the second round of advanced demonstration projects and summarizes the results found by NREL in its 2013 summary of hydrogen fuel cell electric buses. The cost of the second-phase hybrid hydrogen fuel cell buses remained high at over $2.2 million each (Gladstein, Neandross, and Associates, 2014, p. 10) compared to just over $450,000 for a natural gas bus. However, performance improvements were demonstrated in the second phase. After one year of the second phase, the fuel cell buses were shown to be some 50 percent as reliable and cost about $3.41 per mile more to operate than a natural gas bus. The report by NREL (2015) provides the latest update. There are 11 active fuel cell electric bus demonstration programs currently ongoing within the United States, which includes some 24 total buses deployed (Table 7-1).

NREL began collecting data on two more projects in 2014: a Proterra fuel cell electric bus (FCEB) in operation at Capital Metro in Austin, Texas, and an EV America FCEB in operation at Birmingham-Jefferson County Transit Authority in Birmingham, Alabama. Both these FCEBs have battery-dominant fuel cell propulsion systems. The status of FCEBs relative to DOE targets is summarized in Table 7-2.

However, NREL finds these systems to not yet be ready for commercialization, and considers them on a Technology Readiness Level (TRL) 7.

TABLE 7-1 U.S. Fuel Cell Bus Demonstrations in Calendar Year 2015

	Bus Operator	Location	Total Buses	Active Buses[a]	Technology Description
1	AC Transit, Zero Emission Bay Area,[b]	San Francisco Bay Area, CA	13	12	Van Hool bus and hybrid system integration, US Hybrid fuel cell
2	SunLine Transit Agency, AT FCEB[b]	Thousand Palms, CA	1	1	New Flyer bus with Bluways hybrid system and Ballard fuel cell
3	SunLine Transit Agency, American Fuel Cell Bus (AFCB) (prototype)[b]	Thousand Palms, CA	1	1	ElDorado/BAE Systems/Ballard next-generation advanced design to meet "Buy America" requirements
4	SunLine Transit Agency, AFCB	Thousand Palms, CA	3	3	ElDorado/BAE Systems/Ballard updated AFCB design
5	Birmingham-Jefferson County Transit Authority[b]	Birmingham, AL	1	1	EVAmerica bus with Embedded Power hybrid system and Ballard fuel cell
6	Capital Metro[b]	Austin, TX	1	1	Proterra composite body with a next-generation battery-dominant hybrid system and a Hydrogenics fuel cell
7	University of California at Irvine	Irvine, CA	1	1	ElDorado/BAE Systems/Ballard updated AFCB design
8	Flint MTA, Nutmeg[b]	Flint, MI	1	1	Van Hool bus and hybrid system integration, US Hybrid fuel cell
9	Flint MTA, AFCB	Flint, MI	1	1	ElDorado/BAE Systems/Ballard updated AFCB design
10	University of Delaware (Phase I and 2)	Newark, DE	2	1	Ebus battery-dominant plug-in hybrid using Ballard fuel cells (22 ft)
11	Greater New Haven Transit District	New Haven, CT	1	1	Ebus battery-dominant plug-in hybrid using Ballard fuel cells (22 ft)
		Total	26	24	

[a]Total buses in actual service as of August 2015.
[b]Project received funding through the Federal Transit Administration's National Fuel Cell Bus Program.
SOURCE: NREL (2015).

7.1.2.11 Nikola Hydrogen Fuel Cell Class 8 Tractor and Day Cab

In December 2016, Nikola Motor Company announced the introduction of a hydrogen fuel cell–powered Class 8 tractor which charges a 320 kWh battery pack for over-the-road service (see Table 7-3 for more technical specifications). According to the company, the battery pack is comprised of some 32,000 individual lithium-ion cells. By comparison, a 2015 Tesla Model S with an 85 kWh battery utilizes over 7,000 individual cells (see Chapter 8).

The Nikola 1 subsystems are described in Box 7-1. A fixed production date has not been announced at the time of this writing. However, deposits have been taken for future production vehicles.[10] Nikola Motor Company has also announced plans for a Class 8 day-cab version for urban use.

[10] Nikola Motor Company, December 1, 2016, "Nikola One Truck Revealed Tonight @ 7:00 p.m. MST," https://dxtn4vayafzin.cloudfront.net/nikolamotor/uploads/press_release/pdf/8/December_1_Release.pdf (accessed March 21, 2017).

BOX 7 1
Description of Subsystems of Nikola One (Nikola Motor Company, 2017)

1. **Front Radiator Assembly.** Electric fans with electronic speed control eliminating the need for mechanical PTOs.
2. **Nikola™ Motor Gearbox/Steering Front End.** Front end which incorporates two electric motors and two gearboxes allowing torque vectoring.
3. **Power Electronics.** The electronics are used to manage energy flow between the fuel cell stack, battery pack, and drive motor/generators.
4. **Nikola™ Battery Storage System.** A reported 320 kWh battery comprised of over 32,000 individual lithium-ion cells welded together on the Nikola One™.
5. **Chiller and Air Tanks.** The 800 V chiller provides cold water to cool the batteries, motors, and power electronics.
6. **Fuel Cell.** A hydrogen fuel cell which reportedly outputs nearly 300 kW of electricity which is fed directly to the 320 kWh battery pack.
7. **Hydrogen Fuel System.** The hydrogen storage system is reported to be sufficient for a total driving distance of 1,200 miles of assumed highway driving.
8. **Rear Motor Gearbox Housing and Independent Suspension.** The rear motor gearbox housing should hold both electric motors and gearboxes.
9. **Fifth Wheel.** The HOLLAND FWAL aluminum fifth.

At the time of this writing, pricing for the Nikola One Class 8 tractor is identified as a lease plan which includes hydrogen fuel. Due to limited hydrogen fueling infrastructure in the United States, the extent of potential market penetration remains to be seen. However, the concept of a fuel cell power source coupled to a large battery pack represents a very expensive combination, such that significant cost reduction would be necessary for such a solution to become commercially viable.

7.1.2.12 Hydraulic Hybrid Powertrain Vehicles

Due to the level of power which is potentially required to brake or propel a medium- or heavy-duty hybrid vehicle, the size, cost, and weight associated with batteries, battery packs, electric motors, and other related equipment can become challenging. Especially for duty cycles in which there are a significant number of braking and subsequent acceleration events, hydraulic hybrid powertrain systems have demonstrated high regenerative braking efficiencies (>70 percent),[11] at high power (100 to 200 kW or greater).

In a hydraulic hybrid powertrain vehicle (HHV), the energy storage device is a hydraulic accumulator, which uses high-pressure hydraulic fluid to move a piston or expanding elastomeric bladder to compress a gas, normally nitrogen. Unlike batteries, there is no need for electrochemical energy transfer, such that very high power levels can be achieved with high in/out energy storage efficiency. Hydraulic pumps/motors replace electric generators/motors in the HHV system and are very compact for the level of power which can be generated. Operating pressures for such systems are normally around 5,000 pounds per square inch (psi) (340 bar) although work is ongoing to increase the operating pressures to reduce the size of the high-pressure accumulators. There is also a low-pressure reservoir that stores hydraulic fluid after it passes through the hydraulic motor piston assembly and transfers the high potential energy of the fluid into torque and power across the motor to drive the wheels. During regenerative braking, a hydraulic pump, or the same pump/motor system acting as a pump, pressurizes the system, which in turn compresses the gas in the high-pressure accumulator.

[11] U.S. Environmental Protection Agency, "Hydraulic Hybrid Vehicles," https://www.epa.gov/cati/hydraulic-hybrid-vehicles (accessed May 18, 2017).

TABLE 7-2 Comparison of Fuel Cell Electric Bus Performance with U.S. DOE Targets

	Units	Current Status[a] (Range)	2016 Target	Ultimate Target
Bus lifetime	years/miles	0.25-4.9/7,900-117,000[b]	12/500,000	12/500,000
Power plant lifetime[c]	hours	660-20,000[b,d,e]	18,000	25,000
Bus availability	%	40-92	85	90
Fuel fills[f]	per day	1	1 (<10 min)	1 (<10 min)
Bus cost[g]	$	2,100,000-2,400,000	1,000,000	600,000
Road call frequency (bus/fuel cell system)	miles between road calls	1,800-6,800/ 9,000-104,000	3,500/15,000	4,000/20,000
Operation time	hours per day/days per week	7-21/5-7	20/7	20/7
Scheduled and unscheduled maintenance cost[h]	$/mile	0.54-1.33	0.75	0.40
Range	miles	240-340[i]	300	300
Fuel economy	miles per diesel gallon equivalent	5.56-7.71	8	8

[a]The summary of results in this report represents a snapshot from the included demonstrations: data from August 2014 to July 2015.
[b]Accumulated totals for existing fleet through July 2015; these buses have not reached end of life.
[c]For the DOE/Federal Transit Administration targets, the power plant is defined as the fuel cell system and the battery system. The fuel cell system includes supporting subsystems such as the air, fuel, coolant, and control subsystems. Power electronics, electric drive, and hydrogen storage tanks are excluded.
[d]The status for power plant hours is for the fuel cell system only; battery lifetime hours were not available.
[e]The highest-hour power plant was transferred from an older-generation bus that had accumulated more than 6,000 hours prior to transfer.
[f]Multiple sequential fuel fills should be possible without an increase in fill time.
[g]Cost targets are projected to a production volume of 400 systems per year. This production volume is assumed for analysis purposes only and does not represent an anticipated level of sales.
[h]Excludes mid-life overhaul of power plant.
[i]Based on fuel economy and 95 percent tank capacity.
SOURCE: NREL (2015).

TABLE 7-3 Additional Specifications Reported for Nikola One

Motor type	800V AC motors
Battery capacity and chemistry	320 kWh lithium ion
Transmission	Two-speed automated with direct drive and low-noise gears
Drive system type	6×4 four-wheel drive
Active descent control	Regenerative braking
Cooling	Biodegradable non-electrically conductive liquid coolant

SOURCE: Nikola Motor Company.

Figure 7-10 shows the layout of a series HHV where, in common with a series HEV, the hydraulic pump is driven by the engine, which either charges the high-pressure accumulator or drives the hydraulic pump/motor operating in motor mode to drive the wheels.

During a regenerative braking event, the hydraulic pump/motor switches into pump mode and recharges the high-pressure accumulator. Additional braking is required once the accumulator is fully charged; just as in an HEV, the friction brakes are used to stop the vehicle. However, due to the high amount of braking energy that is absorbed by the hydraulic system, due to its high efficiency (typically >70 percent), significant maintenance cost savings can be realized by reducing the friction brake replacement frequency by a factor of 2 to 4. The low-pressure reservoir completes the hydraulic circuit, either charging the hydraulic pump during the engine power transfer mode, accepting the low-pressure discharge from the drive motor which is connected to the wheels, or charging the pump/motor acting as a pump during regenerative braking.

A hydraulic pump/motor/accumulator system can transfer high power with high efficiency in a short period of time. Battery systems, on the other hand, tend to require increasingly large batteries in order to transfer sufficient power generated during medium- and heavy-duty vehicle events, due to the large mass of such vehicles and the power needed to brake or accelerate them. However, the specific energy (amount of energy stored per unit of volume or mass [weight]) is relatively low for hydraulic propulsion systems, compared to batteries. Therefore, the time during which a medium- or heavy-duty HHV can be driven by the accumulator system is relatively short, typically not much longer than an acceleration from stop to 40 to 60 mph.

HHVs can also be configured as a parallel system. In such a layout, the size of the high-pressure accumulator is typically limited to store the energy of a single regenerative braking cycle. As such, this parallel system is also sometimes called a "launch assist" configuration. Figure 7-11 shows the layout of a typical parallel HHV system. Parallel HHVs are typically used for applications where there is a significant amount of operation under repeated braking and acceleration cycles, combined with a reasonable portion of the duty cycle that is related to normal on-highway use. Refuse trucks that require significant highway driving to reach their intended route, followed by a long distance to the disposal site, exemplify such an application.

The decision to use a series or parallel HHV for particular applications is primarily influenced by the efficiency of the propulsion system while driving on the highway. The regenerative braking and launch assist efficiency of series or parallel systems can be very similar (>70 percent). However, while operating under extended periods of highway driving, a series hybrid will cycle the engine on and off quite frequently due to the relatively low-energy storage capability of the accumulator (specific energy). In theory, the engine can operate at its peak efficiency (load and speed point) as it cycles on and off, thereby minimizing fuel consumption. On the other hand, in a parallel system, the engine can drive the wheels through a mechanical transmission with gear-drive efficiencies greater than 90 percent. Therefore, during on-highway operation, there is a trade-off between the frequently cycling series system and the continuous high-efficiency-transmission parallel system which may be driven by an engine whose operating points will be below its optimum efficiency curve. As in all medium- and heavy-duty hybrid vehicles,

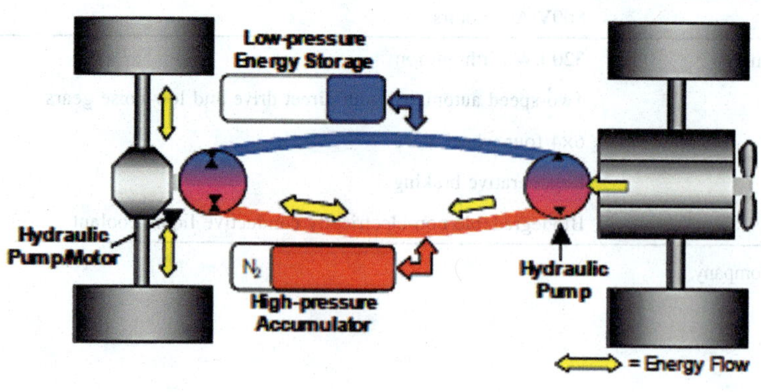

FIGURE 7-10 Series hydraulic hybrid architecture.
SOURCE: NRC (2010a).

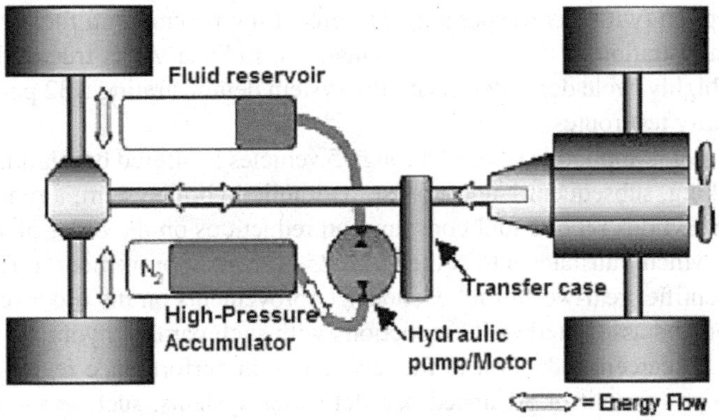

FIGURE 7-11 Parallel hydraulic launch assist hybrid architecture.
SOURCE: NRC (2010a).

the operational duty cycle is extremely important in determining the benefits which may be derived in real-world use and therefore commercial purchase decisions.

Currently, there are at least three HHV systems offered for commercial vehicles. Parker Hannifin has offered its "RunWise" system, which is installed in refuse truck applications. According to Parker[12] there are some 200 refuse trucks currently using the RunWise system in commercial service and Parker claims up to 47 percent reduction in fuel consumption for refuse truck applications with a very high level of acceleration and braking cycles during normal use. This type of service, similar to city bus operating cycles, represents perhaps the best use of HHV technology because of the high efficiency of hydraulic systems during high power transfer for relative short periods of time during acceleration and subsequent braking to a stop. This allows the size of the high-pressure accumulator to be limited to the energy necessary for a braking cycle with subsequent high load acceleration.

The RunWise system represents a series hybrid drive system that incorporates a three-speed power drive unit that allows the engine to operate at peak efficiency. According to Parker, the "drive system is work brake compatible, has integrated traction control and end-user diagnostic capability" and "integrates with the existing ABS system and has a single point electrical interface to the chassis."[13] Parker is also introducing a hydraulic power-split transmission with regenerative braking and launch assist for application in vehicles ranging from delivery vans to transit buses (Parker Hannifin Corporation, 2012). According to Parker, depending on the speed range of the engine, a reduction gear assembly can be incorporated to maintain a maximum transmission speed of 2,600 rpm. A two-speed gearbox on the input shaft is offered as an option to allow additional initial torque for improved launch performance.

A second system applied to commercial vehicles, in this example refuse trucks, is offered by Stored Energy Solutions, Inc. (SES).[14] This system is a parallel, mechanical-hydraulic break energy recovery and launch assist system that is installed together with the production refuse truck transmission. Unlike the Parker RunWise system, which replaces the production vehicle transmission, the SES system essentially reduces the launch and drive torque on the driveshaft of the truck, thereby reducing fuel consumption and lowering GHG emissions. There may also be benefits in reducing NO_x and particulate emissions due to lower transient torques during vehicle acceleration. According to an Association for the Work Truck Industry publication,[15] the SES system demonstrated a 16.3 percent

[12] Parker Hannifin, September 21, 2015, "Parker Hannifin Hydraulic Transmission Outperforms Proposed EPA Emissions Regulations for Heavy-duty Trucks: RunWise® hydraulic technology reduces fuel consumption by an average of 43 percent," http://www.parker.com/portal/site/PARKER/menuitem.31c35c58f54e63cb97b11b10237ad1ca/?vgnextoid=c0edc90375584510VgnVCM100000e6651dacRCRD&vgnextchannel=9104fbdc71fd7310VgnVCM100000200c1dacRCRD&vgnextfmt=EN&newsroom=Y&vgnextcat=News Release Details (accessed May 18, 2017).

[13] Parker Hannifin, "Runwise Advanced Series," http://ph.parker.com/us/en/runwise-advanced-series (accessed May 18, 2017).

[14] Further information is available at http://ses-hydraulichybrid.com/intellidrive-2/.

[15] NTEA Work Truck Show, February 2016.

reduction in fuel consumption (with a corresponding 21 percent improvement in fuel economy) while participating in a refuse truck demonstration in the City of Indianapolis, Indiana. As is true with all hybrid systems, the results were shown to be highly cycle dependent, with the system demonstrating a 32 percent improvement in fuel economy on one of the daily test routes.

A third HHV system that is applied to Class 3 through 5 vehicles is offered by Lightning Hybrids. This system is also a regenerative braking, subsequent launch assist, hydraulic hybrid system, a rendering of which is shown in Figure 7-12. Lightning Hybrids reports fuel consumption reductions on the order of 15 to 20 percent for stop-and-go bus-cycle testing, which translates into some 20 to 25 percent improvement in fuel economy.[16] Lightning Hybrids on its website identifies real-world fuel economy improvements on the order of 20 to 30 percent.

The fuel consumption and associated GHG reductions with such parallel hydraulic launch assist systems all appear to be in the 15 to 25 percent reduction in fuel consumption performance region. As the level of technology increases, to include power-split or advanced parallel-series systems, such as the Parker RunWise system, improvements in fuel consumption are seen to achieve levels of 35 to 40 percent, in addition to reduced criteria pollutants such as NO_x and CO, and significant improvements in brake system life, due to greatly reduced loads on the foundation brakes on commercial vehicles fitted with regenerative braking.

7.1.2.13 Full Series Hydraulic Hybrids

For a period of about 15 years, ending in 2013, EPA supported a major research and development effort to demonstrate the fuel consumption reduction and GHG emission benefits that could be gained by a true series hydraulic hybrid system. Demonstration programs conducted with prototype delivery vans[17] and 30-person buses demonstrated improvements in fuel economy that exceeded 100 percent, or doubling of the fuel economy, which represents over a 50 percent reduction in fuel consumption (EPA, 2016).

The system was a series configuration, with a pump/motor which drives the wheels through a three-speed transmission which receives high-pressure hydraulic fluid from the high-pressure accumulator, or from the high-pressure hydraulic pump that is connected to the combustion engine. High efficiency is achieved by careful matching of the variable displacement traction motor to the required operational characteristics of the vehicle. This hydraulic motor also acts as a high-efficiency pump during regenerative braking cycles, replenishing the high-pressure fluid in the high-pressure accumulator.

As the system is of the series design, a variable displacement pump is driven by the engine in the vehicle, which allows the engine to operate near its optimum efficiency point while it is running, to provide high-pressure hydraulic fluid to drive the traction motor, recharge the accumulator, or both. The engine shuts down during regenerative braking, when the traction pump motor recharges the high-pressure accumulator. During a subsequent launch the stored higher-pressure fluid in the accumulator flows through the traction motor, driving the vehicles. When the quantity of high-pressure fluid is depleted, the engine starts, then runs near optimum efficiency while driving the high-pressure supply pump. This particular series hydraulic hybrid technology, developed under EPA funding, is still in the developmental stage and has not yet been released for series production (Kargul et al., 2015).

7.1.2.14 Hydraulic Hybrid System Costs

In the case of a series/parallel type of system, similar to that currently offered by Parker Hannifin, the anticipated net price (after some credit for replacing the normal transmission) of the total hybrid system is estimated to be on the order of $45,000 to $55,000 in production volumes up to about 10,000 units per year. Such a system includes a high-pressure accumulator and associated control valves, high-pressure hydraulic lines, two high-pressure hydraulic pumps/motors, a low-pressure reservoir system, a two- or three-speed gearbox, electronic controls, and in some cases expanded use of a hydraulically actuated accessory drive in order to operate a power steering pump and air conditioner compressor while the engine is off. In some extremely high-frequency stop-start operating

[16] "Lightning Hybrids: The Solution," https://www.nrel.gov/workingwithus/assets/pdfs/2015-igf-lightning-hybrids.pdf (accessed May 18, 2017).

[17] EPA, "Hydraulic Hybrid Vehicles," https://www.epa.gov/cati/hydraulic-hybrid-vehicles (accessed May 18, 2017).

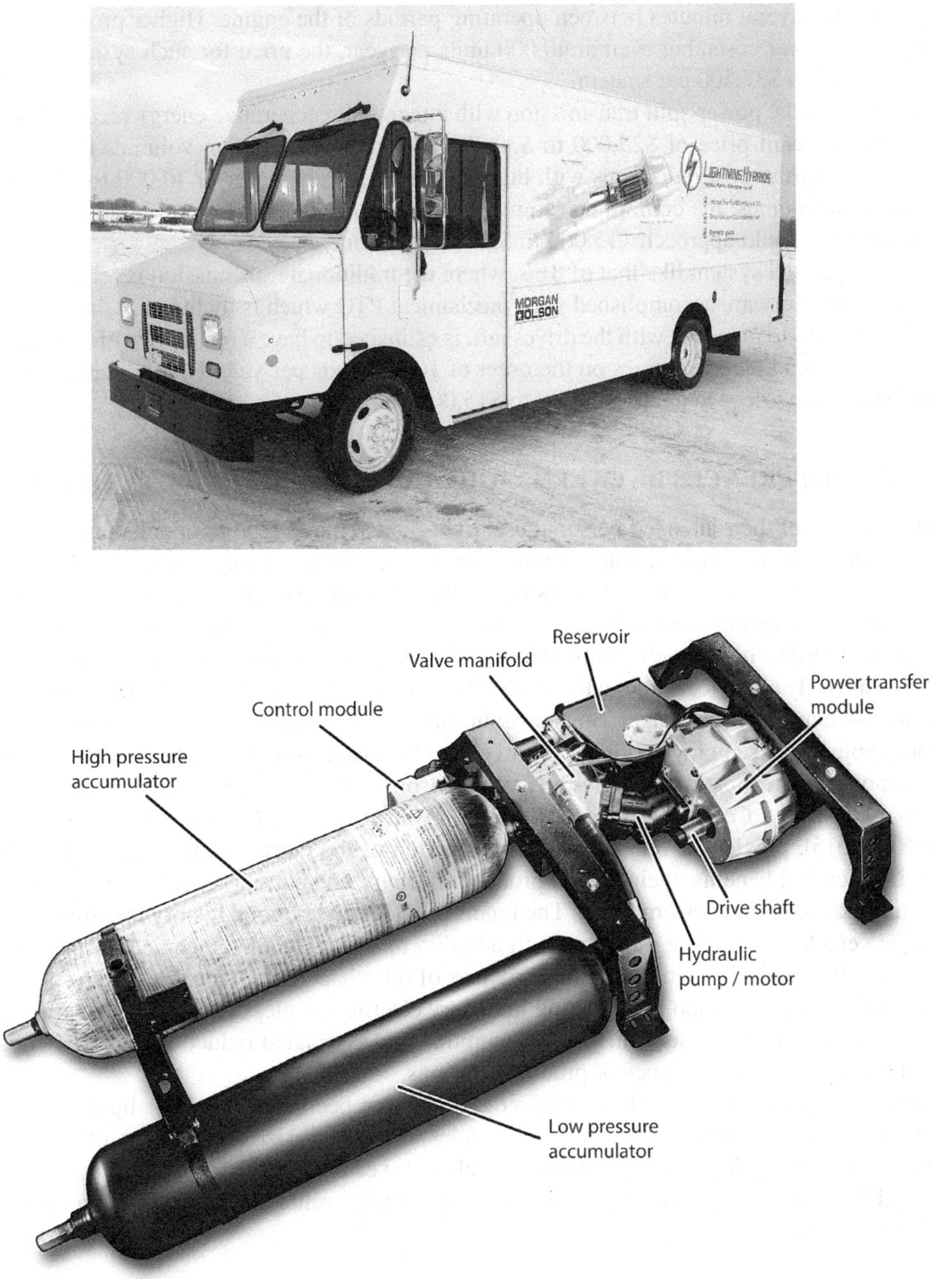

FIGURE 7-12 Parallel, hydraulic hybrid regenerative braking and launch assist system for Class 3 through 6 commercial vehicles.
SOURCE: Image used by permission of Lightning Hybrids, LLC.

cycles, there may be several minutes between operating periods of the engine. Higher production volumes could result in somewhat lower costs, but even at 40,000 units per year, the price for such systems would likely be in the range of $35,000 to $37,500 per system.

However, a hydraulic power-split transmission with integrated regenerative energy recovery and launch assist, would likely have a retail price of $25,000 to $30,000 per unit in production volumes under 10,000 units per year, but could see greater cost reductions with higher volumes, on the order of 40,000 to 50,000 units per year. At such production volumes, the committee estimates the retail price for these regenerative braking, power-split transmission systems would approach $15,000 to $20,000 per unit.

The cost of a parallel system like that of SES, where the traditional transmission is retained but parallel drive and brake energy recovery are accomplished via a mechanical PTO which is included in the existing transmission or through an auxiliary drive in-line with the driveshaft, is estimated to have a retail price of approximately $15,000 to $20,000, again in production volumes on the order of 10,000 units per year. Higher volumes (40,000 units per year) could lower these prices to about $12,500 to $15,000 per unit.

7.2 DIFFERENCES BETWEEN LIGHT-DUTY AND HEAVY-DUTY HYBRIDS

Due to the nature of their intended personal use, light-duty vehicles tend to operate under relatively low load conditions, except during acceleration, trailer towing conditions, or other heavier load cycles. This is not necessarily true for certain light-duty trucks, vans, and SUVs which are used in commercial or industrial applications. However, across the normal passenger car and light-duty truck, the clear majority of vehicles are driven for personal use.

Since the mid-1970s there have been regulatory and other efforts aimed at reducing criteria pollutants and emissions that adversely affect local air quality (NRC, 2010b). Over the past decade, there has been very significant investment across almost all passenger car and light-duty truck manufacturers, as evidenced by new product offerings (see Table 8-1), to employ varying levels of electric hybridization, to reduce the smog-forming emissions, but also in response to the most recent light-duty vehicle Corporate Average Fuel Economy and GHG emissions.

A recent NRC study, *Cost, Effectiveness, and Deployment of Fuel Economy Technologies for Light-Duty Vehicles* (NRC, 2015), provides a very thorough explanation of the current state of electrification in light-duty vehicles, their functional benefits, their current product offerings, and current associated costs and projections for future potential (both cost and performance). The report provides a very good history and summary of hybrid and battery electric vehicles in the light-duty space. In addition to a very detailed technical discussion, which shows the similarities to the systems identified in the early part of this chapter, the report summarizes product offerings by different levels of electrification, culminating in battery electric vehicles. The report also provides a very good summary of the relative improvements in fuel consumption and associated reductions in GHG emissions through the implementation of the various types of product offerings.

Although the general types of hybrid technology that have been developed for light-duty vehicles can, in principle, be applied to medium- and heavy-duty vehicles, the overall size and weights of commercial vehicles, combined with the significantly different operating cycles, makes the application of light-duty hybrid technology much more difficult to employ, challenging to meet commercial vehicle lifetimes, and sometimes difficult to achieve cost effectiveness.

7.3 MAPPING TECHNOLOGIES TO DUTY CYCLES

The fuel consumption and CO_2 benefits that can be derived from the use of HEVs are highly dependent on the type of vehicle, its size and weight, and, most importantly, its intended duty cycle. Use cases involving a high frequency of braking cycles, such as refuse trucks or urban buses, can gain significant benefits through regenerative braking, where the size of the energy storage system can be somewhat limited in size to recover primarily only the braking energy and then launch the vehicle.

Other applications, such as urban delivery trucks, may choose to incorporate larger energy storage systems and optimize total route operational efficiency with a combination of regenerative braking and periods of operating the ICE and higher load levels, to minimize carbon-based fuel consumption.

The NRC Phase One Report, *Technologies and Approaches to Reducing the Fuel Consumption of Medium- and Heavy-Duty Vehicles* (NRC, 2010a), provides a detailed explanation of the operational cycles to be considered and the relative fuel consumption and GHG benefits that may be derived.

Table 7-4 is presented below to list the relative benefits in application of hybrid technology for different classes of vehicles and the type of duty cycle they typically perform. The table shows a range of hybridization including simple 12-volt stop-start; a higher voltage (48 V) integrated starter/generator mild hybrid with launch/regeneration capability; and a high voltage (>200 V) strong hybrid with limited all-electric operation. The costs associated with these systems are shown in Table 7-6 below and the cost effectiveness for different use cases is summarized in Chapter 12.

7.4 CURRENT MANUFACTURERS AND PRODUCT RANGES

A summary of several commercially available hybrid commercial vehicles (both electric and hydraulic) available at the time of this writing is included in Table 7-5. Attempting to spur early market introduction of a technology that would otherwise not be cost effective for the purchaser, various federal, state, and local financial incentives are available to subsidize the cost of many hybrid vehicle and powertrain system suppliers. Especially in the case of transit buses, these subsidies can be significant. Within the 2030 time frame of this report focus,

TABLE 7-4 Relative Fuel Consumption Benefits of Electric Hybrid Components by Class

Vehicle Class	Hybrid Type	Vehicle Use	Duty Cycle	Approx. FC Reduction (%)
Class 2b-3	12 V start/stop (S/S) belt-starter-generator (BSG)	Heavy-duty pickup and van	Urban + highway	3.5
Class 2b-3	Integrated S/S with regen/launch assist	Heavy-duty pickup and van	Urban + highway	10
Class 2b-3	Parallel strong hybrid	Heavy-duty pickup and van	Urban + highway	20
Class 4-5	12 V S/S BSG	Vocational vehicles	Urban + highway	3.5
Class 4-5	Integrated S/S with regen/launch assist	Vocational vehicles	Urban + highway	16
Class 4-5	Parallel strong hybrid	Vocational vehicles	Urban + highway	20
Class 4-5	12 V S/S BSG	Delivery trucks	Urban delivery	3.5
Class 4-5	Integrated S/S with regen/launch assist	Delivery trucks	Urban delivery	20
Class 4-5	Parallel strong hybrid	Delivery trucks	Urban delivery	25
Class 6-7	12 V S/S BSG	Vocational vehicles	Urban + highway	3.5
Class 6-7	Integrated S/S with regen/launch assist	Vocational vehicles	Urban + highway	16
Class 6-7	Parallel strong hybrid	Vocational vehicles	Urban + highway	20
Class 6-7	12 V S/S BSG	Delivery trucks	Urban delivery	3.5
Class 6-7	Integrated S/S with regen/launch assist	Delivery trucks	Urban delivery	20
Class 6-7	Parallel strong hybrid	Delivery trucks	Urban delivery	25
Class 8	12 V S/S BSG	Vocational refuse truck	Urban accel/brake	3.5
Class 8	Integrated S/S with regen/launch assist	Vocational refuse truck	Urban accel/brake	22
Class 8	Parallel strong hybrid	Vocational refuse truck	Urban accel/brake	30

TABLE 7-5 Hybrid Commercial Vehicles Currently Available (2017)

Manufacturer	Product Example	Application
Hino	Class 5 cab-over electric hybrid	Delivery Trucks
Allison	Heavy heavy-duty (HHD) parallel hybrid	Transit Bus
BAE	HHD series or parallel hybrid	Transit Bus
XL	Class 3-4 mild electric hybrid	Shuttle Bus
Crosspoint Kinetics	Class 3-7 mild electric hybrid	Delivery trucks, shuttle buses
Lightning Hybrids	Class 2-5 hydraulic hybrid	Delivery trucks
Parker Hannifin	Medium heavy-duty hydraulic hybrid	Delivery trucks
Freightliner Custom Chassis	MHD hydraulic hybrid	Delivery trucks
Morgan Olson	MHD hydraulic hybrid	Delivery trucks
Autocar-Parker	RunWise hydraulic hybrid	Refuse trucks
Eaton (1) China Application	HHD parallel electric hybrid	Trucks and buses
Odyne	Plug-in electric hybrid, E-PTO	Utility trucks
Efficient Drivetrains Inc.	Plug-in electric hybrid, E-PTO	Utility trucks

such subsidies may or may not exist. Therefore, in the cost analyses presented in the following chapter, the cost benefit is evaluated without subsidy.

7.5 ADVANCES IN TECHNOLOGY AND COST REDUCTION OF ELECTRIC MOTORS AND POWER ELECTRONICS FOR HYBRID AND BATTERY ELECTRIC VEHICLES

As identified in the preceeding sections of this chapter, significant progress has been made in the reduction of battery cell and pack costs for use in light-duty hybrid and battery electric vehicles. A rather detailed discussion on battery technology and projected improvements is included in Chapter 8.

While many years of research and development by the automotive industry, battery suppliers, and governmental research laboratories has yielded significant battery advancements, similar efforts on the rest of the electrified powertrain components have been somewhat later in developing. However, it appears that much more attention is now being placed on these non-battery electrified powertrain systems, especially within the EERE of DOE (Rogers and Boyd, 2016).

As shown in Figure 7-13 below, the remaining parts of the electrified powertrain include the electric motor (which may also function as a generator during regenerative braking), the power inverter(s), any on-board battery charger(s), and voltage converter(s). The complete system also includes low-voltage wiring, high-voltage cables, cooling circuits (including pumps), electrical disconnect circuits, and various microprocessor control modules.

The Electric Drive Technologies (EDT) program is a joint research and development program which includes the DOE, various national laboratories, vehicle original equipment manufacturers (OEMs), Tier 1 suppliers, and various other tiered suppliers, other government agencies, universities, and research organizations. Much like the U.S. Automotive Battery Consortium (USABC), specific size, weight, specific energy, and cost targets are established and monitored annually in program merit reviews.

The stated goal of the EDT program is to approach similar cost and convenience targets for plug-in electric vehicles as their gasoline-fueled counterparts by the year 2022. To accomplish this aggressive goal, the electrified powertrain motors and power electronics would need to demonstrate a combined unit cost of $8/kW propulsion power, with a system efficiency of 94 percent, while achieving a specific energy of 1.4 kW/kg and an energy density of 4.0 kW/L (Rogers and Boyd, 2016). Figure 7-14 shows that to achieve these aggressive targets, compared to 2012 electrified drivetrain production systems, the cost will need to be reduced by a factor of 4, with a 35 percent reduction in size, 40 percent reduction of weight, and elimination of 40 percent of electrical, fluidic, and frictional losses in the system.

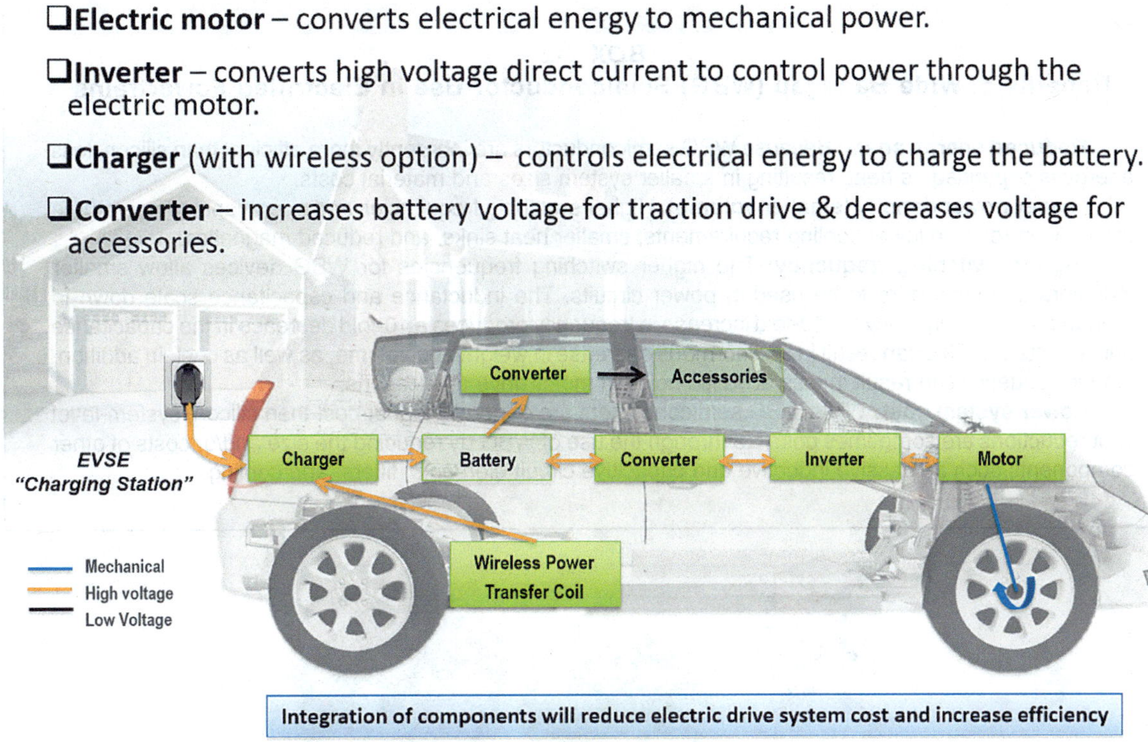

FIGURE 7-13 Schematic representation of electrified powertrain components.
SOURCE: Rogers and Boyd (2016).

As a gauge for individual subsystem cost reduction challenges, the 2015 production costs for the electric traction motors and power electronics for a typical passenger car (3,500 to 4,000 pounds vehicle weight) are approximately $7/kW and $5/kW, respectively, for a total combined cost of $12/kW. In order to achieve the 2022 combined cost of $8/kW, Figure 7-15 shows that the power electronics cost must achieve about $3.30/kW and the traction motor must approach $4.70/kW (Rogers and Boyd, 2016).

The 2016 roadmap to achieve these aggressive cost reduction targets, which was developed in cooperation with OEMs and Tier 1 suppliers, is shown in Figure 7-16. The reduction in combined power electronics and traction motor cost, from about $30/kW in 2012 to $12/kW in 2015, was primarily accomplished by improvements in silicon-based power electronics and reduction in rare-earth-metal content in the permanent magnet motors typically used in electrified powertrains. However, to achieve the 2022 goal of $8/kW, a shift in technology application will be necessary.

As Figure 7-16 depicts, further improvements in power electronics cost, energy density, and specific energy will likely require the production development of wide-bandgap (WBG) systems. The benefits of WBG electronics, compared to silicon-based systems, are summarized below (Box 7-2), which originates in the DOE document on WBG electronics (DOE, 2015b).

Under the EDT program, General Motors is developing a next-generation inverter which, when scaled to its highest power configuration is expected to achieve the $3.30/kW cost target by 2022. According to General Motors, the highly consolidated design incorporates advanced, vertically integrated manufacturing processes. Under the high-power rating, the traction motor inverter is also projected to achieve the specific energy and energy density targets of 14.1 kW/kg and 13.4 kW/L, respectively. Figure 7-17 shows the inverter exterior configuration.

Another major effort and technical challenge is the reduction in cost of electric traction motors for use in hybrid and battery electric vehicles. As identified in Figure 7-18, a key to achieve the 2022 cost target of $4.70/

BOX 7-2
Benefits of Wide Bandgap (WBG) Semiconductor Use in Electrified Powertrains

Reduced energy costs: Because WBG semiconductors are inherently more efficient than silicon, less energy is expended as heat, resulting in smaller system sizes and material costs.

Higher power density (smaller volume): Higher switching frequencies and operational temperatures than silicon result in lower cooling requirements, smaller heat sinks, and reduced magnetics.

Higher switching frequency: The higher switching frequencies for WBG devices allow smaller inductors and capacitors to be used in power circuits. The inductance and capacitance scale down in proportion to the frequency—a 10-fold increase in frequency produces a 10-fold decrease in the capacitance and inductance. This can result in an enormous decrease in weight and volume, as well as cost. In addition, higher frequency can result in less acoustic noise in motor drive applications.

Lower system cost: While WBG semiconductors are generally higher cost than silicon, system-level cost reductions are sometimes possible through the use of WBG by reducing the size and/or costs of other components such as passive inductive and capacitive circuit elements, filters, and cooling.

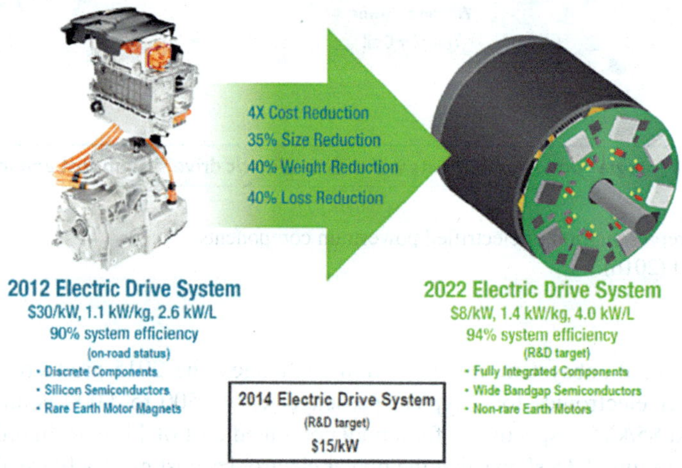

FIGURE 7-14 DOE Electric Drive Technology Program electrified powertrain technical and cost targets.
SOURCE: DOE (2015c).

kW is a very large reduction in, or elimination of, rare-earth material use in the motors. Under the EDT program, General Electric is developing next-generation high-performance motors with non–rare-earth materials (Rogers and Boyd, 2016).

The challenges to achieve the motor cost goals include

- High efficiency over wide speed and load ranges;
- High power density and high coolant inlet temperature;
- High speed, which poses mechanical challenges;
- Non–rare-earth permanent magnets; and
- Low-cost targets based on 100,000 units per year.

To achieve the high energy and power densities, the motors will need to operate at relatively high speeds, around 14,000 rpm. However, with high speed comes high frequencies, higher stresses, and higher stator core and rotor losses. The technical goals for this next generation of non–rare-earth motors are shown in Figure 7-18 (Rogers and Boyd, 2016).

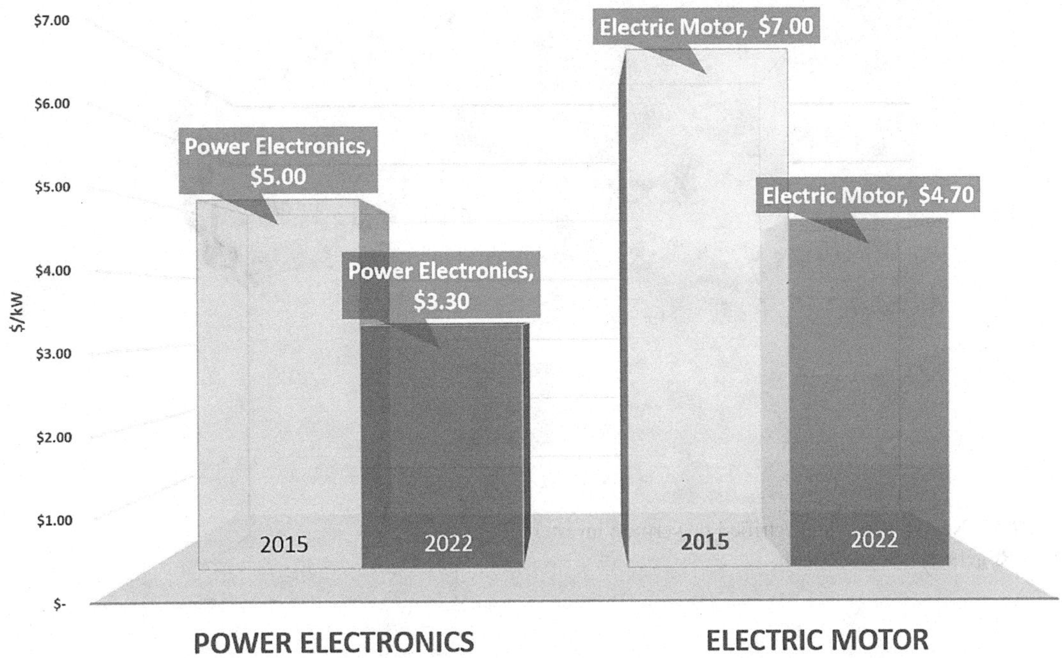

FIGURE 7-15 Cost reduction targets for electrified powertrain power electronics and traction motors from 2015 to 2022.
SOURCE: Rogers and Boyd (2016).

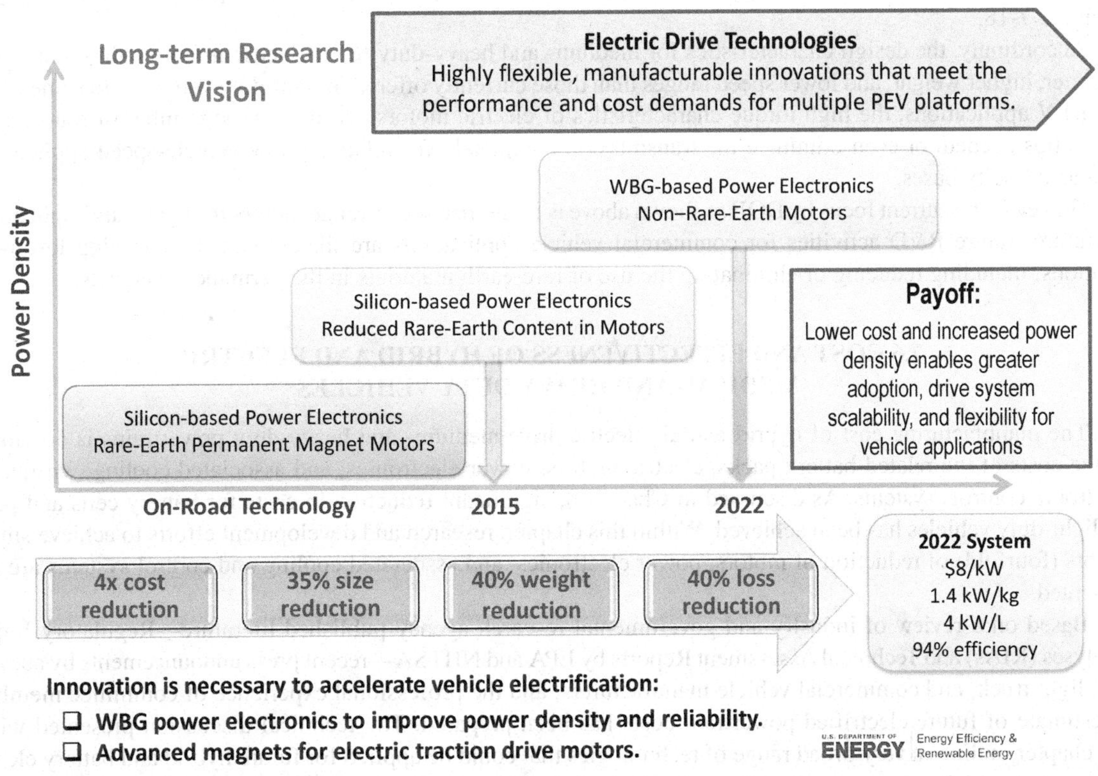

FIGURE 7-16 DOE Electric Drive Technologies program roadmap for cost reduction.
SOURCE: Rogers and Boyd (2016).

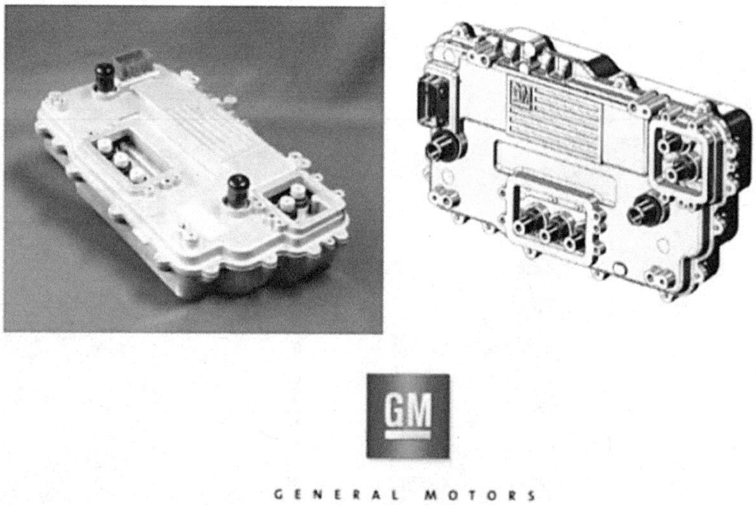

FIGURE 7-17 Next-generation electrified powertrain inverter.
SOURCE: Rogers and Boyd (2016).

Although the overall power requirements for the use of electric motors in large passenger cars and light-duty trucks are similar to those in medium-duty and some heavy-duty vehicles (approximately 250 to 350 hp maximum), the torque requirements for commercial vehicles are significantly higher. However, the potential to generate high torque at low speeds from electric motors is excellent, as shown by the increasing torque with decreasing speed in Figure 7-18.

Accordingly, the design characteristics for medium- and heavy-duty truck electric motors will require larger diameter, higher weight, and lower speed ranges than those currently offered in light-duty vehicles. In some PHEV and BEV applications, the high-torque characteristics of electric motors can reduce the number of transmission gear ratios needed, or even eliminate the transmission completely for relatively low-vehicle-speed applications, such as intracity buses.

However, the current focus by DOE as shown above is for higher-speed-range motors for light-duty applications. Therefore, future R&D activities for commercial vehicle applications are likely needed to develop lower-cost solutions, including reducing or eliminating the use of rare-earth materials in the permanent magnets.

7.6 COST AND EFFECTIVENESS OF HYBRID AND ELECTRIC MEDIUM- AND HEAVY-DUTY VEHICLES

The manufacturing cost of hybrid and all-electric drive medium- and heavy-duty powertrains is dominated by the costs of the related battery packs, electric motors, power electronics, and associated cooling, wiring, and electronic controls systems. As discussed in Chapter 8, significant reduction in costs for battery cells and packs for light-duty vehicles has been achieved. Within this chapter, research and development efforts to achieve similar factors (fourfold) of reduction of motors, power electronics, and associated cooling and control systems are also presented.

Based on a review of industry and governmental research agency published literature—Regulatory Impact Analyses (RIAs) and Technical Assessment Reports by EPA and NHTSA—recent press announcements by automotive, light truck, and commercial vehicle manufacturers, and the professional experience of committee members, an estimate of future electrified powertrain costs has been prepared. The technical discussion presented within this chapter outlines a very broad range of technologies that could be applied for future hybrid and battery electric commercial vehicles. The committee has reduced the technology applications and range of vehicle classes to those

Items	Specification
Max. Speed	14,000rpm
Peak Power	55kW @ 20% speed for 18sec
Maximum Current	400Arms
Cont. Power	30kW @ 20~100% speed @ Vdc=325
Efficiency	Refer to target efficiency map
Operating Voltage	200~450V (325V nominal)
Back EMF	<600Vpk line-to-line @ 100% speed
Torque Pulsation	<5% of Peak Torque @ any speed
Characteristic Current	< Maximum Current
Weight	≤35kg
Volume	≤9.7L
Cost @100k	≤$275
Ambient (outside housing) Operating Temperature	-40~140°C
Coolant inlet	105°C, <10LPM, 2psi drop, <20psi inlet
Minimum isolation impedance-phase terminal to GND	1Mohm

- Build and test final 55kWpk non-rare earth motor to meet DOE specifications

Figure 1. Motor Efficiency Targets

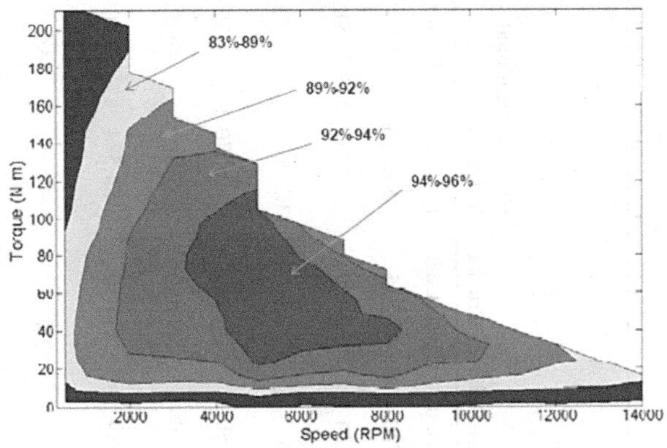

FIGURE 7-18 Technical targets for low-cost electric motors with non–rare-earth materials.
SOURCE: EL-Refaie (2016).

the committee believes have a reasonable probability of success, based on relative cost, vehicle class, potential duty cycle, and the potential for effective service and warranty support.

Table 7-6 presents a projected hybrid system distribution across Class 2b and 3 trucks and vans, Class 4 vocational vehicles and Class 4 delivery trucks, Class 5 vocational vehicles and Class 5 delivery trucks, Class 6 and 7 vocational vehicles and Class 6 and 7 delivery trucks, and Class 8 vocational vehicles (refuse trucks). The levels of hybridization range from stop-start systems (12 V) with no regeneration capability, to integrated stop-start (48 V) with limited launch assist and regenerative energy recovery, to fully integrated high-voltage hybrid systems with full tractive power to operate the vehicle for an extended time under electric power drive only.

The percent reduction in fuel consumption for each technology category is estimated based on relative duty cycle and is compared to a model year (MY) 2015 typical diesel-powered vehicle. The relative improvement from a reduction in energy storage is assumed to be the same for MY 2022 and MY 2030. The difference in cost reflects further development and increasing production quantities which may lead to reduced production cost.

The relative fuel consumption improvement values were derived from several sources including the first medium- and heavy-duty fuel economy report (NRC, 2010a), published results from the EPA and NHTSA RIAs, and the analyses conducted for the committee by Southwest Research Institute.

The intent of Table 7-6 is to show the relative improvements in fuel consumption which can be obtained through the application of hybridization in medium- and heavy-duty vehicles. It is important to note that the improvements, relative to typical MY 2015 diesel-powered vehicles, are dependent on the size of the vehicle, the type of hybridization, and the typical duty cycle under which the vehicles may be operated.

Due to the state of battery, power electronics, and electric motor technology, production costs for these vehicles will be higher than for light-duty vehicles. The committee has attempted to estimate the production costs for these major components as they would be applied to commercial vehicles, recognizing that initial production volumes

TABLE 7-6 Hybrid System Cost Estimates (High and Low Cases)

		Light Heavy-Duty (LHD)									
Vehicle Class	Year or or % Reduction in FC	Class 2b and 3 Van and Pickup		Class 4 Vocational Vehicles		Class 4 Delivery Trucks		Class 5 Vocational Vehicles		Class 5 Delivery Trucks	
Technology		Low Cost	High Cost	Low Cost	High Cost	Low Cost	High Cost	Low Cost	High Cost	Low Cost	High Cost
Non-hybrid	2015	$1,190	$1,210	$1,710	$1,725	$1,710	$1,725	$1,710	$1,775	$1,710	$1,775
Enhanced accessories	2022	$477	$540	$724	$800	$724	$800	$744	$850	$744	$850
12 V stop-start	% Reduction in FC	3.50%	3.50%	3.50%	3.50%	3.50%	3.50%	3.50%	3.50%	3.50%	3.50%
	2030	$422	$481	$650	$719	$650	$719	$670	$739	$670	$739
RIA integrated	2015	$5,150	$5,175	$9,800	$9,850	$9,800	$9,850	$9,800	$9,900	$9,800	$9,900
Mild hybrid regen/ launch	2022	$2,208	$2,435	$4,584	$5,010	$4,584	$5,010	$4,584	$5,060	$4,584	$5,060
With stop-start	% Reduction in FC	10%	10%	16%	16%	22%	22%	16%	16%	22%	22%
	2030	$2,005	$2,201	$4,240	$4,602	$4,240	$4,602	$4,260	$4,652	$4,260	$4,652
Parallel strong hybrid	2015	$12,250	$12,350	$17,100	$17,300	$17,100	$17,300	$17,100	$17,500	$17,100	$17,500
SHEV (18-22% pickup and van)	2022	$5,094	$5,730	$7,426	$8,320	$7,426	$8,320	$7,426	$8,520	$7,426	$8,520
Electric drive	%Reduction in FC	20%	20%	20%	20%	25%	25%	20%	20%	25%	25%
		$4,640	$5,178	$6,785	$7,562	$6,785	$7,562	$6,785	$7,762	$6,785	$7,762

TABLE 7-6 Continued

Vehicle Class	Year or % Gain	Medium Heavy-Duty (MHD) Class 6 and 7 Vocational Vehicles		Class 6and 7 Delivery Trucks		Heavy Heavy-Duty (HHD) Class 8 Vocational Vehicles (Refuse)		Class 8 Tractor-Trailers
Technology		Low Cost	High Cost	Low Cost	High Cost	Low Cost	High Cost	
Non-hybrid	2015	$2,150	$2,180	$2,150	$2,180	$3,300	$3,350	
Enhanced accessories	2022	$891	$1,000	$891	$1,000	$1,389	$1,565	
12 V stop-start	(% gain)	3.50%	3.50%	3.50%	3.50%	3.50%	3.50%	
	2030							
RIA integrated	2015	$17,100	$17,225	$17,100	$17,225	$21,000	$21,900	
Mild hybrid regen/launch	2022	$7,850	$8,625	$7,850	$8,625	$10,080	$11,700	
With stop-start	(% gain)	16%	16%	22%	22%	22%	22%	
	2030							
Parallel strong hybrid	2015	$22,000	$22,250	$22,000	$22,250	$34,500	$35,000	
SHEV (18-22% pickup&van)	2022	$10,020	$11,100	$10,020	$11,100	$14,780	$16,500	
Electric drive	(% gain)	20%	20%	25%	25%	30%	30%	
	2030							

will be low and costs will likely come down in the 2022 and 2030 time frames. Chapter 12 evaluates the cost effectiveness of these potential applications and their commercial payback periods.

Although there is significant development ongoing for current and future introduction of PHEVs and full BEVs for passenger cars and light trucks and sport-utility vehicles (SUVs), the committee believes that the market penetration for these technology concepts for medium- and heavy-duty trucks and commercial vehicles will be rather limited over the next 10 years. This is based on the unique duty cycle, commercial vehicle durability, rough service requirements, and large power transfer requirements which differ from light-duty applications and the challenging cost effectiveness (payback period for the higher cost incurred) due to rather low projected energy and fuel costs over the next decade. Lack of high-power charging infrastructure may also limit deployment and use of MHD BEVs, for those that would need to be charged outside of a fleet facility. A detailed discussion of this cost-effectiveness assessment is included in Chapter 12.

As can be seen in Table 7-6, significant reductions in total system cost are estimated, starting from published costs extracted from various sources for 2015, and then projected by the committee to 2022 and 2030. The table also provides low and high cost estimates based on uncertainty associated with achieving cost reduction targets and the challenges related to the transfer of technology developments from light-duty passenger vehicles to much larger commercial vehicles.

Due to the much higher vehicle weights for commercial vehicles compared to passenger cars, the relative size of battery packs, electric motors, power electronics, and the rest of the hybrid-related system is much larger. As a result, the total system costs, related by cost per kilowatt of power or the necessary kilowatt-hour levels, are much larger, as shown in the table.

Table 7-6 also shows the percentage improvement in fuel consumption which was estimated by the committee, based on the vehicle class and the intended service. For vocational vehicles used in job-site-type applications, the potential fuel consumption improvement is less than delivery truck–type of usage where a high percentage of use is braking and subsequent acceleration driving. Therefore, the cost effectiveness, or fuel savings cost versus system cost, is dependent on the predominant driving cycle of the vehicle.

As stated above, Table 7-6 does not show projections for PHEV or BEV versions of commercial vehicles because the cost-effectiveness assessment does not show likelihood of commercialization. However, regulatory developments in foreign markets or local financial incentives in certain U.S. cities or states could promote earlier commercialization than would otherwise occur. International regulations under consideration to outlaw combustion engine use in commercial vehicles used in highly populated cities could justify an alternative set of financial boundary conditions, which could cause increased introduction of PHEV and BEV vehicles in certain U.S. markets.

Table 8-8 is presented at the end of Chapter 8 to show the projected cost breakdown assumptions which the committee used to develop the estimated total hybrid system costs. The table shows that the committee believes that the future cost reduction projections will be very significant, but will not achieve the very low costs estimated from passenger car applications. As discussed earlier in this chapter, the unique requirements for commercial vehicles will not allow complete carryover from light-duty vehicles. The table also shows the breakdown by battery pack, traction motor, power electronics, and other components which make up the "rest of system." The relative sizes of battery pack (in kilowatt-hours), traction motor (in kilowatts), and method of hybridization are broken down by vehicle class and low or high cost estimates, and are projected to 2022 and 2030.

7.7 FINDINGS

Finding: Due to significant progress in the reliability improvement and cost reduction of lithium-ion battery systems in light-duty vehicles, meaningful application of hybridization will occur in Class 2b through 6 vehicles over the next 10 to 15 years, especially in those with urban stop-and-go duty cycles.

Finding: Certain subsystems developed as part of complete hybrid systems may individually be applied to Class 8 vehicles for cost-effective benefits, for example, engine decoupled coasting, stop-start, electrification of accessories, and auxiliary power units.

Finding: Because the commercial vehicle industry is so cost driven, compressed natural gas technology was adopted in many urban and centralized fueling applications as a means to reduce criteria pollutants and fueling costs. As hybrid technology, especially full hybrid and plug-in hybrid, costs continue to decline, it will compete as the lowest-cost option for urban and centralized fleets.

Finding: Light-duty dedicated hybrid and electric vehicles are being designed and manufactured in normal plant cycles to maximize synergy in the overall vehicle system, such as the Chevrolet Volt and Bolt. Currently, medium- and heavy-duty hybrid vehicles are produced by "up-fitting" or converted by higher cost by low-volume producers. Future dedicated hybrid designs will result in additional cost savings.

Finding: As purchase costs, maintenance costs, and operating downtime are reduced, the business case for hybrid applications in vocational vehicles with significant job-site and stationary operations will increase and market penetration will be significant.

Finding: Current R&D activities being undertaken by DOE and its industry partners are focused on the cost reduction of electric motors/generators for use in light-duty hybrid and electric vehicles. Due to the much higher torque requirements for medium- and heavy-duty drivetrains, the technical approaches for light-duty electric motor cost reduction may not directly carry over to the commercial vehicle space.

Recommendation 7-1: NHTSA and EPA should coordinate with DOE, other government agencies (Department of Defense, Office of Naval Research, the Defense Advanced Research Projects Agency, etc.), and private industry to develop a strategy for electric motor cost reduction R&D that is applicable to the high torque requirements for medium- and heavy-duty vehicles.

7.8 REFERENCES

DOE (U.S. Department of Energy). 2015a. 2015 Annual Progress Report. Hydrogen and Fuel Cells Program. https://www.hydrogen.energy.gov/annual_progress15.html (accessed December 19, 2019).

DOE. 2015b. Wide Bandgap Power Electronics Technology Assessment. Available at https://energy.gov/sites/prod/files/2015/02/f19/QTR%20Ch8%20-%20Wide%20Bandgap%20TA%20Feb-13-2015.pdf (accessed May 18, 2017).

DOE. 2015c. Quadrennial Technology Review 2015: Chapter 8: Advancing Clean Transportation and Vehicle Systems and Technologies: Technology Assessments. Available at https://www.energy.gov/sites/prod/files/2016/01/f28/QTR2015-8E-Plug-in-Electric-Vehicles.pdf (accessed August 31, 2019).

EL-Refaie, A. 2016. "Alternative High-Performance Motors with Non-Rare Earth Materials." Presentation for DOE Peer Review. June 7. Department of Energy, Washington, DC. Available at https://www.energy.gov/sites/prod/files/2016/06/f32/edt045_elrefaie_2016_o_web.pdf.

EPA (Environmental Protection Agency). 2016. "Demonstration Vehicles" Office of Transportation and Air Quality. February 20. https://archive.epa.gov/otaq/technology/web/html/demonstration-vehicles.html (accessed December 20, 2019).

Gladstein, Neandross, and Associates. 2014. *Equivalent Strategies for the ARB Zero Emission Bus Regulation.* Irvine, CA: Gladstein, Neandross, and Associates. June.

Hino Motor Sales Australia. 2014. Hino 300 Series Hybrid: Cleaner, More Efficient Trucks in the Light-Duty Class. Available at http://www.hino.com.au/300/hybrid/.

Kargul, J., A. Moskalik, K. Newman, D. Barba, and J. Rockwell. 2015. Design and demonstration of EPA's integrated drive module for commercial series hydraulic hybrid trucks and buses. *SAE International Journal of Commercial Vehicles* 8(2):549-567.

NASA. 2017. Definition of Technology Readiness Levels. Available at https://esto.nasa.gov/files/TRL_definitions.pdf.

NHTSA (National Highway Traffic Safety Administration) and EPA. 2016. *Regulatory Impact Analysis, Final Rule for Greenhouse Gas Emissions and Fuel Efficiency Standards for Medium- and Heavy-Duty Engines and Vehicles, Phase II.* Washington, DC: NHTSA and EPA.

NRC (National Research Council). 2010a. *Technologies and Approaches to Reducing the Fuel Consumption of Medium- and Heavy-Duty Vehicles.* Washington, DC: The National Academies Press.

NRC. 2010b. *Hidden Costs of Energy: Unpriced Consequences of Energy Production and Use.* Washington, DC: The National Academies Press.

NRC. 2015. *Cost, Effectiveness, and Deployment of Fuel Economy Technologies for Light-Duty Vehicles.* Washington, DC: The National Academies Press.

NREL (National Renewable Energy Laboratory). 2015. Fuel Cell Buses in U.S. Transit Fleets: Current Status 2015. NREL/TP-5400-64974.

Odyne Hybrid Systems LLC. 2014. "Optimization of Hybrid Systems for Combined Driving/Stationary Duty Cycles." Presentation at Battery Show. Available at https://www.odyne.com/wp-content/uploads/2019/02/odynehybridbatteryshowpresentation09182014.pdf (accessed September 17, 2019).

Parker Hannifin Corporation, Hybrid Drive Systems Division. 2012. Parker RunWise Advanced Series Hybrid Drive Brochure.

Rogers, S., and Boyd, S. 2016. "Overview of the DOE VTO Electric Drive Technologies R&D Program," Presentation at the 2016 DOE Vehicle Technologies Office and Hydrogen and Fuel Cells Program Annual Merit Review and Peer Evaluation Meeting about Electric Drive Systems. June 7. Department of Energy, Washington, DC.

XL Hybrids, Inc. 2018. XLH Hybrid Electric Uplift Ford F-59 Brochure. Available at https://www.xlfleet.com/content/assets/Uploads/XL-XLH-F59-Flyer-8.5x11-LR.pdf (accessed September 17, 2019).

8

Battery Technology for Medium- and Heavy-Duty Hybrid and Electric Vehicles

8.1 INTRODUCTION

As identified in Chapter 7, there has been a significant increase in the number of hybrid and battery electric vehicles being offered in passenger cars and light-duty trucks within the past few years. Stimulated by research and development funding by the U.S. Department of Energy (DOE), there has been significant progress relative to the three major hurdles to wide-scale commercialization of hybrid and battery electric vehicles: (1) battery cell energy density and useful life; (2) battery pack reliability, thermal management, and safety; and (3) battery cell and resulting pack cost.

Another factor driving increased battery research and development is increasing regulatory pressure to eliminate the use of internal combustion engines, especially diesel engines, in large, densely populated cities. The mayors of Paris, Mexico City, Madrid, and Athens have announced that they will stop the use of all diesel-powered cars and trucks by 2025 in order to improve air quality.[1]

8.1.1 Evolution of Automotive Battery Chemistries

For the first 80 years or so of the automotive and commercial trucking industry, lead acid was the most common choice for battery chemistry due to its relatively low cost, useful level of its state of charge (SOC), and the ability to recycle the core elements of the battery. They were employed to provide 12-volt (V) power to starter motors, with the ability to provide rather large current levels for the time necessary to start spark-ignited, primarily gasoline-fueled engines in passenger cars and light trucks. For use in commercial trucks and buses, the voltage remained at 12 V, but the cold cranking amperage typically increased to 1,400 amperes (A) or more due to the higher power and torque needed to start diesel engines. Lead-acid batteries typically have energy densities of approximately 30 watt-hours per kilogram (Wh/kg).

With the arrival of the first electric car prototypes and limited production vehicles (GM EV1) in the early 1990s, battery chemistry migrated to nickel-cadmium (Ni-Cd) to increase energy density. Such Ni-Cd batteries typically exhibited energy densities of about 50 Wh/kg. When the first hybrid electric vehicles were introduced in the late 1990s (e.g., the Honda Insight and Toyota Prius), there was a need for even greater energy density, resulting in the introduction of nickel-metal hydride (NiMH). The energy density for NiMH increased approximately from 60 to

[1] See http://www.bbc.com/news/science-environment-38170794 (accessed january 2, 2017).

70 Wh/kg, which, combined with its relative safety and thermal management requirements, saw very significant market penetration and is still in use today.

In the ongoing quest for higher energy density, especially for application in full hybrid and battery electric vehicles, lithium-ion (Li-ion) chemistries have seen increasing use in vehicular applications. Li-ion batteries underwent much research and development to support the tremendous growth of the portable computer, tablet, and cell phone industry. Although the basic chemistry may be appropriate for use in transportation, the application of such batteries to large-power battery packs has been the focus of the automotive and its Tier 1 supplier industry for the past 15 to 20 years. Advances in electrical control and thermal management have allowed increasing penetration into most current and future hybrid electric vehicles (HEVs), plug-in hybrid electric vehicles (PHEVs), and battery electric vehicles (BEVs). This is due to the much higher energy density than that of NiMH, with Li-ion packs demonstrating 120 to 180 Wh/kg, depending on the particular electrolyte and anode/cathode chemistry employed. Lithium-polymer batteries, for example, utilize a semisolid gel-type electrolyte. Figure 8-1 shows the range of specific energies for different battery chemistries compared to specific power (W/kg).

8.1.2 Breadth of Battery Systems Offered in Light-Duty Hybrid Vehicles

There has been a significant increase in the number of hybrid and battery electric passenger cars and light trucks sold in the United States between 2010 and 2014. In 2015, the report *Cost, Effectiveness, and Deployment of Fuel Economy Technologies for Light-Duty Vehicles* identified some 68 total hybrid and electric vehicle models offered in 2014, which included 46 different hybrid vehicle models and an additional 22 plug-in hybrids or battery electric vehicles (NRC, 2015a). Total sales numbers for model year 2014 were approximately 455,000 HEVs and 129,000 PHEVs and BEVs.

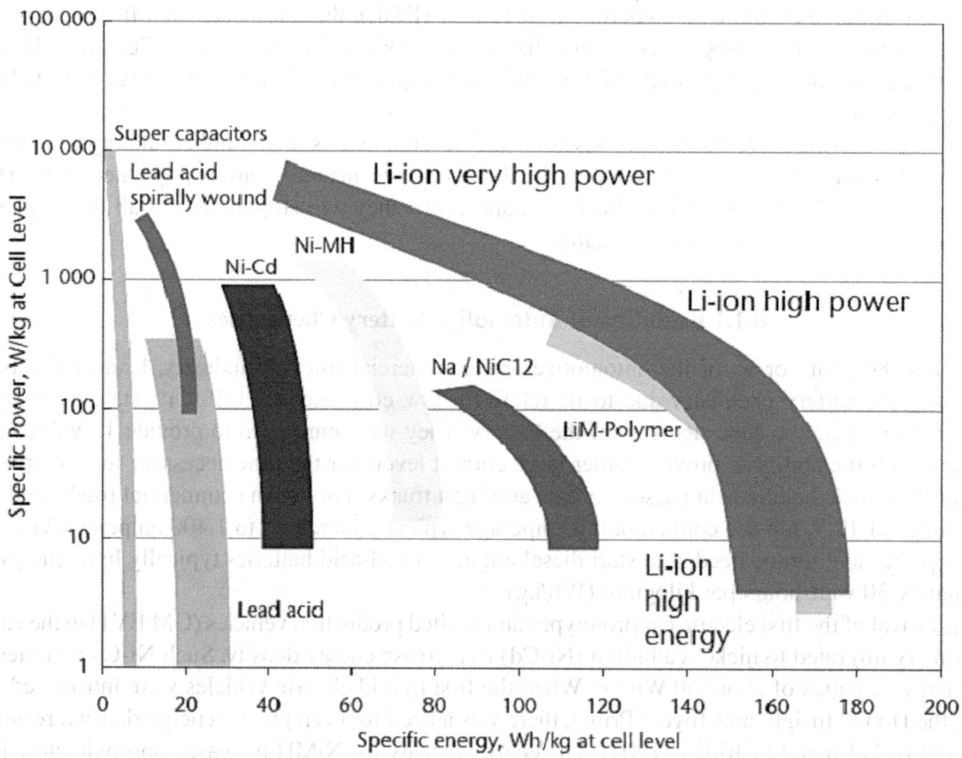

FIGURE 8-1 Comparison of specific power and associated specific energy for varying types of battery chemistry.
SOURCE: Pesaran et al. (2013).

Table 8-1 is reproduced from the 2015 National Research Council (NRC, 2015a) report to give the reader a better impression of the breadth of vehicles offered, which requires an equally broad development of battery cells and packs with varying chemistries, different thermal management and electronic control approaches, unique operating cycle optimization, and many different pack geometries, power requirements, and energy capacity needs.

8.1.3 Basic Working Principles of Li-Ion and Other Battery Chemistries

Batteries are designed to store energy and deliver it at needed rates, producing the power required to move the vehicle or power accessories. There are trade-offs in choice of battery chemistry and battery component design to maximize the energy or power performance of battery systems. For example, batteries that are designed to contain as much charge as possible, as used in BEVs, are designed for higher energy storage capacity. In contrast, batteries that must survive a high number of battery charging and discharge cycles, such as those used in HEVs, are designed with the ability to transfer electrical energy at higher power rates. In order to better understand the challenges associated with the use of high-energy-density Li-ion batteries in automotive applications, and especially in medium- and heavy-duty vehicles, an abbreviated explanation of working principles, application trade-offs, and influences on commercialization is presented below.

Most simply described, a battery contains at least two electrochemical cells that convert chemical energy into electrical energy in the form of current, voltage, and power (based on energy transfer over time). Figure 8-2 shows a very basic representation of an electrochemical battery.

During electrical discharge, current flows from the cathode (positive terminal) to the anode (negative terminal), both of which are isolated by a separator. This occurs because the cathode of a charged battery has a higher electrical potential than the anode, by definition. It is important to note that the direction of current flow is opposite that of electron flow. Therefore, current, by definition, flows through a circuit from the cathode to the anode, while the electrons flow from the anode to the cathode. When a battery is connected to an external circuit, the electrolyte within the battery supports the flow of ions created by the chemical reactions as they transfer between the cathode and anode. Under discharge, cations (positive ions) migrate to the cathode, where they can combine with the electrons, which flow from the anode to the cathode. If anions (negative ions) exist, as in a lead-acid battery, they migrate toward the anode. Therefore, the relative movement of the ions supports corresponding electron flow and associated current flow, such that the battery produces electrical energy over time to perform work.

As previously discussed, Li-ion battery chemistries offer a higher level of specific energy than lead-acid, NiMH, and Ni-Cd types. Figure 8-3 provides additional details about the electrochemical function of Li-ion batteries.

Li-ion batteries *store* energy by intercalation of charged lithium ions within the anode, which may be carbon/graphite, titanate, silicon based, or some other metal oxide. The charger drives electrons from the cathode side of the battery to the anode, which in turn causes the positively charged Li+ to migrate through the separator to the anode. The chemical reactions, in the case of a graphite anode and lithium iron phosphate, are as follows:

Li-ion Battery Charging Mechanism

Reaction at Cathode:
$LiFePO_4 \rightarrow FePO_4 + Li^+ + e^-$

Electron flow is driven by the charger and creates Li^+.

Reaction at Anode:
$Li^+ + e^- + C_6 \rightarrow LiC_6$

Li^+ ions pass through the separator to combine with electrons.

As shown in the figure, as electrons are stripped and transferred, an equal number of Li^+ ions are created in order for the process to remain balanced. During charging, these Li^+ ions transfer within the electrolyte, through

TABLE 8-1 Hybrid Models in Light-Duty Vehicles in 2014

Manufacturer	Model	Sales CY 2014	U.S. Hybrid Share 2013	Architecture
Toyota	Prius Liftback	122,776	0.2715	PS
Toyota	Prius C	40,570	0.0897	PS
Toyota	Camry Hybrid	39,515	0.0874	PS
Ford	Fusion Hybrid	35,405	0.0783	PS
Toyota	Prius V	30,762	0.0680	PS
Hyundai	Sonata	21,052	0.0466	P2
Ford	C-Max Hybrid	19,162	0.0424	PS
Lexus	CT200h	17,673	0.0391	PS
Toyota	Avalon Hybrid	17,048	0.0377	PS
Lexus	ES Hybrid	14,837	0.0328	PS
Honda	Accord Hybrid	13,997	0.0310	i-MMD
Kia	Optima Hybrid	13,776	0.0305	P2
Lincoln	MKZ	10,033	0.0222	PS
Lexus	RX 400 / 450 h	9,351	0.0207	PS
Subaru	XV Crosstrek Hybrid	7,926	0.0175	Other
Buick	Lacrosse Hybrid	7,353	0.0163	MH w IMA
Honda	Civic Hybrid	5,070	0.0112	MH w IMA
Honda	Insight	3,965	< 0.01	MH w IMA and CVT
Toyota	Highlander Hybrid	3,621	< 0.01	PS
Honda	CR-Z	3,562	< 0.01	MH w IMA and CVT or M6
Infiniti	Q50 Hybrid	3,456	< 0.01	PS
Nissan	Pathfinder Hybrid[a]	2,480	< 0.01	P2
Volkswagen	Jetta Hybrid	1,939	< 0.01	P2
Infiniti	QX 60 Hybrid[a]	1,678	< 0.01	P2 with CVT
Chevrolet	Malibu Hybrid	1,018	< 0.01	SS
Buick	Regal Hybrid	662	< 0.01	MH
Porsche	Cayenne Hybrid	650	< 0.01	P2
Chevrolet	Impala Hybrid	565	< 0.01	SS
Acura	ILX Hybrid	379	< 0.01	P2
Audi	Q5 Hybrid	283	< 0.01	P2
Lexus	GS 450h	183	< 0.01	PS
Infiniti	Q70 Hybrid	180	< 0.01	
Mercedes	E400H	158	< 0.01	P2
BMW	Active(335ih)	151	< 0.01	Other
Acura	RLX Hybrid	133	< 0.01	
BMW	Active (535ih)	112	< 0.01	Other
Chevrolet	Tahoe Hybrid	65	< 0.01	2-Mode
Lexus	LS 600h	65	< 0.01	PS
BMW	7-Series Hybrid	45	< 0.01	Other
Cadillac	Escalade Hybrid	41	< 0.01	2-Mode
GMC	Yukon Hybrid	31	< 0.01	2-Mode
Volkswagen	Touareg Hybrid	30	< 0.01	P2
Chevrolet	Silverado Hybrid	24	< 0.01	2-Mode
Mercedes	ML450H	20	< 0.01	
Mercedes	S400HV Hybrid	10	< 0.01	P2
GMC	Sierra Hybrid	6	< 0.01	2-Mode
Overall Hybrid Share of LDV Market			2.75%	

[a]PHEV/BEV breakdown unknown.

NOTE: CVT, continuously variable transmission; IMA, integrated motor assist; i-MMD, intelligent multi-mode drive; MH, mild hybrid; PS, power split; SS, stop-start.

SOURCE: NRC (2015a).

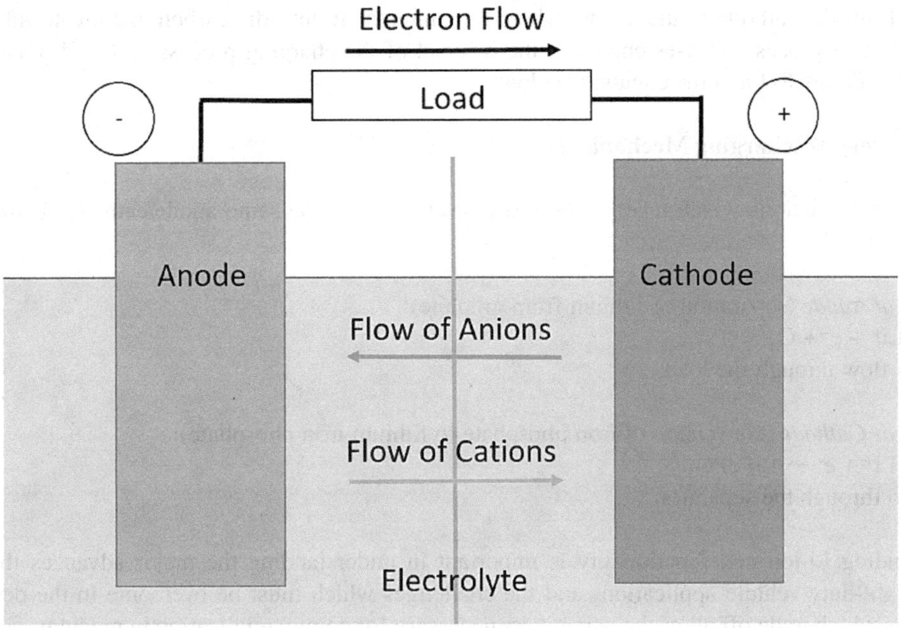

FIGURE 8-2 Basic battery structure and working principle showing electron and ion flow, and major battery components of the anode, cathode, separator, electrolyte, and load.

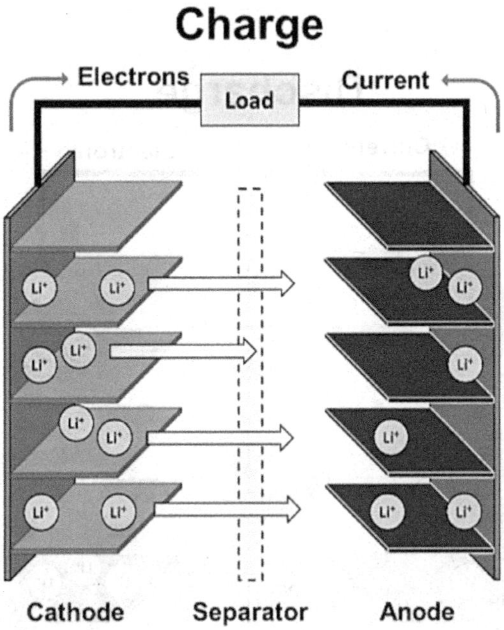

FIGURE 8-3 Lithium-ion structure and battery working principle illustrating charging of the battery by electron flow from the cathode to the anode through the charger, resulting in lithium separation from the cathode, and flow across the separator and intercalation into the anode material.
SOURCE: Wessler (2018).

the separator from the cathode to the anode, where they intercalate into the carbon and create lithiated carbon, LiC_6. The discharge process releases energy in the reversal of the charging processes. The discharge process is characterized in Figure 8-4 and the equations below.

Li-ion Battery Discharging Mechanism

The chemical reactions which take place during discharge, for the same anode/cathode chemistries, are as follows:

Reaction at Anode (separation of lithium from graphite):
$LiC_6 \rightarrow Li^+ + e^- + C_6$
Electrons flow through the load.

Reaction at Cathode (conversion of iron phosphate to lithium iron phosphate):
$FePO_4 + Li^+ + e^- \rightarrow LiFePO_4$
Li ions go through the separator.

Understanding Li-ion cell functionality is important in understanding the major advances that have been achieved in light-duty vehicle applications and the challenges which must be overcome in the development of battery designs which trade off all of the parameters necessary for commercialization in medium- and heavy-duty trucks. Although the examples above are simplified, optimizing all of the parametric factors that must be done for best performance in particular applications is very complex. For example, the graphite-electrode, lithium iron phosphate cell chemistry is only one of many that are currently being commercialized.

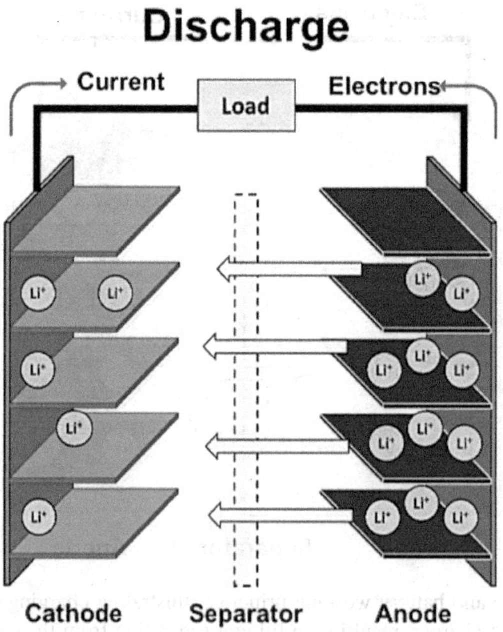

FIGURE 8-4 Lithium-ion battery working principle illustrating discharge of the battery, with electrons flowing from anode to cathode, powering a load and lithium ions separating from the anode and crossing through the electrolyte and separator to combine with the cathode material.
SOURCE: Wessler (2018).

Many anode chemistries are being evaluated and commercialized depending on the intended duty cycle, operating voltages, durability requirements, cost constraints, and safety goals. A group of about a dozen chemistries are the primary focus of attention. Current anode systems include, but are not limited to, carbon graphite (most popular, cost effective), silicon based (improved life and capacity), lithium titanate ($Li_4Ti_5O_{12}$); 1.8-2.8 V (graphite on cathode), and metal oxides (Spotnitz, 2005). Graphite is relatively inexpensive and is used almost exclusively for most light-duty vehicle applications. However, graphite has a charge rate that is limited by lithium plating. In addition, graphite tends to be calendar-life limited due to the growth of solid electrolyte interphase, which forms a coating over time on the graphite that impedes Li^+ transfer into the graphite anode. Lithium titanate oxide (LTO) technology is a derivative of traditional Li-ion batteries which adds additional lithium titanate nanocrystals to the surface of the carbon anode. As a result, the anode has a surface area of around 100 meters squared per gram (m^2/g), which is some 30 times more than the 3 m^2/g of conventional graphite, which supports electrons transferring to or from the anode much more quickly.[2] As a result of this much larger surface area, recharging of the LTO battery is much faster than a traditional graphite anode battery. Furthermore, the increased surface area significantly improves the LTO cells' general stability and safety during charging. LTO also exhibits excellent power at low temperatures.

Likewise, various electrolyte chemistries are being developed and optimized for commercial application, including salts such as $LiPF_6$ and $LiBF_4$ added, dimethyl or ethylene carbonates, ionic liquids, polymer electrolytes, and various solid electrolytes (Spotnitz, 2005). The liquid electrolyte solution permits the transport of Li^+ ions between the cathode and anode. The most common electrolyte employs lithium salt, such as $LiPF_6$, dissolved in an organic solution. In addition to lithium salt, other additives are necessary to achieve the desired properties of the electrolyte solution. These additives improve the stability of the solution by opposing dendritic formation and degradation of the solution. The electrolyte formulation will vary depending on the specific anode and cathode materials being used.[3]

In addition, much research and development are ongoing in cathode chemistry. Accordingly, different higher-performance, higher-capacity cathode materials including lithium cobalt oxide (LCO), spinel-based lithium manganese oxide (LMO), lithium nickel cobalt manganese oxide (NMC), lithium iron phosphate (LFP), and lithium nickel cobalt aluminum oxide (NCA). The cathode materials (Table 8-2) are constructed of cobalt, nickel, and manganese in the crystal structure forming a multi-metal oxide material to which lithium is added. This family of batteries includes a variety of products that cater to different user needs for high energy density and/or high load capacity.

Figure 8-5 compares the specific energy of lead- and nickel-based chemistries with those that are lithium based. Although Li-Al (NCA) demonstrates the highest specific energy, Lithium manganese (LMO) and lithium phosphate (LFP) show higher specific power and thermal stability. Lithium titanate (LTO), on the other hand, has relatively low specific energy but demonstrates excellent calendar life and performs the best in cold temperatures. As we look to medium- and heavy-duty vehicles, safety and calendar and cycle life may be more important than just specific energy or total energy capacity.

Table 8-3 provides a summary of different battery chemistries and construction methods which are provided in hybrid and electric vehicles available in model year 2015. Of interest is the variation in chemistries, the cell count associated with pack topology, and specific energy. This suggests that no single chemistry is either optimum or preferred.

8.2 COMPARISON OF DIFFERENT LITHIUM BATTERY CHEMISTRIES, PERFORMANCE, AND APPLICATIONS

Figure 8-1, presented in the beginning of this chapter, showed a graphic representation of specific energy capacity versus specific power of different basic battery chemistries. However, the Li-ion curve showed a broad range of possibilities. In order to better understand in what applications particular chemistry combinations are most preferred, a trade-off in performance metrics must be done for each application.

---

[2] See http://investorintel.com/technology-metals-intel/lithium-titanate-battery-technology-bigger-better/ (accessed December 1, 2019).

[3] See http://www.targray.com/li-ion-battery/electrolyte (accessed December 1, 2019).

TABLE 8-2 Comparison of Typical Cathode Materials Used in Lithium-Ion Batteries

Chemistry	Nominal V	Charge V Limit	Charge and Discharge C-Rates	Energy Density (Wh/kg)	Applications
Cobalt	3.60 V	4.20 V	1C limit	110-190	Cell phone, cameras, laptops
Manganese (spinel)	3.7-3.80 V	4.20 V	10C cont. 40C pulse	110-120	Power tools, medical equipment
NCM (nickel cobalt manganese)	3.70 V	4.10 V[a]	~5C cont. 30C pulse	95-130	Power tools, medical equipment
Phosphate	3.2-3.30 V	3.60 V[a]	35C cont.	95-140	Power tools, medical equipment

[a]Higher voltages provide more capacity but reduce cycle life.
SOURCE: Targray Technology International, Inc. (2017).

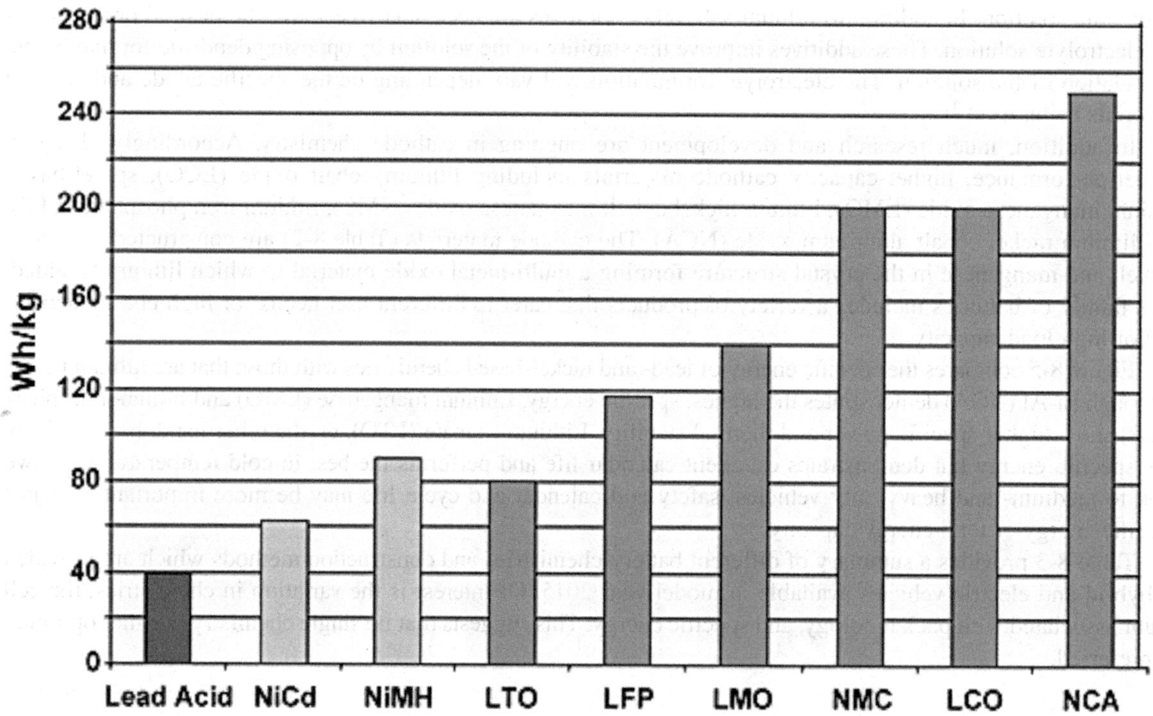

FIGURE 8-5 Comparison of specific energy densities for common battery chemistries.
SOURCE: Battery University (2016), batteryuniversity.com (accessed August 20, 2019).

TABLE 8-3 Summary of Li-Ion Battery Chemistries and Geometries Utilized in Model Year 2015 Electric and Hybrid Electric Light-Duty Vehicles

EV	Cathode	Anode	Battery Supplier	Type of Cell	Number of Cells	Electric Energy (kWh)	Power (kW)	Specific Energy Density (Wh/kg)	Electric Range (mi)
Tesla Model S	NCA (layered)	Carbon (layered)	Panasonic	Cylindrical	>7,000	85	270	116 (pack)	265
Chevy Volt	LMO (spinel)	Carbon (layered)	LG Chem	Prismatic	>200	16	111	88 (pack)	40
Nissan Leaf	LMO (spinel)	Carbon (layered)	Nissan NEC JV	Prismatic	192	24	90	140 (pack)	84
Honda Fit	NCM (layered)	LTO (layered)	Toshiba Corp.	Prismatic	432	20	92	100 (cell)	82
BYD E6	LFP (olivine)	Carbon (layered)	BYD	Prismatic	96	48	75	–	186

SOURCE: NRC (2015a).

The six primary battery performance metrics which must be traded off are

1. Specific energy capacity (Wh/kg or watt-hours per liter [Wh/l]),
2. Specific and total power (watts per kilogram [W/kg]),
3. Cell and pack safety (abuse tolerance),
4. Cell and pack life (total cycle and/or calendar life),
5. Total battery system cost (dollars per watt-hour [$/Wh]), and
6. Cell performance.

The specific energy or capacity of a battery pack is directly related to the total operational time before recharging is necessary. In the case of hybrid and electric vehicles, this corresponds to electric mileage or electric range. The specific power of the battery cells or the pack corresponds to the ability of the system to deliver high levels of current, which, at a given voltage over time, determines the total power that can be generated. Cell and pack safety are associated with the ability of the system to withstand harsh operational conditions or abuse. Protection against thermal runaway or other thermal events is also a factor considered in pack safety. The lifetime of cells and packs is evaluated not only on number of charging and discharging cycles, but also on calendar life or longevity, which is influenced by growth of deposits that will degrade lithium-ion transfer. Cell, battery pack, and total electrical system costs are very important parameters which are particularly important in commercial vehicle application where capital and operational costs are of the utmost importance. The component and total system costs are shown in Table 8-8 at the end of this chapter. Pack and cell performance, under both cold and hot conditions, is also important to consider. Factors such as self-discharge percentage, capacity (in percentage), and internal resistance, all evaluated as functions of the total number of charging/discharging cycles, are used to evaluate performance, which varies by cell chemistry. There are many other factors which must be considered, such as ease of maintenance, component commonality across different vehicle applications for a single fleet, fast-charge capabilities, and propensity to self-discharge and shelf life—all characteristics which must be considered in product commercialization. Table 8-4 shows a summary of many key features and functional parameters associated with seven leading Li-ion battery chemistries. These give the reader an understanding of the breadth of variation in performance functionality of batteries which together are all classified under "lithium-ion batteries."

Figure 8-6 shows spider diagrams of the six major Li-ion cell chemistries which are currently being developed and/or produced by the battery industry. The six axes on the spider diagram represent relative achievements of each battery type in successfully accomplishing the metric being evaluated: (1) specific energy, (2) specific power, (3)

TABLE 8-4 Voltage, Specific Energy, and Volumetric Energy for Several Battery Chemistries

Material	Voltage (Average vs Li/ Li^+)	Capacity (mAh/g)	Crystal Density (g/cm^3)	Tap Density (g/cm^3)	Specific Energy (Wh/kg)	Volumetric Energy (Wh/L)
$LiCoO_2$	3.8	150	5.10	2.9	570	2,907
$LiNi_{1/3}Mn_{1/3}Co_{1/3}O_2$ (NMC)	3.7	170	4.75	2.5	629	2,988
$LiNi_{0.8}Co_{0.15}Al_{0.05}O_2$ (NCA)	3.7	185	4.85	2.5	685	3,322
$LiMn_2O_4$	4.0	110	4.31	2.5	440	1,896
$LiFePO_4$	3.4	160	3.60	1.5	544	1,958
$0.5Li_2MnO_3{\cdot}0.5LiMO_2$[a]	3.6	250	4.30	1.8	828	3,870
$LiMn_{1.5}Ni_{0.5}O_4$[a]	4.7	135	4.40	2.0	635	2,794

[a]Not yet commercialized.

NOTE: Crystal densities are theoretical values, while densities represent typical, practical values, determined experimentally as the actual weight per unit volume occupied by a given material. Volumetric energy densities are calculated using the crystal densities for comparison because the optimum, final electrode densities will vary among materials. M=Mn, Ni, or Co in $LiMO_2$. Capacity and voltage targets for $0.5Li_2MnO_3{\cdot}0.5LiMO_2$ are based on DOE's end-of-life goals for composite materials; crystal density is calculated as an average of Li_2MnO_3 and $LiMn_{0.5}Ni_{0.5}O_2$.

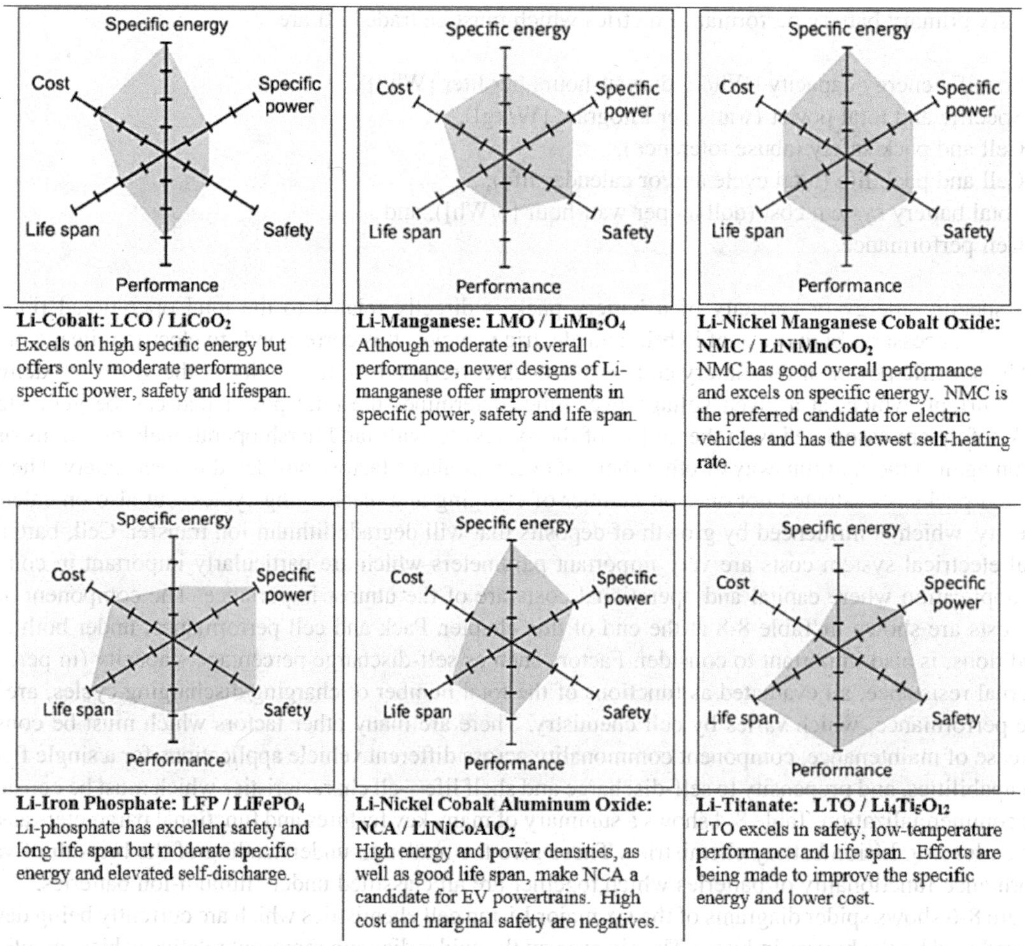

FIGURE 8-6 Relative performance, functional and safety characteristics of six leading Li-ion battery chemistries.
SOURCE: Battery University (2016), batteryuniversity.com (accessed August 20, 2019).

safety, (4) performance, (5) life span, and (6) cost. These diagrams, which are adapted from information provided by Cadex Electronics through the Battery University website, provide a convenient way to compare the key performance metrics when evaluating application for vehicles.[4] A review of the diagrams suggests that lithium nickel manganese cobalt oxide ($LiNiMnCoO_2$, known as NMC) displays excellent values in most of the key metrics, and that is why it is a leading battery chemistry for light-duty hybrid and electric vehicles. NCA also demonstrates high specific energy and very reasonable specific power, which has made it the choice of Tesla-Panasonic (Table 8-7 at the end of this chapter). However, the cobalt concentration in NCA causes it to be more expensive than NMC. Lithium titanate demostrates relatively low specific energy and specific power, but its very long life span, excellent cold weather performance, and very high safety have made it the choice for some commercial applications (UPS) and certain lightweight electric vehicles (iMEV, Honda Fit EV).

As can be seen in Figure 8-6, the selection of basic battery chemistry has many different features and parameters which must be optimized for particular applications. However, other equally important issues that must be considered are the physical features of the cells, the complexity of electrical connection between the cells, the method of cooling (liquid versus air), available geometric space in the vehicle, and many more.

8.2.1 Li-Ion Battery Physical Characteristics

Certain key parameters which influence successful operation and commercialization are outlined below. The first feature to consider is Li-ion cell geometric and design configuration. There are two major options being pursued, which are *cylindrical* and *prismatic*. Cylindrical cell configurations can be divided into "jellyroll" or hard can types. For prismatic designs, they can be configured in wound or stacked layers which are encased within a soft pouch or hard can. Figures 8-7 and 8-8 show examples of these different configurations. Table 8-7 lists battery offerings from various manufacturers and describes their configurations and other characteristics.

Cylindrical cells have the advantage that the same cell geometry may be used in many consumer electronics applications, such that economies of scale tend to reduce the individual cell costs. An example of such dual use is the cells used in the Tesla Model S, where over 7,000 Panasonic 18650 cells are integrated into the battery pack.

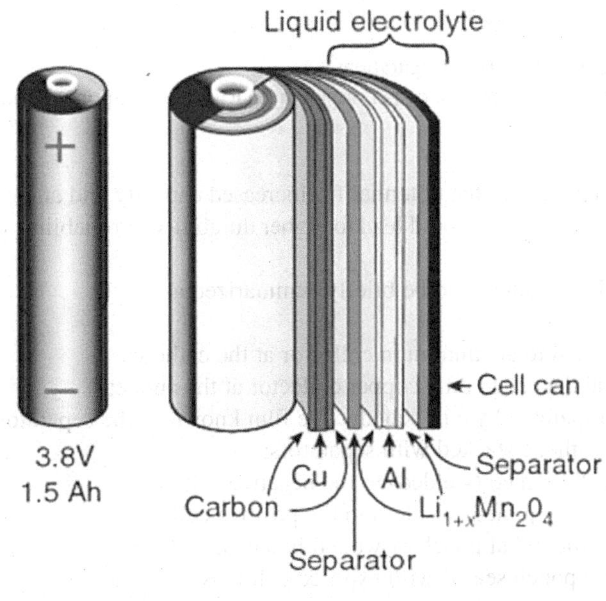

FIGURE 8-7 Examples of cylindrical battery configurations.
SOURCE: Tarascon and Armand (2001), reprinted by permission from RightsLink: Springer Nature, Tarascon & Armand, 2001.

[4] See http://batteryuniversity.com/learn/article/types_of_lithium_ion (accessed August 20, 2019).

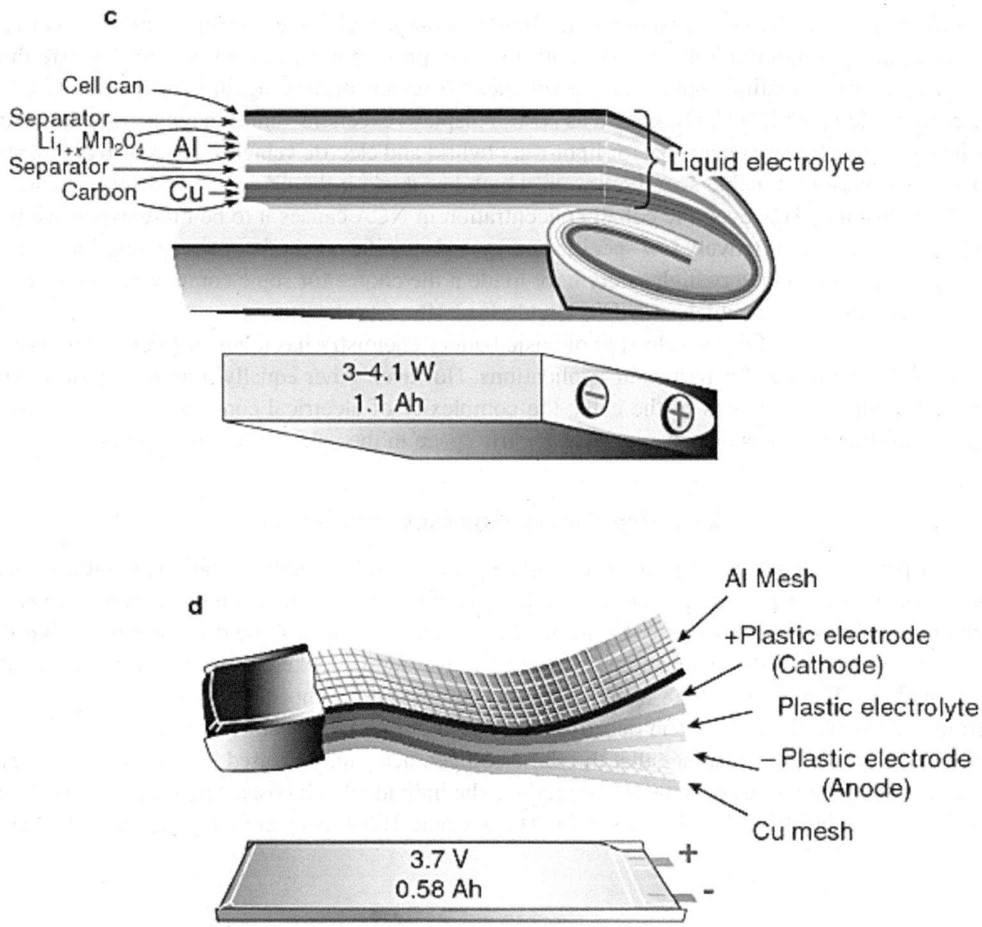

FIGURE 8-8 Examples of prismatic battery configurations.
SOURCE: Tarascon and Armand (2001), reprinted by permission from RightsLink: Springer Nature, Tarascon & Armand, 2001.

Prismatic cells, on the other hand, have the potential for increased capacity and amperage for each unit that must have external electrical connections. This could lead to higher durability or reliability, especially in the application for commercial vehicles.

The composition of a Li-ion battery can be briefly summarized as

- Active material laminated to an aluminum collector at the cathode;
- Carbon-based material deposited on a copper collector at the anode;
- Anode and cathode, separated by a nonconductive film known as the separator;
- Layers of anode and cathode stacked with separators;
- Anode layers electrically joined (welded) to the negative cell tab;
- Cathode layers electrically joined (welded) to the positive cell tab;
- Stack inserted in a plastic/metal pouch or wound in a can; and
- Electrolyte poured and pouch sealed with exposed cell tabs.

This basic construction applies to both cylindrical and prismatic cells.

8.2.2 Commonly Used Terms Important in Battery Assessment

In order to better understand the ability of battery technology to fulfill future requirements for medium- and heavy-duty applications, it is important to understand commonly used terms that are important when evaluating battery cell and pack characteristics. Key terms include the following:

- **C-Rate**
  - Charging and discharging of batteries is governed by the C-rate, a measure of the discharge current. C-rate is the amount of constant current at which a battery cell can be discharged (or charged) for 1 hour.
  - Example: A battery cell which is capable of providing 10 A for one hour has a 1C rate of 10 A.
- **Capacity**
  - Coulomb counting of current (in amperes) is capacity.
  - The above-mentioned battery has a capacity of 10 amp-hours (Ah).
  - Discharging the same cell at a 2C rate would theoretically mean withdrawing 20 A for 0.5 h.
  - Putting such two cell in parallel would increase the C-rate of the battery to 20 A and capacity to 20 Ah.
  - A higher C-rate typically reduces capacity owing to higher internal losses.
  - It is not uncommon to discharge Li-ion batteries at 10C rates for a very short duration of time.
- **State of Charge (SOC)/Depth of Discharge (DOD):** SOC is the percent of charge or energy remaining in the battery. DOD is the percent of charge or energy used. If the SOC is 80%, the DOD is 20%.
- **Resistance to DC Current (DCR):** Measure of a battery's internal resistance.
  - Charge/discharge resistance depends on SOC, temperature, and C-rate.
  - Internal resistance of a battery grows over the life span of the battery.
- **High Self-Discharge:** Ni-Cd and NiMH have the highest self-discharge.
- **Cell Balancing**
  - As cells charge and discharge, their SOC levels deviate from one another due to variation in internal resistances. Periodically, cells in a battery pack need to be balanced to gain maximum usage.
  - Battery cells are typically balanced at top of charge (PHEVs and electric vehicles [EVs]) or bottom of the charge (typically HEVs).
- **Power:** Power in simple terms equates to acceleration or the ability of an EV to climb a hill. Power = Voltage × Current (P = VI).
- **Energy:** Energy equates to vehicle range or capacity to do work. It is measured in watt-hours.
- **Cycle Life:** The cycle life is the number of cycles the battery can undergo before it fails to meet performance criteria. A complete cycle is defined as one deep charge/discharge cycle at a given C-rate.
- **Calendar Life**
  - Dependency on temperature, SOC, and time.
  - Battery resistance increases with calendar life; thus, retained capacity decreases. Optimum storage temp is below 30°C. The calendar life of Li-ion batteries is far superior to that of other types of batteries.
- **State of Health (SOH):** SOH is used to predict range, state of charge, etc.
  - Measure of DCR, capacity, alternating current resistance etc.
- **WH Throughput:** Total energy in and out.
- **Hybrid Pulse Power Characterization (HPPC):** HPPC is a test to measure a battery's power capability at various SOC levels.
- **Charge Neutrality:** A pulse cycle which discharges and charges the battery to maintain charge neutrality (100 A out/100 A in). SOC will still very slowly deplete.

8.2.3 Battery Pack Design Considerations

Once the basic priority for cell performance is defined, the cells must be combined into battery packs that much achieve a significant number of electrical, thermal, structural, and weight targets in order to have successful installation into a vehicle.

- Cells can be joined in various parallel and series configurations to achieve required power and energy.
  - o Cells in parallel increase capacity.
    - —Manufacturing/assembly puts limitations on how many cells can be put in parallel.
  - o Cells in series increase pack voltage.
    - —Cell tabs are joined with low resistance methods (welding).
- Cells are put together in modules (arrays).
  - o Electromechanical construction comprises welding cell tabs, cell constraining mechanism, and cooling considerations.
  - o Provide means for cell voltage and temperature monitoring.
- Various numbers of modules are put together to make a battery pack.

There are several major components within battery packs that are common across most applications. These include the following:

- **Modules (Arrays)**
  - o Cells in parallel/series combination. Modules are usually stand-alone components structurally.
- **Battery Management System (BMS):** The BMS is low voltage and
  - o Controls all the devices internal to the battery.
  - o Limits power to and from the battery if cells are operated outside their recommended parameters.
  - o Can shut down the battery completely if needed.
  - o Communicates externally to the vehicle control module on the Controller Area Network bus.
  - o Monitors cell voltages and temperature for optimum life.
- **Distribution Box**
  - o Contains switches, pre-charge resistor, current sense module, etc. Electronics are powered by low voltage.
- **Service Disconnect**
  - o A manual switch which can be removed when working on the battery packs.
- **Fuse**
  - o Safety fuse which is rated at a current somewhat higher than application current.
- **Vent**
  - o Battery pack mounted inside the cabin has cell venting strategy to eject vent gases outside.

In addition to the battery pack components, there are other major parts to the electric drive system, which include the following:

- **DC-to-DC Converter**
  - o Reduces batteries DC voltage to required DC voltage to run electronics/accessories.
- **DC-to-AC Inverter**
  - o Converts battery's DC voltage to AC voltage to accept charge and send to the traction motor.
- **Traction Motor**
  - o Typically a multi-phase AC motor.
  - o Provides torque to wheels.
  - o Acts as a generator during braking (regenerative power).

- o Some PHEVs have a separate generator for charging the battery while running the engine at its optimum efficiency.
- o Typically liquid cooled.

8.3 INFLUENCE OF MEDIUM- AND HEAVY-DUTY USAGE ON BATTERY PERFORMANCE

In some cases, experience gained in light-duty usage sheds some light on the challenges which may be experienced in commercial vehicle applications. The report *Overcoming Barriers to Deployment of Plug-in Electric Vehicles* (NRC, 2015b) evaluates the current state of light-duty plug-in electric vehicles, their technologies, costs, and challenges for large-scale commercialization. Although the focus of the 2015 report is on smaller, light-duty passenger vehicles, the assessment of future challenges and opportunities has some direct bearing on medium- and heavy-duty applications.

Concerns about the durability and performance of the current lithium-ion batteries at extremely high and low temperatures could be a barrier widespread adoption. As more experience is gained with a larger number of vehicles and increasing miles driven, effects on durability are becoming clearer, which is an extremely important factor in commercial vehicle use (Steffke et al., 2008).

One study that evaluated a PHEV with a 20 kWh battery showed that a hot climate accelerates the normal degradation of battery capacity with time (Figure 8-9) (Pesaran et al., 2013). Reports on shorter battery life for

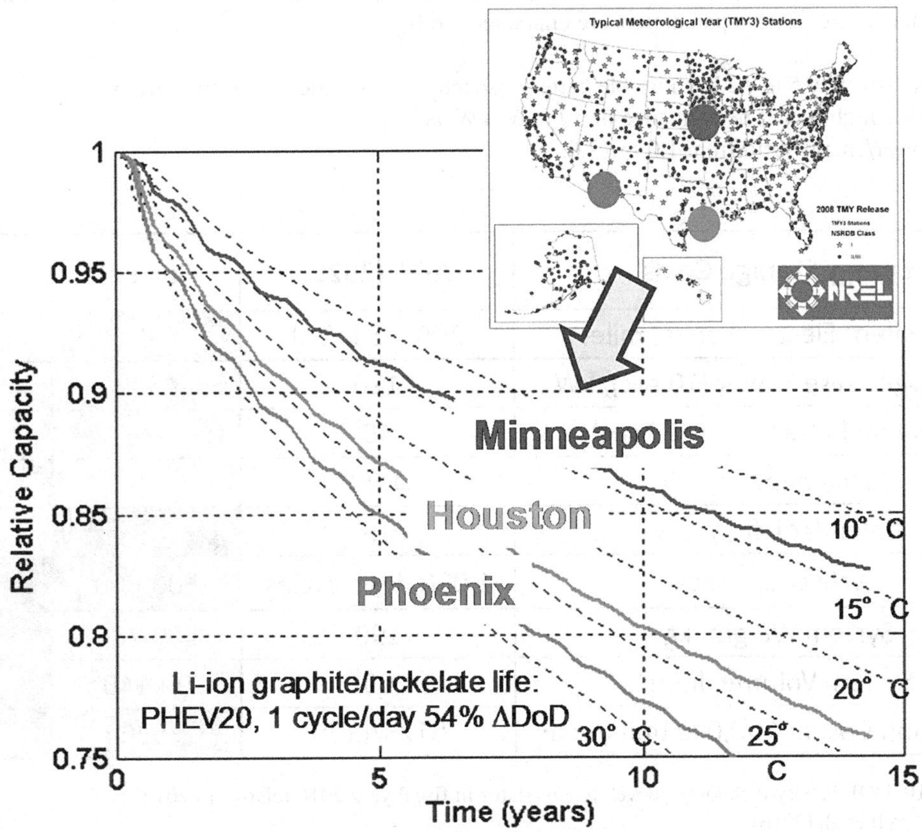

FIGURE 8-9 Effect of ambient temperature on battery capacity for a 20 kWh battery in a PHEV.
NOTE: DoD, depth of discharge; PHEV20, plug-in hybrid 20 mile range.
SOURCE: Pesaran et al. (2013).

Nissan Leafs in Arizona seem consistent with that observation (Gordon-Bloomfield, 2013). As a result, Nissan has tested new battery pack designs to address the observed problem (Gordon-Bloomfield, 2013).

Although Figure 8-10 illustrates preliminary results of studying the effect of temperature on battery capacity, battery life depends also on cycling at various depths of charge, rate of charge, and discharge, which are prevalent in commercial vehicle applications. Further experience in hot climates may be necessary to determine additional approaches for commercial vehicle use.

There are many requirements for battery use in medium- and heavy-duty commercial vehicles, compared to passenger cars and light-duty trucks, which must be addressed, including the following:

1. Larger size with accompanying manufacturing complexity (number of welds),
2. Higher power and associated cooling requirements,
3. More demanding duty cycle,
4. Higher annual mileage than typical passenger cars,
5. Higher hours of usage than passenger cars,
6. Higher number of charge/discharge cycles than light-duty vehicles,
7. Harsher environment (temperatures, weather elements, shock/vibration, etc.),
8. Higher sensitivity to total cost of ownership, and
9. Higher manufacturing variability due to up-fitters, various bodies, and in-use requirements.

According to the National Renewable Energy Laboratory (NREL), the energy efficiency of PHEV and BEV passenger cars is approximately 200 to 400 Wh/mile, based upon vehicle weight, performance, aerodynamics, rolling resistance, etc.[5] These parameters are characterized by

- Approximately 5 to 12 kWh usable battery capacity for 30 miles of electric driving,
- A 2-sec high-power performance of 30 to 60 kW, and
- A power/energy ratio from 1.0 to 12.0.

Energy Storage Goals	AEV (2020)	Current	2012 EV Battery
Equivalent Electric Range, miles	200 (45 kWh)	✓	
Discharge Pulse Power (10 sec), kW	120	✓	
Regenerative Pulse Power (10 sec), kW	40	✓	
Recharge Rate, kW	1.20	✓	
Calendar Life, years	15	tbd	tbd
Cycle Life, cycles	1,000 deep cycles	500-600	200
System Weight, kg	190	200-280	500
System Volume, liters	90	90-140	200
Production Cost at 100,000 units/year	$125/kWh	< $268	$600

FIGURE 8-10 DOE battery technology development status in fiscal year 2016 relative to 2012.
SOURCE: Howell et al. (2016).

[5] K. Smith, S. Santhanagopalan, A. Pesaran, and K. Kelly, NREL, "Battery Technologies for Heavy Duty Electric Vehicles," webinar to members of APTA, August 19, 2015.

By comparison, a medium- or heavy-duty vehicle will require energy efficiencies in the range of 1,000 to 2,000 Wh/mile. Likewise, the associated parameters would be

- More than 30 to 60 kWh usable battery capacity for 30 miles of electric driving;
- A 2-sec high-power performance of 60 to 150 kW or more;
- That volume, weight, and costs for such large systems are challenging; and
- That thermal management for high-power systems operating in commercial vehicles, especially in vocational vehicles, is also challenging.

A 40-ft bus for transit operation could have as much as a 320 kWh (total energy) battery system. Table 8-5 identifies some of the technical goals that have been guiding advanced battery technology development for light-duty vehicles in recent years. Significant progress has been made in achieving performance targets, while reducing the level of cooling and still achieving acceptable safety and durability. Figure 8-10 depicts the progress that has been made in several technical areas under the U.S. Department of Energy Vehicle Technologies Program related to batteries.

However, the technical focus has been directed toward increasing specific energy density and capacity of cells and packs associated with PHEV and BEV applications, including increasing the acceptable discharge level. This increases the available range of such plug-in vehicles, which is one of the major barriers to broad acceptance of such vehicles. This "range anxiety" was identified as major disadvantage in the report *Overcoming Barriers to Deployment of Plug-in Electric Vehicles* (NRC, 2015b).

Another very important factor for commercialization of both light-duty and commercial vehicle application of electrified propulsion systems is cost. The costs associated with traction motors, power electronics, and electrical accessories were discussed in Chapter 7 of this report.

Figure 8-11 shows data concerning the evolution of Li-ion battery pack costs over the past 10 years. Also shown are future projections from various publications out to 2030. The figure references earlier targets identified by the U.S. Automotive Battery Consortium (USABC) (Table 8-6), which were about \$150/kWh as a long-term goal for commercialization. However, more recent projections by USABC suggest that a cost of \$125/kWh is achievable at the pack level (\$100/kWh at the cell level). These targets are shown in Nykvist and Nilsson (2015).

TABLE 8-5 Technical Goals of Recent Light-Duty Vehicle Advanced Battery Technology Development

		PHEV (2015)		
DOE Energy Storage Goals	HEV	PHEV-10	PHEV-40	EV (2020)
Equivalent electric range (miles)	N/A	10	40	200-300
Discharge pulse power: 10 sec (kW)	25	45	38	80
Regen pulse power: 10 sec (kW)	20	30	25	40
Recharge rate (kW)	N/A	1.4	2.8	5-10
Cold crank power: -30°C/2sec(kW)	5	7		N/A
Available energy (kWh)	0.3	3.4		40
Calendar life (year)	15	10+		10
Cycle life (cycles)	300,000 (shallow)	3,000-5,000 (deep)		1,000 (deep)
Maximum system weight (kg)	40	60	120	300
Maximum system volume (l)	32	40	80	133
Operating temperature range (°C)	-30 to +52	-30 to +52		-40 to +85

SOURCE: Howell and Elder (2011).

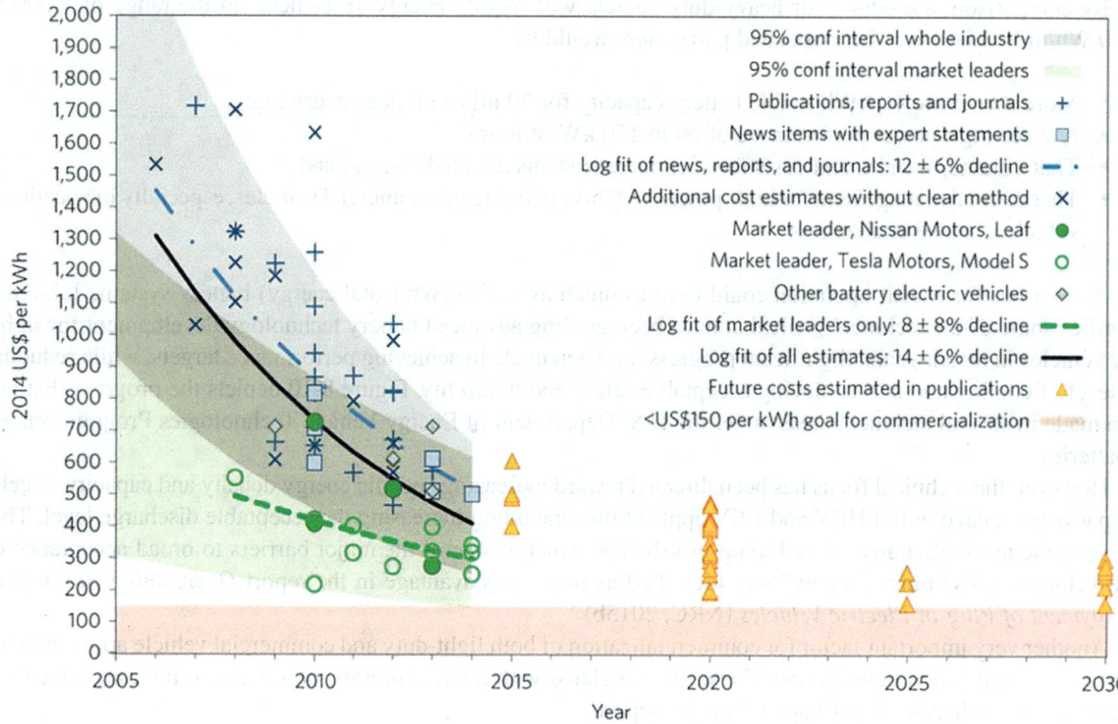

FIGURE 8-11 Battery pack cost per kWh vs time, showing rapidly falling costs of battery packs for light-duty electric vehicles.
SOURCE: Nykvist and Nilsson (2015), reprinted by permission from RightsLink: Springer Nature, Nykvist & Nilsson, 2015.

Figure 8-12 shows some technical breakdown of how the Department of Energy (DOE) and its research contractors envision steps to achieve further cost reduction, without compromising safety, reliability, useful life, or performance. However, for medium- and heavy-duty commercial vehicle applications, further development to address the power, functional, environmental, reliability, and cost issues for such applications is necessary. One major topic of interest for the development of commercial vehicle hybrid and electric propulsion systems is future battery chemistries which may address the functional and safety requirements for such vocational vehicles. Figure 8-13 provides a graphical representation of potential improvements in cell voltage and energy capacity which may be accomplished with new cathode and anode materials.

Figure 8-14 shows an exemplar cost breakdown for different advanced battery chemistries which could be utilized for further cost reduction, to levels of $100/kWh.

When parameters that are especially important for commercial vehicle application are considered, including safety, performance, reliability, and durability, graphite-LFP and LTO-NMC chemistries represent viable choices for commercial EVs in the foreseeable future, through about 2020. Figure 8-15 depicts the relative status of various aspects of different battery chemistries and their potential application in commercial vehicles, while Table 8-7 shows battery configurations currently offered by suppliers.

Graphite-LFP demonstrates a good compromise between these important parameters with high safety at a very competitive cost. LTO-NMC demonstrates the highest durability, excellent safety, and the highest charging rate and power, all of which are very important for commercial vehicle application. Therefore, despite a lower specific energy and higher cost, the durability, performance, and safety make LTO-NMC a viable option for commercial vehicles.

TABLE 8-6 USABC Goals for Advanced Batteries for 2020 EVs

End-of-Life Characteristics at 30°C	Units	System Level	Cell Level
Peak discharge power density, 30 sec pulse	W/L	1000	1500
Peak specific discharge power, 30 sec pulse	W/kg	470	700
Peak specific regen power, 10 sec pulse	W/kg	200	300
Usable energy density @ C/3 discharge rate	Wh/L	500	750
Usable specific energy @ C/3 discharge rate	Wh/kg	235	350
Usable energy @ C/3 discharge rate	kWh	45	N/A
Calendar life	Years	15	15
DST cycle life	Cycles	1000	1000
Selling price @ 100K units	$/kWh	125	100
Operating environment	°C	-30 to +52	-30 to +52
Normal recharge time	h	<7 h, J1772	<7 h, J1772
High rate charge	min	80% ΔSOC in 15 min	80% ΔSOC in 15 min
Maximum operating voltage	V	420	N/A
Minimum operating voltage	V	220	N/A
Peak current, 30 s	A	400	400
Unassisted operating at low temperature	%	>70% usable energy @ C/3 discharge rate at -20°C	>70% usable energy @ C/3 discharge rate at -20°C
Survival temperature range, 24 h	°C	-40 to +66	-40 to +66
Maximum self-discharge	%month	<1	<1

SOURCE: USABC (2017).

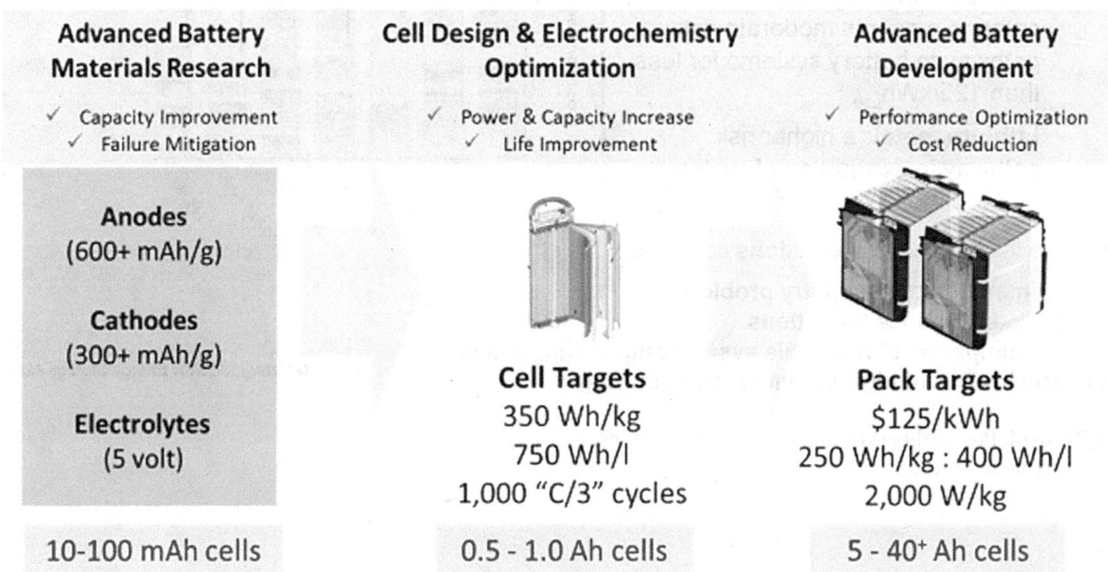

FIGURE 8-12 Cost reduction approach for energy storage systems.
SOURCE: Howell (2014).

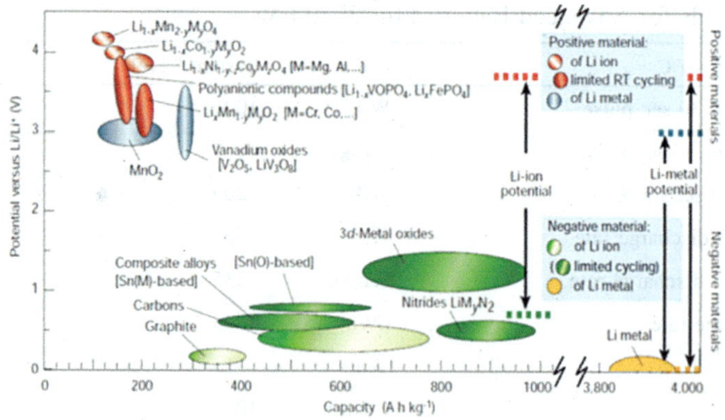

FIGURE 8-13 Future battery chemistries to achieve higher cell voltages and energy capacity.
SOURCE: Tarascon and Armand (2001), reprinted by permission from RightsLink: Springer Nature, Tarascon & Armand, 2001.

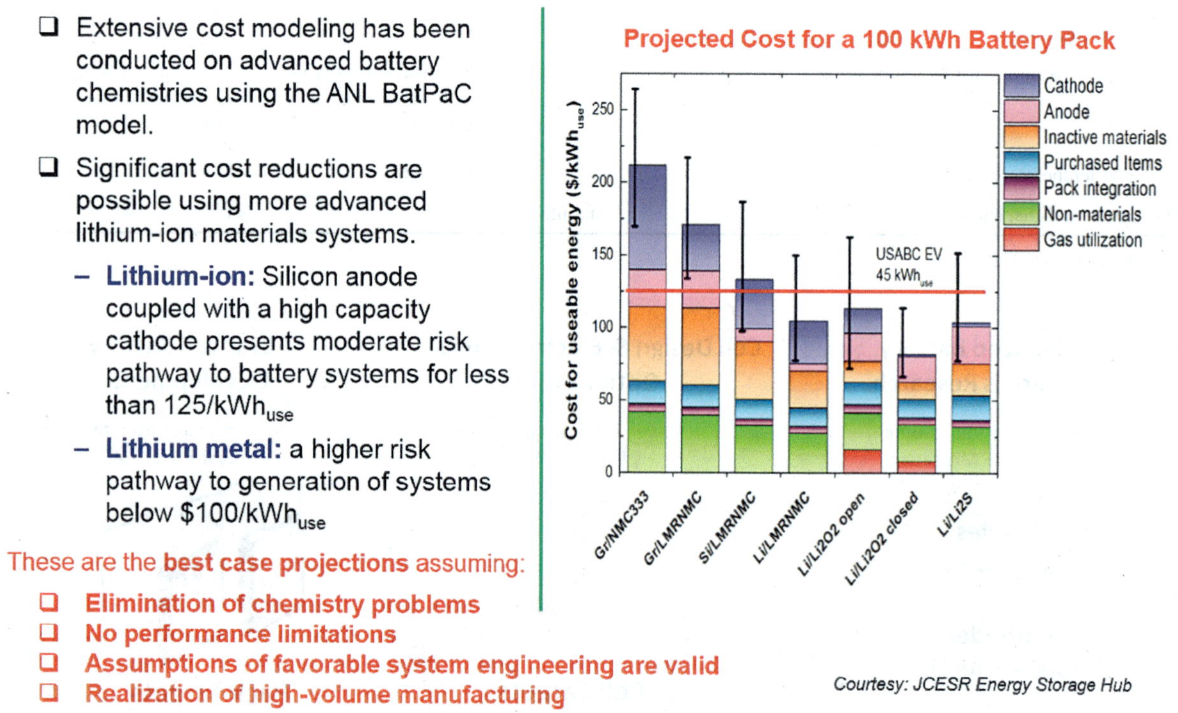

FIGURE 8-14 Potential cost of future battery chemistries.
SOURCE: Duong (2015).

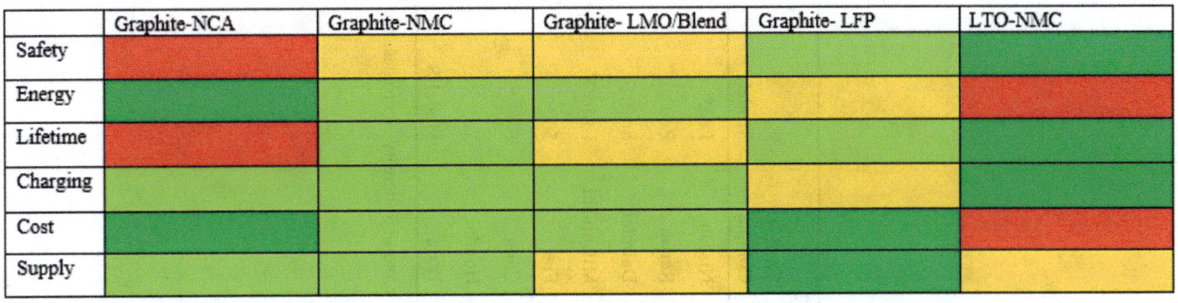

FIGURE 8-15 Li-ion battery characteristics related to commercial vehicle application.
NOTE: red = poor, yellow = satisfactory, light green = good, and dark green = excellent.
SOURCE: Adapted from Kukkonen (2014).

8.4 FINDINGS

Finding: Spurred by increasing regulatory pressure for combustion-free transportation in many highly populated cities around the world, very significant research and development is being done to reduce the cost, increase the specific energy, and demonstrate sufficient safety of batteries over their useful life in PHEV and BEV passenger vehicles.

Finding: Several international passenger vehicle manufacturers and Tier 1 battery suppliers are projecting costs for PHEV/BEV battery packs that will achieve \$120/kWh by 2020 and \$100/kWh or less by 2025. At these cost levels, significant market growth with increasing levels of light-duty vehicle electrification can be expected worldwide by 2030.

Finding: As a result of the rugged use, high power, long life, and reliability requirements for commercial vehicles, the direct carryover of battery systems from light-duty vehicles will be limited. Additional research and development, beyond that which is being done for light-duty applications, will be necessary and could delay the market penetration of electrified commercial vehicles.

Finding: Due to the duty cycle, service life, and other boundary conditions for commercial vehicles, the battery-system costs are expected to decline significantly by 2030, but will remain somewhat higher than light-duty battery systems.

Finding: International regulatory incentives to promote combustion-free powertrain vehicle use in highly populated cities could support earlier introduction of electrified commercial vehicles than would normally occur in the technology and manufacturing cost development cycle.

8.5 RECOMMENDATIONS

Recommendation 8-1: EPA and the National Highway Traffic Safety Administration (NHTSA) should recognize the challenging performance, reliability, and safety requirements for commercial vehicles and coordinate with DOE, commercial battery suppliers, and international organizations to promote a timeline for development of battery chemistries and pack technologies which will allow the safe and efficient introduction of hybrid and all-electric drive of commercial vehicles, especially in highly-populated urban centers.

Recommendation 8-2: In recognition of the challenges associated with electrification of commercial vehicles, EPA and NHTSA should coordinate with U.S. and other international governmental bodies to assess the performance and safety progress of battery systems for use in commercial vehicles in an attempt to prevent forced early adoption which could result in poor reliability and safety performance and delayed market penetration.

TABLE 8-7 Battery Configurations Offered by Various Manufacturers

Cell Maker	Chemistry Anode/Cathode	Capacity (Ah)	Configuration	Voltage (V)	Weight (kg)	Volume (l)	Energy Density (Wh/l)	Specific Energy (Wh/kg)	Used in Company	Model
AESC	G/LMO-NCA	33	Pouch	3.75	0.80	0.40	309	155	Nissan	Leaf
LG Chem	G/NMC-LMO	36	Pouch	3.75	0.86	0.49	275	157	Renault	Zoe
Li-Tec	G/NMC	52	Pouch	3.65	1.25	0.60	316	152	Daimler	Smart
Li Energy Japan	G/LMO-NMC	50	Prismatic	3.7	1.70	0.85	218	109	Mitsubishi	i-MiEV
Samsung	G/NMC-LMO	64	Prismatic	3.7	1.80	0.97	243	132	Fiat	500
Lishen Tianjin	G-LFP	16	Prismatic	3.25	0.45	0.23	226	116	Coda	EV
Toshiba	LTO-NMC	20	Prismatic	2.3	0.52	0.23	200	89	Honda	Fit
Panasonic	G/NCA	3.1	Cylindrical	3.6	0.048	0.018	630	233	Tesla	Model S

Tesla-Panasonic's current cell offers specific energy 50 percent higher than competition. This is primarily due to the use of highly reactive NCA cathodes and high-density electrodes. The gap will shrink to 20 to 25 percent in the next 3 years.
SOURCE: Anderman (2014).

TABLE 8-8 Component Costs of Hybrid Systems

					Light Heavy Duty (LHD)					Medium Heavy Duty (MHD)		Heavy Heavy Duty (HHD)	
	Battery Cells	Battery Pack	Power Electronics	Electric Motor	Class 2b-3 Specs	Class 2b-3 Cost	Class 4-5 Specs	Class 4 Cost	Class 5 Cost	Class 6-7 Specs	Class 6-7 Cost	Class 8 Specs	Class 8 Cost
Micro-Hybrid with Stop-Start Low Cost													
2015 $/kWh	$420.00	$80.00			1.5	$750.00	2.0	$1,000.00	$1,000.00	2.5	$1,250.00	3.5	$1,750.00
2022 $/kWh	$130.00	$36.00			1.5	$249.00	2.0	$332.00	$332.00	2.5	$415.00	3.5	$581.00
2030 $/kWh	$110.00	$25.00			1.5	$202.50	2.0	$270.00	$270.00	2.5	$337.50	3.5	$472.50
2015 $/kW			$12.00	$18.00	10	$300.00	15	$450.00	$450.00	20	$600.00	35	$1,050.00
2022 $/kW			$3.60	$5.20	10	$88.00	15	$132.00	$132.00	20	$176.00	35	$308.00
2030 $/kW			$3.30	$4.70	10	$80.00	15	$120.00	$120.00	20	$160.00	35	$280.00
Rest of System						$140.00		$260.00	$280.00		$300.00		$500.00
2015 Total Cost						$1,190.00		$1,710.00	$1,710.00		$2,150.00		$3,300.00
2022 Total Cost						$477.00		$724.00	$744.00		$891.00		$1,389.00
2030 Total Cost						$422.50		$650.00	$670.00		$797.50		$1,252.50
2015 $/kWh	$420.00	$80.00			1.5	$750.00	2.0	$1,000.00	$1,000.00	2.5	$1,250.00	3.5	$1,750.00
2022 $/kWh	$140.00	$40.00			1.5	$270.00	2.0	$360.00	$360.00	2.5	$450.00	3.5	$630.00
2030 $/kWh	$125.00	$25.00			1.5	$225.00	2.0	$300.00	$300.00	2.5	$375.00	3.5	$525.00
2015 $/kW			$12.00	$18.00	10	$300.00	15	$450.00	$450.00	20	$600.00	35	$1,050.00
2022 $/kW			$4.50	$6.50	10	$110.00	15	$165.00	$165.00	20	$220.00	35	$385.00
2030 $/kW			$4.00	$5.60	10	$96.00	15	$144.00	$144.00	20	$192.00	35	$336.00
Rest of System						$160.00		$275.00	$325.00		$330.00		$550.00
2015 Total Cost						$1,210.00		$1,725.00	$1,775.00		$2,180.00		$3,350.00
2022 Total Cost						$540.00		$800.00	$850.00		$1,000.00		$1,565.00
2030 Total Cost						$481.00		$719.00	$739.00		$897.00		$1,411.00

SOURCE: Committee generated estimates from the synthesis of its information gathering.

TABLE 8-8 Continued

	Battery Cells	Battery Pack	Power Electronics	Electric Motor	Light Heavy Duty (LHD) Class 2b-3 Specs	Class 2b-3 Cost	Class 4-5 Specs	Class 4 Cost	Class 5 Cost	Medium Heavy Duty (MHD) Class 6-7 Specs	Class 6-7 Cost	Heavy Heavy Duty (HHD) Class 8 Specs	Class 8 Cost
Mild Hybrid with Regen/Launch and Stop-Start Low Cost													
2015 $/kWh	$420.00	$80.00			5	$2,500.00	8	$4,000.00	$4,000.00	15	$7,500.00	20	$10,000.00
2022 $/kWh	$130.00	$36.00			5	$830.00	8	$1,328.00	$1,328.00	15	$2,490.00	20	$3,320.00
2030 $/kWh	$110.00	$25.00			5	$675.00	8	$1,080.00	$1,080.00	15	$2,025.00	20	$2,700.00
2015 $/kW			$12.00	$18.00	60	$1,800.00	120	$3,600.00	$3,600.00	200	$6,000.00	200	$6,000.00
2022 $/kW			$3.60	$5.20	60	$528.00	120	$1,056.00	$1,056.00	200	$1,760.00	200	$1,760.00
2030 $/kW			$3.30	$4.70	60	$480.00	120	$960.00	$960.00	200	$1,600.00	200	$1,600.00
Rest of System						$850.00		$2,200.00	$2,200.00		$3,600.00		$5,000.00
2015 Total Cost						$5,150.00		$9,800.00	$9,800.00		$17,100.00		$21,000.00
2022 Total Cost						$2,208.00		$4,584.00	$4,584.00		$7,850.00		$10,080.00
2030 Total Cost						$2,005.00		$4,240.00	$4,260.00		$7,225.00		$9,300.00
2015 $/kWh	$420.00	$80.00			5	$2,500.00	8	$4,000.00	$4,000.00	15	$7,500.00	20	$10,000.00
2022 $/kWh	$140.00	$40.00			5	$900.00	8	$1,440.00	$1,440.00	15	$2,700.00	20	$3,600.00
2030 $/kWh	$125.00	$25.00			5	$750.00	8	$1,200.00	$1,200.00	15	$2,250.00	20	$3,000.00
2015 $/kW			$12.00	$18.00	60	$1,800.00	120	$3,600.00	$3,600.00	200	$6,000.00	200	$6,000.00
2022 $/kW			$4.50	$6.50	60	$660.00	120	$1,320.00	$1,320.00	200	$2,200.00	200	$2,200.00
2030 $/kW			$4.00	$5.60	60	$576.00	120	$1,152.00	$1,152.00	200	$1,920.00	200	$1,920.00
Rest of System						$875.00		$2,250.00	$2,300.00		$3,725.00		$5,900.00
2015 Total Cost						$5,175.00		$9,850.00	$9,850.00		$17,225.00		$21,900.00
2022 Total Cost						$2,435.00		$5,010.00	$5,060.00		$8,625.00		$11,700.00
2030 Total Cost						$2,201.00		$4,602.00	$4,652.00		$7,895.00		$10,820.00

continued

TABLE 8-8 Continued

	Battery Cells	Battery Pack	Power Electronics	Electric Motor	Light Heavy Duty (LHD) Class 2b-3 Specs	Class 2b-3 Cost	Class 4-5 Specs	Class 4 Cost	Class 5 Cost	Medium Heavy Duty (MHD) Class 6-7 Specs	Class 6-7 Cost	Heavy Heavy Duty (HHD) Class 8 Specs	Class 8 Cost
HEV with Full Performance Low Cost													
2015 $/kWh	$420.00	$80.00			10	$5,000.00	15	$7,500.00	$7,500.00	20	$10,000.00	40	$20,000.00
2022 $/kWh	$130.00	$36.00			10	$1,660.00	15	$2,490.00	$2,490.00	20	$3,320.00	40	$6,640.00
2030 $/kWh	$110.00	$25.00			10	$1,350.00	15	$2,025.00	$2,025.00	20	$2,700.00	40	$5,400.00
2015 $/kW			$12.00	$18.00	180	$5,400.00	220	$6,600.00	$6,600.00	250	$7,500.00	300	$9,000.00
2022 $/kW			$3.60	$5.20	180	$1,584.00	220	$1,936.00	$1,936.00	250	$2,200.00	300	$2,640.00
2030 $/kW			$3.30	$4.70	180	$1,440.00	220	$1,760.00	$1,760.00	250	$2,000.00	300	$2,400.00
Rest of System						$1,850.00		$3,000.00	$3,000.00		$4,500.00		$5,500.00
2015 Total Cost						$12,250.00		$17,100.00	$17,100.00		$22,000.00		$34,500.00
2022 Total Cost						$5,094.00		$7,426.00	$7,426.00		$10,020.00		$14,780.00
2030 Total Cost						$4,640.00		$6,785.00	$6,785.00		$9,200.00		$13,300.00
2015 $/kWh	$420.00	$80.00			10	$5,000.00	15	$7,500.00	$7,500.00	20	$10,000.00	40	$20,000.00
2022 $/kWh	$140.00	$40.00			10	$1,800.00	15	$2,700.00	$2,700.00	20	$3,600.00	40	$7,200.00
2030 $/kWh	$125.00	$25.00			10	$1,500.00	15	$2,250.00	$2,250.00	20	$3,000.00	40	$6,000.00
2015 $/kW			$12.00	$18.00	180	$5,400.00	220	$6,600.00	$6,600.00	250	$7,500.00	300	$9,000.00
2022 $/kW			$4.50	$6.50	180	$1,980.00	220	$2,420.00	$2,420.00	250	$2,750.00	300	$3,300.00
2030 $/kW			$4.00	$5.60	180	$1,728.00	220	$2,112.00	$2,112.00	250	$2,400.00	300	$2,880.00
Rest of System						$1,950.00		$3,200.00	$3,400.00		$4,750.00		$6,000.00
2015 Total Cost						$12,350.00		$17,300.00	$17,500.00		$22,250.00		$35,000.00
2022 Total Cost						$5,730.00		$8,320.00	$8,520.00		$11,100.00		$16,500.00
2030 Total Cost						$5,178.00		$7,562.00	$7,762.00		$10,150.00		$14,880.00

8.6 REFERENCES

Anderman, M. 2014. The Tesla Battery Report. Oregon House, CA: Total Battery Consulting, Inc. https://totalbatteryconsulting.com/industry-reports/Tesla-report/overview.html (accessed December 20, 2019).

Battery University. 2016. BU-205: Types of Lithium-Ion. Available at http://batteryuniversity.com/learn/article/types_of_lithium_ion (accessed July 25, 2019).

Duong, T.Q. 2015. "Overview and Progress of the Advanced Battery Materials Research (BMR) Program," presentation for DOE Merit Review, June 8. Washington DC.

Gordon-Bloomfield, N. 2013. "Nissan Testing New Battery Pack to Address Problems with LEAF in Hot Weather." Plug-in Cars, August 26. http://www.plugincars.com/nissan-testing-new-battery-pack-leaf-128088.html (accessed December 20, 2019).

Howell, D., and R. Elder. 2012. "Electrochemical Energy Storage Technical Team (EESTT)," presentation to the National Research Council's Committee on Review of the U.S. DRIVE Partnership, January 26. Washington, DC.

Howell, D. 2014. "Overview of the DOE Advanced Battery R&D Program" presentation given by U.S. Department of Energy at 2014 DOE Hydrogen and Fuel Cells Program and Vehicle Technologies Office Annual Merit Review and Peer Evaluation Meeting. June 16. https://www.energy.gov/sites/prod/files/2014/09/f18/battery_rd_amr_plenary_june_2014_final.pdf (accessed December 20, 2019).

Howell D., B. Cunningham, T. Duong, and P. Faguy. 2016. "Overview of the DOE VTO Advanced Battery R&D Program," presentation for DOE Merit Review, June 6. Washington, DC.

Kukkonen, S. 2014. "Current Trends in Battery Technology," presented at Electric Commercial Vehicles National Seminar, September 24.

NRC (National Research Council). 2015a. *Cost, Effectiveness, and Deployment of Fuel Economy Technologies for Light-Duty Vehicles*. Washington, DC: The National Academies Press.

NRC. 2015b. *Overcoming Barriers to Deployment of Plug-in Electric Vehicles*. Washington, DC: The National Academies Press.

Nykvist, B., and M. Nilsson. 2015. Rapidly falling costs of battery packs for electric vehicles. *Nature Climate Change* 5:329-332.

Pesaran, A., S. Santhanagopalan, and G. Kim. 2013. "Addressing the Impact of Temperature Extremes on Large Format Li-Ion Batteries for Vehicle Applications." Presented at the 30th International Battery Seminar, March 11-14, Fort Lauderdale, FL. Available at https://permanent.access.gpo.gov/gpo41672/58145.pdf (accessed September 23, 2019).

Spotnitz, R. 2005. "Advanced EV and HEV Batteries." Presented at the 2005 IEEE Vehicle Power and Propulsion Conference, September, Chicago, IL. Available at http://ieeexplore.ieee.org/stamp/stamp.jsp?tp=&arnumber=1554579 (accessed August 25, 2019).

Steffke, K., S. Inguva, D. Van Cleve, and J. Knockeart. 2008. "Accelerated Life Test Methodology for Li-Ion Batteries in Automotive Applications." SAE Technical Paper 2013-01-1548, doi:10.4271/2013-01-1548.

Tarascon, J.M., and M. Armand. 2001. Issues and challenges facing rechargeable lithium batteries. *Nature* 414:359-367.

Targray Technology International, Inc. 2017. Targray Cathode Materials for Battery Technologies. Available at http://www.targray.com/li-ion-battery/cathode-materials (accessed July 30, 2019).

USABC (U.S. Advanced Battery Consortium). 2017. USABC Goals for Advanced Batteries for EVs—CY 2020 Commercialization. Available through USCAR News center: https://www.uscar.org/guest/article_view.php?articles_id=85 (accessed September 23, 2019).

Wessler, G. (Duke SciPol). 2018. "Science Module: Energy Storage" available at http://scipol.duke.edu/content/science-module-energy-storage (accessed March 30, 2019).

9

Freight Operational Efficiency

This chapter discusses methods and technologies for improving the efficiency of freight operations that, loosely speaking, do not directly involve improving the efficiency of the vehicle powertrain. These methods entail improvements to freight efficiency through changes to trucking operations, such as those that involve the driver or through changing limits on truck size and weight. Also included is discussion of the impact on freight efficiency through use of other transportation modes, vessels, and conveyances.

9.1 METHODS TO IMPROVE MOVEMENT OF FREIGHT

Drivers face many challenges with respect to efficient movement of freight. A brief summary of these mechanisms includes the following:

- Driver's Hours of Service per Federal Motor Carrier Safety Administration (FMCSA) requirements: At the end of 2015, FMCSA issued its proposed rule requiring the use of electronic logging devices. This will increase the already substantial penetration of on-board computers and will likely lead to operational improvements.
- Safety Inspections: State agencies and the FMCSA perform safety inspections on a random basis. Depending on the type of inspection, this can take from several minutes to over an hour to complete. If the vehicle can wirelessly provide information on its safety status, the number of physical inspections can be reduced while still accomplishing the needed safety audits. The PrePass system, a wireless intelligent transportation system service, has been enabling safety and weigh station inspection bypassing for well over a decade. FMCSA had a program for Wireless Roadside Inspection; however, it was ended, incomplete, in 2017.
- Weigh Stations: Various states monitor the weight of commercial vehicles on roads at weigh stations, frequently placed on major highways where traffic enters a state. PrePass and Drivewyze provide electronic verification, allowing bypassing of such stations via transponders or cell phones.
- Toll Booths: Eliminating the need to slow down, stop, or wait to pay tolls for bridges and tolled highways can improve average load-specific fuel consumption. E-ZPass is a network of automated collection of electronic tolls for the northeastern United States. Many such networks have been developed, and PrePass is now fully integrated with all E-ZPass roads.

- Border Crossings: Several sources provide real-time estimates and average wait times on border crossing delays. While the average delays to enter and leave Canada are typically less than 20 minutes, several of the United States–Mexico border crossings have average delays of over an hour (Calit2, 2016).
- Congestion: Congestion slows traffic down, often to stop-and-go, for miles. The amount of fuel used to accelerate and stop commercial vehicles can be minimized by reducing the amount of congestion that a commercial vehicle encounters by minimizing overall congestion, rerouting, and driving at alternative times.
- Accidents: The National Highway Traffic Safety Administration (NHTSA) tracks accident statistics, primarily as it relates to injuries and deaths. More recently, the impact of accidents on traffic congestion and fuel consumption has become important. Traveler information systems are providing information on accidents to eliminate additional accidents, relieve stress in drivers, and suggest alternate routes.
- Loading and Unloading: Additional driving hours are possible when the time to load and unload a vehicle is minimized. While there are limits on the number of hours for driving, there are also limits on the amount of time a driver is on duty. In some cases, driving time is impacted by waiting. Fleets work to minimize this lost time. Properly loading a trailer to maximize the load, in terms of both overall weight and volume, is a difficult challenge. Furthermore, it is constantly changing as new products and packaging are created. A primary requirement is to maintain legal weights on the axles that are properly spaced for whatever road requirements exist in the states being traveled. A second need is to maximize the amount of cargo in the trailer. Dry van trailers are used to transport items as diverse as square boxes on standardized pallets to axles for trucks at assembly plants, to full vehicles, such as race cars. A third requirement relates to how the product is configured on the trailer for loading and unloading. Some manufacturing operations expect the load on the trailer to be arranged in the sequence that the material will be used.

9.2 METHODS TO REDUCE DEADHEADING

Deadheading is the term used for a commercial vehicle moving without a load. After a load is delivered, the vehicle often has to move to another location to obtain a load to move. This is a significant number of miles in many cases. According to the American Trucking Associations (ATA), professional truck drivers cover more than 275 billion miles in a year (ATA, 2015a). Of those miles, 18 percent are non-revenue-generating miles according to surveys conducted annually by the American Transportation Research Institute (ATRI, 2014, 2015). While not new, improvements in load matching services may lead to a reduction in some of these miles.

Finding: The government and fleets are looking at numerous ways to keep commercial vehicles moving at highway speeds, while ensuring safety.

Recommendation 9-1: NHTSA should work more closely with the other parts of DOT to focus efforts on minimizing interruption of freight movements and encourage optimized loading.

9.3 METHODS TO IMPROVE DRIVER-VEHICLE INTERACTION

As discussed earlier in the section on transmissions, the driver plays a role in determining the fuel consumption of the vehicle. Fuel can be lost through unwanted manipulations of the throttle, unnecessary movements of the steering wheel, poor management of speed that causes excessive and hard braking, fast acceleration, and poor selection of gear for the powertrain and vehicle conditions, to name a few. Numerous suppliers of telematics and other driver interaction devices for commercial vehicles have done their own work to substantiate savings through eco-driver feedback. An example of one such system is SMARTDRIVE. Company information claims that the technology will significantly reduce collisions and will reduce fuel use by 10 percent. Recent studies by University of California, Riverside researchers found that providing real-time information to drivers can reduce fuel consumption by 3.5 to 7.5 percent (Jin et al., 2016; Scora et al., 2015). Driving more slowly is another method of improving fuel economy, and the anticipated FMCSA/NHTSA speed limiter regulation is expected to have a positive impact on this factor, removing it from the driver's control to some extent.

As the FMCSA rule for electronic logging devices takes effect, the feedback to drivers on fuel consumption will improve. Furthermore, fleets will begin to have more real-time information for judging the effects of fuel. Some suppliers are including real-time information on loads, grades, road conditions, and weather to provide more reliable, achievable goals on fuel consumption improvements the driver can influence. Vehicle original equipment manufacturers (OEMs) are offering, via instrument clusters and in on-board computer systems, eco-driver feedback information to help them control fuel consumption.

The vehicle OEMs are working to minimize the effect of the driver on fuel consumption. They are implementing methods to control fuel during acceleration, increase the use of cruise control, and creating fuel-saving predictive cruise control products that anticipate fuel usage in order to optimize it. Semiautonomous and autonomous vehicles further decrease the driver influence on fuel consumption as well as impacting safety. The biggest gain in this area of driver-vehicle interactions may stem from idle reduction. In addition to state laws restricting idling, strong information campaigns linked with technology advances such as auxiliary power units, fuel-operated heaters, and electrification of accessories and appliances are significantly reducing idling.

Finding: Immediate feedback to the driver has been shown to be effective in getting fuel-conscious drivers to improve fuel consumption. Fleets provide economic incentives to drivers to get them to respond to the prompts.

Recommendation 9-2: NHTSA should work with FMCSA to complete additional research on driver influence on fuel consumption, taking into account the newer regulations for electronic logging devices, speed limiters, anti-idling technologies, and the introduction of technologies to reduce the driver influence on fuel consumption.

9.4 COUNTERVAILING FORCES TO FUEL CONSUMPTION REDUCTION

While the commercial vehicle freight market has a strong focus on improving costs and reducing fuel consumption, there is also a focus on customer satisfaction. This occurs especially in the areas of on-time delivery, just-in-time delivery, and line sequencing for assembly operations, to name a few. These countervailing forces tend to increase the number of trucks needed and to increase the amount of fuel used. For instance, just-in-time delivery often causes fleets to use trailers as warehouses to hold inventory close to an assembly operation. It also causes fleets to make sure they arrive early and that a driver is waiting to make the delivery. This can take away from valuable hours of service for the drivers. Furthermore, in some industries, the product on a trailer has to be loaded in an exact order to fit with the planned production at the consignee, such as an automotive assembly plant. This can reduce the amount of cargo the fleet is able to put on the trailer, to optimize the operation of the plant versus maximizing the load to be carried.

Several organizations are now pushing for faster delivery, sometimes as short as 1 to 2 hours in urban areas. These are small deliveries of restaurant food or consumer items, so the vehicles impacted will be medium-sized ones. However, this new delivery model increases the number of vehicles needed and those vehicles are far from optimally loaded. Drone deliveries have been proposed for these types of small deliveries in dense, urban areas. However, there are many issues to be addressed by the Federal Aviation Administration. Amazon, UPS, Workhorse, and many other companies are conducting tests of combination van and drone services, although it is not clear when or if these will be allowed.

Finding: Freight of all sorts is delivered to consignees for purposes that drive the timing and business model of shippers and brokers.

Recommendation 9-3: NHTSA should encourage the relevant government agencies, such as the Department of Transportation, the Environmental Protection Agency, and the Department of Energy, to monitor and assess new models for freight delivery that could impact fuel consumption and safety.

9.5 INTERMODAL SYSTEMS

"Intermodal" refers to the movement of freight among different modes of transportation, generally truck, rail, ocean, and inland waterways. While international intermodal is dominated by maritime-truck movements, domestic intermodal is almost exclusively truck-rail. It most often refers to movement of goods by rail and truck. Compared to historic truck-only supply chains and in particular tractor-trailer for longer routes or trucks for shorter ones, intermodal requires a major mode shift. This shift causes a movement of the freight itself or the container containing the freight to be moved onto a separate cargo-carrying vehicle. The advantage is that different, more efficient modes can be used for portions of the freight's journey: ships for long sea travel, rail for longer land trips, and trucks on each end to most effectively reach the start and end of the route, a dock at a manufacturing site or a retail store on the other end. Many technologies, such as Global Positioning Systems (GPS) and freight tracking, help enable more efficient intermodal transportation of goods.

International intermodal refers to goods movement lanes that include ocean transportation and a port that transfers freight from a container ship to land transport, either directly onto rail or via truck drayage (very short-haul truck movements to the railyard) or to a longer truck move. Domestic intermodal, which has had and will likely continue to have higher growth rates, is one where the entire supply chain is in North America and where the growth comes from the addition of mode shifting from more traditional long-haul sleeper tractor single-mode moves. Goods movement in the United States over the past 60 years has been dominated by long-haul sleeper tractor moves as long as full coast-to-coast journeys. With higher fuel costs over the past 10 years, making part of this journey via rail has been considered an efficiency and an opportunity to cut costs.

9.5.1 Truck-Rail Intermodal Transport

Typical truck-rail intermodal movements involve freight moved by truck at the beginning and end of its trip and by rail over a long distance in the middle. Movement by rail is efficient due to its high ton-miles per gallon. A single locomotive can move a great deal of freight via many carloads with less overall energy, even though diesel fuel is predominately consumed. According to the Association of American Railroads, rail fuel efficiency in 2014 was 479 miles per gallon per ton (AAR, 2015). Historically, rail has been the primary solution for goods movement in the United States for bulk commodities such as grain, coal, and crude oil. These bulk commodities are sometimes referred to as "non-economically sensitive" because of their low value and also because deliveries are typically not time sensitive. As more goods have been hauled longer distances throughout the past 50 years, the movement of general freight—goods other than bulk commodities—has been moved more economically by rail for distances exceeding 2,000 miles. This was expanded with the development of containers in sizes commensurate with the needs of ocean and truck capabilities.

The most common challenges for shipping by rail are related to the fact that a mode shift must occur, as necessitated by the fact that railways do not generally go to the origin and destination of the freight. Technologies for rail intermodal moves have evolved over time and include trailer-on-flat-car moves in which a Class 8 trailer is placed on top of a rail car, container-on-flat-car, and increasingly double stacked containers, usually in well cars that are deep and offer stability, efficiency, and also security benefits because the well prevents the bottom trailers from being opened en route.

During the 1990s, just-in-time delivery of goods became a common and exploited inventory practice in the United States. The time sensitivity of freight became more important, a state of affairs that has created a head wind for movement to more domestic intermodal transport. Just-in-time manufacturing systems, such as the Toyota Manufacturing System, and warehousing infrastructure limit the ability to lengthen supply chain times. This has been a force keeping intermodal shipments low, though the size of the market has continued to expand slowly as technologies for managing the mode shifts have been improved.

Given domestic intermodal transport replaces part of the freight's journey with rail in lieu of truck, the common decision-making aspect of this mode shift is the savings associated with moving part of the trip with more fuel efficient means—rail—with the increased cost of a mode shift, from truck on one end to another truck on the other and the risk of delay. Generally rail-truck delivery increases the time for the overall shipment. International and domestic intermodal lanes are being served by expanding inland terminals (sometimes called inland ports or dry ports) in Atlanta, Chicago, Dallas, Kansas City, and other U.S. cities.

Capacity is an issue too. For example, the increases in oil and gas shipments have caused many delays over the past few years (Morris et al., 2014; Nixon, 2014). Adding capacity can be very expensive, requiring new tracks or modifying tracks, for example, lowering tracks along lines so that doubles can make it under bridges and into tunnels for capacity). Other infrastructure challenges are due to community and state involvement approvals, and long lead times for orders on new locomotives and rail cars. Rail regulations are also an issue. Adding weight for safer tanker cars, potentially lower speed limits, and increasing crew numbers are all under consideration and would have a negative effect on capacity.

Normal economic cycles may stall investment and actions to rationally exploit the opportunities for rail freight in the long-haul sector, which has historically favored trucking. Many shippers see rail as unreliable for time-sensitive general freight. Inland terminals in high-congestion cities, such as Atlanta and Chicago, present challenges which delay truck traffic in and out of the terminals. But the industry is responding by investing in new locomotive technologies, railcars, and other rail infrastructure in an effort to raise capacity and efficiency. However, capitalizing on these investments will require significant time and money. The industry is also responding to a reduction in demand for coal movements by increasing services for moving natural gas and crude oil. Trucking companies such as JBHunt, Schneider, CR England, FedEx, and UPS continue to increase their presence in intermodal transport through ownership of containers and chassis and improvements in logistics management. Costs appear to be improving with the movement of some shorter hauls being moved to intermodal transport. The rail industry claims that less than 1,000 miles and even down to 500 miles will become efficient for a shift to rail. However, with the movement to lower miles, the mode-shift cost accounts for a much higher portion of the overall time and labor cost for intermodal freight.

The truck driver shortage will continue to limit long-haul trucking capacity increases in the future. (Interestingly, there is evidence that drivers of less-than-load have far lower turnover.) (O'Marah, 2016) Moving freight to rail would reduce the need for truck drivers which will be increasingly important because many older drivers are retiring and the newer drivers are less willing to spend days or weeks on the road. The ATA estimates that there was a shortage of 48,000 drivers in 2015 and that the number could be as high as 175,000 by 2024 (ATA, 2015b).

Efficiencies can be improved at intermodal terminals. GPS and autonomous movement of containers by cranes, trucks, and dollies can improve times. Tracking of freight, containers, chassis, and pallets and individual orders can help with the reliability of the intermodal moves. Eastern U.S. ports are increasing their capacity by dredging to accept bigger ships that will be allowed with the expansion of the Panama Canal. Increasing capacity of lifts and moves within all terminals continues.

Finding: A long-term collaborative effort to optimize intermodal movements could help reduce transportation costs and fuel consumption. This is a system issue, though, and one that requires dedication to a long-term strategy with many stakeholders, many of whom consider each other competitors more than collaborators.

9.5.2 Truck-Truck Intermodal Transport

A solution that has been considered for nearly half a century, but which is gaining popularity in the United States and abroad, is the development of inland ports (often called freight villages) which allow for the transfer of loads from long-distance vehicles to smaller regional or urban vehicles (see, for example, Russo et al., 2015; Viscelli, 2014; Wu and Haasis, 2013).

These moves are truck-truck domestic intermodal moves. Improvements to the efficiency of long-haul sleeper tractors and trailers via aerodynamics, rolling resistance, and powertrains optimizes them for highways. Without the requirement to enter congested cities, technologies such as aggressive aerodynamics, platooning, and others become more realistic. The congestion in urban areas causes trucks to sit at low speeds around cities, where they operate inefficiently. Truck parking for mandatory rest periods can be more difficult to find in urban areas as well. As a result, trucks often overnight outside the urban area and subsequently travel into these areas during congested periods. Allowing drivers to drop their trailers on the outskirts of large cities and head back out on the open road allows these tractors to operate more efficiently. Inside the city, the freight is moved by more highly maneuverable tractors, alternatively fueled, hybridized, etc., more specifically designed for a stop-start urban environment. Conflicts between urban freight facilities and other land uses have also been encountered. Therefore, the planning and development of these types of facilities needs to address both the opportunities and the challenges.

Finding: The development of freight transfer facilities near urban areas will increase the use of more agile, fuel efficient, and less polluting vehicles for "last-mile" freight movements.

9.5.3 Inland Waterway Transport

Inland water transportation is used mainly for bulk commodities such as petroleum and chemicals, coal, grain, and soybeans. While an important component of the overall freight transportation system, the share of freight moving on rivers will continue to be very small.

9.5.4 Drayage

Drayage is the movement of goods short distances to enable a mode shift in international or domestic intermodal transport. Historically, these moves were made with older trucks that have been taken out of long-haul service. However, the Environmental Protection Agency and local air quality regulations have recently required newer and much cleaner trucks to be adopted for drayage (for example, in the Ports of Long Beach and Los Angeles). A few active development, test, and deployment projects have emerged to support technologies to limit the emissions around these drayage routes. The 710 corridor set of projects in the Long Beach, California, area includes technologies such as all-electric trucks, catenary trucks (Impullitti, 2012), and other zero emission ideas.

The costs of implementing these technologies remain very high, and constraints on capacity in high-volume freight corridors limit the extent to which right of way can be dedicated to clean corridor projects. And the cost to produce a tractor specific for the routes could price these as much as an order of magnitude higher than the older Class 8 tractors currently in use.

Finding: The successful implementation of clean corridors will require incentives at the federal, state, and local levels for development of the necessary technologies.

9.6 SIZE AND WEIGHT CHANGES (VEHICLE-MILES-TRAVELED REDUCTION)

The discussion in this section was adapted, with permission of the author, from a recent Transportation Research Board paper (Woodrooffe, 2016) which features a much more detailed discussion and analysis.

Size and weight regulation defines the upper limit of both low- and high-density freight vehicle efficiency. In doing so, these regulations have direct influence on how trucks use infrastructure and contribute to the wear and tear of infrastructure, use fuel, produce emissions, and influence crash frequency. There is considerable evidence that the reduction in vehicle miles traveled that would result from more closely aligning U.S. regulations with those of our North American Free Trade Agreement (NAFTA) partners would more than make up for any additional danger to passenger car drivers and vulnerable road users that might result from longer and heavier vehicles. For the magnitude size and weight increase required to harmonize the NAFTA region, there is no evidence that such vehicle units pose additional threat to public safety (OECD, 2010).

Size and weight policy is not uniform across the United States. Both the federal and state governments have their own policies that apply to roads under their respective authority. For example, state roads are governed by individual state policies while Interstate and National Network facilities are governed by federal law. However, the federal law is not absolutely uniform in terms of vehicle weight. While the stated allowable gross vehicle weight (GVW) on the Interstate System is restricted to 80,000 pounds (36,300 kilograms [kg]), grandfather rights were given to states that had higher weights than federal law at the time of creation of the Interstate Highway Program in 1956. Michigan is a case in point where trucks can operate up to 164,000 pounds (74,400 kg) on the Interstate

System provided they have a sufficient number of axles to comply with Michigan size and weight policy (FHWA, 2016).

More than a dozen states have gross vehicle weight limits on state roads that exceed the limits on the Interstate and other national truck network routes, and many more states have commodity- or industry-specific exemptions that allow even higher truck weights on specific routes. This patchwork of truck size and weight limits presents challenges for interstate trucking operations as well as for public regulation of the trucking industry. Economies are regional and normally require operations across multiple state boundaries. For example, sugar beet growers operate trucks daily across the border between Minnesota and the Dakotas, but Minnesota has lower truck weight limits than the Dakotas, so carriers must break their loads into smaller units when they cross the state border into Minnesota. This drives up the cost of operations for truckers and contributes to suboptimal transport efficiency.

The Surface Transportation Assistance Act (STAA) of 1982 effectively froze size and weight limits on the National Network, limiting vehicle weight to 80,000 pounds (36,300 kg) and limiting so-called longer combination vehicles (LCVs) to twin 28 ft (8.5 m) trailers known as STAA doubles.[1] The freeze was put in place at a time when there was great uncertainty regarding size and weight policy reform and at the time legislators were understandably cautious. However, since then, other countries have increased size and weight limits in what may be seen as a move to balance infrastructure consumption with vehicle productivity and safety, thereby making substantial improvements in transportation efficiency.

9.6.1 International Comparison

Compared with other industrialized countries, national weight limits in the United States are significantly lower. Table 9-1 provides detail on axle weights and GVW permitted for the common tractor semitrailer (OECD, 2010; Woodrooffe et al., 2010). U.S. federal size and weight regulations do not recognize the tridem axle group required for the six-axle tractor semitrailer, shown in Figure 9-1. Therefore, productivity comparisons were made based on the U.S. 80,000 pound (36,360 kg) five-axle tractor semitrailer against other countries' six-axle tractor semitrailer. These vehicles represent the so-called workhorse tractor semitrailer of each country. Productivity was determined by subtracting an estimated tare (empty) weight of 36,400 pounds (16,500 kg) for six-axle tractor semitrailers from the maximum GVW to resolve the allowable cargo weight and then comparing this to the five-axle U.S. tractor semitrailer assuming a lower tare weight of 34,200 pounds (15,500 kg) to account for one less axle.

Uniformity of size and weight regulations in countries separated by oceans, such as the European Union and North America, is not as critical as those sharing a continent. Figure 9-2 shows the diversity of permitted vehicle weights for the signatories of NAFTA: Canada, Mexico, and the United States. The 80,000 pounds (36,300 kg)

TABLE 9-1 Comparative International Tractor Semitrailer Axle Weight Limits, GVW, and Productivity

Country	Steer lb (kg)	Drive lb (kg)	Tridem lb (kg)	GVW lb (kg)	Productivity Advantage
Mexico	14,300 (6,500)	43,000 (19,500)	49,600 (22,500)	106,900 (48,500)	53%
Denmark	17,600 (8,000)	35,300 (16,000)	52,900 (24,000)	105,800 (48,000)	51%
Canada	12,100 (5,500)	37,500 (17,000)	52,900 (24,000)	102,500 (46,500)	44%
Australia	14,300 (6,500)	37,500 (17,000)	49,600 (22,500)	101,400 (46,000)	41%
UK	14,300 (6,500)	35,300 (16,000)	47,400 (21,500)	97,000 (44,000)	32%
USA	12,000 (5,500)	34,000 (15,400)	34,000[a] (15,400)[a]	80,000 (36,300)	--

NOTE: Most of the axle weight values listed above are legal limits but some are approximate as higher axle weight may be permitted but the vehicle GVW is limited. This is the case for the UK and Mexico.

[a]The United States does not have weight standards specified beyond tandem axles, so the tridem weight limit is the same as for the tandem, 34,000 pounds.

SOURCE: OECD (2010), Woodrooffe et al. (2010).

[1] 97th Cong., P.L. 97-424 (1983).

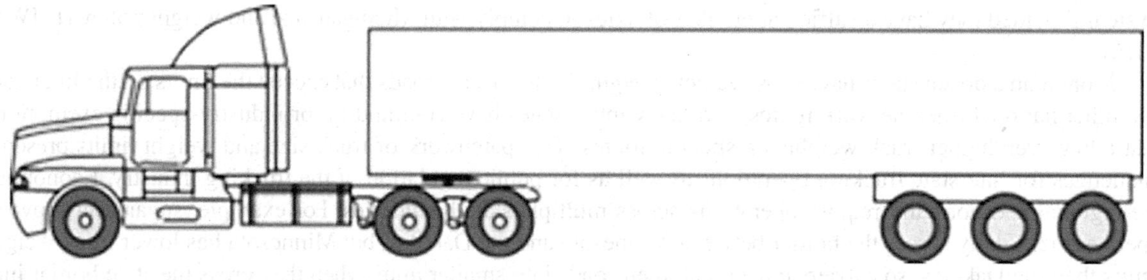

FIGURE 9-1 Six-axle tractor semitrailer workhorse vehicle.

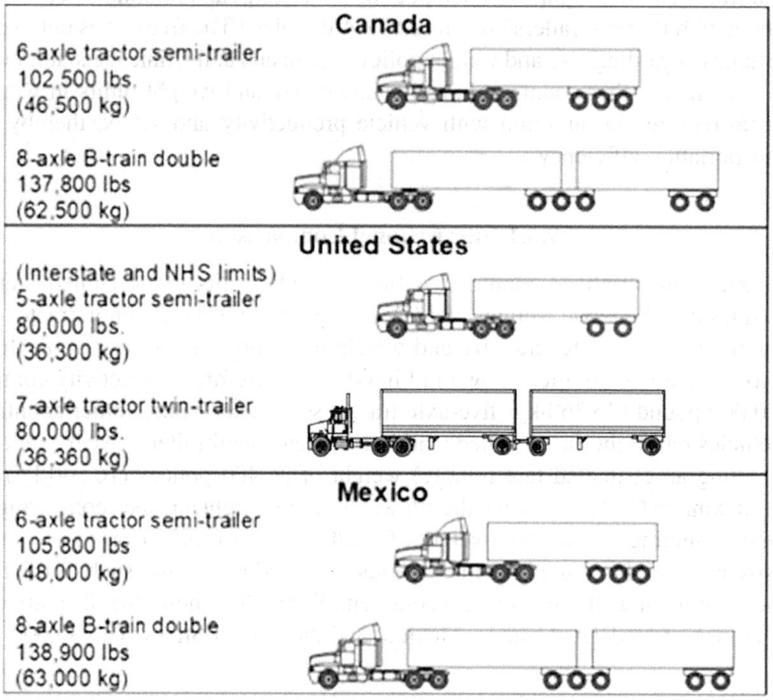

FIGURE 9-2 Comparing NAFTA region size and weight uniformity.

U.S. federal size and weight limit greatly restricts the ability to move mass-limited freight through the region efficiently. For tractor semitrailers, U.S. carriers are disadvantaged by 44 to 53 percent based on available cargo mass capacity, and for double trailer combinations, they are disadvantaged by a factor of 2. The 80,000 pound (36,360 kg) U.S. GVW limit represents a bottleneck for mass-limited freight moving within the NAFTA region, which has freight efficiency and societal cost implications for all NAFTA partners.

9.6.2 Longer Combination Vehicles

Transportation of low-density freight benefits from longer trucks with multiple trailers (LCVs). An example of an LCV combination is a tractor pulling two 53-foot (16.2 m) trailers. Because of their length, these vehicles are often restricted to certain road classes having geometric design characteristics that are compatible with LCVs. In some jurisdictions, LCVs operate under a special permit program governed by strict operating conditions. The structure and enforcement mechanisms engender a level of safety consciousness within the LCV fleet that far exceeds that found in other vehicle classes (Woodrooffe et al., 2004). Canada has developed unique policies with respect to the operation of LCVs and select provinces use a performance-based standard to approve candidate vehicles. The provinces of Alberta and Ontario have developed policy governing LCV movements so as to reduce high-risk travel of the LCV fleet (Alberta Government, 2015). This is done by restricting movement during peak congestion hours, public holidays, and inclement weather. The safety performance of the Alberta LCVs was found to be on the order of three to five times better than the standard tractor-semitrailer fleet operating on identical roads (crashes, fatalities, and injuries per distance traveled) (Montufar et al., 2007; Woodrooffe et al., 2010).

9.7 SOCIETAL VALUE OF FREIGHT EFFICIENCY IMPROVEMENT

To obtain a rough estimate of the societal opportunity costs from improvements in freight efficiency, a recent study (Woodrooffe, 2014) used aggregated fuel use by vehicle class and applied carbon value analysis to estimate the savings that would accrue from an improvement in freight efficiency that results in reduced total truck travel distance. The study noted that the contributing strategies that influence truck transport efficiency include improved loading of the existing truck fleet; increased use of other modes, particularly rail transport, though increased intermodal capacity; and, most importantly, reforming truck size and weight policy with the intended goal of reducing total truck travel distance through consolidation of the freight task. The degree to which truck transportation efficiency can be improved was found to depend on many factors, including the amount by which policy makers might wish to extend truck size and weight limits. This makes forecasting freight efficiency improvement difficult and somewhat speculative. Therefore, rather than projecting what the magnitude of freight efficiency improvement might be for any number of options and assumptions, the study looked at what a goal of 10 percent reduction in truck travel distance from transport efficiency improvements would generate in societal benefit. Estimates of benefits associated with safety, emissions, and fuel are provided in Table 9-2. The total estimated benefit for a 10 percent reduction in truck miles traveled is approximately $16 billion annually.

A further category of benefits that might accrue from reducing truck miles traveled is the reduction in fuel and CO_2 emissions. In the United States, single-unit trucks consume approximately 54.1 billion liters of diesel fuel and produce approximately 144 million metric tons of CO_2 annually. Articulated vehicles, largely five-axle tractor-semitrailers, use approximately 106 billion liters of fuel per year and generate approximately 283 million metric tons of CO_2 per year (FHWA, 2015a). Both the single-unit truck and the articulated vehicles will benefit from improved loading and reduced logistics costs. Articulated vehicles will also benefit from size and weight regulation reform while single-unit trucks are less likely to; however, there would be some diversion for single-unit truck freight to more productive tractor-semitrailers.

Finally, improvements in freight efficiency through various means including size and weight reform would likely have an added net safety benefit because the reduced amount of truck travel through improved truck cargo capacity would result in proportionately fewer truck crashes and that reduction would likely dominate negative safety consequences per distance traveled associated with increased vehicle size and weight. The recently completed U.S. size and weight study was unable to predict, with confidence, the likely safety outcome associated with particular larger heavier vehicle scenarios due to a lack of data (FHWA, 2015b). However, direct international experience with such vehicles has not found a substantive negative relationship between GVW and crash rate provided the vehicles comply with regulations (OECD, 2010; Woodrooffe, 2001). Insight into the potential truck efficiency improvement that could be gained from U.S. size and weight policy reform in line with NAFTA partners is that the improvement in freight efficiency would likely be in the range of 10 to 15 percent (Woodrooffe et al., 2009).

Finding: Higher weight limits and longer combination vehicles could significantly improve the productivity and therefore reduce the overall distance traveled in the heavy-vehicle long-haul transportation sector.

Recommendation 9-4: NHTSA and the U.S. Department of Transportation should consider urging Congress to adopt size and weight regulations that are in line with those of other North American trade partners.

TABLE 9-2 Estimated Reduction in the Number of Victims, Fuel Use, Greenhouse Gas Emissions, and Associated Opportunity Cost When Truck Exposure Is Reduced by 10 Percent

Benefits Study Variable	Injury Severity	Reductions Assuming 10% Reduction in Exposure	Estimated Annual Benefits ($US billion)
Estimated safety benefits attributed to a 10% reduction in truck travel distance	No apparent injury	21,562	0.20
	Possible injury	2,929	0.44
	Evident injury	2,724	0.68
	Disabling injury	1,453	0.87
	Killed	330	2.54
	Total safety cost saving attributed to 10% reduction in exposure		4.73
Estimated fuel and emissions benefits attributed to a 10% reduction in truck travel distance	Category	Quantity saved	Annual cost saving ($US billion)
	Diesel fuel reduction	10.6 billion liters	10.60
	CO_2 reduction	28.3 million metric tons CO_2	0.680
Combined benefits	Total estimated annual savings		16.01

SOURCE: Woodrooffe et al. (2009).

9.8 REFERENCES

AAR (Association of American Railroads). 2015. Rail Intermodal Keeps America Moving. https://www.aar.org/BackgroundPapers/Rail%20Intermodal.pdf (accessed June 5, 2016).

Alberta Government. 2015. Conditions for the Operation of Long Combination Vehicles. Version 5.05. http://www.transportation.alberta.ca/Content/docType276/Production/lcv.pdf (accessed December 16, 2019).

ATA (American Trucking Associations). 2015a. Trucking Moves America Forward: Professional Truck Drivers. http://www.trucking.org/ata%20docs/what%20we%20do/image%20and%20outreach%20programs/misc%20documents/pro%20Truck%20Drivers_final.pdf (accessed December 16, 2019).

ATA. 2015b. Truck Driver Shortage Analysis. http://www.trucking.org/ATA%20Docs/News%20and%20Information/Reports%20Trends%20and%20Statistics/10%206%2015%20ATAs%20Driver%20Shortage%20Report%202015.pdf (accessed December 16, 2019).

ATRI (American Transportation Research Institute). 2014. An Analysis of the Operational Costs of Trucking: A 2014 Update. http://www.atri-online.org/wp-content/uploads/2014/09/ATRI-Operational-Costs-of-Trucking-2014-FINAL.pdf (accessed December 16, 2019).

ATRI. 2015. An Analysis of the Operational Costs of Trucking: A 2015 Update. http://atri-online.org/wp-content/uploads/2015/09/ATRI-Operational-Costs-of-Trucking-2015-FINAL-09-2015.pdf (accessed December 16, 2019).

Calit2. 2016. Best Time to Cross the Border. http://traffic.calit2.net/border/border-crossing-wait-times.php# (accessed April 4, 2016).

FHWA (Federal Highway Administration). 2015a. Modal Shift Comparative Analysis Technical Report, Comprehensive Truck Size and Weight Limits Study. http://ops.fhwa.dot.gov/freight/sw/map21tswstudy/technical_rpts/mscanalysis.pdf (accessed December 16, 2019).

FHWA. 2015b. Highway Safety and Truck Crash Comparative Analysis Technical Report, Comprehensive Truck Size and Weight Limits Study. http://ops.fhwa.dot.gov/freight/sw/map21tswstudy/technical_rpts/hstcanalysis.pdf (accessed December 16, 2019).

FHWA. 2016. Compilation of Existing State Truck Size and Weight Limit Laws. http://ops.fhwa.dot.gov/FREIGHT/policy/rpt_congress/truck_sw_laws/app_a.htm#mi (accessed May 9, 2016).

Impullitti, J. 2012. Mobile Source Zero Emission Catenary Truck Project. South Coast Air Quality Management District Clean Fuels Program Advisory Group, August 29, Diamond Bar, CA. http://www.aqmd.gov/docs/default-source/technology-research/clean-fuels-program/clean-fuels-program-advisory-group---august-29-2012/mobile-source-zero-emission-catenary-truck-project.pdf?sfvrsn=6 (accessed April 6, 2016).

Jin, Q., K. Boriboonsomsin, and M. Barth. 2016. Energy and Emissions Benefits of a Real-Time Driving Speed Advisory System for Heavy-Duty Trucks, No. 16-6008. In *Transportation Research Board Annual Meeting Compendium of Papers*. Transportation Research Board, Washington, DC, January 10-12.

Montufar, J., J. Regehr, G. Rempel, and R.V. McGregor. 2007. Long Combination Vehicle (LCV) Safety Performance in Alberta: 1999-2005. http://www.transportation.alberta.ca/Content/docType61/production/LCVFinalReport2005.pdf (accessed December 16, 2019).

Morris, B., J. Bunge, and J.W. Miller. 2014. Surge in rail shipments of oil sidetracks other industries. *Wall Street Journal*, March 13. http://www.wsj.com/articles/SB10001424052702304914904579437680173044774 (accessed May 9, 2016).

Nixon, R. 2014. Grain piles up, waiting for a ride, as trains move North Dakota oil. *New York Times*, August 25. http://www.nytimes.com/2014/08/26/us/grain-piles-up-waiting-for-a-ride-as-trains-move-north-dakota-oil.html (accessed May 9, 2016).

OECD (Organisation for Economic Co-operation and Development). 2010. Moving Freight with Better Trucks. International Transport Forum, Leipzig, Germany, May 26-28.

O'Marah, K. 2016. Truck driver shortage is a shortage of imagination. *Forbes*, September 1. https://www.forbes.com/sites/kevinomarah/2016/09/01/truck-driver-shortage-is-a-shortage-of-imagination/#73a1fde41d7e (accessed November 8, 2017).

Russo, F., C. Rindone, P. D'Agostino, C. Lanciano, and T. Scattarreggia. 2015. A city logistics system for long distance freight transport provisioning. In *Urban Transport XXI*, edited by C.A. Brebbia and J.L. Miralles I Garcia, pp. 279-290. Southampton, UK: WIT Press.

Scora, G., K. Boriboonsomsin, and M. Barth. 2015. Value of eco-friendly route choice for heavy-duty trucks. *Research in Transportation Economics* 52:3-14.

Viscelli, S. 2014. Urban Truck Ports, White Paper. http://media.wix.com/ugd/10984e_c38fde52fd804389bd4e94a55abe34a0.pdf (accessed December 16, 2019).

Woodrooffe, J. 2001. Long Combination Vehicle Safety Performance in Alberta 1995 to 1998. http://www.transportation.alberta.ca/Content/docType61/production/LCVSafetyPerformanceReport.pdf (accessed December 16, 2019).

Woodrooffe, J. 2014. Reducing Truck Fuel Use and Emissions: Tires, Aerodynamics, Engine Efficiency, and Size and Weight Regulations. University of Michigan Transportation Research Institute Report 2014-27. Ann Arbor, MI.

Woodrooffe, J. 2016. Opportunity cost for society related to US truck size and weight regulation: Freight efficiency. *Transportation Research Record: Journal of the Transportation Research Board* 2547:25-31. doi: 10.3141/2547-04.

Woodrooffe, J., D. Anderson, and L. Ash. 2004. The Influence of Policy on Crash Rates of Long Combination Vehicles. Presented at the 8th International Symposium on Heavy Vehicle Weights & Dimensions, Muldersdrift, Gauteng, South Africa, March 14-18.

Woodrooffe J., B.M. Belzowski, J. Reece, and P.F. Sweatman. 2009. Analysis of the Potential Benefits of Larger Trucks for U.S. Businesses Operating Private Fleets. University of Michigan Transportation Research Institute Report 2009-11. Ann Arbor, MI.

Woodrooffe, J., K.P. Glaeser, M. Bereni, A. Germanchev, P. Eady, B. Jacob, and P. Nordengen. 2010. Safety, Productivity, Infrastructure Wear, Fuel Use and Emissions Assessment of the International Truck Fleet-A Comparative Analysis. OECD International Transport Forum, Leipzig, Germany, May 26-28.

Wu, J., and H.-D. Haasis. 2013. Converting knowledge into sustainability performance of freight villages. *Logistics Research* 6(2-3):63-88.

10

Intelligent Transportation Systems and Automation

10.1 INTRODUCTION

In this chapter we review the intelligent transportation systems (ITSs) and automation technologies that appear poised to have the greatest impact on the fuel efficiency and environmental impacts of medium- and heavy-duty vehicles (MHDVs), including

- Platooning, a promising technology that is being heavily tested and even implemented;
- Intelligent truck parking, another fairly straightforward and very important technology; and
- Connected vehicles (sometimes referred to as cooperative intelligent transportation systems (C-ITSs) or ad hoc vehicular networks), which are a precursor to automation or partial automation. Though it does not seem likely that fully autonomous driving will have a major impact on the fuel economy of MHDVs in the near term, it may well be that in certain situations it will be important. In the near term these will be fully grade-separated operations such as large ports and agricultural centers. In the longer term these systems will have an enormous impact on the efficiency of freight and fleet systems.

While the most important near-term impacts of adoption of connected vehicle technologies for MHDVs will be primarily safety related, fuel consumption and environmental impacts will also be impacted.

10.2 TRUCK PARKING TECHNOLOGY

One area in which ITS technologies are sure to have an immediate impact is in the area of truck parking. While the primary reason for these systems is a pressing need to improve rest opportunities for drivers, the combination of a lack of available parking spaces and a lack of information about those spaces results in drivers wasting fuel and generating pollution, either while searching for parking or because they then park in spaces without appropriate parking facilities, using engine idling for power. Therefore, these systems will have an immediate impact on heavy vehicle fuel economy.

Simple sensor and communication technologies are being developed in order to provide drivers with information on available parking spaces in their vicinity and will also enable reservation systems which will guarantee parking to drivers upon their arrival at a parking facility. Although such systems have been discussed for many years (for a review of earlier research see Garber et al., 2004), only in the past few years have they begun to see widespread testing across the United States and internationally. It seems likely that there will be large-scale adoption of these systems in developed countries within the next 5 years.

In several U.S. states and European countries variable message signs are already providing real-time information to truck drivers about available spots, and reservation systems are being tested (see, for example, Figure 10-1). Many studies have been performed regarding the sensor or imaging technologies that are appropriate for cataloging parking inventories. These technologies include video imaging (including stereovision and computer vision technologies), which is very promising (Chachich and Smith, 2011; Cook et al., 2014; Deruytter et al., 2012; Gertler and Murray, 2011; Modi et al., 2011); inexpensive, but apparently insufficiently accurate, magnetometers (Fallon and Howard, 2011); in-pavement sensors (Thinking Highways, 2014); human-in-the-loop closed-circuit television monitoring systems; and simple systems to monitor entrance and exit from truck parking facilities (Easyway, 2012; Woodrooffe et al., 2016). While most of the pilot studies have focused on roadside variable message systems for communicating with drivers, information is increasingly made available via standard web interfaces, phone apps, or instant messaging (see, for example, Thinking Highways, 2015).

A number of reservation systems have also been proposed and several have been tested. These range from simple web- or phone-based reservation systems (Mbiydzenyuy et al., 2012) to more sophisticated multi-agent-based negotiation platforms (Garcia et al., 2014).

10.3 TRUCK PLATOONING

Interest in truck platooning has been strong for more than 25 years now. Tests of prototype systems were conducted in the United States, Europe, and Japan in the late 1990s (Franke et al., 1995; Fritz, 1999), but the technologies required to make these systems operational were rather slow to mature. In the past 3 years, connected vehicle technologies have advanced to the extent that these systems are being widely tested and, in a few isolated cases, becoming operational. Bergenhem et al. (2012) details tests in the United States, several European countries, and Japan. Tests have also been conducted in Australia and several other Asian countries.

Though estimates vary, the consensus is that platooning reduces fuel consumption by about 7.5 percent. In 2016, NACFE conducted a study for potential fuel consumption savings versus a separation distance of two tractor-trailers in platooning mode. For a separation distance of 40 to 50 feet, the average fuel savings for the trailing truck in a platoon is about 10 percent and for the lead truck is about 4 percent (NACFE, 2016). While platoon testing in a controlled environment can produce 7.5 percent fuel consumption savings, it is not practical to expect this level

FIGURE 10-1 Changeable message road signs identify the vacancy level at respective rest areas and exits.
SOURCE: Woodrooffe et al. (2016).

in the real world (NACFE, 2016). Considering that not all miles can be in platooning mode and that the platoon will be disturbed and affected by traffic surrounding the two trucks, the real-world fuel consumption savings would be closer to 4 percent, which is still a significant amount for fuel economy savings (NACFE, 2016).

In order to realize these savings along significant parts of any heavy-duty vehicle trip, coordinated scheduling of trips within and among companies would be required. Larsson et al. (2015) define an optimization problem that maximizes platooning opportunities for all trucks in a network and present heuristics that work well with problems involving 200 trucks. They point out that real-world problems of this type would include several hundreds or even thousands of trucks in a network. Van de Hoef et al. (2015) introduced an optimization "framework for a fuel-optimal coordination where the trucks adapt their speeds in order to form platoons during their journeys. The influence of speed on the fuel consumption is explicitly considered and trucks are guaranteed to arrive at their destinations by their arrival deadlines." The opportunity to improve safety while also saving fuel and lowering greenhouse gases is evident in platooning. A recent study (Mihelic et al., 2015) examined the relative merits of platooning compared with longer combination vehicles (LCVs), concluding that LCVs provide more benefit in economics and fuel and emissions reduction compared with platooning.

Finding: Automated operation that enables truck platooning has the opportunity to save fuel and greenhouse gas emissions achieved through decreases in drag of both trucks.

Figure 10-2 shows platooning trucks next to longer combination vehicles.

10.4 CONNECTED VEHICLES/COOPERATIVE-INTELLIGENT TRANSPORTATION SYSTEMS

Connected vehicle technologies, also known as cooperative intelligent transportation systems (C-ITS) technologies, are seeing rapid development in all major automotive manufacturing companies and original equipment manufacturers (OEMs). We include technologies enabling semiautonomous or assisted driving in this category even though in some cases those technologies do not involve cooperation or communication between vehicles.

FIGURE 10-2 A longer combination vehicle, platooning vehicles, and a single vehicle.
SOURCE: Mihelic et al. (2015).

As mentioned above, the fuel economy impacts realized by platooning can be significant. The fuel economy benefits of eco-cooperative cruise control, which employs vehicle-to-vehicle (V2V) communication to determine the preceding vehicle's speed, acceleration, and distance to compute the most fuel efficient vehicle trajectory using additional information such as roadway characteristics and vehicle configuration, should also be significant. A recent research project (Kamalanathsharma and Hesham, 2012) estimates the potential fuel savings for connected vehicles in the vicinity of signalized intersections to be 30 percent. While those savings apply less to long-distance trucking operations, they will impact the performance of urban delivery, transit, and vocational vehicles.

Another variant of cooperative adaptive cruise control (CACC) is the infrastructure-to-vehicle to vehicle-to-infrastructure (V2I) systems in which the roadway infrastructure, through its traffic management center and roadside devices, provides recommended speeds to the vehicle speed control systems. This infrastructure-based information could be static (providing average best speeds) or dynamic (providing variable suggested speeds). Shladover et al. (2015) and Nowakowski et al. (2015) provide definitions and operational concepts related to CACC.

Other technologies aimed at eco-driving could also have an impact on the fuel economy of MHDVs. Some of these systems provide drivers with more information about the real-time performance of their vehicles and recommended speeds, while others could take control over the driving speed of vehicles in order to optimize fuel efficiency.

Impediments to near-term success of these technologies include standards and interoperability, issues that must be worked out by automotive manufacturers and OEMs. And, security is a key issue. Recently, several researchers demonstrated how vulnerable these systems are to various kinds of attacks (Markey, 2015; Wilonsky, 2015; Zhang et al., 2014), although some credible sources have questioned the reliability of those reports (see, for example, Newcomb, 2015; Pogue, 2016).

10.4.1 Vehicular Ad Hoc Networks

Vehicular Ad Hoc Networks (VANETs) operate with little or no permanent infrastructure and are characterized by (1) high mobility, (2) fixed road networks, (3) predictable speed and traffic patterns in congested conditions, and (4) very few power constraints or storage limitations. Unlike other communication systems, in which the primary goal is to achieve high message throughput, VANETs aim primarily for communication reliability and fast dissemination.

Communication pathways include

- V2V, whereby messages are transmitted between neighboring vehicles. This includes "one-hop" and "multi-hop" messaging scenarios in which vehicles communicate directly with other vehicles or through intermediary vehicles.
- V2I, whereby messages are transmitted between vehicles and road-side units located on nearby arterial road intersections or highway on-ramps.
- Vehicle-to-pedestrian (V2P), whereby messages are transmitted between vehicles and pedestrians who send and receive messages via their phones or other wireless devices.

These vehicular communication networks can help improve safety, the environment, and mobility. For example, V2V and V2I systems use information on acceleration and braking behaviors of nearby vehicles to track dangers beyond a driver's line of sight, helping to prevent collisions. V2P systems will improve the safety of pedestrians crossing at intersections and facilitate carpooling and ridesharing.

The primary impediments to large-scale implementation of VANETs are related to communication challenges, reliability issues, security and privacy issues, and societal issues such as liability and jurisdiction/cost allocation and control of public-sector components of such systems. And, as originally imagined, these systems will likely be rendered obsolete by private-sector investments automated vehicles, though they can be considered precursors to those technologies.

It is critical that vehicular communications networks have robust and fault-tolerant software that has the ability to recover gracefully from connection downtime, system errors, and unforeseen scenarios. For V2V and

V2I applications, a few seconds' delay in downtime may be the deciding factor in a traffic accident. And unlike consumer electronics that may involve frequent hardware turnover (for example, smartphones or tablets), the computing hardware for vehicular communication networks (on-board units, vehicle sensors, etc.) needs to have a much longer life span, closer to the decade or more operating life of the vehicle. In addition, future repair or maintenance may or may not happen according to manufacturers' guidelines. The reliability of both the computing software and the hardware components remain major technical obstacles that must be overcome in order for these systems to achieve their promise.

Finally, issues related to liability must be addressed. For a recent detailed discussion of all of these issues please see Regan and Chen (2015).

10.4.2 Automated Vehicles

It seems likely that in the distant future there will be a considerable number of fully automated (referred to as "independent operation") MHDVs. In the short and mid term, trucks will at a minimum operate by driver assistance, but wil be "cooperatively operated" by the equipment.

In September 2015, the Technology and Maintenance Council of the American Trucking Association issued a white paper, "Automated Driving and Platooning Issues and Opportunities" (ATA/TMC, 2015). The paper details a number of automated operations opportunities for trucks, including level 1 and greater automation. These include

- Truck platooning,
- Highly automated trucks,
- Traffic jam assist,
- Automated trailer backing,
- Highway pilot, and
- Automated movement in queues (ports and other intermodal facilities).

The report details some important tasks for the trucking industry to undertake in order to facilitate automation:

- "Prepare, as needed, model legislation for states regarding operation of near-term truck AV [autonomous vehicle] systems.
- Assess current insurance approaches for suitability for increasing levels of automation. Work with insurers to develop new models as needed. Address the particular needs of self-insured fleets.
- Assess and facilitate public acceptance of near-term truck AV systems.
- Address specific technical and scientific evaluation tasks that are needed to adequately verify performance of these systems.
- Conduct research to determine what technology is needed to control the platooning engagement/disengagement process such that corporate policies are so enforced, if so desired. Current platooning models assume the driver will make the decision as to whether to platoon or not. However, some companies may wish to restrict this decision-making to a preferred list of carriers/operators.
- Conduct research on developing criteria for joining a platoon based on vehicle configuration, loads, weather conditions, routes, internal company cost, customer contracts, etc. This, because not all runs are profitable and not all platooning may result in economic savings even if fuel costs are reduced.
- Assess the impact that local, state and federal government may have on regulating the distance between platooned vehicles, since size and weight regulations are often addressed at a state or local level."

There will be many benefits to such systems, including improved safety, around-the-clock operation, and fuel and freight efficiency improvements, but many technology and public policy issues must be worked out before widespread adoption can be achieved.

For large-scale autonomous driving to be a success, issues of reliability, security, and privacy are gravely important (see, for example, Kalra et al., 2009; Marchant and Lindor, 2012). Automakers and public-sector officials are all concerned that these issues must be properly addressed before these systems are fully operational.

10.5 SAFETY-IMPROVING TECHNOLOGIES

While the goal of safety-improving technologies is to reduce the negative impacts of accidents—deaths, injuries, congestion, and costs due to damage to vehicles and the environment—many of these technologies provide opportunities for improving the overall efficiency of freight and fleet movements. Additionally, these safety technologies are the key to the success of driver-assisted or driverless vehicles. The primary technologies available are electronic stability and roll stability control, forward collision and mitigation systems, adaptive cruise control, and lane-keeping systems.

10.6 REFERENCES

ATA/TMC (American Trucking Association/Technology and Maintenance Council). 2015. Future Truck Program Position Paper 2015-3: Recommendations Regarding Automated Driving and Platooning Systems. http://www.trucking.org/ATA%20Docs/About/Organization/TMC/Documents/Position%20Papers/Future%20Truck%20Position%20Papers/FT_PP_2015_3.pdf (accessed December 16, 2019).

Bergenhem, C., H. Pettersson, S. Shladover, E. Coelingh, C. Englund, S. Tsugawa. 2012. "Overview of platooning systems." Proceedings of the 19th ITS World Congress, October 22-26, Vienna, Austria. Available at http://publications.lib.chalmers.se/records/fulltext/174621/local_174621.pdf (accessed September 23, 2019).

Chachich, A., and S. Smith. 2011. Smart Park: Truck Parking Field Operation Test Results, Volpe National Transportation Systems Center. http://ntl.bts.gov/lib/43000/43000/43029/Chachick_Smart_Park.pdf.

Cook, D.J., T. Morris, V. Morellas, and N. Papanikolopoulos. 2014. An automated system for persistent real-time truck parking detection and information dissemination. Pp. 3989-3994 in *IEEE International Conference on Robotics and Automation.* IEEE, Hong Kong, China, May 31-June 7.

Deruytter, M., K. Maddelein, W. Favoreel, D. Tsishkou, S. Bougnoux, and R. Bendahan. 2012. Stereovision truck parking occupancy detection. In *19th ITS World Congress, Vienna, Austria, 22 to 26 October 2012.* ITS America, Washington, DC.

Easyway. 2012. Freight & Logistics Services Intelligent and Secure Truck Parking Deployment Guideline. http://www.rits-net.eu/uploads/media/EW-DG-2012_FLS-DG01_IntelligentAndSecureTruckParking_02-00-00.pdf (accessed November 12, 2015).

Fallon, J., and K. Howard. 2011. Smartpark Truck Parking Availability System: Magnetometer Technology Field Operational Test Results. FMCSA-RRT-10-041. Federal Motor Carrier Safety Administration, Washington, DC. http://ntl.bts.gov/lib/51000/51300/51359/SmartPark-Magnetometer.pdf.

Franke, U., F. Bottiger, Z. Zomotor, and D. Seeberger. 1995. Truck platooning in mixed traffic. Pp. 1-6 in *Proceedings of the IEEE Intelligent Vehicles '95 Symposium.* IEEE, Detroit, MI, September 25-26.

Fritz, H. 1999. Longitudinal and lateral control of heavy duty trucks for automated vehicle following in mixed traffic: Experimental results from the CHAUFFEUR project. Pp. 1348-1352 in *Proceedings of the 1999 IEEE International Conference on Control Applications, Volume 2.* IEEE, Kohala Coast, HI, August 22-27.

Garber, N., H. Teng, and Y. Lu. 2004. A Proposed Methodology for Implementing and Evaluating a Truck Parking Information System, Research Report No. UVACTS-15-5-86. Charlottesville: Center for Transportation Studies, University of Virginia.

Garcia, J.F., V.R. Tomas, L.A. Garcia, and J.J. Martinez. 2014. An autonomic system for intelligent truck parking. Pp. 810-916 in *Proceedings of the IEEE 11th International Conference on Informatics in Control, Automation and Robotics (ICINCO), Volume 2.* IEEE, Vienna, Austria, September 1-3.

Gertler, J., and J. Murray. 2011. Smartpark Truck Parking Availability System: Video Technology Field Operational Test Results. FMCSA-RRT-10-002. Federal Motor Carrier Safety Administration, Washington, DC. http://ntl.bts.gov/lib/51000/51300/51360/SmartPark-Video.pdf.

Kalra, N., J. Anderson, and M. Wachs. 2009. Liability and Regulation of Autonomous Vehicle Technologies. California PATH Research Report, UCB-ITS-PRR-2009-28. University of California Institute of Transportation Studies, Berkeley, CA.

Kamalanathsharma, R.K., and R. Hesham. 2012. Agent-based modeling of Eco-Cooperative Adaptive Cruise Control systems in the vicinity of intersections. Pp. 840-855 in *Proceedings of the 15th International IEEE Conference on Intelligent Transportation Systems (ITSC).* IEEE, Anchorage, AK, September 16-19.

Larsson, E., G. Sennton, and J. Larson. 2015. The vehicle platooning problem: Computational complexity and heuristics. *Transportation Research Part C: Emerging Technologies* 60:258-277.

Marchant, G.E., and R.A. Lindor. 2012. The coming collision between autonomous vehicles and the liability system. *Santa Clara Law Review* 52(4):1321-1340.

Markey, E. 2015. Tracking & Hacking: Security & Privacy Gaps Put American Drivers at Risk. http://www.markey.senate.gov/imo/media/doc/2015-02-06_MarkeyReport-Tracking_Hacking_CarSecurity%202.pdf (accessed December 16, 2019).

Mbiydzenyuy, G., J.A. Persson, and P. Davidsson. 2012. Proposed core services for the deployment of intelligent truck parking. In *19th ITS World Congress, Vienna, Austria, 22 to 26 October 2012.* ITS America, Washington, DC.

Mihelic R., J. Smith, and M. Ellis. 2015. Aerodynamic comparison of tractor trailer platooning and A-train configuration. *SAE International Journal of Commercial Vehicles* 8(2):740-746.

Modi, P., V. Morellas, and N. Papanikolopoulos. 2011. Counting Empty Parking Spots at Truck Stops Using Computer Vision. CTS 11-08. Intelligent Transportation Systems Institute, Center for Transportation Studies, University of Minnesota, Minneapolis, MN.

NACFE (North American Council for Freight Efficiency). 2016. Confidence Report on Two-Truck Platooning. https://nacfe.org/technology/two-truck-platooning/ (accessed September 23, 2019).

Newcomb, D. 2015. "Congress, 60 Minutes exaggerate threat of car hacking." *Forbes*, February 9. http://www.forbes.com/sites/dougnewcomb/2015/02/09/60-minutes-joins-car-hacking-hype/#5fdb8cb04f4d (accessed May 10, 2016).

Nowakowski, C., S.E. Shladover, X.Y. Lu, D. Thompson, and A. Kailas. 2015. *Cooperative Adaptive Cruise Control (CACC) for Truck Platooning: Operational Concept Alternatives.* California Partners for Advanced Transportation Technology, University of California, Berkeley, CA. http://eprints.cdlib.org/uc/item/7jf9n5wm (accessed May 10, 2016).

Pogue, D. 2016. "Why car hacking is nearly impossible." *Scientific American*, February 22. http://www.scientificamerican.com/article/why-car-hacking-is-nearly-impossible/ (accessed May 10, 2016).

Regan, A.C., and R. Chen. 2015. Ad hoc vehicular networks. Pp. 29-36 in *Vehicular Communications and Networks*, edited by W. Chen. Cambridge, UK: Elsevier.

Shladover, S.E., C. Nowakowski, X.Y. Lu, and R. Ferlis. 2015. Cooperative adaptive cruise control: Definitions and operating concepts. *Transportation Research Record: Journal of the Transportation Research Board* 2489:145-152.

Thinking Highways. 2014. "Nedap deploys smart truck parking systems in Denmark." http://thinkinghighways.com/nedap-deploys-smart-truck-parking-system-in-denmark/ (accessed June 4, 2015).

Thinking Highways. 2015. "Truck Parking Europe now available as a web portal." http://thinkinghighways.com/truck-parking-europe-now-available-as-a-web-portal/ (accessed June 4, 2015).

van de Hoef, S., K.H. Johansson, and D.V. Dimarogonas. 2015. Fuel-optimal centralized coordination of truck platooning based on shortest paths. Pp. 3740-3745 in *American Control Conference (ACC).* IEEE, Chicago, IL, July 1-3.

Wilonsky, R. 2015. "Dallas attorney sues Ford, Toyota and GM, claiming their cars are 'susceptible to hacking'." *Dallas Morning News,* March 10. http://bizbeatblog.dallasnews.com/2015/03/dallas-attorney-sues-ford-toyota-and-gm-claiming-their-cars-are-susceptible-to-hacking.html/ (accessed May 10, 2016).

Woodrooffe, J., D. Blower, and J. Sullivan. 2016. Evaluation of Michigan DOT Truck Parking Information and Management System. University of Michigan Transportation Institute Report 2016-15. Ann Arbor, MI.

Zhang, T., H. Antunes, and S. Aggarwal. 2014. Defending connected vehicles against malware: Challenges and a solution framework. *IEEE Internet of Things Journal* 1(1):10-21.

11

Manufacturing Considerations

11.1 INTRODUCTION

Previous National Research Council and agency reports and the section "Projecting Total Capital Costs" in Chapter 12 of this report have provided extensive discussion on the parameters of manufacturing to be considered for the technologies and approaches that may be employed for the reduction of fuel consumption. See section 11.3 "Annex" for a synopsis of those reports and their findings. Significant discussion in those reports is directed toward estimating the financial factors and influences for projecting future manufacturing costs, including such items as direct manufacturing cost, indirect costs, retail price equivalents, indirect cost multipliers, economies of scale, time- and volume-based learning effects, and stranded capital.

This report chapter does not intend to repeat what has already been thoroughly reported and debated. The approach of this report for manufacturing considerations will focus primarily on the emerging technologies for the latter part of the next decade that may have significant influence over fuel consumption reduction. As stated in Chapter 12, several technologies with strong fuel consumption improvement potential also carry the burdens of high current and projected manufacturing costs, posing challenging pathways through the stages of low volume growth and rational consumer choice to reach the volumes necessary to achieve improved, acceptable cost effectiveness.

This issue is further exacerbated by the fact that U.S. fuel prices are currently low and projected to remain relatively so for the next decade at a minimum. With fuel demand reducing as a result of existing and promulgated regulations for light, medium, and heavy vehicles through the middle of the next decade, there will be continued downward pressure on fuel prices, making the rational consumer's business case to purchase fuel-consumption-reducing technologies more difficult. This is clearly demonstrated by the analysis on hybrids shown in Chapter 12. It underscores the fact that greater emphasis needs to be placed on reducing the cost of the most effective fuel-consumption-reducing technologies. Those manufacturers pursuing them are concentrating on system cost reductions that are gained in incremental stages along volume- and time-dependent learning curves, along with substantial step improvements in cost provided by design cycles of learning. A good example would be in the case of waste-heat recovery (WHR) systems for heavy-duty diesels in line-haul trucks. Whereas the costs for WHR reported in the Phase II proposed rule (EPA and NHTSA, 2015a) and Regulatory Impact Analysis (RIA) (EPA and NHTSA, 2016) were based on systems in consideration prior to Phase I, recent and substantial changes to the system design have the potential for significant reduction in manufacturing and implementation cost.

While design and system changes alone often provide the largest and most obvious opportunities for cost reduction, what is often elusive and overlooked is the potential for new manufacturing technologies, in synergetic combination with design iterations, to serve as enablers for even further cost reduction. Therefore, the primary focus of this chapter is to identify a few of the relevant known and emerging manufacturing technologies and considerations that offer promising opportunities for reduction of the cost of the high impact of fuel-consumption-reduction technologies. This will improve their rates of adoption and further enable naturally achieved economies of scale and improved rates of learning. This will include the manufacturing technology opportunities for reducing the costs of hybrids, WHR, and mass reductions.

This chapter will conclude by recommending additional actions that should be considered between now and the introduction date for a possible Phase III of regulations (taking effect after 2027) to increase the potential for manufacturing technology to further improve fuel-consumption-reduction business cases.

11.2 ADVANCED MANUFACTURING TECHNOLOGIES CONSIDERED

There are many advanced manufacturing technologies presently under development, and with the increasing capabilities in advanced computational methodologies, greater application of more accurate and representative mathematical models and simulations, and a broader global collaborative approach many more will be generated before the Phase III regulations are in effect. To cover every known advanced manufacturing technology would be exhaustive, so this report will only highlight a few of those that currently appear to have highly relevant potential for this report's primary directions.

It is important to note up front that this is only a snapshot at this point in time, and the list of technologies will evolve continuously in the many years between now and Phase III regulations. The present snapshot will indicate where additional development emphasis will be beneficial to improving fuel consumption, but it must be recognized that the pace of new developments in these areas is already brisk and many of the identified challenges will be solved, only to be replaced by emerging new challenges between now and the time Phase III rulemaking commences. So the manufacturing technologies presented at this juncture should serve primarily to exemplify some trending items with high potential trajectories, if properly guided, but there is high likelihood that many additional high-potential manufacturing technologies will emerge between now and Phase III. This indicates the strong need for a continuing coordinated and focused effort to identify and prioritize resources such that the most promising technological directions with the greatest potential and impact can be sufficiently developed to have major presence and effect on Phase III compliance efforts.

11.2.1 Additive Manufacturing

Currently, the most widely discussed advanced manufacturing technology is additive manufacturing. It is considered by some to be a principal driver of a new manufacturing age, offering the promise of dramatic reductions in raw material, manufacturing capital equipment and tooling, energy costs, machining by-products, handling, and shipping costs. It also has demonstrated potential for freeing physical designs from many current manufacturing constraints, allowing production of more optimal designs to satisfy the given part or system performance objectives.

Fundamentally, additive manufacturing refers to processes that build up three-dimensional objects by fusing successive layers of materials that are defined by numerical database representations of the geometry. The layers are created and shaped by controlling ultraviolet light, focused intense laser beams, electron beams, or nozzles based on computer-aided design files describing each layer of the part, to harden a resin, melt and fuse a powdered metal, extrude a material onto a layer, or otherwise control material deposition and bonding in an accurately described fashion. Material removal (subtractive manufacturing) through machining is minimized or eliminated entirely, depending on dimensional accuracy and surface finish considerations. Initiated in the early 1980s with resinous compounds that hardened under exposure to ultraviolet laser light (stereolithography), the field has rapidly expanded to a variety of plastics extrusion additive manufacturing processes (fusion deposition modeling, fused filament fabrication, etc.), metal powder bed fusion processes (selective laser sintering, selective laser melting

[SLM], direct metal laser sintering, direct metal laser melting [DMLM], electron-beam melting [EBM], etc.), laminated solids processes, and wire-fed processes (electron-beam freeform fabrication [EBF3]). More variations will undoubtedly emerge in the ensuing period. Hundreds of companies are now developing more advanced additive manufacturing equipment and processes, and new applications of the technology are developed daily.

Additive manufacturing applications have expanded rapidly, from making small plastic models of parts used primarily for physical representation in space claim mockups, to making functional parts and assemblies to demonstrate proof of concept, to making metallic prototype parts, low- and medium-volume tooling and molds, and low-volume-production metallic parts. Additive manufacturing is advancing quickly in the medical and dental fields, from creation of dental implants and prosthetic devices to 3D printing of life-like human tissue.[1] In the industrial sectors, the leading usages of additive manufacturing are in aerospace, where low-volume complex, lightweight structures that must operate in harsh environments and intricate manufacturing jigs and fixtures can be produced faster and better than by conventional methods. Successively more challenging applications are expanding in the aerospace industry, such as gas turbine engine blades under development by Siemens[2] and CFM International's LEAP engine fuel nozzles (Figure 11-1) at intended production volume of 35,000 to 40,000 per year, replacing a 20-piece welded assembly and saving 25 percent of the weight.

Automotive industries have long used additive manufacturing for physical space claim, display, and mockup parts; engine intake and exhaust flow models; rapid prototyping; and experimental tooling. Metallic and nonmetallic special tools and functional prototype metallic assemblies are increasingly being made using the technology. Precision additive manufacturing at automotive volumes and quality with reasonable cost is still a challenge, currently limiting the application of the technology to low-volume prototypes and engineering development models. However, as additive manufacturing equipment cost and throughput are rapidly improving, low-volume (<50,000/year) parts such as electrical connectors are starting to be produced by additive manufacturing.[3] Clearly, additive manufacturing usage has grown and is forecast to continue to grow exponentially in many manufacturing sectors (Figure 11-2).

FIGURE 11-1 GE LEAP gas turbine engine fuel nozzle made by additive manufacturing.
NOTE: CFM International is a 50/50 joint company between GE and Safran Aircraft Engines.
SOURCE: CFM International.

[1] See http://additivemanufacturing.com/2016/02/10/engineers-3d-print-a-new-lifelike-liver-tissue-for-drug-screening/ (accessed December 20, 2019).

[2] See https://press.siemens.com/global/en/pressrelease/siemens-achieves-breakthrough-3d-printed-gas-turbine-blades (accessed December 20, 2019).

[3] See https://3dprint.com/136839/carbon-clip-delphi/ (accessed December 20, 2019).

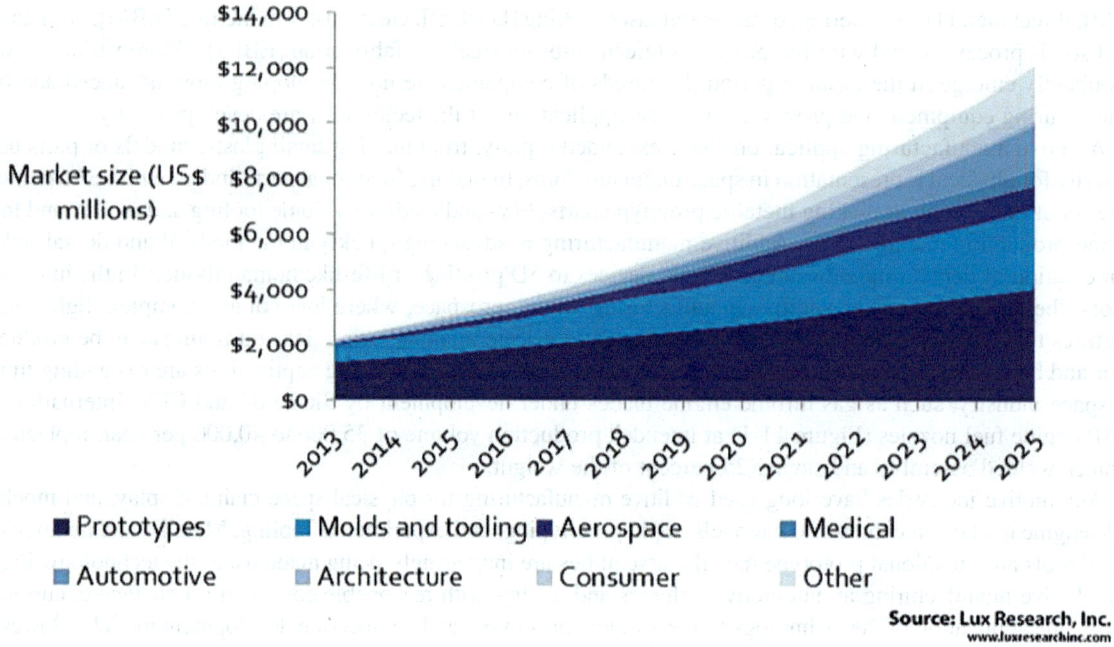

FIGURE 11-2 Growth of additive manufacturing measured in market size for the aerospace, medical, automotive, electronics and consumer product industries.
SOURCE: Vicari et al. (2014).

The many advantages of additive manufacturing are dependent on the specific fields of application. Most of the waste associated with subtractive manufacturing, material removal machining, is eliminated.[4] Additive manufacturing can provide additional benefits to the nearer-term technologies considered in this report. One example of this would be in the improvement of engine thermal efficiency. Chapter 4 details the progress and prognosis for engine thermal efficiency improvements, recognized as one of the most significant factors in vehicle fuel consumption. Thermodynamic cycle losses in the form of heat transfer to coolant and exhaust have eluded rapid improvement for decades, partly because of limitations in the way engine parts need to be designed to be economically manufactured by traditional, predominantly subtractive manufacturing processes. Additive manufacturing can allow new geometries in pistons, cylinder heads, and cylinder blocks to more optimally control heat transfer, through the design and specific placement of heat dams, cooling channels, entrained air pocket insulators, and other part geometries previously not possible. By retaining more of the combustion energy in the cylinders to produce more piston work, engine thermal efficiency can be further improved. Chapter 6 clearly shows the strong benefit of improving engine thermal efficiency on vehicle fuel consumption.

Another example would be in the application of additive manufacturing to electric machines used in hybrid propulsion systems. Chapter 8 describes the various hybrid architectures available and in development. As indicated in Chapter 12, Table 12-2, copied from the Phase II RIA, the penetration of hybrids in vehicle Classes 2b and 3 is required to increase steeply to provide fuel consumption improvement beyond 2.5 percent per year, yet the private benefits of hybrids at the currently projected cost of fuel yield excessive payback periods to recover the hybrid system capital costs. A consumer-driven growth of this important technology is therefore challenged, unless the capital cost of the technology can be reduced by over 30 percent, per the analysis.

[4] In the more reaching scenarios posed by popular futurists, the need to consume so much of our transportation industry fuel in the shipping of raw materials, finished parts, and machining by-products (chips, cutting oils, packaging, etc.) will be dramatically reduced, enabled by additive manufacturing centers widely distributed to consumers who will be able to download and 3D print whatever they need from a locally optimized supply of basic elemental powders and ultraviolet (UV) light-sensitive resins. This state, if realized someday, would probably have the greatest effect on transportation fuel consumption.

The largest incremental cost drivers in hybrid systems are the energy storage system, the electric machines, and the power electronics. The electric machines are complex geometries designed to optimize mechanical and electromagnetic performance efficiency, controllability, space claim, weight, and cost. Manufacturability considerations heavily influence how motors are designed, and often impact resulting performance and cost of materials such as rare-earth magnets and copper alloys. With additive manufacturing technology, new higher-performance electric machine geometries could be possible, allowing mixed-material assemblies to be manufactured with optimal configuration parts and minimum use of high-cost materials. While the ultimate reduction in system cost is impossible to accurately project, the likely benefit to the current payback dilemma that would otherwise hinder a more consumer-driven growth in penetration rate for hybrids underscores why additive manufacturing application to hybrid motors needs to be emphasized.

Battery costs remain the dominant concern influencing the economic attractiveness of hybrid systems, although economies of scale and learning cycles on cathode, anode, and electrolyte materials have allowed battery costs to be reduced to less than half per kilowatt-hour (kWh) of energy storage than what they were just 5 years ago, with predictions of $100/kWh cell cost by the end of this decade or slightly thereafter.[5] Additive manufacturing technologies, applied to the intricate shapes, housings, control, and protection circuitry, and mixed-material assemblies found in mobile energy and power batteries could supplement this trend, further reducing the cost while improving energy storage density.

As mentioned earlier, WHR systems for heavy-duty diesel engines have evolved rapidly in the past decade and show promising fuel-consumption-reduction potential (4 to 5 percent) in the Class 8 vehicle segment. The current- (fourth-)generation WHR system from Cummins utilizes a low–global warming potential refrigerant that transfers exhaust heat energy to a mechanical turbine expander made by Cummins Turbo Technologies, shown in Figure 11-3.

The complex WHR turbine expander rotor and blade geometries could be candidates for additive manufacturing processing, as are the turbocharger compressor and turbine wheels in the conventional turbochargers, and can be considered an extension of what the aerospace industry is doing for gas turbine engine blades to achieve higher-performance geometries with materials impractical for conventional manufacturing processes.

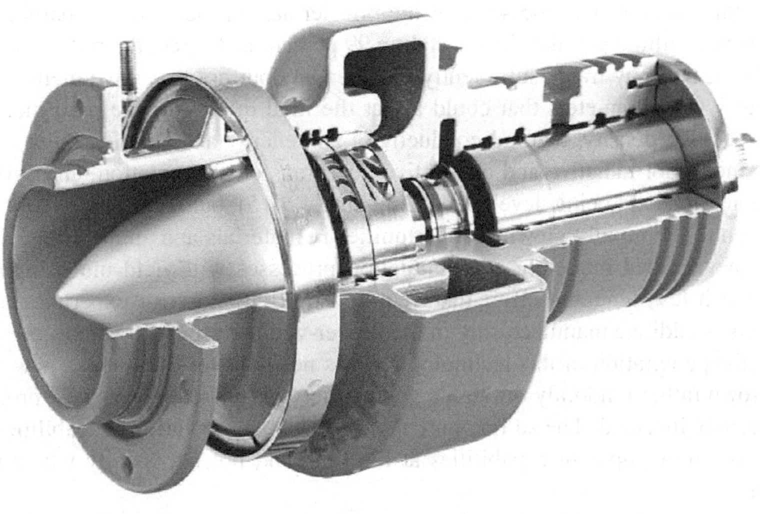

FIGURE 11-3 Cummins Turbo Technologies mechanical turbine expander.
SOURCE: Cummins Inc.

[5] See http://www.hybridcars.com/tesla-projects-battery-costs-could-drop-to-100kwh-by-2020/ (accessed December 20, 2019) and http://www.hybridcars.com/gm-ev-battery-cells-down-to-145kwh-and-still-falling/ (accessed December 20, 2019).

Vehicle weight reduction is also shown in Chapters 5 and 12 to be an opportunity for improving vehicle fuel consumption. However, material substitution (replacing ferrous materials with aluminum, magnesium, or composite materials) initiatives by themselves are generally considered to yield marginal incremental returns on investment (ROI) at best, due to the cost of the substituting materials. Again following the aerospace industry, the use of additive manufacturing processes to enable much more material-efficient, lattice-like design configurations for the heavier structural components in vehicles without compromising strength, rigidity, or durability is an alternative approach to material substitution. Lightening of parts currently made by conventional casting and forging processes by hollowing out material that is only there because of current process constraints (draft angles, casting core support and rigidity requirements, directional solidification considerations, casting venting, etc.) is another approach enabled by additive manufacturing processes. Less material will ultimately be used in the vehicle, so material costs will be lower. Ferrous alloys can continue to be used for cost and/or performance needs, or lighter alloys can be substituted with minimum cost impact due to the capability to create intricate shapes and structures to provide the required part strength and rigidity with minimum material.

These examples barely scratch the surface in highlighting additive manufacturing's potential for enabling the higher-impact fuel-consumption-reduction technologies. While they paint an optimistic picture of what may be possible, there are several significant challenges with additive manufacturing processes that must be overcome before economical high-volume production of consistently high-quality metallic parts can be realized.

Confidence in the quality and durability of the parts made by additive manufacturing to operate satisfactorily in the challenging automotive environments for hundreds of thousands of miles needs to be improved. With annual volumes in the thousands and millions, the production rates and capacities need to increase by several magnitudes over current levels. Machine-to-machine variation, machine throughput and speed, and education of engineers and designers to the capabilities and limitations of additive manufacturing need to improve. Major current roadblocks exist with new material invention, characterization, and qualification, build envelope size limitations, and required post-processing to achieve required surface finish, stress relief, hardness, and net shape.

The highest strength and density metallic parts made by additive manufacturing are currently made by laser or electron-beam fusion metal processes (SLM, DMLM, EBM, EBF3). Confidence in the ability of these additive manufactured parts to succeed in the challenging automotive environments can be gained through understanding of the physics of the processes. Modeling and simulation, employing high-performance computing and iterative verification through factorial-based designs of experiments on physical samples, will increase this confidence level. Modeling and simulation at both the scale of the powder and the scale of the parts need to be done. For SLM applications, where resulting part density exceeding 99 percent is desired, the process needs to be controlled to melt the powder layer, track by track, sufficiently for adequate but not excessive melting of the powder. The process involves around 130 parameters that could affect the final quality of the part, including such items as powder morphology, packing density, thermal conductivity and energy absorptivity, laser wavelength, intensity, geometry and speed, melt pool radiative and convective heat transfer, vaporization, and surface tension effects. These models and simulations are being developed (King et al., 2015), but much more work needs to be done to expand their application to the rapidly growing opportunities in materials and applications.

Clearly, the use of validated modeling and simulation processes will yield much higher confidence parts and processes, with much less time and effort than required by trial-and-error experimentation. This will help accelerate the adoption of additive manufacturing in the higher-volume ground transportation industry. However, the pathway for how the penetration of this technology grows needs better focus. As a new technology, additive manufacturing has grown rather randomly through a wide variety of applications as new processes and materials were being simultaneously invented. The ad hoc purchase of machines of various capabilities led to applications being selected based on the equipment capabilities at the time and not necessarily where the ultimate benefit would be the greatest.

As the range of equipment has matured and process understanding improves, the product needs should more clearly define the pathway for increased penetration of this technology. In this context, the priority directions to reduce transportation industry fuel consumption should identify from the thousands of parts and tools used in a vehicle which ones are the highest-yield target opportunities, and the output of the ever-improving additive manufacturing modeling and simulation tools should identify which materials, processes, and equipment would be capable for application of additive manufacturing to improve performance at least cost. With this more-focused pull on the right pathways, growth of this technology will further accelerate in the direction to benefit the ROI of fuel consumption reduction. Ideally, a neutral, readily accessible entity should serve in this "clearing house" role, collaboratively matching product needs to advancing process and equipment capabilities, and thereby driving future research agendas and efforts. The National Additive Manufacturing Innovation Institute (NAMII), a node within the National Network for Manufacturing Innovation (NNMI) within the Department of Commerce, is a public-private partnership entity that should fulfill this role. The National Highway Traffic Safety Administration (NHTSA) and the Environmental Protection Agency (EPA) should strongly pursue cross-agency coordination of projects and priorities to accelerate the prime additive manufacturing technologies for benefiting fuel consumption improvement.

Finding: Additive manufacturing technologies appear to offer opportunities for improving the ROI of high-effect fuel-consumption-improving technologies, such as hybrids, WHR, and mass reduction.

Finding: Challenges affecting the rate of additive manufacturing application to high-volume transportation vehicles with required quality can be addressed efficiently by advanced modeling and simulation techniques, employing high-performance computational methods.

Finding: The application of additive manufacturing to transportation systems for fuel consumption improvement needs prioritization, focus, and emphasis such that progress on the right pathways proceeds in coordination with regulatory timetables.

Recommendation 11-1: NHTSA and EPA should strongly pursue cross-agency coordination of projects and priorities to accelerate the prime additive manufacturing technologies for benefiting fuel consumption improvement. This should be done through NNMI and NAMII within the Department of Commerce.

11.2.2 Joining Technologies

As fuel-consumption-reduction technologies such as weight reduction and hybrid propulsion systems need to increase penetration in the medium- and heavy-duty vehicle classes, the need for manufacturing process innovation and refinement in selected joining processes increases. The most direct approach to vehicle mass reduction is material substitution, with the replacement of ferrous materials with aluminum materials being the main approach due to the very large difference in material densities. This trend is apparent in light-duty vehicles today and is expected to increase as light-duty-vehicle regulations become more stringent. The Ford F-150 truck is an outstanding recent high-volume example of manufacturing innovation enabling fuel consumption reduction, where the entire former steel body has been replaced with high-strength aluminum for a savings of over 700 pounds. Other manufacturers are increasing the use of aluminum stampings in major body parts. Joining processes and equipment are very material specific, so this has given rise to the need for innovation in how aluminum stampings can be joined to each other and to other nonstamped, dissimilar material parts. The switch to aluminum required Ford engineers to undertake a massive plant retooling effort. Traditional steel resistance spot-welding equipment was replaced with state-of-the-art technology that allows the truck to be assembled with self-piercing rivets, flow-drill screws, and structural adhesives, in addition to clinching and laser welding. Significant manufacturing engineering innovation was employed to advance these processes for the high-volume usage in this vehicle's assembly.

Use of aluminum joining methods gives rise to several new challenges. Laser-welded aluminum seams are subject to hot cracking with certain aluminum alloys containing magnesium and/or silicon. As was discussed with additive manufacturing, the use of sophisticated modeling and simulation employing high-performance computing allows this sensitivity to be understood and resolved much faster and more effectively than trial-and-error experimentation.

Resistance spot welding (RSW) of aluminum body and frame panels could offer the lowest-cost joining alternative, but due to the different properties of aluminum compared to steel, it has not been commonly used due to poor perceived robustness, variability, high current requirements, large flange requirements, and poor high-cycle fatigue properties.

An excellent example of how advanced manufacturing technology has provided a solution to this is General Motors' (GM's) patented multi-ring domed (MRD) electrode design, which creates high radial plastic strain in the electrode-sheet interface, breaking up the surface oxide layer, which reduces contact resistance and greatly improves control over heat generation.

As a result of this innovation, GM has been able to convert from steel to aluminum body and frame panels with less manufacturing capital equipment cost and lead time, since existing steel RSW equipment can be simply retooled to increase weld gun stiffness, add transformers where required for the higher current draw, and refit with the new MRD electrodes and dressing tools. The retooling cost is negligible in comparison to converting an RSW-intensive body shop over to self-piercing riveting.

GM has further evolved this technology to apply it successfully to various thickness sheets, extrusions, and die cast materials. A recent article discusses ongoing product evaluation of the technology to allow joining of aluminum to steel by RSW, which would further reduce cost, capital equipment, and lead time for material conversion initiatives to reduce vehicle weight.[6]

With the trends in vehicle weight reduction by both direct material substitution and metal thickness reductions, the generally agreed-upon eventual outcome will be the mixed-material vehicle body and frame, where each segment of the structure is made from the optimal material for that specific location. Gage thicknesses will change, as will the specific metallic and composite materials, as required across the vehicle. While many of the current joining processes may still suffice, the trends for usage of structural adhesives in combination with them will continue to increase. Structural adhesives, usually consisting of one-part epoxies, offer advantages by stiffening joints to either offset the effects of thinner-gauge materials used for weight reduction or improving noise and vibration by raising resonant frequencies of the structure. They also block galvanic corrosion between dissimilar metals, supplement joint strength, improve vehicle crashworthiness, and can provide sealing capabilities where desired.

The manufacturing technologies for usage of structural adhesives has developed rapidly, but additional innovations in adhesive strength and toughness, application parameters, process variables, long-term durability, cost, debond capabilities (for recycling), and many other requirements will be necessary as the range of joined material combinations continues to increase. Innovation in this field will be a key enabler for future vehicle mass reduction.

Finding: The drive for vehicle mass reduction will create the need for effective, low-cost, and investment joining technologies for aluminum, mixed materials, and a wider range of material thicknesses. Innovation in aluminum resistance spot welding has allowed conversion of existing steel body shops to aluminum with less investment, and retains this low-cost joining process as an alternative to self-piercing rivets, flow-drill screws, and adhesives.

Finding: Advanced modeling and simulation techniques using high-performance computing will continue to assist in the efficient development of joining methods as the range of materials used in vehicle design becomes more diverse.

11.2.3 Materials Technologies

Focused application of emerging materials technologies on vehicle components and systems holds promise for greatly improving the performance and reducing the cost of several fuel-saving technologies. The field of nanoscale materials is growing exponentially, with many automotive applications already in use, including chip-resistant paints, dirt-resistant coatings, low-friction oils, nano-enhanced thermal fluids, lightweight nanocomposite components, wear-resistant tires, high-performance thermal interface materials, flexible displays, acoustic absorbers, and many others.

Returning to the premise stated in the introduction to this chapter, the priority areas for greater manufacturing technology emphasis are those that would reduce the most fuel consumption if penetration rates can be increased,

[6] See http://www.autonews.com/article/20160229/OEM06/302299967/gms-sharks-hunt-for-innovations (accessed December 20, 2019).

with consumers naturally pulling on the technology's growth because of good ROI on their capital cost to acquire the technology. The priority areas include hybrids for vehicle Classes 2b and 3, and Classes 4 through 6 in vocations that have high stop-start duty cycles, WHR for Class 8 line-haul tractors, and, to a lesser extent, mass reduction. Therefore, the use of new materials, especially nanostructured materials that can reduce the investment and/or cost of these technologies, will encourage greater penetration with less dependence on mandates and incentives. Much is already being done in applying nanostructured materials to batteries for electric and hybrid vehicle energy storage systems, and battery costs have reduced by half in the past 5 years. More needs to be done, however, to approach the \$100/kWh battery cell cost objective that many claim will be the tipping point for attractive ROI with this technology at the current and projected cost of fuels. Use of nanotechnologies to lower the cost or improve the performance of WHR systems should also be a priority. For example, in this arena, employing thermal fluids and solids enhanced with nanoscale particles may provide an opportunity to improve heat transfer from the exhaust to the organic Rankine cycle expander.

Another attractive emerging manufacturing technology utilizing nanoparticle technology is in catalyzed exhaust aftertreatment systems. Most aftertreatment catalysts use platinum group metals (PGMs) in an active-layer washcoat on the catalytic converter substrate to cause the conversion of criteria exhaust emissions to less harmful compounds. This washcoat is traditionally manufactured from the elemental PGMs by wet chemistry methods. New technology companies, such as SDC Materials, Inc., have developed processes to manufacture nanoscale catalytic compound particles and disperse them into very stable, effective washcoat layers, where the high surface-to-volume ratio attribute of nanotechnology enables excellent thrifting of the PGM necessary for the desired catalytic action. PGM savings in excess of 50 percent have been demonstrated with this technology. Alternatively, aftertreatment performance can be enhanced at the same PGM loading, as shown qualitatively in Figure 11-4.

The need for more effective, lower-cost aftertreatment systems is heightened by desired changes to engine calibrations to optimize the trade-offs between CO_2 and other emissions, such as NO_x. The value of this manufacturing technology will increase as manufacturers strive to meet more stringent emissions regulations with more effective catalysts using a minimum amount of PGM in their washcoats.

As is the case with additive manufacturing, the field of nanoscale materials is advancing so rapidly that accurately projecting the exact technologies for the Phase III regulation timing is futile at this point in time. However, these and many other examples underscore the potential impact this technology can have on improving the attractiveness of fuel-saving components and systems, if properly focused, prioritized, and resourced in the intervening period. Many forums and collective efforts already exist. The National Nanotechnology Initiative (NNI) is a U.S. government research and development initiative consisting of the individual and cooperative nanotechnology-

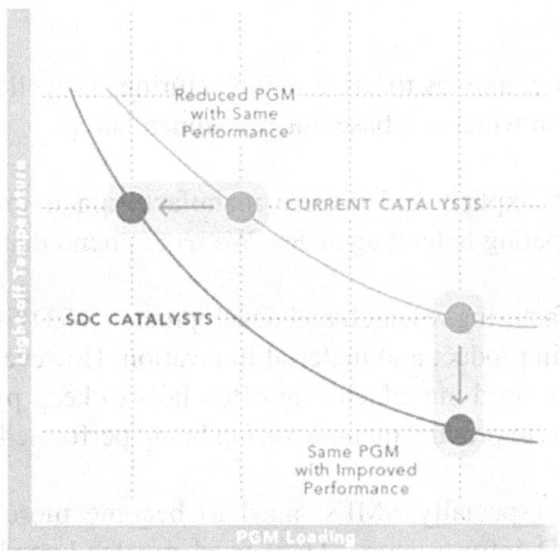

FIGURE 11-4 SDC nano-enhanced catalyst PGM capability measured in light-off temperature vs PGM loading.
SOURCE: SDC Materials, Inc.

related activities of federal agencies with a range of research and regulatory roles and responsibilities. It is managed within the framework of the National Science and Technology Council, the Cabinet-level council under the Office of Science and Technology Policy at the White House.

Finding: The rapidly expanding field of nanoscale materials development has produced many successful existing applications to vehicles, and many more are under development that, with proper priority, can provide fuel consumption and greenhouse gas (GHG) reduction benefits.

Recommendation 11-2: Emphasis needs to continue on developing applications for reducing the cost of high-impact fuel-consumption-reducing technologies, including hybrid and electric propulsion batteries, heat transfer systems for WHR, mass reduction, and exhaust aftertreatment catalysts. A pathway to ensure appropriate priority and coordination of nanotechnologies for fuel consumption should be determined by NHTSA and EPA working through the councils and committees associated with the NNI.

11.2.4 Conclusion

Manufacturing innovation has been shown to offer strong potential for improving the cost and effectiveness of product technologies in general, and this chapter has highlighted only a sampling of those emerging and rapidly expanding technologies that presently appear capable of markedly improving the cost of some of the highest-potential fuel consumption reduction technologies, thereby improving their natural penetration rate increase in the nation's fleets. It should be pointed out that, given the long time between this report and Phase III rulemaking, the trajectories for these technologies are not easily projected, and the emergence of additional new manufacturing innovations will undoubtedly occur. However, with proper emphasis, focus, and support, the development and introduction of the most effective technologies in the right time frame can be maximized.

To do this, identifying the correct entities and pathways for prioritizing, emphasizing, focusing, encouraging, incentivizing, and continually refreshing the advanced manufacturing technology agenda should be undertaken. With the need for fuel consumption improvement a high national priority in a climate of low fuel prices, a cross-agency effort within the nation's governmental entities is necessary. The general need for a coordinated approach is underscored by the National Manufacturing Council,[7] under the U.S. Department of Commerce, in its October 2014 recommendations, excerpted here:

> There's an important nexus between manufacturing and innovation. More than half of the STEM-related PhDs in the U.S. are employed in manufacturing, focused on research, development and advanced manufacturing practices.
>
> To ensure that innovation continues to drive manufacturing, as well as reverse, we have identified the following key findings upon which we base our recommendations:
>
> 1. The U.S. needs a clear, explicit and diverse manufacturing innovation strategy that is visibly integrated across participating federal agencies. We recommend that the Department of Commerce lead this effort.
> 2. U.S. spending on manufacturing research and development (R&D) is critically important to realize value from investments in product and material innovation. However, research in enabling process and technologies for advanced manufacturing often fails to keep pace with product development research. This contributes to product manufacturing being performed in other countries and the loss of U.S. jobs.
> 3. Many U.S. companies, especially SMEs, need to become more deeply engaged in the U.S. manufacturing and R&D infrastructure. They need greater knowledge and increased visibility

[7] U.S. National Manufacturing Council, letter to Secretary of Commerce The Honorable Penny Pritzker, October 15, 2014.

within the range of existing government programs related to procurement, innovation, research and development, and further exposure to manufacturing technology, processes and innovation.
4. Manufacturers need a widely recognized single mechanism that could provide visibility of current U.S. manufacturing capabilities among public and private sector procurement officials.

The council offers the following recommendations, which are based on these findings and support a strategy to strengthen American manufacturing.

1. Designate the federal manufacturing innovation programs become an Inter-agency Science and Technology Initiative, by executive decree.
2. Foster research and development that enables manufacturing processes and technologies to be expanded and optimized by evolving the role of national labs and strengthening the National Network for Manufacturing Innovation (NNMI) program with continued funding and coordination.
3. Improve the framework for collaboration and teambuilding within and between the public and private sectors.
4. Facilitate early supply chain creation for new innovation.
5. Expand connections for collaboration and commercialization of technologies.

With the adoption of these recommendations, the fuel-consumption-reduction strategies and priorities can be properly emphasized in better coordination with the regulatory agenda, and key advanced manufacturing innovations that offer the greatest potential for impacting total fuel consumed across the nation's medium- and heavy-duty vehicle fleets can be identified and nurtured.

The status and, if necessary, mid-course correction of the national advanced manufacturing strategy, as it applies to enabling fuel consumption improvement, should be included within the interim evaluation recommended in Chapter 2 of this report.

Finding: The pace of advanced manufacturing innovation suggests a need for a constant, continual review and adjustment of the strategy and priorities to focus development projects on high-impact fuel-consumption-reduction initiatives in coordination with the timing of the regulatory agenda.

Recommendation 11-3: The recommendations identified in the National Manufacturing Council's October 2014 report to the Department of Commerce should be vigorously pursued.

Recommendation 11-4: The status and, if necessary, mid-course correction of the national advanced manufacturing strategy, as it applies to enabling fuel consumption improvement, should be included within the interim review recommended in Chapter 2 of this report.

Recommendation 11-5: NHTSA and EPA, through their representation in or coordination with the relevant entities within the Department of Commerce, the Department of Energy, and the Office of Science and Technology Policy, should prioritize development and application of additive manufacturing, materials joining processes, nanostructured materials, and other yet-to-be-identified promising manufacturing innovations on those parts and systems that offer the greatest potential for fuel consumption and GHG reduction. The progress of these technologies, their prioritization status, and prognosis for effective commercialization within the Phase III regulatory period should be included in the interim evaluation proposed by this report.

11.3 ANNEX

This section summarizes recent reports by the National Research Council (NRC) and federal contractors on the subject of manufacturing, with an emphasis on those reports with findings related to reduction of vehicle fuel consumption. The reports described below covered both known issues with current manufacturing processes for vehicles but also discussed new manufacturing techniques.

The recent study by NRC (2015) of fuel economy of light-duty vehicles considered both financial factors that enable projections of future manufacturing costs (i.e., direct manufacturing costs and indirect costs). The study opined that tear-down analyses are the most reliable albeit the most expensive and further noted that this technique provides the best window into cost differences between technologies. Other techniques, including cost quotes, have limitations or include the effect of confounders such as supply-demand fluctuations.[8]

The NRC (2015) study further noted that, at least for auto manufacturers, the product development process has been accelerating, with one driver being the need to implement new technologies faster to meet fuel economy standards. This has the potential to increase stranded manufacturing capital.

The study also considered changes to the manufacturing process incident to the introduction of new technologies for better fuel consumption and focused on lightweighting and on powertrain technologies. The study found that changes to the vehicle requiring a higher level of integration, such as lightweighting, would need to be synchronized with the corresponding new vehicle program. Specifically, the study noted that "forming parts out of aluminum instead of steel may require greater engineering lead time and resources than simply changing the shape of a part made of the same material." In contrast, low-rolling-resistance tires would be an example of a technology that could probably be introduced during minor upgrades.

The NRC considered the stages of manufacturing that would be influenced by the introduction of new material grades such as occurs in lightweighting. Introducing lightweighting in automobiles, they found, would necessitate change in several points along the manufacturing value chain including the extensive cold-stamping infrastructure. Other affected areas included the assembly tools designed specifically for steel and the paint shop, recognizing that different materials paint differently and may have different thermal expansion coefficients (which can lead to novel complications when mixing materials).

The study described the impact on automation and gave several examples. Automation frequently relies on magnets to pick up (steel) parts, with obvious implications were aluminum-intensive parts to be introduced. Similarly, robot welding relies on the conductive properties of the material, which again would change with new material grades. The study did, however, note that flexible automation, introduced recently into production lines, can in some instances allow for faster introduction of innovations.

With respect to changes to the powertrain for reducing fuel consumption, the study found, for example, that reducing friction calls for "tighter machining, balancing, assembly tolerances, and bearing specifications, which all have a potential to increase manufacturing cost." The report also noted that standardized engine features can enable manufacturing flexibility, such as engines with varying numbers of identical in-line cylinders being manufactured on the same line as are V-6 and V-8 engines. The report identified this as a major trend among original equipment manufacturers. Turning to turbochargers, an oft-used strategy for improving fuel consumption, the report found that advanced concepts such as twin scroll and variable geometry placed exceptional demands on the manufacturing process.

11.4 REFERENCES

Cummins, Inc. 2009. Comments in Response to Greenhouse Gas Emissions and Fuel Efficiency Standards for Medium- and Heavy-Duty Engines and Vehicles—Phase 2. EPA-HQ-OAR-2014-0827 and NHTSA-2014-0132.

EPA (Environmental Protection Agency) and NHTSA (National Highway Traffic Safety Administration). 2015. Greenhouse gas emissions and fuel efficiency standards for medium- and heavy-duty engines and vehicles—Phase 2. *Federal Register* 80:40137-40765. July 13.

EPA and NHTSA. 2015b. *Commercial Medium- and Heavy-Duty Truck Fuel Efficiency Technology Cost Study.* Washington, DC: EPA and NHTSA.

EPA and NHTSA. 2016. *Regulatory Impact Analysis, Final Rule for Greenhouse Gas Emissions and Fuel Efficiency Standards for Medium- and Heavy-Duty Engines and Vehicles, Phase II.* Washington, DC: EPA and NHTSA.

[8] A complete review of techniques for determining direct costs was included in a previous report by the NRC (2011).

King, W.E., A.T. Anderson, R.M. Ferencz, N.E. Hodge, C. Kamath, S.A. Khairallah, and A.M. Rubenchik. 2015. Laser powder bed fusion additive manufacturing of metals; physics, computational, and materials challenges. *Applied Physics Reviews* 2:041304. doi: 10.1063/1.4937809.

NRC. 2011. *Assessment of Fuel Economy Technologies for Light-Duty Vehicles.* Washington, DC: The National Academies Press. https://doi.org/10.17226/12924.

NRC. 2015. *Cost, Effectiveness, and Deployment of Fuel Economy Technologies for Light-Duty Vehicles.* Washington, DC: The National Academies Press.

Vicari, A., R. Kozarsky, and M. Holman. 2014. "How 3D Printing Adds Up: Emerging Materials, Processes, Applications, and Business Models." State of the Market Report. Available at https://members.luxresearchinc.com/research/report/14139 (accessed September 23, 2019).

12

Costs and Benefits

12.1 ESTIMATES OF COSTS AND BENEFITS

12.1.1 Introduction

This chapter examines the costs and benefits associated with fuel-saving technology. The chapter begins with a review of the benefits and costs of meeting the Phase II standards as calculated by the Environmental Protection Agency (EPA) and the National Highway Traffic Safety Administration (NHTSA). The chapter uses these agencies' analysis as a base for examining increases in standards beyond the Phase II rule. The analysis of this chapter then focuses on the marginal cost of increases in the fuel efficiency standards beyond those of Phase II regulations. The approach taken in this chapter does not look at every possible technology application, but rather focuses on technologies central to making advances in fuel efficiency beyond where the industry is likely to be in 2027 when the Phase II regulations have been fully implemented.

Following the overall analysis of costs and benefits, the chapter turns to a detailed examination of issues involved in the measurement of social benefits and in the projection of future technology costs. The factors examined with respect to the projection of costs include learning that comes with time and with increased production as well as those technology costs incurred in addition to direct manufacturing expenses such as research and development (R&D).

12.1.2 Agency Estimates of Costs and Benefits

NHTSA and EPA estimate total costs and benefits of several scenarios of increased fuel efficiency standards in the Regulatory Impact Analysis (RIA) (NHTSA and EPA, 2016). These estimates provide a starting point for analyzing the costs and benefits of improving fuel consumption in medium- and heavy-duty trucks. For example, NHTSA and EPA's Phase II rule requires a 2.5 percent reduction in fuel consumption per year in the new truck fleet. In addition, NHTSA and EPA model the effects of a 2 percent, a 3.5 percent, and a 4 percent per year alternative. Table 12-1 is from the Regulatory Impact Analysis and presents the agencies' summary estimates of costs and benefits of the Phase II rule—a 2.5 percent annual reduction in fuel consumption.

The improvement in fuel efficiency and the reduction in the greenhouse gas (GHG) and other emissions resulting from increased standards for medium- and heavy-duty vehicles (MHDVs) and engines provides benefits both to those who own and operate the vehicles and also to others. The benefit to owners or operators of the

TABLE 12-1 Estimates of Lifetime Cost and Benefits of Rule Requiring 2.5 Percent Per Year Reduction in Fuel Consumption for Model Years 2018-2029 (billion $2013)

Discount rate	3%	7%
Private costs	25.4	17
Technology and indirect cost, normal Profit on additional investment	23.7	16.1
Additional routine maintenance	1.7	.9
Private benefits	157.5	84.7
Fuel savings	149.1	79.7
Savings from less frequent refueling	3.0	1.6
Benefits from additional vehicle use	5.4	3.4
Total net private benefits	132	67.7
Social benefits	64.4	49.1
Reduced climate change from greenhouse gas (GHG) emissions	33	33
Reduced health damages from reduced non-GHG emissions	27.2	14.5
Increased energy security	7.3	3.9
Congestion, crashes, fatalities, and noise from increased vehicle use	-3.1	-1.8
Total net benefits (private + social)	196.4	117

NOTE: The baseline used in the above table assumes Phase I standards are met and some additional technology is adopted in the fleet if it can pay back the capital through fuel savings in 6 months or less. In all cases examined by NHTSA and EPA, a 3 percent discount rate was used to value the benefits of reduced GHG emissions.
SOURCE: NHTSA and EPA (2016, Table 8-30, p. 8-73).

vehicles takes the form of a reduced cost for fuel consumption, which is considered their *private* (direct) benefit. They bear a cost in the form of a higher purchase (and perhaps maintenance) cost for vehicles complying with the increased standard, which is considered a private (direct) cost to them. The private firms gain benefits in the form of lower fuel costs, less time spent in refueling, and economic gains from increased vehicle use that follows from the lower operating costs associated with improved fuel efficiency. In Table 12-1, private benefits and costs are distinguished from *external*, or *social*, benefits and costs, which are those that accrue to *other* people besides the owners or operators of the vehicles subject to the increased standards. These external benefits include the effects of avoided emissions to society at large, including emissions of both greenhouse gases and of criteria pollutants, and also the national security benefits of reduced reliance on imported fuels. A negative consequence associated with increased vehicle use is increased congestion and accidents, constituting a negative externality or social cost.

Table 12-1 shows the benefits and costs of the Phase II rule as estimated by NHTSA and EPA using two different discount rates.[1] Both discount rate cases show a positive net benefit from the rule, ranging from $117 billion to $196 billion, depending on the interest rate used to discount benefits and costs. Interestingly, the bulk of the benefits are private benefits. Net private benefits amount to between 59 and 67 percent of *total net benefits*, depending upon the discount rate used.[2] The reduced climate change impacts are valued at $33 billion, ranging from 50

[1] In the Regulatory Impact Analysis NHTSA and EPA also present results measured against a slightly different baseline, reflecting different assumptions about technology deployment in response to the Phase I standards. The results differ slightly. The results shown here reflect what NHTSA and EPA call the "dynamic" baseline, which assumes more aggressive migration of light-truck technology to the heavy-duty (HD) segments. For a discussion of these two baselines, see NHTSA and EPA (2016, p. 10-120). There are other scenarios considered in the Regulatory Impact Analysis including a low and a high oil price scenario. As expected, total benefits are lower or higher depending on the fuel price largely since the value of fuel savings to private actors changes accordingly.

[2] Normally, we would expect these potential net private savings to result in firms making the investment in fuel-saving technology. NHTSA and EPA discuss why they think that private firms react insufficiently to these potential private gains. The agencies argue that this represents a market failure. The evidence as to whether this does constitute a market failure is ambiguous, and is discussed in Chapter 13 of this report.

to 67 percent of total *social* benefits, depending on the discount rate. In the next section we build on the modeling done by NHTSA and EPA and discuss what lies behind these estimates of costs and benefits to approximate the marginal cost and benefits of going beyond the 2.5 percent improvement in fuel consumption.

12.1.3 Identifying the Marginal Cost of Fuel Efficiency in 2027: Heavy-Duty Pickups and Vans (Classes 2b and 3)

Other aspects of the analysis done by NHTSA and EPA provide insight into the "marginal cost" and "marginal benefits" of going beyond the rule. Consistent with the task given to the committee to examine post–Phase II standards we assume that the Phase II rule takes effect and that the agencies are considering a possible Phase III rule for the period 2027 and beyond. Under this assumption, the required 2.5 percent per annum improvement would already have been met and NHTSA and EPA would be considering further required improvements. In order to meet the higher standard, additional technology will need to be applied to the new vehicle fleet. The question is what technology will be relied upon to achieve these more aggressive standards.

Table 12-2 is also from the Regulatory Impact Analysis (NHTSA and EPA, 2016). The table shows the rates at which specific technologies need to be deployed in the heavy-duty pickup and van segments in order to meet the fuel-saving standards under various hypothetical alternative rules. Recall that the 2.5 percent is the actual Phase II rule. The 3.5 percent rule would achieve the same level of stringency (15.6 percent) as the 2.5 percent rule, only it would achieve it 2 years earlier. The 4 percent rule achieves a greater reduction in fuel consumption (17.9 percent), also 2 years earlier than the reductions achieved by the 2.5 percent rule.

The greater fuel consumption reductions consistent with the 4 percent rule are achieved, according to the results shown in Table 12-2, by the more widespread use of electrification technologies—electronic power steering (EPS) and accessories, 12 volt (V) stop-start, and strong hybrids in the segments analyzed.[3] For example, the penetration of stop-start technology rises from 80 percent in the 2.5 and 3.5 percent annual improvement cases to 96 percent in the 4.5 percent per annum case. Similarly, the utilization of strong hybrids in the fleet rises from 2 to 7 percent. Other technologies either plateau at high rates such as the case with eight-speed automatic transmission (AT) or turbocharging. Stop-start and strong hybrids appear to be a key incremental technology necessary to move beyond the proposed rule and can be used to measure the marginal cost—the incremental cost of increasing fuel economy beyond the proposed rule—as well as the incremental benefit of compliance. Mass reduction and aerodynamic improvements also play a role in moving from a 2.5 to a 4.0 percent annual reduction in fuel consumption.

12.1.4 Identifying the Marginal Cost of Fuel Efficiency in 2027: Vocational Segments

NHTSA and EPA expect that by 2027 mild hybrids will be deployed in 14 percent of vehicles certified in the multipurpose and urban subcategories (6 percent integrated and 8 percent non-integrated). Hybrids are seen to play an increasing role in the vocational fleet as the Phase II rule is implemented, rising from 0 percent in 2021, suggesting that further stringency in the standards would rely on further penetration of hybrid technology.

Another view of the role of technology at the margin is provided in a study done by the California Hybrid, Efficient and Advanced Truck Research Center (CalHEAT) for the California Energy Commission (CalHEAT, 2013). Hybrid technology also plays a significant role in the CalHEAT study of the "roadmaps" to CO_2 emissions reduction in the California truck fleet. This study looked at what actions were necessary to reduce truck-related petroleum use and CO_2 emissions to 1990 levels by 2020 and by further reductions post-2020 so that by 2050 emissions levels are reduced by 80 percent from 1990 levels. The CalHEAT analysis takes into account the effects of Phase I, but goes beyond the reductions to be achieved by the Phase II standards. Figure 12-1, taken from that study, shows the important role of hybridization in meeting the California targets. The use of hybridization

[3] In the Preliminary Regulatory Analysis, using an earlier version of the Department of Transportation's Corporate Average Fuel Economy model, even greater penetration rates were estimated for stop-start and mild integrated starter/generator (ISG) and strong hybrids. These results are shown in the final Regulatory Impact Analysis (NHTSA and EPA, 2016, p. 10-122). These results are referred to as "Method B" analysis. This earlier analysis showed a much higher penetration rate for the mild and strong hybrids of 51 percent in the case of the 4 percent annual fuel consumption improvement. Stop-start technology is used in 13 percent of the fleet under the same scenario.

TABLE 12-2 NHTSA and EPA Summary of the Impact of Various Alternatives for HD Pickups and Vans Versus the Dynamic Baseline (Alternative 1b)

Alternative	2	3	4	5
Stringency of Standards				
Annual increase in stringency beginning in model year (MY) 2021	2.0%	2.5%	3.5%	4.0%
Increases until	MY 2025	MY 2027	MY 2025	MY 2025
Total increase in MY 2030 stringency relative to final Phase I standards[a]	9.6%	15.6%	15.6%	17.9%
Estimated Average Fuel Economy (miles per gallon)				
Required in MY 2030	19.03	20.37	20.38	20.95
Achieved in MY 2030	19.20	20.47	20.45	20.98
Average Fuel Consumption (gallons/100 miles)				
Required in MY 2030	5.25	4.91	4.91	4.77
Achieved in MY 2030	5.21	4.88	4.89	4.77
Estimated Average Greenhouse Gas Emissions (grams per mile)				
CO_2 required in MY 2030	494	462	462	450
CO_2 achieved in MY 2030	490	460	460	449
Technology Penetration in MY 2030 (percent)				
Variable valve timing and/or variable valve lift	56	56	56	56
Cylinder deactivation	4	4	4	4
Direct injection engine	17	27	26	29
Turbo charged engine	59	69	68	68
Eight-speed AT	77	95	94	95
EPS, accessories	52	80	80	96
12V stop-start	0	0	3	11
Strong hybrid	0	2	2	7
Aeodynamic improvements	46	80	80	98
Mass Reduction (vs. No Action)				
Mass reduction (lb)	28	240	24	289
Mass reduction (percent of curb weight)	0.43	3.6	3.7	4.3
Technology Costs (vs. No-Action)				
Average vehicle ($)	$500	$1470	$1480	1890
Payback period (m)[b]	19	30	31	33

[a] This increase in stringency is based on the estimated percentage change in fuel consumption (gallons per 100 miles) stringency projected by the model for the MY 2030 fleet under the final Phase II standards relative to the continuation of the Phase II standards.

[b] Here payback period is calculated using estimated undiscounted retail fuel savings and the initial technology costs for MY2030.

SOURCE: NHTSA and EPA (2016).

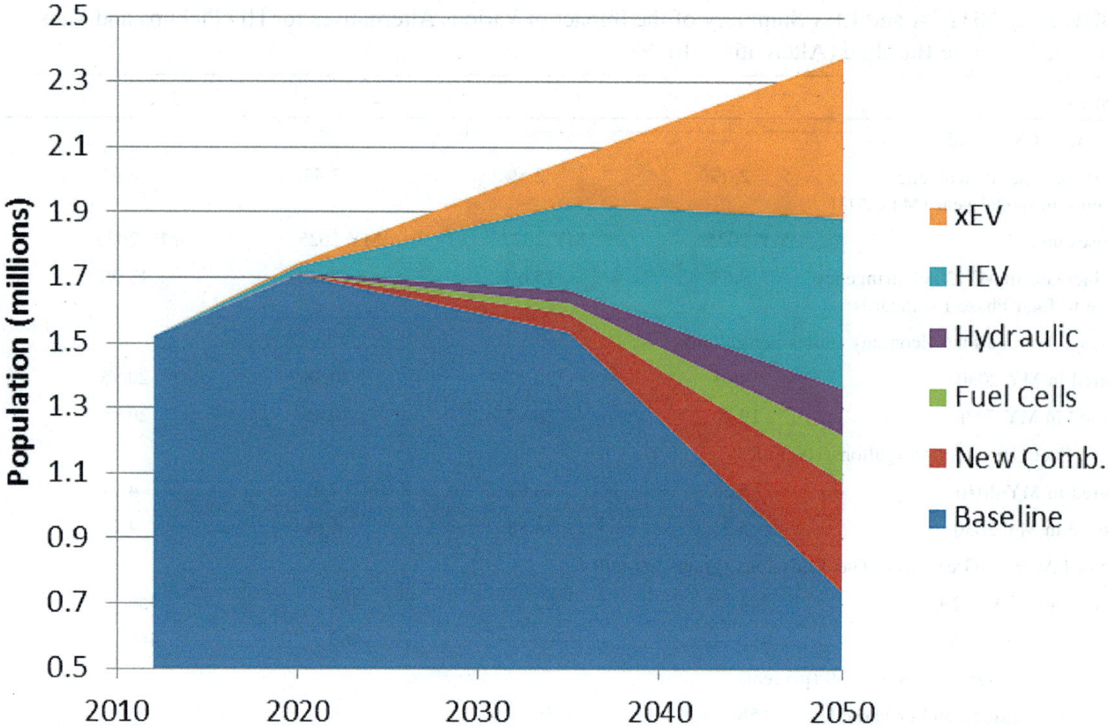

FIGURE 12-1 Technology adoption across all truck categories of CalHEAT roadmap action items by technology group, shown by number of vehicles affected, 2010 through 2050. By 2020, the roadmap action items could result in efficiency and emission improvements in approximately 1.7 million trucks, and by 2050, this impact could increase to 2.4 million trucks.
NOTE: xEV refers to electric trucks, plug-in hybrids, and electrified corridors; HEV includes hybrid electric, electrified auxiliaries, and electric power take-off; hydraulic refers to hydraulic hybrids; new comb. includes alternative power plants and combustion cycles.
SOURCE: CalHEAT (2013).

increases substantially in the post-2020 period, under these assumptions of large reductions in CO_2 emissions, but they are consistent with the results of the RIA in that hybridization begins to play a large role in the post-2020 period (CalHEAT, 2013, p. 28).

The CalHEAT study sees hybridization and electrification applied pervasively across a wide range of vehicle classes, with the exception of over-the-road Class 8 vehicles (CalHEAT, 2013, p. 20).

Given the large role that hybridization is currently expected to play in improving fuel efficiency in the post-2027 world, we treat it as representative of the marginal technology for the next round of efficiency improvements in a Phase III regulation. We assume either that hybridization is this marginal (incremental) technology or that the marginal technology has the same magnitude of cost and fuel efficiency impact as hybridization. Therefore, in the remainder of this section, we analyze the costs and benefits of a Phase III fuel efficiency target using, as a proxy, data on costs and related efficiency improvement possible by hybridization.

12.1.5 Estimating the Marginal Cost of Fuel Efficiency Improvements

Table 12-3 shows estimates of the cost of full and partial hybridization in the heavy-duty van and pickup segments as well as other vocational segments. The capital costs are from Chapter 7 of this report. The capital costs include the direct manufacturing costs as well as the indirect costs such as R&D costs, corporate overheads, and sales and distribution costs. Allowance is also made for a return on capital. In addition, the capital costs are projected forward to the 2027 to 2030 period, taking into account expected learning effects over time as volumes increase.

TABLE 12-3 2027-2030 Estimated Costs, Payback Period, Break-even Fuel and Capital Cost for Hybrid Technology, Waste-Heat Recovery Technology, and Trailer Skirts across Different Vehicle Classes (2013 dollars)

Vehicle Type	Technology	Duty Cycle	Capital Cost[a] (2030)	Mpg[b] (2027)	Efficiency Gain[c]	VMT[d]	Break-even Fuel Price[e]	Break-even Capital Cost[e]	Payback (years)
Class 2b pickup	Stop-start	Urban	425	20.6	4.5%	14,000	$1.98	$619	4.8
		Multipurpose	425	20.6	4.5%	14,000	$1.98	$619	4.8
	Mild hybrid	Urban	2,000	20.6	17.5%	14,000	$2.39	$2,406	5.8
		Multipurpose	2,000	20.6	12.5%	14,000	$3.35	$1,719	8.2
	FHEV	Urban	4,650	20.6	22.5%	14,000	$4.33	$3,093	10.6
		Multipurpose	4,650	20.6	17.5%	14,000	$5.57	$2,406	13.6
Class 4 delivery van	Stop-start	Urban	700	8.4	4.5%	14,000	$1.33	$1,854	2.7
		Multipurpose	700	8.4	4.5%	14,000	$1.33	$1,854	2.7
	Mild hybrid	Urban	4,600	8.4	17.5%	14,000	$2.25	$7,211	4.5
		Multipurpose	4,600	8.4	13.5%	14,000	$2.91	$5,563	5.8
	FHEV	Urban	7,550	8.4	23.5%	14,000	$2.74	$9,683	5.5
		Multipurpose	7,550	8.4	17.5%	14,000	$3.69	$7,211	7.4
Class 8 refuse truck	Stop-start	Urban	1,400	6	4.5%	25,000	$1.06	$4,636	2.1
	Mild hybrid	Urban	10,825	6	16.5%	25,000	$2.24	$16,997	4.5
	FHEV	Urban	14,875	6	23.5%	25,000	$2.16	$24,208	4.3
Urban bus	Stop-start	Urban	1,400	6	4.5%	34,000	$0.78	$6,304	1.6
	Mild hybrid	Urban	10,825	6	16.5%	34,000	$1.65	$23,116	3.3
	FHEV	Urban	14,875	6	23.5%	34,000	$1.59	$32,923	3.2
Class 8 tractor-trailer	WHR	Highway	7,025	6.4	2.5%	120,000	$2.13	$11,589	4.3
	Skirts	Highway	900	6.4	2.5%	40,000	$0.82	$3,863	1.6

[a]Capital cost of technologies: Chapter 7 of this report. The estimates reflect the high-cost estimates in Chapter 7, Table 7-6, for all technology/vehicle combinations except for Class 2b. It is assumed that the experience gained in light vehicles will inform the Class 2b segment and the lower cost estimates will be achieved. The estimates have been rounded to the nearest $25 interval.

[b]Initial mpg:

a. Class 2b pickups—from (Lutsey, 2015, table 1): estimate for MY 2010 of 16.4.
b. Class 3—15.1 (Lutsey, 2015): estimate for MY 2010.
c. Vocational vehicles—from Lutsey (ICCT, June 2010): 6 mpg (from table 5, Lutsey estimate for 2027 gasoline vehicles; note: table 1 in the source document gives an estimate of 2014 mpg for various classes with gasoline and diesel combined).

[c]Efficiency: Midpoint estimate from Chapter 7.

[d]VMT:

a. Class 2b and 3: from NRC Phase I report (2010a).
b. Refuse truck: from Gordon et al. (2003); Greening Garbage Trucks: New Technologies for Cleaner Air, Inform, Inc. 2003. Cited in www.afdc.energy.gov/data/ (accessed December 20, 2019).
c. Urban bus: www.afdc.energy.gov/data (accessed December 20, 2019).
d. Class 8 tractor-trailer: same as b and c.
e. VMT for trailer skirts reflects an estimate of three trailers used per tractor.

[e]Forecasted fuel prices used in payback calculations and in break-even capital cost calculation are from Annual Energy Outlook, 2017, expressed in 2013 dollars.

The capital cost of the technology is compared to the value of the fuel savings. Several measures are presented to analyze the value of the technology. The first, and most commonly used in the industry, is the simple payback period. This is the capital cost divided by annual fuel savings, at an assumed fuel price. The committee has frequently heard from industry representatives that truck manufacturers and purchasers use the payback period as a heuristic for the value of new technology. The payback period captures, in a rough approximation, the cost of capital and the uncertainty associated with the adoption of new technologies. These uncertainties include the benefits of the new technology under real-world and particular duty cycles, volatility in fuel prices, unanticipated maintenance costs, etc. The committee has heard from manufacturers and purchasers that they look for 1.5- to 2-year paybacks or, in other cases, for a payback period that is half the expected ownership period of the first owner of the vehicle.[4] For the Class 2b pickups the payback periods in the table are calculated using the gasoline price projected for 2027 by the Department of Energy in its Annual Energy Outlook (AEO) 2017 of $2.88 per gallon. For the other vehicle classes, the payback periods are calculated using the forecast for the diesel fuel price of $3.52 per gallon.[5]

The second measure shown in the table is the "break-even" fuel price. This measure calculates what the fuel price would have to be in order for the present discounted value of the fuel savings (over 10 years) to equal the initial capital cost. If the forecasted fuel price exceeds the "break-even price," the technology offers savings to a firm adopting the technology. The final measure is the "break-even capital cost," that is, the capital cost that would just equal the present value of the fuel savings at the relevant assumed fuel price. Again this is done for the projected 2027 fuel prices.

The "break-even capital cost" can also be interpreted as the cost any fuel-saving technology would have to meet to be attractive to a private firm for the vehicle segment, duty cycle, and efficiency gain shown. It is therefore a proxy for the targets any technology must meet to be attractive to private manufacturers and purchasers.

The calculations in the table represent the typical use in the segment, as measured by the average vehicle miles traveled (VMT) in the segment. Consistent with the Regulatory Impact Analysis, a 7 percent discount rate for private investment is assumed. The life of the vehicle is assumed to be 10 years.[6] Using these metrics for evaluating the net benefits of a fuel-saving technology are only approximations since they exclude other potential costs such as maintenance, variability in duty cycle, and driver convenience. Nevertheless, they are broadly suggestive of the net economic benefits of the technologies.

Table 12-3 suggests that payback periods for hybrid applications are generally long. The payback periods for technologies other than stop-start, sometimes called micro hybrid, range from 3.2 years in the case of a full hybrid deployed on an urban bus to more than 13 years for a full hybrid on a Class 2b pickup used in a multipurpose duty cycle. Stop-start technology, however, has relatively low paybacks in many applications shown in Table 12-3. Based on what the committee has heard in presentations from industry representatives and survey evidence cited above, other than for stop-start, the estimated paybacks would likely not be attractive to most private purchasers of medium- and heavy-duty trucks or the manufacturers. In other words, to achieve fuel efficiency gains beyond the improvements offered by stop-start technology, the larger gains from mild and strong hybrids likely would not be sufficient by themselves to drive widespread diffusion of the technology, given that the adoption decisions would be made based on relatively short payback period requirements.

The break-even fuel price metric used here calculates what price of fuel would produce fuel savings such that the present value of those savings over a 10-year period equals the stated capital cost. Importantly, this metric

[4] Klemick et al. (2015) confirms this observation in Chapter 13 of this report, citing Roeth et al. (2013), and points out that first ownership periods, while lengthened during the recession of 2008 and 2009, have come down recently to about 4 years, suggesting an acceptable payback period in the 2-year range.

[5] These are prices converted from dollars of 2016 to constant dollars of 2013. The cost estimates in Chapter 7 used here assume that hybrid technology is added to a gasoline-fueled vehicle for Class 2b pickups. For the other vocational vehicle classes shown in Table 12-3, hybrid technologies are assumed added to diesel-fueled vehicles. Therefore, the break-even fuel price is compared to gasoline prices for Class 2b vehicles and to diesel fuel prices for the other vehicle classes shown.

[6] The assumption of a 10-year life, coupled with the assumption of a constant VMT over the 10-year period, also biases the estimates in favor of the technologies shown in Table 12-3. In the case of hybrid vehicles, battery life is the relevant period for calculating the fuel savings. These calculations assume, in essence, that the life of the battery is 10 years and, after that, significant additional capital expenditure would be necessary.

assumes away the uncertainties that private firms face and that they proxy by their use of the simple payback period. In contrast to the payback criterion, the calculated break-even fuel price metric suggests that if the projected capital costs and efficiency gains are consistently realized, the application of mild and full hybrids in the 2027-2030 period will offer cost-efficient fuel savings in many vehicle classes and duty cycle combinations.

Consistent with its relatively quick payback, *stop-start* technology in most applications would make economic sense at fuel prices far below the prices projected by the Energy Information Agency. Comparing the calculated break-even fuel prices to the AEO forecasts of $2.88 and $3.52 for gasoline and diesel fuel, respectively, suggests that *mild hybrids* in most cases also would be economically justified. Only in the case of the Class 2b pickup when used in a multipurpose duty cycle does the break-even price for the mild hybrid exceed $2.88 per gallon.

Again, using the break-even fuel price metric, *full or strong hybrids* present a mixed case. Full hybrids appear to be uneconomic in Class 2b pickups, with break-even fuel prices substantially exceeding the AEO price forecast. In a Class 4 delivery van, the break-even price is below the forecasted diesel price of $3.52 per gallon in the urban duty cycle, but slightly exceeds it in the multipurpose cycle. Full hybrid electric vehicles (FHEVs) appear to meet that economic test in the heavier vocational vehicles, where VMT is high and the vehicles are utilized in an urban duty cycle.

The contrast between the results using payback periods and those using break-even fuel prices highlights a debate in the energy economics literature, often referred to as the "energy paradox." Some would argue that private trucking firms and manufacturers using relatively short payback periods are "leaving money on the table" by not adopting cost-effective fuel-saving technology. Others would argue that engineering analysis does not capture all the costs involved in utilizing the new technology or that the break-even fuel price measures use too long a vehicle life and do not reflect the varieties of duty cycles as the vehicle ages.[7] To the extent that the use of short payback periods, however, represents uncertainties about the costs and performance of new technologies, one would expect private firms to lengthen the acceptable payback period over time if the technologies demonstrate consistent cost and vehicle performance.

The break-even capital cost provides an estimate of how much cost reduction would be necessary to produce a favorable economic case for the technologies and applications shown in Table 12-3. For example, in the case of the full hybrid deployed in Class 2b pickups, capital costs would have to decline by approximately 35 to 50 percent to equate the capital cost to the present discounted value of the fuel savings. In contrast, in the stop-start case in the Class 2b vehicle, the estimated capital cost is about two-thirds of the break-even level, indicating a favorable economic evaluation of the technology. Similarly, many of the other vehicle technology–duty cycle combinations shown in Table 12-3 have estimated capital costs below the break-even capital cost level.

As VMT increases, the value of the fuel savings goes up and the technology becomes more economically attractive. Figure 12-2 shows the sensitivity of the break-even period and the break-even fuel price to VMT. As the graphs show, even at a VMT of 20,000 miles, strong and mild hybrid applications still have relatively high payback periods in the Class 2b and Class 4 van segments, again suggesting that private firms using this metric would not favor rapid adoption of these technologies.[8] When capturing 10 years of potential fuel savings, however, the break-even price of fuel, at 20,000 VMT, falls below the forecasted price of fuel of $3.52 for mild and strong hybrid applications in Class 4 vans. In the case of Class 2b pickups at 20,000 VMT, mild hybrids would be economic at the forecasted fuel price, while break-even fuel prices for full hybrids would still exceed the forecasted fuel price.

In contrast, also on the chart are the curves representing waste-heat recovery (WHR) technology in Class 8 vehicles and trailer skirts for trailers that lower aerodynamic drag. The WHR technology is still in development and has not yet been deployed on vehicles. NHTSA and EPA expect that this technology will gain increased application beginning in 2021 and rise to 25 percent of tractor engines in 2027 (NHTSA and EPA, 2016, p. 2-82). In the post-2027 period, engine manufacturers as well as the committee expect that WHR technology will be deployed (Cummins, 2015). The curve shown in Figure 12-3 therefore represents a view of its potential cost and efficiency, once the obstacles are overcome and the technology is in production. Given the large number of vehicle

[7] This issue is discussed further in Chapter 13.

[8] There is clearly a relationship between VMT per year and life of vehicle. At 20,000 annual miles traveled, the 10-year life of vehicle is less likely an appropriate assumption. The estimates here do not adjust for that and thus are biased toward favorability of the technology.

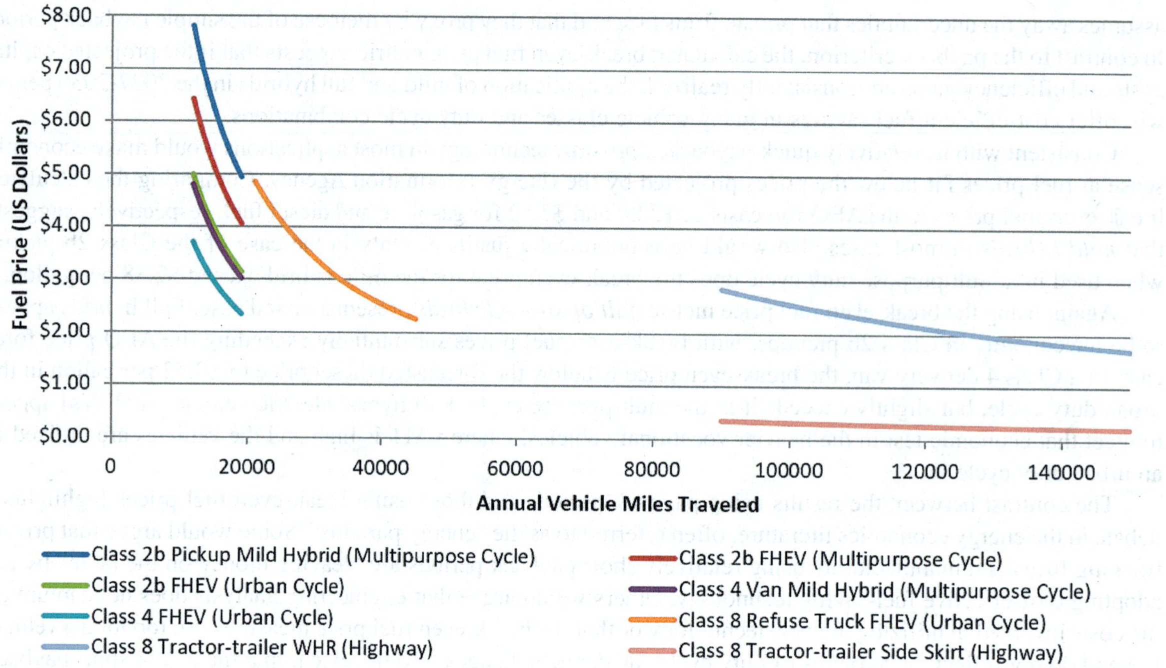

FIGURE 12-2 Break-even fuel price sensitivity to vehicle miles traveled for different vehicle technologies implemented on class 2b-8 vehicles.
SOURCE: Committee calculations based on Table 12-3.

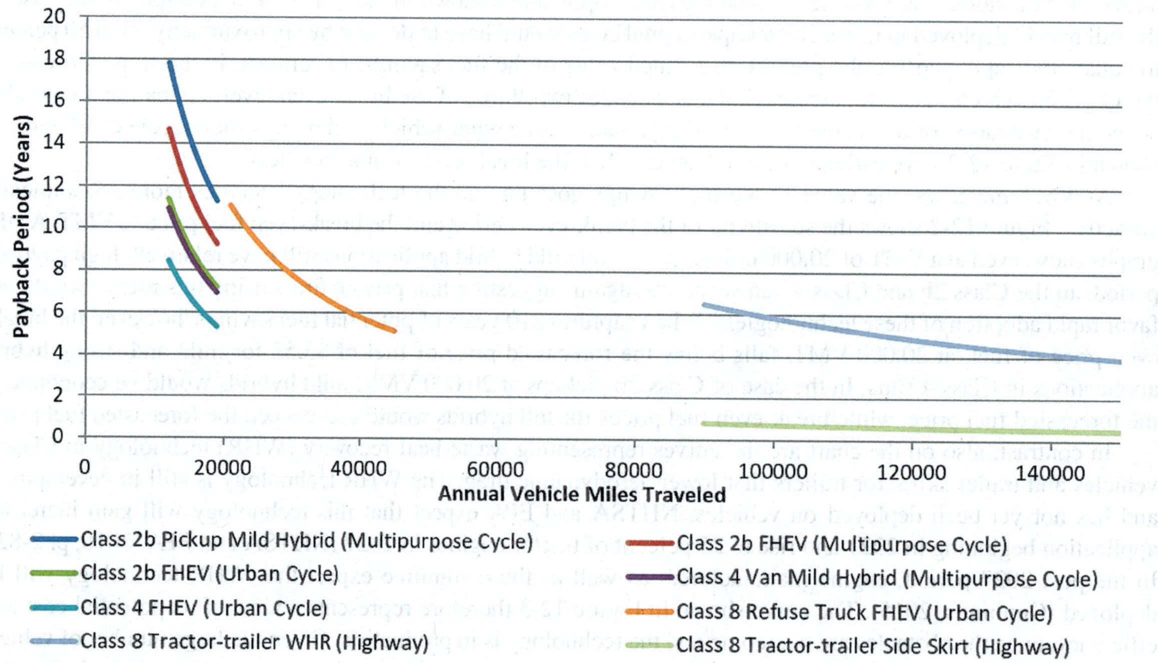

FIGURE 12-3 Payback period sensitivity respect to vehicle miles traveled for different vehicle technologies implemented on class 2b-8 vehicles.

miles traveled by Class 8 long-haul vehicles, this technology does offer potentially large absolute fuel savings per vehicle as well for the industry as a whole. The large fuel savings per vehicle, in turn, lead to a low break-even fuel price and a break-even capital cost significantly above the estimated potential cost. Nevertheless, at almost 5 years, the simple payback is still higher than the acceptable payback period to private firms, as discussed above. Trailer skirts, on the other hand, are widely deployed on trailers today and have a very low break-even price of fuel and a very quick payback.

As discussed above, the private fuel savings are only part of the benefits to society. In addition to these private savings, reductions in fuel use lower the environmental, health, and national security externalities. In order to reflect these externalities their cost can be added to the forecasted price of fuel. If the actual price of fuel plus the value of the externalities equals or is greater than the break-even price of fuel, the technology is socially efficient. Even if the private decision maker would not benefit from adopting the technology, society would still gain by forcing the technology adoption through regulation or through taxation.[9]

The following sections of this chapter consider how to value the externalities associated with fuel production and consumption. The median estimated value of reducing carbon emissions, based on U.S. government estimates and as discussed in the section "The Benefits of Reduced CO_2 Emissions" below, is the equivalent of about $0.50 per gallon in 2013 dollars, using a 3 percent social discount rate.[10] The value of the avoided health damages associated with lower criteria pollutant emissions offset partially by increased congestion and noise costs, based on the estimates in the RIA, is approximately $0.31[11] using a 3 percent discount rate. Using a 7 percent discount rate the value is approximately $0.16. The national security cost as discussed in Section 12.2 below has an estimated median value of about $0.17 per gallon. The break-even price of fuel can be compared to the social cost of the fuel by adding the value of the externalities to the forecasted price of fuel. Taking the median value of the cost of carbon of $0.50 per gallon, the median estimate of the national security premium of $0.17 per gallon and the estimate of health benefits using the 3 percent discount rate of $0.31 per gallon results in a total external cost of $0.98 per gallon. The forecast of fuel prices of $2.88 and $3.52 for gasoline and diesel fuel, respectively, results in social costs of fuel of approximately $3.86 and $4.50.

A comparison of the estimates of break-even fuel prices in Table 12-3 to the estimated social cost of fuel, suggests that hybridization, is likely to be the "next up" technology in the post-2027 time frame in all the vehicle segments and duty cycles shown in Table 12-3, if the vehicles attain at least the level of VMT shown in the table. Only in the case of full hybridization in the Class 2b pickup does the break-even fuel price exceed the estimated social cost of fuel. It is important to restate that this analysis ignores the variation in duty cycles within each vehicle segment shown and that the estimates of social costs, as discussed in the sections that follow, vary considerably. Nevertheless, the results here are suggestive of the technology likely to be used to meet any potential increase in fuel efficiency standards.

---

[9] As noted earlier in this section, this analysis uses average duty cycles and abstracts from some potential costs such as increased maintenance that society may incur by forcing the technology. The variability in duty cycles also means that forcing the technology does risk higher net social costs in some uses.

[10] Section 12.2 discusses how the social cost of carbon is measured. Shown in Table 12-4 is the median estimate of the Interagency Working Group (IWG) of $46 per tonne in 2025 and $50 in 2030. The midpoint of $48 is used here to proxy for the cost in 2027. These estimates were in 2007 dollars, so they are converted to 2013 dollars using a factor of 1.18. The cost per tonne of CO_2 is converted to an equivalent cost per gallon of gasoline using the converter found at http://www.icbe.com/carbondatabase/dollarspertontocentspergallon.asp. Note also that the social cost of carbon is the global cost. In other words, the cost of emitting one tonne of carbon leads to the estimated damage globally. The effect on the United States is estimated by the IWG to be from 7 to 23 percent of the global cost. It is, as discussed in Section 12.2, a policy choice whether to consider the global cost or just the impact on the United States. In the last rulemaking NHTSA and EPA chose to use the global cost. A March 28, 2017, executive order issued by President Trump rescinds the Obama administration guidance on the use of the social cost of carbon in rulemaking and relies instead on previous guidance. One result of this order will be to estimate costs and benefits on a U.S.-only basis.

[11] The value of the other non-CO_2 pollutants, discussed in Section 12.2.1, is calculated most appropriately on a per-VMT basis. A proxy measure is calculated here by using the total value of avoided health effects offset by congestion and noise effects as estimated in NHTSA and EPA's RIA (Table 8-30). The net present discounted value of these non-CO_2 externalities ranges from $12.2 billion at a 7 percent discount rate to $24.1 billion at a 7 percent discount rate using NHTSA and EPA's dynamic baseline. Dividing these totals by the 77.7 billion gallons of fuel saved yields an estimate of average non-CO_2 externalities of $0.16 to $0.31 per gallon of fuel saved (RIA, Table 11-20). (A slightly different baseline yields a range of $0.18 to $0.34 per gallon.) In Section 12.3 of this report the national security premium is discussed. As shown in that section, the median estimate by NHTSA and EPA in the Phase II rulemaking was the equivalent of about $0.17 per gallon.

12.1.6 Findings

Finding: Waste-heat recovery used in Class 8 over-the-road vehicles potentially offers significant cost-effective fuel savings in the post-2027 period. Expected progress in WHR technology suggests that in the post-2027 period this technology will see increasing penetration in Class 8 tractor-trailers.

Finding: In 2027 and beyond, stop-start technology applications are expected to have payback periods that would make them attractive in many applications to private firms. The greater fuel savings, however, that are possible with the application of mild and strong hybrids are expected to be less attractive to private firms, given their relatively long payback periods and the weight placed on payback periods by private firms.

Finding: If the anticipated cost reductions and efficiency gains are in fact consistently achieved, in the post-2027 period increases in fuel efficiency beyond the Phase II rule are likely to be met with increasing applications of hybrid technologies including stop-start, mild, and strong hybrids. The technology penetration will vary depending on vehicle class and duty cycle.

Finding: Considering 10-year fuel savings, break-even fuel prices are below the social cost of fuel in most hybrid applications. These results suggest that hybridization is a technology that is "next up" for meeting more stringent post–Phase II standards. Only in Class 2b does the break-even fuel price for full hybrids exceed the forecasted social cost of fuel.

Recommendation 12-1: Since hybrid technologies and WHR potentially could play increasing roles in achieving reductions in fuel consumption in the post-2027 period, developments in the cost and efficiency of these technologies should be monitored and be included in a formal interim review of fuel consumption standards.

12.2 THE BENEFITS OF REDUCED CO_2 EMISSIONS

CO_2 and most other GHGs cause harm not through direct exposure of humans or wildlife to the pollutant but, rather, because their accumulation changes the energy balance in the atmosphere and the chemistry of the ocean. The change in the earth's energy balance leads to changes in climate worldwide, including changes in temperature and precipitation, and to melting of sea ice sheets and sea level rise. In the ocean, their accumulation leads to ocean acidification. These changes have many consequences that affect the well-being of humans and natural ecosystems worldwide. In some cases, the effects are beneficial; often they are harmful.

Because GHGs emitted anywhere on the earth mix globally in the atmosphere and are then introduced into the ocean through interface with the near-surface atmosphere, they are different from other air pollutants including criteria air pollutants: The emission of a GHG anywhere on earth affects global climate and ocean acidification. A second distinctive feature of GHGs is that they are what is known as *stock* pollutants: The harm comes from the accumulated stock of GHGs in the atmosphere and the ocean. Conventional air pollutants are generally *flow* pollutants: The harm from them comes, essentially, contemporaneously with the release of emissions and they do not linger. With GHGs, as a stock pollutant by contrast, harm continues from the time of emission until the stock of pollutant has dissipated, which can take centuries in the case of CO_2.[12] This has two implications for the assessment of the damages. First, for an emission in any given year, the effect on global climate and the resultant consequences play out over time until that emission has entirely decayed. Second, the emission interacts in the atmosphere and the ocean with *subsequent* emissions and its impact on climate is influenced by those later emissions. Therefore, accounting for the consequences of GHG emissions requires dynamic modeling of the accumulation and dissipation of GHGs in the atmosphere and the ocean as well as the resultant evolution of the climate system. This is done through what are known as Integrated Assessment Models (IAMs). These are mathematical models that track, in sequence, the emission of GHGs as a function of economic activity; the accumulation and

[12] The persistence of GHGs varies greatly. The warming impacts of ozone or contrails last only days or months. Those of methane last for 20-30 years. Those of CO_2 persist for hundreds of years. Some GHGs persist even for thousands of years.

decay of GHGs in the atmosphere, the surface (or upper ocean), and the deep ocean; the link from atmospheric concentration to global temperature via radiative forcing;[13] and the impact of temperature change on economic activity and human well-being, measured in monetary terms as the equivalent reduction in the gross domestic product (GDP).

Reduced CO_2 emissions are valued using the *social cost of carbon* (SCC). This is a monetary measure of the discounted present value of the net damage resulting over time from the emission of an additional unit of CO_2 in a given year, estimated by means of an IAM. We use here the estimate of the SCC developed by the federal government's IWG in 2010 (IWG, 2010), as subsequently updated in 2013 (IWG, 2013).[14]

The IWG ran the three main IAMs used in the literature[15] and averaged the results across them. The models were run over the period through 2300.[16] Rather than using the emissions projection native to each IAM, the IWG standardized the emission projections across the IAMs by using a common set of projections taken from an external source.[17,18] Five alternative projections were used. Four assume business-as-usual (BAU) emissions through 2100, with a leveling off and downturn in emissions between 2100 and 2200; the fifth alternative assumes that emissions are reduced ("down-scaled") during the period 2000-2100 to meet a target of 2 degrees Celsius (°C) warming by 2100 and then remain roughly constant thereafter. The results with each of the emissions projections were weighted equally, thereby assigning a 20 percent probability to the low (2°C) projection and 80 percent collectively to the BAU projections.

Given a trajectory of emissions, a key parameter in translating these emissions to subsequent changes in temperature is the *climate sensitivity*. This parameter measures the long-run equilibrium increase in the global annual average surface temperature resulting from a doubling of atmospheric CO_2 relative to the pre-industrial level.[19] It reflects both the direct warming from the increase in radiative forcing and also the warming resulting indirectly from feedback processes triggered by that warming, such as melting of the permafrost.[20] Uncertainties in the value of the climate sensitivity have received considerable attention in the scientific literature.[21] To account for this, the IWG employed a probability distribution for the value of the climate sensitivity, which had the effect of making the

---

[13] In tracking the linkage from emissions to changed global temperature, the IAMs offer a simplified and reduced-form representation of the vastly more detailed analysis found in Global Climate Models (GCMs). GCMs project precipitation as well as temperature. Because of their highly simplified representation of the carbon cycle the IAMs project only temperature. The GCMs project temperature on a spatial scale consisting of grids of about 175 km. The IAMs project temperature on the spatial scale of large groups of countries (FUND, PAGE) or the entire globe (DICE). Also, while the GCMs operate on something like an hourly time scale, the IAMs operate on an annual time scale.

[14] There was subsequently a small correction to the 2013 analysis in IWG (2015).

[15] These are the DICE model, developed by Professor William Nordhaus in 1990; the PAGE model, developed by Chris Hope in 1992; and the FUND model, developed by Richard Tol in the early 1990s. For its 2010 analysis, the IWG used the 1999 version of DICE (Nordhaus and Boyer, 2000), the 2002 version of PAGE (Hope, 2006), and version 3.5 of FUND (Anthoff and Tol, 2012). Following updating of these models by their developers, the 2013 IWG analysis used DICE 2007 (Nordhaus, 2007), PAGE 2009 (Hope, 2011), and FUND 3.8 (Anthoff and Toll, 2014).

[16] The native time horizon in PAGE is 2200; that in DICE is 2595; that in FUND is 3000. The IWG made 2300 a common time horizon across all three models. The IWG did not alter the IAM's native time steps, which are annual in FUND and decadal in DICE; the time steps in PAGE are decadal through 2060 and 20 years thereafter.

[17] The source was a model intercomparison exercise conducted by the Stanford Energy Modeling Forum, EMF-22 (Clarke et al., 2009), designed to assess the economic cost (and feasibility) of meeting a target of 2°C by 2100. The models compared were all versions of an IAM but, with one exception, they stop at the projection of future warming and they lack a damage function. The IWG used the four best known of the models participating in the EMF exercise.

[18] While FUND and PAGE use exogenous projections of emissions, emissions are endogenous in DICE. DICE is a modified model of optimal economic growth in which, in each period, there is choice of how much of what is produced in that period to consume, how much to invest (i.e., to allocate to capital formation for use in future production), and how much to allocate to emission abatement. The allocation to investment determines the future level of production which, in turn, determines future emissions. The IWG disabled the optimizing allocation of consumption and investment in DICE and made emissions exogenous, as in FUND and PAGE.

[19] This is distinct from the *transient* climate response, which determines the response of temperature within a fixed short-term period, and which also is represented in the IAM carbon cycle.

[20] The climate sensitivity includes feedback processes that occur over the short to medium term (up to 100 to 200 years), but it does not include long-term feedbacks which occur on a time scale of many hundreds to thousands of years.

[21] A significant part of the uncertainty arises from the feedback processes and the warming that they generate.

SCC estimate probabilistic.[22] The IWG used a well-known probability distribution for the climate sensitivity, that of Roe and Baker (2007), and applied it across all three IAMs.[23] Ten thousand draws of the value of the climate sensitivity were taken from this distribution and input into each of the three IAMs. Given an emission scenario, for each draw the IAM was then run through year 2300, generating a particular trajectory of warming and damages. With three IAMs and five emission scenarios, this generated 150,000 trajectories of warming and damages.

The IWG estimated the SCC associated with an increment in emissions occurring in 2010, 2020, 2040, and 2050.[24] In each case, the procedure was to introduce a one-time pulse in emissions in the reference year, leaving later emissions unchanged, rerun the model, and, for each period following the pulse, calculate the additional damages in that period compared to the base run of the same version of the model without the pulse in emissions. For each reference year this generated 150,000 trajectories of incremental damages associated over time with the pulse in emissions. The trajectories of incremental damages were converted to a discounted present value (an SCC value) using three alternative discount rates: 2.5, 3.0, and 5.0 percent. The result was 450,000 estimates of the SCC, corresponding to the three IAMs, the 10,000 values of the climate sensitivity, the five alternative emissions scenarios, and the three alternative discount rates. The results were summarized by reference year and by discount rate, for the three IAMs combined.[25] The IWG tabulated the average SCC value at each of the three discount rates, together with the 95th percentile value at the 3 percent discount rate (the results are reproduced in Table 12-4). For its central estimate, the IWG used the average SCC value across the three IAMs at the 3 percent discount rate.

The IWG's 2010 and 2013 estimates of the SCC have been used in numerous rulemaking proposals by federal agencies, some of which provided opportunities for public comment. In addition, following the publication of the IWG's 2013 estimate, the Office of Management and Budget (OMB) put out a formal request for comments on the methodology used by the IWG. In July 2015, the IWG issued a formal response to these comments (IWG, 2015).

The issues raised by the comments fall mainly into two groups: modeling choices that should be considered as policy decisions being made by the U.S. government, and decisions that represent an exercise of judgment with regard to technical details of the empirical analysis.

TABLE 12-4 Revised Social Cost of CO_2, 2010-2050 (in 2007 dollars per metric ton of CO_2)

Discount Rate Year	3.0% Avg
2010	31
2015	36
2020	42
2025	46
2030	50
2035	55
2040	60
2045	64
2050	69

SOURCE: IWG (2015, p. 3).

[22] The climate sensitivity is modeled as a random variable in FUND and PAGE, along with several other parameters. DICE, by contrast, is entirely deterministic.

[23] This probability distribution has a median value of 3.0 and a mean of 3.5. With this distribution, there is only a 1.3 percent chance that the value of the climate sensitivity is below 1.5°C, and only a 10 percent chance that is it above 5.86. This is broadly consistent with the values used for DICE (a fixed value of 3.0) and for PAGE and FUND (both probabilistic with a mean of 3.0), and it is consistent with the findings of both the Fourth and Fifth Intergovernmental Panel on Climate Change Assessment Reports.

[24] The IWG used linear interpolation to generate SCC estimates for the in-between years of 2015, 2025, 2035, and 2045.

[25] Compared across reference years, the SCC estimate increases (in constant dollars) over time because, as time passes, global temperature is higher and there is a greater accumulation of CO_2 in the atmosphere. This causes both the warming impact of a pulse in emissions to be larger and the extra damage from a given increment in warming to be larger.

The policy judgments include the decision to consider global impacts rather than just impacts in the United States, the decision to consider impacts through 2300, and the choice of 2.5, 3, and 5 percent as discount rates when calculating the SCC.

The IWG estimated that damages in the United States might amount to between 7 and 23 percent of the global climate impact reflected in its SCC estimate (IWG, 2010, p. 11). The IWG justified its adoption of a global perspective because of the global externality caused by U.S. emissions, which contribute to damages around the world. Consequently, "to address the global nature of the problem, the SCC must incorporate the full (global) damages caused by GHG emissions" (IWG, 2010, p. 10). It noted that accounting for the global impacts can encourage reciprocal action by other nations. It further observed that adverse impacts of climate change on other countries can have spillover effects on the United States, "particularly in areas of national security, international trade, public health and humanitarian concerns." Moreover, "if all countries acted independently to set policies based only on the costs and benefits of carbon emissions, it would lead to an economically inefficient level of emission reductions which could be harmful to all countries, including the US" (IWG, 2015, p. 32).

The IAMs used by the IWG have different default time horizons: 2200 (Policy Analysis of the Greenhouse Effect [PAGE]), 2595 (Dynamic Integrated Climate-Economy Model [DICE]), and 3000 (Climate Framework for Uncertainty, Negotiation and Distribution [FUND]). The IWG considered the year 2200 too short a time horizon because it could omit a significant fraction of damages, and chose to run the three IAMs through 2300. Some commenters argued for a shorter time horizon, say 2100, due to the uncertainty associated with projecting impacts out to 2300. IWG (2015, p. 29) rejected this suggestion as inappropriate given the long atmospheric lifetime of CO_2.

The longer the time horizon, the greater the role of the discount rate in estimating the SCC.[26] OMB's Circular A-4 identifies 3 and 7 percent as appropriate discount rates for regulatory impact analysis conducted pursuant to Executive Order 12866. The 7 percent rate is an estimate of the before-tax, real opportunity cost of capital for business borrowers in the United States. The 3 percent rate is an estimate of the social rate of time preference (the consumption rate of interest) for consumers in the United States. The IWG found that the consumption rate of interest is the relevant discount rate for the SCC. IWG (2015) notes that most regulatory impact analysis is conducted over a time frame in the range of 20 to 50 years. The SCC involves a far longer time horizon. OMB Circular A-4 recognizes that special ethical considerations arise when comparing benefits and costs across generations. "Although most people demonstrate time preference in their own consumption behavior, it may not be appropriate for society to demonstrate a similar preference when deciding between the well-being of current and future generations. … Estimates of the discount rate appropriate in this case, from the 1990s, ranged from 1 to 3 percent" (IWG, 2015, p. 21). The central value, 3 percent, is consistent with estimates provided in the economics literature and OMB's Circular A-4 guidance for the consumption rate of interest. The upper value of 5 percent represents the possibility that climate damages are positively correlated with market returns, which would suggest a rate higher than the risk-free rate of 3 percent. Additionally, this discount rate may be justified by the high interest rates that many consumers use to smooth consumption across periods. The low value, 2.5 percent, is included to incorporate the concern that interest rates are highly uncertain over time. It represents the average rate after adjusting for uncertainty using a mean-reverting and random walk approach as described in Newell and Pizer (2003), starting at a discount rate of 3 percent. Furthermore, a rate below the riskless rate would be justified if climate investments are negatively correlated with the overall market rate of return. Use of this lower value also responds to the ethical concerns discussed above regarding intergenerational discounting (IWG, 2015, p. 22).

Other issues raised in the comments include the projection of future GHG emissions, the adequacy of the IAM damage functions used to assess the damage from future warming, and the averaging of the SCC estimates across the three IAMs.

Some commenters argued that the emission projections were too pessimistic and understated the future abatement that would occur; others argued that the projections were too optimistic. In response, IWG (2015) acknowledged the uncertainty in these projections: "The chosen scenarios capture a wide range of potential future states of the world, but were not intended to represent a comprehensive accounting of the full range of uncertainty, and therefore it is possible that future outcomes will fall outside of this range" (IWG, 2015, p. 20).

[26] The present value today of $10,000 in 2300, discounted at 2.5, 3, and 5 percent, amounts, respectively, to only $8.78, $2.19, and $0.01.

Some commenters argued that it is not possible to meaningfully assess how climate impacts occurring 200 or 300 years in the future will be valued by society. Some suggested that the IAM damage functions are arbitrary and unreliable. In response, the IWG acknowledged that none of the three IAMs fully incorporates all climate change impacts, whether positive or negative. Using an ensemble of three different models, it pointed out, "was intended to, at least partially, address the fact that no single model includes all of the impacts" (IWG, 2015, pp. 9-10).

Some commenters suggested that the IWG should have used the median rather than the mean of the distribution of SCC estimates, which would produce a lower value. Using the median effectively discards the tails of the distribution. The IWG viewed this as inappropriate in the context of climate change where tail risks—risks associated with low probability but highly damaging outcomes—are of the essence (IWG, 2015, p. 26).

Finding: It is a government policy decision as to which impacts should count toward the social cost of carbon, i.e., the impact of climate change on the United States versus the global impact. It is also a policy decision over what time period the impacts should be considered (through 2300, or later, or earlier), and how impacts occurring at different points in time should be discounted (e.g., at 3 percent or some other rate). The U.S. government's choice of time frame is generally consistent with the time span of the IAMs used. Its choice of a discount rate is consistent with estimates in the literature.

12.2.1 Additional Environmental and Health Costs and Benefits from Different Fuel and Technology Strategies

We discussed above the methods used by NHTSA and EPA to value the climate change externalities associated with the reductions in carbon dioxide emissions resulting from the fuel savings from the use of more efficient MHDV technologies and strategies. Additionally, there are two important environmental costs or benefits associated with the likely changes in technologies, fuels, and strategies that will be induced by the Phase II standards. First, there are changes in emissions from criteria air pollutants—and resulting changes in health and environmental damages induced by those emissions. Second, there are also changes in levels of damages from other greenhouse gases besides CO_2.

12.2.2 Costs and Benefits from Changes in Health and Environmental Impacts from Criteria Air Pollutants

The use of MDHVs powered by fossil fuels (in particular gasoline and diesel) results in emissions of criteria air pollutants, such as nitrous oxides (NO_x), particulate matter (PM), sulfur oxides (SO_x), volatile organic compounds (VOCs), and air toxic emissions. A key distinction between these pollutant emissions and greenhouse gases is that if the source of emissions is removed, the overall concentration of these pollutants decreases rapidly, whereas, as was explained above, the lifetime of GHGs is many decades to centuries and can lead to effects that will be faced by future generations.

In a life-cycle perspective, it is worthwhile pointing out that the upstream emissions associated with the production and transportation of such fuels will also likely change—but over the lifetime of these vehicles, the use phase is by far the largest contributor in terms of emissions. (See, for example, the NRC [2010b] report *Hidden Costs of Energy: Unpriced Consequences of Energy Production and Use*. Note that while the use phase is the largest contributor to emissions for MHDVs, the same is not true for light-duty vehicles [LDVs], where the use phase constitutes about only one-third of the damages.)

The transportation sector overall (including both LDVs and MHDVs) is a very large contributor to some of these criteria air pollutant emissions: the 2015 Transportation Energy Data Book (Davis et al., 2015) reports that in 2014 the transportation sector accounted for 54 percent of total U.S. carbon monoxide emissions, 58 percent of total U.S. NO_x emissions, and 23 percent of total U.S. VOC emissions. The transportation sector is a smaller contributor to overall national PM emissions (accounting for less than 3 percent of national PM_{10} emissions and less than 6 percent of $PM_{2.5}$ emissions) and also a small contributor to SO_2 emissions (about 2 percent of total

U.S. SO_2 emissions are attributable to the transportation sector). Also, just considering the transportation sector, heavy-duty vehicles accounted for about 50 percent of total NO_x highway emissions, 55 percent of PM_{10} highway emissions, and 66 percent of $PM_{2.5}$ highway emissions in 2005.[27]

Depending on where these emissions occur and their exposure pathways, they may, in turn, lead to serious health and environmental impacts, such as increased human mortality and morbidity. According to report *Hidden Costs of Energy* (NRC, 2010b), in year 2005 the vehicle sector overall accounted for \$56 billion in health and other non-climate-change damages, with \$20 billion of these coming from heavy-duty vehicles.

As a result of the mix of strategies, technologies, and fuels used under the Phase II standards the level of emissions from these pollutants and their ambient concentration will change. The emissions of these pollutants have health and environmental effects that are important externalities to consider as part of the overall cost-benefit analysis of the Phase II standards. That task, however, is challenging because—unlike GHG emission damages—the health damages associated with the emissions of criteria air pollutants from MHDVs will be very dependent on where these emissions occur, the ambient baseline concentration of those pollutants, and who is exposed to them. Estimates of the damages associated with these mobile sources generally rely on integrated assessment models that relate the emissions sources to dispersion and reaction models, and to exposure to population, dose-response functions, and a monetized valuation of the health effects. All these modeling and estimation steps have deep uncertainties, and nationally disaggregated data on actual emissions and where these occur are lacking.

Indeed, almost all of the existing studies have focused on estimating the damages associated with the light-duty vehicle fleet, and estimates for the MHDVs are lacking (with the possible exception of NRC [2010b], in which point estimates for the air pollutant damages were provided for MHDVs).

Despite the deep uncertainties, the existing studies often use only point estimates for these damages. This is problematic, as a few key and uncertain assumptions regarding the monetized valuation of the health effects can completely drive the results (for example, the values assumed for the statistical value of life, or the discount rate).

The NRC report *Hidden Costs of Energy* attempted to estimate these damages, and found that for the medium- and heavy-duty vehicles, depending on the vehicle and fuel considered, damages ranged from about \$0.03 to \$0.10 per VMT in 2005, and would be reduced to less than \$0.03 per VMT for all studied vehicle technologies in 2030 in light of ongoing or proposed policies (see Figures 12-4 and 12-5).

The 2016 RIA includes the assessment of changes in damages from carbon monoxide (CO), fine particulate matter ($PM_{2.5}$), SO_x, VOCs, NO_x, benzene, 1,3-butadiene, formaldehyde, acetaldehyde, and acrolein. Importantly, the RIA includes vehicle emissions (i.e., *downstream* emissions) as well as emissions from fuel production and distribution (i.e., *upstream* emissions, using GREET as the source of information for these, updated to account for the most recent National Emissions Inventory). The RIA includes different strategies to model the emissions impacts of the different MHDV classes: for tractor-trailers and vocational vehicles, the EPA Motor Vehicle Simulator model is used, whereas for heavy-duty pickups and vans they used two different methods to estimate the emissions (we refer the reader to the RIA for additional details on these).The RIA then models the emissions consequences in terms of changes in damages using an air quality model, using year 2040 for their analysis where a reference scenario in which no standard is implemented is compared to a scenario with the standards. The overall health and environmental effects from these pollutants associated with the standards are found to be driven by (i) the increased use of auxiliary power units, (ii) a decrease in emissions from the upstream fuel production and distribution, and (iii) a reduction in the damages due to a reduction in emissions associated with "the improved engine efficiency, aerodynamics and tire rolling resistance and absolute changes in average total running weight of the vehicles" (NHTSA and EPA, 2016).

[27] These figures are for year 2005. These estimates are computed using the 2015 Transportation Energy Data Book from the Oak Ridge National Laboratory. More recent data with a breakdown at the medium- and heavy-duty disaggregation level is not reported. Nonhighway emissions are also not reported at the heavy-duty level.

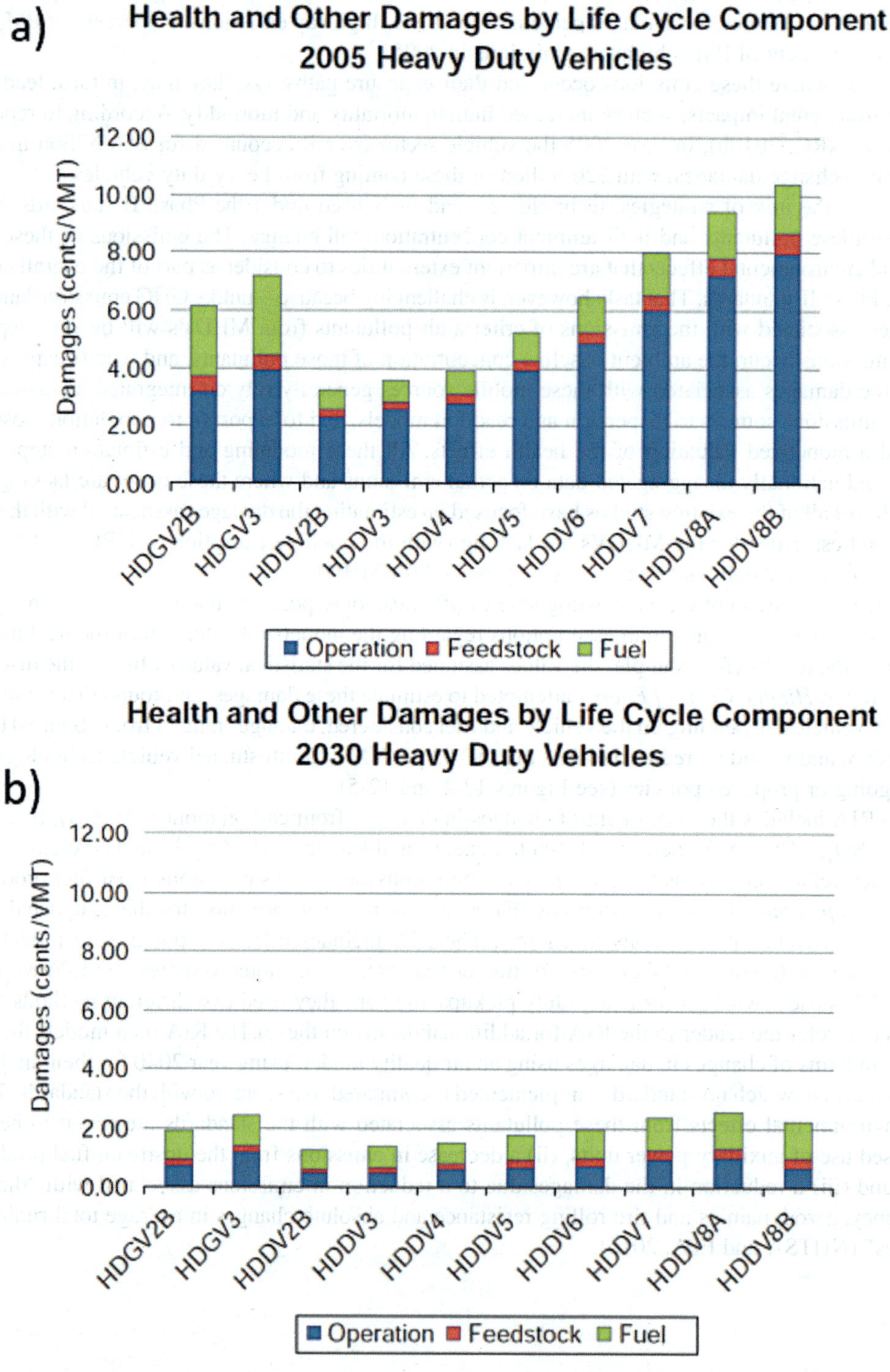

FIGURE 12-4 Aggregate operation, feedstock, and fuel damages of heavy-duty vehicles from air-pollutant emissions (excluding GHGs) in cents per VMT. Panel (a) shows the estimated damages in 2005 and panel (b) shows the estimated damages in 2030.
NOTE: HDDV, heavy-duty diesel vehicle; HDGV, heavy-duty gasoline vehicle.
SOURCE: NRC (2010b).

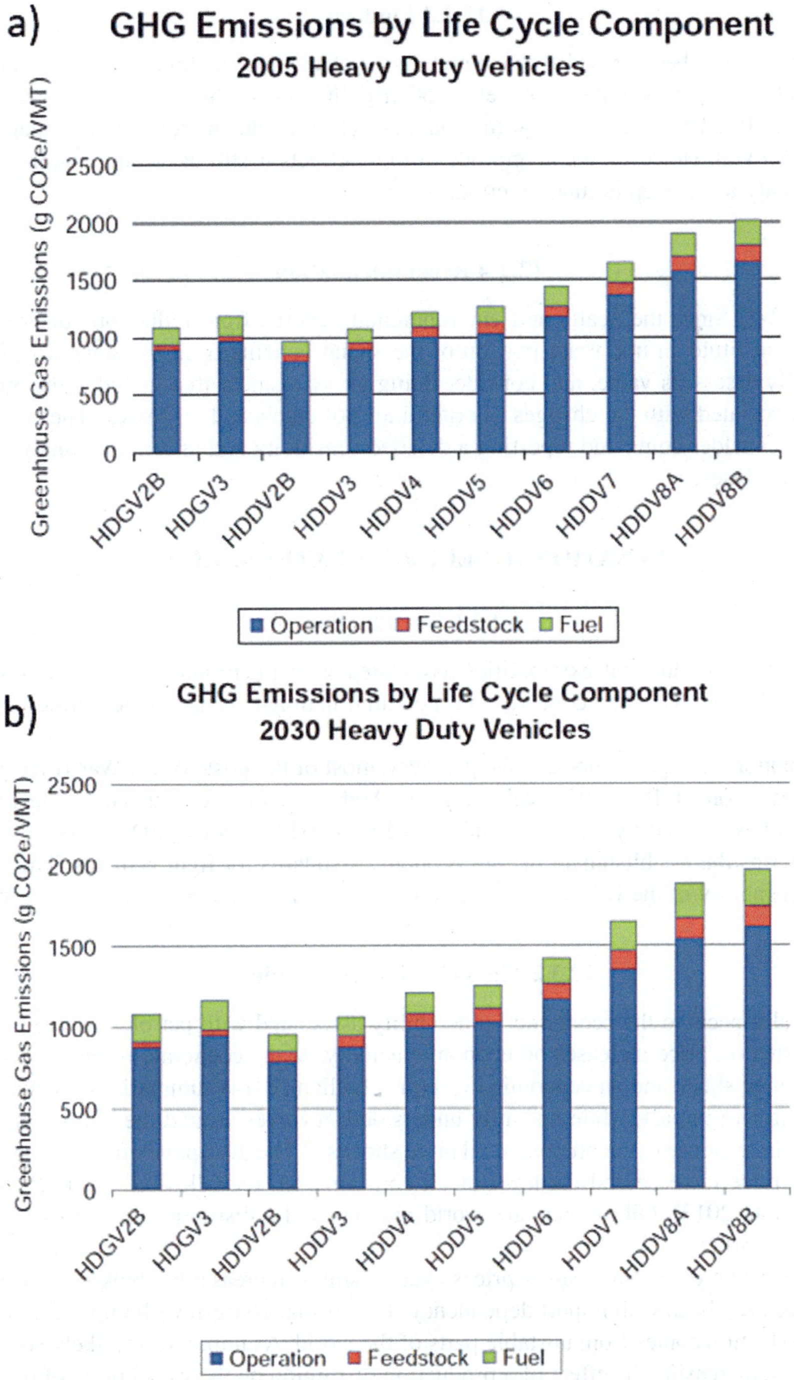

FIGURE 12-5 Aggregate operation, feedstock, and fuel damages of heavy-duty vehicles from GHG emissions (cents per VMT): (a) estimated damages in 2005; (b) estimated damages in 2030.
SOURCE: NRC (2010b).

12.2.3 Finding

Finding: Some attention has been provided to estimating the changes in health and environmental consequences due to criteria air pollutants and air toxics, as well as other pollutants, to the Phase II standards in the 2015 draft RIA and in the final 2016 RIA, suggesting that the changes in health and environmental damages associated with the standards are nontrivial. However, the magnitude of these effects is still uncertain, and the analysis under the final RIA produces only a first step in that direction.

12.2.4 Recommendation

Recommendation 12-2: Since the health and environmental benefits from reductions of criteria air pollutants and air toxics may constitute an important portion of the social benefits from the standards, NHTSA and EPA should more carefully assess its value, and consider doing an assessment that includes the monetization of the costs and benefits associated with the changes in criteria air pollutants and air toxics. The committee also urges NHTSA and EPA to consider doing and reporting a detailed sensitivity and uncertainty analysis for these results of ongoing benefits analyses.

12.3 NATIONAL SECURITY EXTERNALITIES

12.3.1 Introduction

In addition to the environmental externalities associated with oil production and consumption, since the oil crisis of 1973-1974 there has been concern with potential national security externalities associated with oil consumption.

Oil supply disruptions and price shocks have preceded most of the post–World War II recessions (Hamilton, 2009). It was the recession of 1973-1974 following the Arab oil embargo that led to the Energy Policy and Conservation Act (EPCA) passed by Congress and signed by President Ford in December of 1975. Among the provisions of EPCA was the establishment of fuel economy standards for light vehicles. The motivation behind the law was the vulnerability of the U.S. economy to oil supply disruptions originating in the Middle East.

12.3.2 The Oil Security Premium

There are several aspects to the economic vulnerability associated with petroleum use. First is the negative impact of unanticipated oil price increases on economic activity. Many economic researchers have explored the link between an oil price shock and an economic downturn. Killian (2014) summarizes the most recent research. The source of the negative impact, while not fully understood, revolves around the adjustment costs consumers and businesses incur in response to unanticipated oil price shocks.[28] The disruptive effect on the economy depends on the extent of the price increase and the level of oil consumption, regardless of where the oil was produced (Brown and Huntington, 2013). Oil markets are world markets and a disruption anywhere in the world affects prices throughout the market.

In spite of the fact that disruption causes prices everywhere to increase, the bulk of concern over the years since the first oil shock has been with import dependency. The special concern with imports arises because a large fraction of world production comes from unstable parts of the world. As imports rise, likely so does production in these unstable regions, increasing the effect of a production disruption on the world price of oil. Furthermore, the larger is the imported fraction of consumption, the larger is the transfer of economic resources abroad in the event

[28] Killian (2014, p. 144) points out that adjustment costs apply to large price decreases as well as to price increases. However, price decreases result in gains to consumers and resulting increases in consumer spending that serve as at least a partial offset to the adjustment costs and their effect on GDP.

of a price shock. It is for these reasons that some researchers present separate estimates of the security premium for domestic and foreign production (Brown and Huntington, 2013).

It is controversial as to whether the macroeconomic disruption effects are true externalities and represent a market failure. Perry and Darmstadter (2003) point out that to the extent that private actors can take into account future disruption costs and insure against them, there is no market failure. A recent National Research Council Report (NRC, 2010b), in fact, concluded that these disruption costs are not externalities. Other researchers have taken an agnostic point of view and, as pointed out in Perry and Darmstadter (2003), have modeled the external costs parametrically, letting the fraction of instability costs that are anticipated by individuals vary.

Another aspect of oil markets has received attention in the economics literature. The United States as a whole is a large consumer of oil. Were the United States to curtail its consumption, the fall in demand would lead to lower world oil prices. This is known in the economics literature as "monopsony power." The gain from the price reduction would accrue to all U.S. oil imports remaining after the reduction in consumption. The price reduction on nonimported consumption is not a net gain to the United States since the gains to consumers are offset by the losses to domestic producers.

The extent of monopsony power and the value of the reduction has changed over time as the United States has increased its domestic production and reduced imports as a share of consumption, and as other regions of the world have increased their relative proportion of world consumption. The recent decline in oil prices also has lowered the value of the monopsony premium. For example, in the RIA that accompanied the first phase of regulations in the medium- and heavy-duty trucks, NHTSA, based on research by Leiby et al. (1997) and updated in 2008 (ORNL, 2008), estimated the monopsony premium for the year 2020 at $12.07 per barrel (2013 dollars). The RIA accompanying the Phase II rule (NHTSA and EPA, 2016), using the same methodology as the earlier estimate, put the 2030 monopsony premium at $2.21 per barrel (2013 dollars) (NHTSA and EPA, 2010, p. 9-41, 2015, p. 8-77).

It is widely recognized that the monopsony premium does not constitute an externality in oil markets. Rather, the issue is the distribution of gains and losses between oil producers and oil consumers. The gains to the United States from exercising monopsony power would come from losses to oil producers. While not an externality, the potential distributional gains to the United States have led some to call for the inclusion of the monopsony premium in an analysis of benefits to the United States from reducing oil consumption. Whether or not to introduce measures to exploit this power is a policy choice of the U.S. government. In the recent RIA (NHTSA and EPA, 2016), NHTSA and EPA argue that they are taking a global view when establishing policy to deal with climate change. The agencies argue that climate change is a global phenomenon and calculate policy benefits on a global basis. As a result, it would be inconsistent, in their view, to calculate monopsony benefits on a regional basis. In contrast, a recent National Research Council report argues that the United States should count the monopsony benefits as part of the benefits to fuel economy regulations while maintaining a global focus for calculating climate benefits (NRC, 2015).[29]

Finally, it has been observed by many (e.g., NRC, 2010b) that there are potential externalities associated with military expenditures and foreign policy goals associated with protecting oil production and shipping lanes. Nevertheless, most observers conclude that it is very difficult to isolate the increased military expenditures or to quantify the value of the political constraints oil consumption may place on the United States. The NRC (2010b, p. 336) pointed out that "a 20 percent reduction in oil consumption, for example, would probably have little impact on the strategic positioning of military forces in the world."[30]

---

[29] The climate benefits to the United States have been estimated to be between 7 and 23 percent of total benefits (see Gayer and Viscusi, 2013, p. 252). The March 28, 2017, executive order issued by President Trump rescinds the Obama administration guidance on the use of the social cost of carbon in rulemaking and relies instead on previous guidance. One result of this order will be to estimate costs and benefits on a U.S.-only basis. It remains to be seen how this will affect the treatment of the monopsony premium in future rulemakings (Dudley, 2017).

[30] This opinion is shared by NHTSA and EPA: "US military costs are excluded from the analysis performed by ORNL because their attribution to particular missions or activities is difficult and because it is not clear that these outlays would decline in response to incremental reductions in US oil imports" (NHTSA and EPA, 2015, p. 8-85).

12.3.3 Estimates of the Security Premium

There have been many attempts at estimating the security premium associated with oil (see Brown and Huntington [2013] for a list of previous studies). The value of the premium depends on the probability and size of potential disruptions; the impact of the disruption on the world price of oil, which in turn depends on the elasticity of demand and the elasticity of supply of the nondisrupted oil; the elasticity of GDP with respect to oil price increases; the level of the price of oil; and the level of imports. These factors vary over time as do changes in oil market institutions so the premium also varies over time.

Recent developments in oil markets likely have lowered the security premium. The most significant of these developments is the rapid expansion of oil production in the United States from tight formations—"fracking." This expansion has lowered the share of imports in total consumption and has increased the elasticity of supply in the United States.[31,32] These production developments together with attempts of the Organization of the Petroleum Exporting Countries (OPEC) to maintain a particular level of world market share led to a dramatic decrease in the level of oil prices, falling from over $100 per barrel in 2008 to below $30 in early 2016. Also over time the intensity of oil use in the United States economy has declined, likely lowering the impact of any unanticipated increase in price on overall economic activity. Recent research on the relation of oil price changes to GDP changes has also shown a reduction in the impact of oil price changes (Killian, 2014). Of course, oil prices have been volatile over the past decades and looking ahead it is reasonable to expect many of these factors to change.

Table 12-5 shows estimates of the premium made by various authors expressed in 2013 dollars. The table also shows the date the estimate was made and the range of the estimates. Where possible, separate premiums are shown for domestically produced oil and for imports. In today's circumstances the domestic premium is particularly relevant. This is because reductions in consumption in today's circumstances are as likely to result in the reduction of domestic production as in the reduction of foreign production. OPEC and, in particular, Saudi Arabia have shown determination to maintain production in the face of demand reductions. Looking to the future, there is no guarantee this will continue so the relevant premium in, say, 2025 would be some weighted average of the domestic and import premiums calculated at the market conditions prevailing at that time.

All estimates are converted to 2013 dollars using the GDP deflator. Table 12-5 demonstrates the wide band of uncertainty that surrounds these estimates. The estimates depend on estimates of various elasticities that vary over time, as well as the price of oil. More recent research, for example, suggests the effect on GDP of oil shocks is smaller than previous estimates that relied on the experience of 1973-1974 and 1979 (Killian, 2014). Earlier estimates also did not capture the full effect of the large expansion of oil production in the United States and the

TABLE 12-5 Recent Estimates of the National Security Premium (2013 dollars per barrel)

Source	Year Being Forecasted	Domestic Premium (range)	Import Premium (range)	Monopsony Premium (range)
NHTSA and EPA (2010)	2020	NA	7.60 (5.92-12.72)	12.07 (4.13-22.79)
Brown and Huntington (2013)	2013-2014	3.07 (.37-9.33)	5.27 (1.16-15.26)	NA
NHTSA and EPA (2016)	2030	NA	7.26 (3.40-11.73)	2.83 (.83-4.56)

NOTE: The estimates in the NHTSA reports are based on an original model by Leiby et al. (1997), updated to conditions of 2014 by Leiby (2015).

[31] Hydraulic fracturing—"fracking"—technology results in big increases in production in the first year of production. That means that the response to price increases can be more rapid than is the case with conventional production methods. Recent reports of producers that have drilled wells are not producing the oil but waiting for prices to rise (see Mills, 2016).

[32] The import share of total petroleum products supplied is forecast by the 2015 Annual Energy Outlook to decline from 33 percent in 2013 to 14.7 percent by 2025.

dramatic declines in crude oil and petroleum product prices. The last line of the table shows the latest estimates by NHTSA, used in the Phase II rulemaking. The median estimate of $7.26 per barrel is the equivalent of $0.17 per gallon, which is small relative to the volatility in gasoline prices over the past few years.

12.3.4 Findings

Finding: Oil consumption makes the U.S. economy vulnerable to supply disruptions. At least some of the effects of these disruptions are likely to be unanticipated and uninsured against by private firms and consumers.

Finding: To the extent that consumption is satisfied through imports, supply disruptions and consequent price increases cause the transfer of economic resources from U.S. consumers to foreign producers. Therefore, the security premium associated with imports is larger than the premium associated with consumption satisfied by domestic production.

Finding: Recent developments in world oil markets have reduced the vulnerabilities of the U.S. economy to supply disruptions. Most significant is the development of "fracking" technology and the large increase in U.S. domestic oil production. This has increased the ability of the United States to respond to increased prices and reduced the level of imports. Furthermore, OPEC has shown in the recent past a desire to maintain market share and to accommodate the U.S. production increases and the declines in world demand by maintaining their production rates and allowing prices to fall.

Finding: Over the long term it is likely that volatility in oil prices and market institutions will continue to alter the value of the security premium associated with oil consumption and imports.

Finding: The U.S. government has made a policy choice not to consider any monopsony benefits from reducing U.S. oil consumption and, for the Phase II regulations, to take a global view when evaluating the costs and benefits of actions aimed at reducing climate change. In any case, the estimates of the monopsony premium have declined substantially in recent years.

12.3.5 Recommendation

Recommendation 12-3: The committee believes that a security premium reflecting macroeconomic instability and transfers caused by potential supply disruptions are a part of the social cost of consuming and importing oil. It is important to separate the social costs imposed by consumption from the costs imposed by imports. The total cost will be a weighted average of the two premiums and will vary over time with the proportion of consumption satisfied by imports.

12.4 PROJECTING TOTAL CAPITAL COSTS

12.4.1 Indirect Cost Estimation

Cost estimation, whether for regulatory purposes or for other reasons, makes a careful distinction between direct and indirect costs. Estimation of both cost categories is essential both for the manufacturer, which needs to know how to price a product to cover costs and generate an acceptable profit, and for the regulator, who is required to make estimates of the costs and benefits of regulation, including costs of compliance.

The distinction between indirect and direct costs is closely related to the relationships between two different types of inputs in the production process. For some inputs—those exhausted or incorporated into a unit of the product at the time of production and therefore not available to produce another unit—there is a strict, quantifiable relationship between inputs and outputs. Examples in vehicle manufacturing would include the production labor hours, parts and materials, etc. Multiplying the quantity of each direct input by its unit cost and summing provide an estimate of the direct costs of the vehicle.[33]

However, many inputs to the production process, as well as all other corporate activities with costs that must be covered by sales revenues, do not bear this simple or readily observable relationship to outputs, at least not for individual products. These are the inputs that contribute to indirect costs. Foremost among these input categories are the production inputs, which EPA's Office of Transportation and Air Quality (OTAQ) calls production overhead, broken out as follows:

- Production overhead (0.15)
  - Warranty (0.04)
  - R&D (0.05)
  - Depreciation (0.04)
  - Maintenance, repair, and operations (0.02)

As is common in matters of classification, designations of direct or indirect costs are somewhat arbitrary. Warranty costs, for example, may be associated with an individual vehicle, but the relationship is stochastic, not fixed, and the costs are incurred at some indeterminate time after the vehicle is sold.

Indirect costs are not limited to production overhead. The costs of essential nonproduction activities generate no revenue, at least not directly, and must also be covered by sales revenues. The nonmanufacturing categories and subcategories that enter into OTAQ's indirect cost (IC) calculations include the following (the numbers in parentheses are explained below):

- Corporate overhead (0.09)
  - General administrative: personnel, legal, etc. (0.07)
  - Retirement and health care expenses (0.02)
- Sales, including delivery of new vehicles to dealers, and advertising (0.08)
  - Vehicle transport to dealers (0.005)
  - Marketing (0.01)
  - Dealer services, including sales, training of sales personnel, and net income to dealers for new vehicle sales (0.07)
- Net income to manufacturer (0.05)

Of course, *total* indirect costs are reasonably easy to calculate: just identify the costs that are direct and subtract them from total costs (plus net income). Corporate annual and quarterly reports, as well as filings to the Securities and Exchange Commission, generally contain adequate information to do this. The difficulty comes in apportioning the total indirect costs in each of the categories and subcategories above to a multitude of individual products. That job, in turn, requires a vast amount of detailed cost and production data that are generally proprietary and not available to the regulator.

Because of these difficulties, it is common to resort to a simpler calculation of IC that uses readily available data: the "retail price equivalent" (RPE). The RPE is defined as the ratio of total costs (direct plus indirect costs including net income or the cost of capital) to the total direct costs, thus allowing total costs to be estimated from direct costs. The numbers alongside the IC categories and subcategories in the bulleted lists above indicate the contribution of that category to the RPE for medium- and heavy-duty (MHD) trucks. Adding in 1.0 for the production sector, OTAQ's estimate of the total RPE for MHD trucks is 1.36.[34]

According to Rogozhin et al. (2010b), the RPE is commonly used in the motor vehicle industry for this purpose. OTAQ has also estimated the RPE for a variety of light-duty vehicle manufacturers; the average RPE is 1.46, with a

[33] Note that a manufacturer's direct costs may include the indirect costs of parts purchased from suppliers. The fundamental distinction made above still applies (NRC, 2015). To the manufacturer, the cost of purchased parts is precisely known.

[34] The sum of the individual components does not sum to the 1.36 due to rounding.

range from 1.43 to 1.49 (Rogozhin et al., 2010a). It has also been used for heavy-duty vehicles. In the MHD truck industry overall, OTAQ calculated a mean RPE of 1.42, using data from the U.S. Census and Supplier Relations LLC, a data firm serving the industry. OTAQ also estimated RPEs for several engine and truck manufacturers, with averages of 1.28 (engines) and 1.36 (trucks). The engine RPEs tend to be lower because, for the most part, engine manufacturers do not sell directly to the end user, which eliminates part of the retail component of the RPE.

Despite the ease of calculation and wide use of the RPE in the motor vehicle industry, researchers at OTAQ have argued that it has two problematic features. First, OTAQ claims that the RPE is sensitive to the time frame, as many IC activities are short term in nature and do not figure in long-term cost.

Many of the indirect costs are likely to be one-time or short-term activities, such as educating dealers and upgrading mechanics' equipment. These costs will not appear in the long-term IC multipliers. In addition, incremental R&D expenditures will occur over a short period of time, even though they may be amortized over 5 to 10 years. Thus, we expect to see higher indirect costs initially and lower impacts in the long term as companies assimilate the new technologies (Rogozhin et al., 2010a).

For present purposes, however, the time frame is not an issue. We are primarily interested in the incremental long-term costs of regulation, and, in the long run, all costs are incremental.

A more important issue with the RPE, however, is that it implicitly assumes that the ratio of IC to direct cost is identical for every product produced by the firm. While this has to be true on average—the volume-weighted average of the indirect cost multipliers (ICMs) must equal the RPE—since all costs need to be accounted for, nonetheless, it is easy to imagine how a disproportionately outsized demand on overhead could be exerted by products that are new, complicated, or difficult to assemble or, for whatever reason, use up more than their share of management and administrative resources, and similarly, how they could disproportionately affect particular IC subcategories.

To address these issues OTAQ developed a method for producing ICMs that are more flexible than the RPE and in particular more sensitive to both the technology involved and the specific impact on individual corporate subunits. In the end, three levels of technology—low, medium, and high—were examined. Office of Mobile Sources identifies low-level complexity as a technology that introduces only minor changes in component design and does not affect the way various components are linked together. Medium-level complexity either changes the design substantially or changes the way the components are used or fit together. High-complexity technology affects both design and integration of components. Examples of the three complexity levels are single-wide tires, engine turbocompounding, and hybrid-electric powertrains, respectively.

To accomplish these goals, a set of "adjustment factors" was generated, one factor for each combination of IC subcategory and complexity level, to be used to adjust the RPE contribution from that subcategory. The adjustment factors were to be calibrated so that a factor of 0 would indicate that a direct cost would have no effect on IC in that subcategory, while a factor of 1 would indicate an average effect (i.e., the same as the subcategory RPE). Table 12-6 shows the IC multipliers for three types of vehicles—LDVs, MHD engines, and MHD trucks—and three levels of technology complexity.

To estimate the adjustment factors to use with the HDV Fuel Consumption Rule, OTAQ convened a panel of engineers with considerable experience in the vehicle manufacturing industry, and tasked the group with the development of such factors in a group setting. Two different group mechanisms were used: a consensus process where the group had to arrive at a single set of values, and a Delphi process where participants were made aware of the estimates of others in the group and then allowed to change their estimates in response to the opinions of others. The Delphi process was concluded after three iterations. Finally, the adjustment factors coming out of the two processes (consensus and Delphi) were averaged to obtain the final estimates for the low- and medium-complexity technologies. For the high-complexity technologies, the results were different enough that it made sense to report them separately, as High1 (for consensus) and High2 (for Delphi). The exercise was carried out for two engine and four truck manufacturers and averaged within the groups.

The multipliers are consistently lower for MHD engine manufacturers than truck manufacturers, perhaps due to the greater likelihood that engine manufacturers sell directly to end users. Also, for each vehicle category, the ICMs resulting from the expert elicitations conducted by EPA are smaller than RPEs at all technology levels, and they are particularly low for low- and medium-complexity technologies. (In many of the subcategories listed above,

TABLE 12-6 Comparison of RPE and IC Multipliers by Technology Category and Complexity

		IC, by Technology Complexity			
Category	RPE	Low	Medium	High1	High2
LDV	1.46	1.02	1.05	1.26	–
MHD engine manufacturer	1.28	1.04	1.08	1.14	1.24
MHD truck manufacturer	1.36	1.05	1.11	1.22	1.31
MHD sample technology	–	Single-wide tires	Engine turbocompounding	Hybrid-electric powertrains	

in fact, the adjustment factor is zero.) Upon reflection, this seems strange. Indeed, in the RPE-ICM comparisons shown in Table 12-6, each RPE is greater than all of the component ICMs in each row, which implies that the RPE will be greater than the average ICM, regardless of the weights. As noted above, that cannot happen.[35] A more fundamental question is, just what are the entities that ICMs get attached to? It does not seem quite right that it be attached to a "technology," as OTAQ has done, because various applications of the same technology could in principle give rise to different ICMs. Instead, it seems more useful to think of the vehicle (or engine, or whatever) as a bundle of components, which include a mix of technologies of varying complexity, and calculate ICMs that can be attached to tangible products or assemblies of products. For the cost-estimating purposes of this National Academies of Sciences, Engineering, and Medicine committee, it seems safest to continue to use RPEs and ensure that long-run costs are covered. Like the LDV committee, nonetheless, this committee recognizes the cost estimation problem with RPEs that NHTSA and EPA are trying to deal with, and urges them to continue their research on ICMs, especially to develop more solid empirical support for particular ICM values.

12.4.2 Findings

Finding: Accurate calculation of ICs is just as important to regulatory and cost-benefit analysis as is direct cost estimation. While *total* ICs, and the average IC associated with each unit of output, are just as easy to calculate as direct costs are, their calculation on a product-by-product basis can be much more difficult because of the lack of a clear relationship between inputs and outputs. The RPE method, a standard method for calculating ICs, ignores this problem by assigning the same IC multiplier to all products.

Finding: Recently researchers at EPA and NHTSA have been working to develop a new method, which they call ICM, to estimate ICs on a more detailed product- or technology-specific basis. In principle, this makes a good deal of sense, for there are plenty of reasons to expect diversity across products in the use of inputs that give rise to ICs. Considerable progress has been made in developing this new methodology; however, work remains to be done.

Finding: In order to produce more detailed cost breakdowns, the current ICM methodology relies heavily on expert opinion in the absence of product-specific empirical data. In addition, in some circumstances the current ICM methodology gives results that appear to be inconsistent, as shown in the example above where the RPE is calculated to be greater than each of the constituent ICMs.

12.4.3 Recommendations

Recommendation 12-4: The committee encourages NHTSA, in coordination with EPA, to further development of the ICM methodology, which promises to provide more flexible and accurate estimates of IC multipliers. Where

[35] This point was also made by the study of light-duty vehicle fuel economy by the National Academies (NRC, 2015).

possible, these estimates should be validated with observations of actual incremental costs of changing product designs. For particularly important cost regulations or cost estimates, two or more analytical methods could be used to develop a range of cost estimates that could be compared with one another. For example, expert elicitation could be compared with observed cost changes in similar prior regulations or bottom-up engineering analyses. Such parallel analyses would automatically generate a useful sensitivity analysis.

Recommendation 12-5: In order to improve all cost estimates, direct or indirect, NHTSA and EPA should make a habit of commissioning retrospective studies of costs of past regulations. Again, a range of analytical methods is available. One that might be of particular interest for ICs is hedonic estimation, a statistical method that allows bundled cost estimates to be separated into individual component estimates. That property makes it potentially useful for producing estimated IC components across a range of product categories.

12.4.4 Learning Effects on Capital Costs

12.4.4.1 Learning Curves

Repetition drives learning. In business and economics the importance of this fundamental fact is enshrined in the concept of the learning curve. Simply put, the learning curve asserts a declining relationship between the cost of a new product produced by a firm and the firm's cumulative output of the product. Thus, estimation of the costs of regulation, which usually implies adoption of new or modified products or processes, needs to take the learning curve into account.

New and emerging technologies used in MHDV costs and performance will likely evolve over time as a function of several different factors. Policy makers, modelers, and analysts are generally concerned with the effects of transitions to different MHDV technologies and strategies, and how these may influence the outcomes of policies such as the Phase II standards.

To describe how the costs of technologies of strategies will perform in the future, energy economic modelers have frequently used the concept of *learning or experience curves*. These curves can be estimated in the context of retrospective analysis, where the modeler uses historical costs of a technology, a component or a strategy, and assesses how those costs relate to other factors—most commonly the cumulative number of units produced or accumulated service provided. Learning curves are also used in forward-looking modeling efforts, where the modeler assumes specific learning rates for different technologies (often informed by historical learning rates when those estimates exist) to understand how the share and adoption of different technologies is likely to evolve in the future.

The origin of learning curves dates to developments in the psychology of learning in the late 19th century, especially in the work of the German psychologist Hermann Ebbinghaus. Wright (1936) first showed an empirical log-linear relationship between unit costs and cumulative production, in the context of the production of airplanes, and was further supported by experiences in the vast buildup of munitions industries during World War II (Hindle, 2008). Additional development occurred in the early 1960s at the Boston Consulting Group, based on the firm's experience advising a semiconductor manufacturer.

This extraordinarily simple log-linear relationship has shown to hold for a variety of products and technologies. It has been denoted as the "one-factor learning curve," and it is by far the most widely used model in the energy economics literature to understand how unit costs have evolved (Rubin et al., 2015). The one-factor learning curve relates the unit costs of production with cumulative production, using a log-linear functional form:

$$Y = ax^b$$

where Y is the unit cost of production, x is the cumulative experience (often represented by the cumulative number of units produced), a is the unit production cost of the first unit, and b is a constant reflecting the rate of cost reduction (Arrow, 1962).

Intuitively, the one-factor learning or experience curve means that as more units are produced, the firm learns how to reduce the production costs per unit, and empirically it has been shown that this relationship of decreasing costs resulting from learning as the firm produces more units is well described by a log-linear relationship. For example, a 15 percent learning rate (or an 85 percent progress ratio) states that unit costs decline by 15 percent for every doubling of cumulative output.

Studies in various settings have found great variation in the learning rates. A survey by Dutton and Thomas (1984) found average progress ratios of 82 percent (i.e., learning rates of 18 percent), with a range from 55 to 95 percent (learning rates of 5 to 45 percent). More recent work, cited by Taylor and Fujita (2013), found similar results, with an additional finding that within-technology progress ratios varied as much as between-technology ratios. This finding is somewhat unsettling, as it suggests that the accuracy of a learning-curve assumption in a particular context could depend on quite detailed information. It should also be noted that the logarithmic functional form generally employed in learning curves implies that the cost reductions continue indefinitely, approaching zero as an asymptote. As a practical matter, however, the period of learning is in most studies considered finite. For electricity generation, Rubin et al. (2015) reviewed peer-reviewed existing studies on learning curves for 11 electricity-generation technologies and found that there was a wide disparity in one- and two-factor learning rates even within the same technology, time period, or region.

Closely related to the learning-curve concept is that of the *experience curve*. The chief difference between the two is that the learning curve refers only to the relationship between manufacturing costs and the volume of production. The experience curve is broader, relating *total* costs to cumulative sales (Abernathy and Wayne, 1974). For both, the decline in costs is not automatic, but comes only with application of management expertise and attention to the opportunities available. Thus, the potential sources of cost reduction extend far beyond the growth in speed and efficiency that comes with the simple familiarity with the task at hand. As Abernathy and Wayne (1974) note:

> The producer cuts costs with a combination of effects; these include spreading overhead over larger volume, reducing inventory costs as the process becomes more rational and throughput time drops, cutting labor costs with process improvements, achieving greater division of labor, and improving efficiency through greater familiarity with the process on the part of the work force and management.

Dutton and Thomas (1984) lump the categories of improvement into two broad classes. The first is *autonomous* change, resulting from learning by doing. The second category is called *induced* change, resulting from conscious efforts of management, including such measures as worker training and specialization, R&D investments, and streamlining of production processes.

In the one-factor learning curves the number of cumulative units produced is being used as a de facto surrogate for all the factors that influence that technology's cost (Rubin et al., 2015). Indeed, while offering the benefits of a simple model, these simple one-factor learning curves provide no explanation for the underlying factors and processes of technological change or the causality between them (Clarke et al., 2006; Ferioli et al., 2009; Gillingham et al., 2008, Nordhaus, 2009). Modelers have tried to circumvent some of these issues by using more detailed models—multifactor learning models that explicitly incorporate other factors that may be determinants of unit cost changes. Rubin et al. (2015) reviewed the studies for electricity supply technologies and found that previous studies have included factors such as R&D spending (Clarke et al., 2006; Cohen and Klepper, 1996; Jamasb, 2007), knowledge spillovers (Clarke et al., 2006), increased capital investments (Cohen, 1995; Klepper and Simons, 2000), economies of scale (Nemet, 2006; Sinclair et al., 2000; Yeh and Rubin, 2007), changes in input prices (Joskow and Rose, 1985; Nemet, 2006; Söderholm and Sundqvist, 2007), labor costs (Joskow and Rose, 1985), efficiency improvements (Joskow and Rose, 1985; Nemet, 2006), and other public policies (Söderholm and Klaassen, 2007; Söderholm and Sundqvist, 2007). However, lack of data on these explanatory factors often prevents their use in more detailed learning models.

Researchers have found that when R&D is explicitly included as one of the factors contributing to the decrease in unit costs, it is generally a significant explanatory variable (Jamasb, 2007; Klaassen et al., 2005; Kouvaritakis et al., 2000; Söderholm and Klaassen, 2007; Söderholm and Sundqvist, 2007; Watanabe, 1995). Most often, only public R&D is included in such assessments, given the lack of private R&D data. Several studies also explicitly include the enactment of specific policies that may influence costs as controls (most generally as a dummy variable).

Researchers have also used variants of the above-mentioned functional forms. Some authors considered *component-level learning* (i.e., in the case MHDVs, this would mean looking separately at learning for the engine, battery, etc.), or including also factors such as raw materials and labor. Rubin et al. (2007) used the component-based approach for power plants with carbon capture systems. Similarly Weiss et al. (2010) assessed learning effects using component-level analysis for microcogeneration of heat and power. The logic behind this strategy is that for complex multicomponent technologies, different components are likely at different levels of maturity, and therefore have different potential for cost reductions.

Finally, for some technologies, such as nuclear, some researchers have actually found negative learning rates—the unit costs actually increased as the cumulative production increased (see, for example, Cooper [2010] and Grubler [2010]).

Even for those technologies (such wind and solar) where there is a substantial literature on learning rates, a systematic understanding of why different studies achieved fundamentally different estimates for learning, as well as the implications of assuming different learning rate values in policy analysis, is still lacking. This issue is even more pronounced in the MHDV sector due to several factors.

First, the industry is highly segmented and the technologies and components used by different subsectors of the fleet are highly diversified. Indeed, given the diversity of the MHDV fleet it would likely make more sense to think about learning curves at the firm level.

Second, data on historical costs and units produced are lacking for the sector. This makes any estimates of learning rates for MHDVs unreliable.

Third (and this point holds for any prospective use of learning rates), the past may not be a good predictor of the future, namely, in a sector where the technologies and fuels are likely to change over time as a result of the policy. Learning and experience curves are certainly useful *descriptive* tools for describing the reduction in unit costs over time. But can they be considered *predictive*? The long list of potential causes of unit cost reductions, as well as the wide range of learning rates found in empirical studies and the uncertainty surrounding the duration of learning, seems to argue against such a notion (Abernathy and Wayne, 1974; Taylor and Fujita, 2013). Nonetheless, learning and experience curves have been used in RIAs to predict cost reductions of federal regulations for almost 20 years, dating back at least to EPA's 1997 Heavy-Duty Emissions Standards (EPA, 1997). NHTSA began to incorporate a similar but not identical learning-curve analysis into RIAs of fuel economy standards that paired with EPA's greenhouse gas emission standards after 2000 (Taylor and Fujita, 2013).

The predictive value of learning curves may depend not only on the characteristics of the technology, but on the business strategy pursued by the firm. There is, after all, a seeming paradox at the heart of learning-curve analysis. In principle, the learning curve ought to give a huge advantage to a first mover, whose cumulative production is likely to be considerably greater than its rivals, and whose production costs are correspondingly lower. Thus, the first mover in an industry should frequently be the only firm. And yet, most industries contain numerous competitive firms. To resolve the paradox, note that there are other attributes of a firm's products besides costs that consumers care about. By adopting a strategy of pursuing rapid cost reduction above all else, however, the firm may foreclose other strategies that prize innovation and feature product quality and consumer satisfaction over price. The very definition of innovation, after all, is to do something different, not to do the same thing better.

Abernathy and Wayne (1974) illustrate this point by a short case study of the Ford Model T, where the company's exclusive focus on cost reduction of a standardized product put the company in a box where it was producing a very low-cost product that consumers did not want. Once it became fully aware of the problem, the company was forced to expend huge sums and considerable time on designing new products and retooling its plants, in particular causing the giant River Rouge plant to sit idle for more than a year. This debacle almost wrecked the company.

Fourth, as explained in the draft RIA by NHTSA and EPA (2015), studies on learning or experience curves relate costs to cumulative production only after an initial volume threshold is achieved. Indeed, in early stages of technology development, it may be the case that there is an initial increase, rather than a decrease, in unit costs as the first few units are produced (Rubin et al., 2007).

In the 2011 RIA, which accompanied the promulgation of the Phase I rule, EPA's initial approach to learning curves used a learning rate of 20 percent, close to the average found by Dutton and Thomas (1984). The rate is routinely applied only to the long-term direct manufacturing costs of the technology package required by the regulation; its use for indirect costs is at the user's discretion. Thus, it is closer to a learning curve than an experience curve.

When NHTSA and EPA began to collaborate on regulation, a range of learning rates was employed based primarily on the maturity of the technology employed, with new technologies being assigned "steep" learning rates (around 20 percent) and more mature technologies, widely available commercially, "flat" rates in the range of 1 to 3 percent (Taylor and Fujita, 2013).

NHTSA and EPA are now using an algorithm far more complicated than what is generally used in the literature, and arguably with a level of complexity that is not well supported given the overall uncertainties on learning rates. More specifically, NHSTA and EPA are assuming that for emerging and newer technologies, the technology will have what they call a "steep learning/volume-based" curve which will be a 20 percent reduction in cost per year when either there is a doubling of the production capacity *or* after 2 years of production (which reflects the underlying assumption by NHTSA and EPA of a doubling of production every 2 years). After that, the technologies follow what they call a "time-based" or "flat" learning curve which is also applied to the mature technologies. In the time-based or flat learning curve, NHTSA and EPA assume that in the first 5 years there is a 3 percent cost reduction per year, in the following 5 years there is a 2 percent cost reduction per year, and, after that, for the next 10 years, a 1 percent cost reduction per year.

Figure 12-6 shows the year-to-year "learning" cost factors used in the draft RIA. The interpretation of the figure is as follows: when the lines cross the horizontal red line (i.e., the value of 1), that represents NHTSA and EPA's point-cost estimate that the learning was then applied to. So, for example for "Strong Hybrid" and "Low Rolling Resistance Tires-2," the cost estimate from the agencies was done for year 2021, and then the "learning" factors for other years were applied (so that the cost for "Strong Hybrid" in 2016, for example, was assumed to be 1.55 times the 2021 cost).

Recommendation 12-6: NHTSA and EPA should employ a simpler and more transparent method and rationale on how the unit costs of technologies will evolve, either over time or as a function of the production capacity, in their future assessment. Two options the agencies should consider include (1) assuming specific costs per unit over time for each of the technologies under study, informed by manufacturers' and experts' elicitation of likely costs in the future; and (2) applying simple parametric one-factor learning curves, complemented with a sensitivity analysis to assess how high or low these learning factors would need to be for a technology to be more economically appealing than another. NHTSA and EPA also are encouraged to do *ex post* assessments of their forecasts of learning effects.

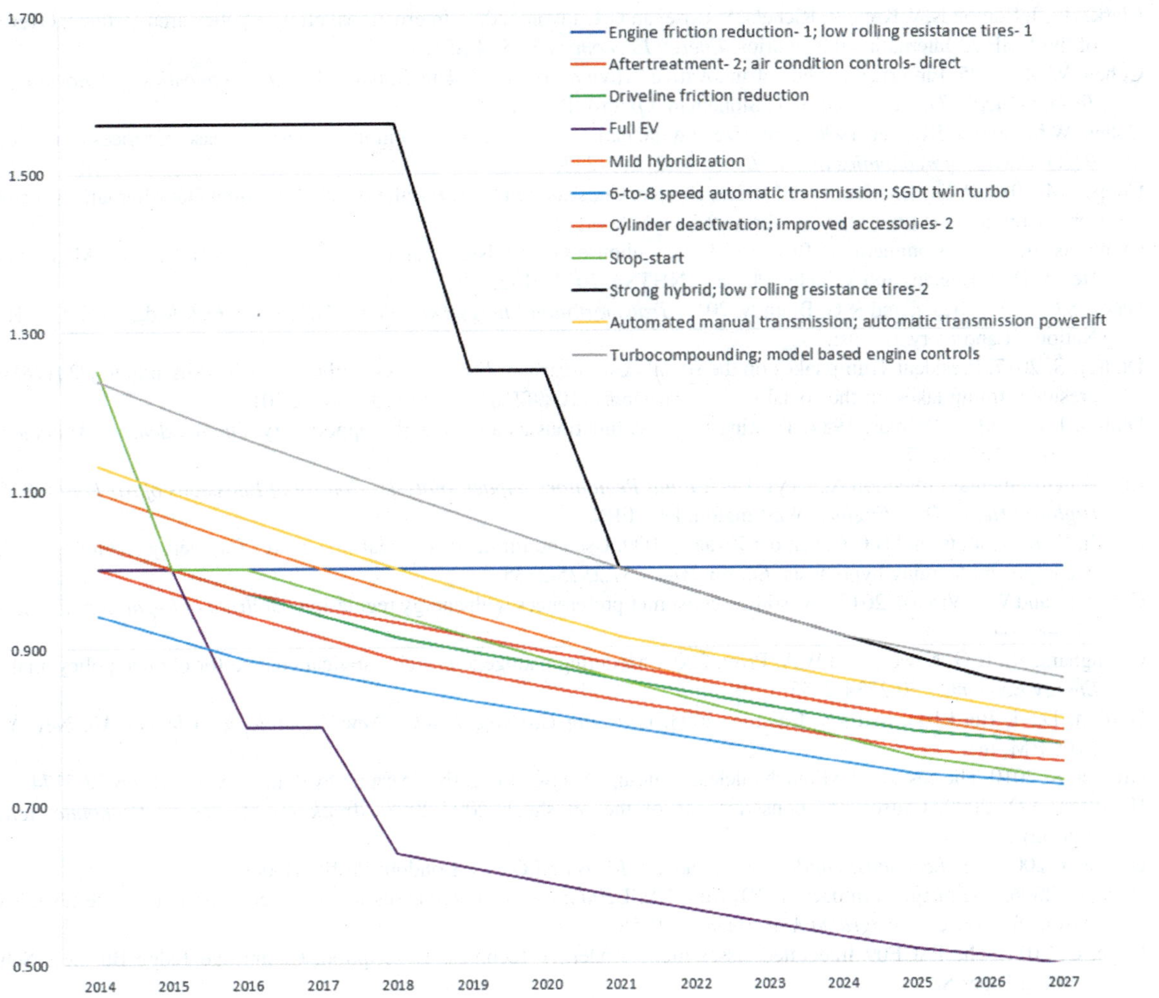

FIGURE 12-6 Cost factors assumed in the Draft RIA. Figure produced by the committee, using the data produced in the RIA.

12.5 REFERENCES

Abernathy, W.J., and K. Wayne. 1974. Limits of the learning curve. *Harvard Business Review* (September).

Anthoff, D., and R.S.J. Tol. 2012. Climate damages in the *FUND* model: A comment. *Ecological Economics* 81:42. https://doi.org/10.1016/j.ecolecon.2012.06.012.

Anthoff, D., and R.S.J. Tol. 2014. The income elasticity of the impact of climate change. In *Is the Environment a Luxury?An Inquiry into the relationship between environment and income.* Ed. by Tiezzi, S. and Martini, C. https://doi.org/10.4324/9781315819594.

Arrow, K.J. 1962. The economic implications of learning by doing. *Review of Economic Studies* 155–173.

Brown, S.P.A., and H.G. Huntington. 2013. Assessing the U.S. oil security premium. *Energy Economics* 38:118-127.

CalHEAT (California Hybrid, Efficient and Advanced Truck Research Center). 2013. Market Transformation Roadmap for Medium and Heavy-Duty Trucks. Prepared for the California Energy Commission. https://calstart.org/libraries-calheat_2013_documents_presentations-calheat_roadmap_final_draft_rev_7-sflb-ashx/ (accessed December 20, 2019).

Clarke, L., J. Weyant, and A. Birky. 2006. On the sources of technological change: Assessing the evidence. *Energy Economics* 28:579-595.

Clarke, L., J. Edmonds, V. Krey, R. Richels, S. Rose, and M. Tavoni. 2009. International climate policy architectures: Overview of the EMF 22 International Scenarios. *Energy Economics* 31:S64-S81.

Cohen, W.M., 1995. Empirical studies of innovative activity. Pp. 182-264 in *Handbook of the Economics of Innovation and Technological Change*, edited by P. Stoneman. Oxford: Blackwell.

Cohen, W.M., and S. Klepper. 1996. Firm size and the nature of innovation within industries: The case of process and product R&D. *Review of Economics and Statistics* 78(2):232-243.

Cooper, M. 2010. Policy Challenges of Nuclear Reactor Construction: Cost Escalation and Crowding Out Alternatives. Institute for Energy and the Environment, Vermont Law School.

Cummins Inc. 2015. Comments in Response to Greenhouse Gas Emissions and Fuel Efficiency Standards for Medium- and Heavy-Duty Engines and Vehicles-Phase 2. NHTSA-2014-0132.

Davis, S.C., S.W. Diegel, and R.G. Boundy. 2015. *Transportation Energy Data Book: Edition 34*. Oak Ridge, TN: Oak Ridge National Laboratory. August.

Dudley, S. 2017. President Trump takes on the social cost of carbon. *Forbes*. www.forbes.com/sites/susandudley/2017/03/30/president-trump-takes-on-the-social-cost-of-carbon/#1110d9f2185e (accessed April 6, 2017).

Dutton, J.M., and A. Thomas. 1984. Treating progress functions as a managerial opportunity. *The Academy of Management Review* 9(2):235-247.

EPA (Environmental Protection Agency). 1997. *Final Regulatory Impact Analysis: Control of Emissions of Air Pollution from Highway Heavy-Duty Engines*. Washington, DC: EPA.

Ferioli, F., K. Schoots, and B.C.C. van der Zwaan. 2009. Use and limitations of learning curves for energy technology policy: A component-learning hypothesis. *Energy Policy* 37:2525-2535.

Gayer, T., and W.V. Viscusi. 2013. Overriding consumer preferences with energy regulations. *Journal of Regulatory Economics* 43:248-264.

Gillingham, K., R.G. Newell, and W.A. Pizer. 2008. Modeling endogenous technological change for climate policy analysis. *Energy Economics* 30:2734-2753.

Gordon, D., J. Burdelski, and J.S. Cannon. 2003. *Greening Garbage Trucks: New Technologies Cleaner Air*. New York: INFORM, Inc.

Grubler, A. 2010. The costs of the French nuclear scale-up: A case of negative learning by doing. *Energy Policy* 38:5174-5188.

Hamilton, J.D. 2009. Causes and consequences of the oil shock 2007-2008. *Brookings Papers on Economic Activity* (Spring):215-261.

Hindle, T. 2008. *The Economist Guide to Management Ideas and Gurus*. London: Profile Books.

Hope, C. 2006. The marginal impact of CO_2 from PAGE2002: An integrated assessment model incorporating the IPCC's five reasons for concern. *Integrated Assessment* 6:19-56.

Hope, C. 2011. The PAGE09 Integrated Assessment Model: A Technical Description. Cambridge Judge Business School Working Paper Series.

ICCT. 2015. *Policy Update: United States Efficiency and Greenhouse Gas Emission Regulations for Model Year 2018-2027 Heavy-Duty Vehicles, Engines and Trailers*. Washington, DC: ICCT.

Interagency Working Group on Social Cost of Carbon (IWG). 2010. Social Cost of Carbon for Regulatory Impact Analysis under Executive Order 12866. February. United States Government. Available at: http://www.whitehouse.gov/sites/default/files/omb/inforeg/for-agencies/Social-Cost-of-Carbon-forRIA.pdf (accessed August 12, 2014).

IWG. 2013. Technical Support Document: Social Cost of Carbon for Regulatory Impact Analysis under Executive Order 12866. United States Government. https://www.epa.gov/sites/production/files/2016-12/documents/scc_tsd_2010.pdf (accessed December 20, 2019).

IWG. 2015. Technical Support Document: Technical Update of the Social Cost of Carbon for Regulatory Impact Analysis under Executive Order 12866. July. United States Government. Available at: https://obamawhitehouse.archives.gov/sites/default/files/omb/inforeg/scc-tsd-final-july-2015.pdf (accessed December 20, 2019).

Jamasb, T. 2007. Technical change theory and learning curves: Patterns of progress in electricity generation technologies. *Energy Journal* 28:51-72.

Joskow, P.L., and N.L. Rose. 1985. The effects of technological change, experience, and environmental regulation on the construction cost of coal-burning generating units. *The RAND Journal of Economics* 1-27.

Killian, L. 2014. Oil price shocks: Causes and consequences. *Annual Review of Resource Economics* 6:133-154. Doi: 10.1146/annurev-resource-083013-114701.

Klaassen, G., A. Miketa, K. Larsen, and T. Sundqvist. 2005. The impact of R&D on innovation for wind energy in Denmark, Germany and the United Kingdom. *Ecological Economics* 54:227-240.

Klemick, H., E. Kopits, K. Sargent, and A. Wolverton. 2014. Heavy-Duty Trucking and the Energy Efficiency Paradox: Evidence from Focus Groups and Interviews. Paper presented to the NRC Committee on Assessment of Technologies and Approaches for Reducing the Fuel Consumption of Medium- and Heavy-Duty Vehicles, Phase Two.

Klemick, H., E. Kopits, A. Wolverton, and K. Sargent. 2015. Heavy-duty trucking and the energy efficiency paradox: Evidence from focus groups and interviews. *Transportation Research Part A: Policy and Practice* 77:154-166.

Klepper, S., and K.L. Simons. 2000. The making of an oligopoly: Firm survival and technological change in the evolution of the U.S. tire industry. *Journal of Political Economy* 108:728-760.

Kouvaritakis, N., A. Soria, and S. Isoard. 2000. Modeling energy technology dynamics: Methodology for adaptive expectations models with learning by doing and learning by searching. *International Journal of Global Energy Issues* 14:104-115.

Leiby, P. 2015. Factors Influencing Estimates of the Energy Security Premium for Heavy Duty Phase II Rule, Oak Ridge, TN: Oak Ridge National Laboratory.

Leiby, P., D. Jones, T.R. Curlee, and R. Lee. 1997. *Oil Imports: An Assessment of Costs and Benefits*. Oak Ridge, TN: Oak Ridge National Laboratory.

Lutsey, N. 2015. *Regulatory Considerations for Advancing Commercial Pickup and Van Efficiency Technology in the United States* Washington, DC: ICCT.

Mills, M.P. 2016. After the carnage, shale oil will rise again. *Wall Street Journal*, January 18.

Nemet, G. 2006. Beyond the learning curve: Factors influencing cost reductions in photovoltaics. *Energy Policy* 34:3218-3232.

Newell, R. and W. Pizer, 2003. Discounting the distant future: How much do uncertain rates increase valuations? *Journal of Environmental Economics and Management* 46(1):52-71. https://doi.org/10.1016/S0095-0696(02)00031-1.

NHTSA (National Highway Traffic Safety Administration) and EPA (Environmental Protection Agency), 2010. Final Regulatory Impact Analysis. *Final Rulemaking for Greenhouse Gas Emissions and Fuel Efficiency Standards for Medium- and Heavy-Duty Engines and Vehicles.* Washington, DC: NHTSA and EPA. Chap. 9.

NHTSA and EPA. 2015. *Proposed Rulemaking for Greenhouse Gas Emissions and Fuel Efficiency Standards for Medium- and Heavy-Duty Engines and Vehicles-Phase 2: Draft Regulatory Impact Analysis.* Washington, DC: NHTSA and EPA.

NHTSA and EPA. 2016. *Regulatory Impact Analysis, Final Rule for Greenhouse Gas Emissions and Fuel Efficiency Standards for Medium- and Heavy-Duty Engines and Vehicles, Phase II.* Washington, DC: NHTSA and EPA.

Nordhaus, W.D. 2007. *The Challenge of Global Warming: Economic Models and Environmental Policy.* New Haven: Yale University.

Nordhaus, W.D. 2009. The Perils of the Learning Model for Modeling Endogenous Technological Change. NBER Working Paper No. 14638. National Bureau of Economic Research.

Nordhaus, W.D., and J. Boyer. 2000. *Warming the World: Economic Models of Global Warming.* Cambridge: The MIT Press. ISBN 0-262-14071-3.

NRC. 2010a. *Technology and Approaches to Reducing the Fuel Consumption of Medium- and Heavy-Duty Vehicles.* Washington, DC: The National Academies Press.

NRC. 2010b. *Hidden Costs of Energy: Unpriced Consequences of Energy Production and Use.* Washington, DC: The National Academies Press.

NRC. 2014. *Reducing the Fuel Consumption and Greenhouse Gas Emissions of Medium- and Heavy-Duty Vehicles, Phase Two: First Report.* Washington, DC: The National Academies Press.

NRC. 2015. *Cost, Effectiveness, and Deployment of Fuel Economies for Light-Duty Vehicles.* Washington, DC: The National Academies Press.

ORNL (Oak Ridge National Laboratory). 2008. The Energy Security Benefits of Reduced Oil Use, 2006-2015. March.

Perry, I.W.H., and J. Darmstadter. 2003. The Costs of US Oil Dependence. Discussion Paper 03-59. Resources for the Future, Washington, D.C.

Roe, G.H., and M.B. Baker. 2009. Why is climate sensitivity so unpredictable? *Science* 318(5850):629-632. DOI: 10.1126/science.1144735.

Roeth, M., D. Kircher, J. Smith, and R. Swim. 2013. Barriers to the Increased Adoption of Fuel Efficiency Technologies in the North American On-Road Freight Sector. Report for the International Council for Clean Transportation. http://www.theicct.org/sites/default/files/publications/ICCT-NACFE-CSS_Barriers_Report_Final_20130722.pdf (accessed December 20, 2019).

Rogozhin, A., M. Gallaher, G. Helfand, and W. McManus. 2010a. Using indirect cost multipliers to estimate the total cost of adding new technology in the automobile industry. *International Journal of Production Economics* 124:360-368.

Rogozhin, A., M. Gallaher, A. Lentz, and W. McManus. 2010b. Heavy Duty Truck Retail Price Equivalent and Indirect Cost Multipliers. Draft report prepared by RTI International for Chi Li, Office of Transportation and Air Quality, USEPA, Ann Arbor, MI.

Rubin, E.S., S. Yeh, M. Antes, M. Berkenpas, and J. Davison. 2007. Use of experience curves to estimate the future cost of power plants with CO_2 capture. *International Journal of Greenhouse Gas Control* 1:188-197.

Rubin, E., I.L. Azevedo, S. Yeh, and P. Jaramillo. 2015. How to improve modelling approaches to represent the future costs of energy technologies? Reviewing the findings from learning curves from energy technologies. *Energy Policy* 86:198-218.

Sinclair, G., S. Klepper, and W. Cohen. 2000. What's experience got to do with it? Sources of cost reduction in a large specialty chemicals producer. *Management Science* 46:28-45.

Söderholm, P., and G. Klaassen. 2007. Wind power in Europe: A simultaneous innovation–diffusion model. *Environmental and Resource Economics* 36:163-190.

Söderholm, P., and T. Sundqvist. 2007. Empirical challenges in the use of learning curves for assessing the economic prospects of renewable energy technologies. *Renewable Energy* 32:2559-2578.

Taylor, M., and K.S. Fujita. 2013. Accounting for Technological Change in Regulatory Impact Analyses: The Learning Curve Technique. Lawrence Berkeley National Laboratory Report LBNL-6195E.

Watanabe, C., 1995. Identification of the role of renewable energy. *Renewable Energy* 6:237-274.

Weiss, M., M. Junginger, and K. Blok. 2010. A review of experience curve analyses for energy demand technologies. *Technological Forecasting and Social Change* 77:411-428.

Wright, T.P. 1936. Factors affecting the cost of airplanes. *Journal of Aeronautical Sciences* 3:122-128.

Yeh, S., and E.S. Rubin. 2007. A centurial history of technological change and learning curves for pulverized coal-fired utility boilers. *Energy* 32:1996-2005.

13

Alternative and Complementary Regulatory Approaches

13.1 INTRODUCTION: WHY CONSIDER ALTERNATIVE APPROACHES

The approach used for regulating the fuel consumption and greenhouse gas (GHG) emissions of medium- and heavy-duty vehicles (MHDVs) under the Phase I and Phase II rules (EPA and NHTSA, 2011, 2016) employs what economists call a quantity instrument (a standard) rather than a price instrument (an economic incentive). Moreover, compared to the quantity instruments used in other regulatory interventions, including light-duty vehicles, the approach used for MHDVs has some significant limitations. Except for the lightest vehicles in the MHDV spectrum, the regulation is implemented though pre-sale certification of the engine (if made by a company that then sells it to the vehicle assembler) or the vehicle—the latter being based on a computerized simulation of the vehicle's fuel use and GHG emissions (i.e., the Greenhouse Gas Emissions Model) but with no direct measurement of those variables. This approach was adopted for Phase I because it was readily implementable, would not mandate the purchase of extensive testing equipment, and was not supplanted by a technologically feasible alternative. Changes made for Phase II significantly improve the sophistication of the engine simulation, and extend its use to simulate the fuel economy of some, but not all, trailers when combined with tractors. But, even with the Phase II changes, the regulation remains ex ante, not ex post; and it relies on pre-sale simulation, not vehicle-level testing, of fuel economy and GHG emissions, the items intended to be controlled. There is a crucial assumption that the ex ante simulation provides a reasonably good approximation of what actually occurs in the real world over the vehicle's lifetime. Without that assumption, there can be no assurance of program effectiveness. With light-duty vehicles, there is pre-sale testing of fuel economy and GHG emissions, and this is complemented by testing of vehicles after sale at the rate of a few thousand vehicles per year to determine if they continue to comply with standards when actually in use. If a test group is determined to be failing, the vehicle model may be recalled provided the problem is deemed repairable. The in-use testing assesses whether light-duty vehicles continue to comply with the laboratory tests that were required pre-sale (DOE and EPA, n.d.). With MHDVs, as noted in Chapter 2, once they are sold, in-use testing is lacking. Even if an on-road test suggested that a particular MHDV exceeded the standards, it is not clear that this could be used as a basis for enforcement since the vehicle was originally certified using a simulation model. The existing regulatory approach, were it to remain a stand-alone approach, would thus have lower efficacy than if it were to be combined with an ex post testing regime.

However, as pointed out in Chapter 3, it would not be easy to apply to MHDVs the in-use testing used with light-duty vehicles because of the major differences in how the two types of vehicles are produced and used. The distinctive feature of the MHDV sector is the tremendous heterogeneity of both the manufacturing companies and the vehicles they produce. While there are many different types of light-duty vehicles, there is a far greater diversity of MHDV types, sizes, and duty cycles. Heavy-duty vehicle production is driven by customer specification of the sort that can lead to far greater variety of pairings between major vehicle components. Unlike light-duty vehicles, the production of an MHDV is typically split between different manufacturers, including the producer of the engine, the producer of the chassis, and another manufacturer that purchases the chassis and adds a body and special equipment. With Class 8 vehicles, the tractor and the trailer are never built by the same company, and they are often not owned by the same company when the truck is operated. On the supply side, the responsibility for an MHDV's fuel consumption is highly fragmented: Quite often, no single company may be responsible for the design and manufacture of the complete vehicle.

Because of these considerations, Chapter 3 observed that (i) simulation represents a more versatile approach than whole-vehicle testing for MHD vehicles and is a reasonable approach in the short and mid term, but (ii) in time, the growing need to reduce fuel use on all types of operation over a full vehicle life span will require in-use methods of compliance. The purpose of this chapter is to consider some alternatives to the current approach used for MHDVs. By the time of Phase III, the current approach will have been in place for well over a decade, and its limitations may have been seen as problematic. It regulates the manufacturer of the engine and the chassis, not the up-fitter who may subsequently modify the body (e.g., into a motor coach or tow truck) and not the owner/operator of the vehicle. And it applies only to new MHDVs and leaves the fleet of older trucks untouched. The narrowness of the approach may eventually become counterproductive. As described below, the experience of another regulatory program, the Clean Water Act, shows that continued narrow-spectrum regulation can generate diminishing returns with regard to attaining regulatory objectives.

As stated in Chapter 2, the problem of reducing MHDV fuel consumption and GHG emissions will ultimately need to be addressed as a societal system optimization and a portfolio of solutions will be needed, including some—perhaps many—for which the National Highway Traffic Safety Administration (NHTSA) and Environmental Protection Agency (EPA) do not have direct authority. The issue will then arise of coordinating the policy instruments used by NHTSA and EPA to promote fuel economy and emissions reduction with the policies and instruments adopted by other actors to achieve those same goals. In that context, a narrow reliance by NHTSA and EPA on pre-sale simulation may become counterproductive.

The chapter is organized as follows. Section 13.2 describes some general principles applicable to the design of regulations of the type being considered in this report. Section 13.3 examines the role that price signals can play to incentivize behaviors needed to meet the Phase III goals. Section 13.4 concludes with a discussion of some regulatory efforts that could complement a possible Phase III regulation issued by NHTSA and EPA.

13.2 SOME GENERAL PRINCIPLES FOR REGULATION

The question of how to frame regulations that shift people's behavior in order to achieve objectives set by the regulator is a topic that has been examined in the economics literature and other policy literatures for at least 50 years. Several general principles have emerged from those literatures. Here we review three such general principles: (1) a regulation that takes the form of a performance standard is generally likely to be superior to other approaches; (2) a regulatory approach that targets a narrow subset of the factors that cause the outcome of interest to the regulator is likely to be less effective, and less economically efficient, than one that targets a broader subset; and (3) any regulatory initiative can have some unintended consequences that diminish its effectiveness or its efficiency but this often comes about because a trade-off has to be made in the choice of a regulatory approach.

13.2.1 The Desirability of a Performance Standard

If a regulation takes the form of a performance standard, it specifies a target outcome, but not the method by which that outcome should be attained. Examples of performance standards for MHDVs could include an annual level of fuel consumption per ton-mile, or an annual level of emissions per ton-mile (these are performance *rate* standards), or, more directly, a level of annual fuel consumption or a level of annual emissions (these are performance *mass* standards). The current approach in Phases I and II sets a rate standard for the engine in pre-sale testing. A performance standard would regulate the actual fuel consumption or actual emissions in operation on the road, which is really what one wants to reduce. With such a performance standard, the regulated entity would be the vehicle owner/operator, whose actions actually determine the fuel consumption and emissions that are to be controlled. The performance standard leaves it up to the owner/operator to decide how to comply with the standard—by changing the age profile of the fleet, making the vehicles more aerodynamic, further reducing the rolling resistance of the tires, increasing the vehicles' cargo capacities, reducing the energy consumption of accessory equipment, changing the routes traveled, investing more in driver education, making greater use of telematics that improve vehicle operation, and so forth. Rather than singling out a specific pathway, the performance standard leaves the owner/operator free to choose any combination of measures that meets the standard. It is a reasonable presumption that owners/operators will choose the most cost-effective approach for their particular circumstances. An equipment mandate or a technology mandate, by contrast, lacks this flexibility and therefore may turn out to be more costly. Moreover, as the National Research Council (NRC, 2010) report *Technologies and Approaches to Reducing the Fuel Consumption of Medium- and Heavy-Duty Vehicles* (henceforth the "NRC Phase One Report") noted, a technology mandate that is not appropriately "tuned" to the operational aspects of a particular class of vehicles may fail to achieve the desired outcome due to incompatibility between the technology and how the vehicle is used. In such cases, especially when characteristics or operations are uncertain, a performance standard offers a surer way of achieving the desired outcome.

The drawback of a performance standard is that it has to be possible to measure the performance outcome. This is a challenge with MHDVs. A pre-sale equipment mandate or a technology mandate is much easier to monitor for compliance, which is why that approach was adopted in Phases I and II of the MHDV regulations. But, the failure to measure performance outcomes is a two-edged sword. It precludes the use of a performance standard and it leaves uncertainty as to whether the equipment or technology mandate is actually achieving the desired regulatory goal over the operating life of the regulated vehicles. The longer this uncertainty persists, the stronger the case for introducing some system of outcome measurement. It is that observation that underpins Chapter 3's conclusion of a need, ultimately, for effective measurement of fuel consumption in use. Below we explore how some form of in-use measurement of fuel consumption could also open the way to a having a performance standard regulation as a complement or substitute for engine or vehicle certification.

Finding: If it can be implemented, a performance standard, which specifies a target outcome but leaves the regulated entity free to choose the best way achieve that outcome, is generally the most efficient form of regulation.

Finding: The current approach to regulating fuel consumption and GHG emissions of MHDVs under the Phase I and Phase II rules is not a performance standard. The policy objective is fuel consumption and/or emissions in the vehicles' use phase. At present these are not measured (except for the lightest MHDVs) and they are not regulated.

Recommendation 13-1: If and when widespread adoption of in-use measurement of fuel use or emissions comes about, serious consideration should be given for future phases of MHDV rulemaking to the adoption of regulation via a performance standard, such as a cap and trade system or fuel taxes, that directly targets fuel consumption or emissions. Such consideration should evaluate the potential gains from a performance standard as well as the practical difficulties and costs on implementing such a system.

13.2.2 The Clean Water Act: A Case Study of Diminishing Returns When Continuing to Adopt a Narrow Approach

The modern law governing surface water pollution in the United States is the Clean Water Act (CWA), enacted in 1972. The 1972 CWA had as its objective "to restore and maintain the chemical, physical and biological integrity of the Nation's waters." To this end, it set the goal that the discharge of pollutants into the waters of the United States be eliminated by 1985. It also set an interim goal of achieving water quality sufficient for "the protection and propagation of fish, shellfish, and wildlife, and [that] provides for recreation in and on the water" ("the *fishable, swimmable* goal") by July 1983.

The Act prohibits the "discharge of any pollutant by any person" into the waters of the United States unless the discharger has obtained a permit and complies with its conditions, including restrictions on which pollutants may be discharged and how much. However, the permit system contains several important exemptions which, in effect, create an opening for the unregulated discharge of water pollutants. The two most important exemptions are that (1) the requirement for a permit is restricted to "point sources" of pollution and not other ("nonpoint") sources and (2) most agricultural point sources became exempt from the requirement for a permit.

Congress made a conscious decision to limit the permit requirement to point sources, defined as a source of water pollution carried by a "discernible, confined, and discrete conveyance." This definition primarily covers discharges by industrial sources (including manufacturing, mining, and oil and gas extraction facilities), government facilities (including publicly owned wastewater treatment plants and military bases), and some agricultural facilities. The Act itself does not define a nonpoint source, but this is usually considered to be pollution "caused by diffuse sources" and associated with rainfall and runoff, including "agricultural, silvicultural and urban runoff, runoff from construction activities, etc."[1]

The 1972 CWA held back from active federal intervention to control nonpoint sources; section 208 left it to the states to decide whether some action was needed for their nonpoint sources and, if so, what. While some states regulated some forms of nonpoint source pollution, in most cases the 208 effort was left to voluntary action. Little was accomplished, and EPA looked on passively (for example, see Gould, 1990, pp. 463, 473-474, 489-491). Because of the perceived ineffectiveness of section 208, when Congress revised the CWA in 1987, in the Water Quality Act (WQA), it created a new section 319. This still left the responsibility with the states, but it directed them to identify waters impaired by nonpoint source pollution, determine the sources of that impairment, and develop Best Management Plans (BMPs) to address the nonpoint source problem. However, most commentators have concluded that section 319, too, has been ineffective (Andreen and Jones, 2008, pp. 27-28; Gould, 1990, pp. 463, 473-474, 489-491; Ruhl, 2000, pp. 298-299).

The original, 1972 CWA defined "pollutant" to include "agricultural waste discharged into water." It specifically included discharges from a "concentrated animal feeding operation" within the definition of a point source. Presumably irrigation return flows, discharged through pipes, ditches, or channels, would also meet the definition of a point source. However, EPA decided to exempt irrigation return flows from the category of point source pollution. In 1975, a federal court held that EPA exceeded its authority in creating this exemption, and EPA moved to include irrigation return flows in the point source permitting program. In 1977, Congress amended the CWA to undo this. For good measure, it added an amendment specifically prohibiting EPA from requiring a permit for discharges "composed entirely of return flows from irrigated agriculture" and, elsewhere, it described irrigation return flows as "agriculturally ... related nonpoint sources of pollution" (Gould, 1990, p. 474; Ruhl, 2000, pp. 293-295). Hence, farms that discharge sediment, animal wastes, fertilizers, or pesticides via their irrigation return flows can do so freely (Ruhl, 2000, p. 295).

For the first two decades of the point source permit program, EPA focused on issuing permits to control discharges from industrial sources and sewage treatment plants. The CWA had also provided EPA with the authority, under certain conditions, to require permits for stormwater discharged through point sources, but EPA did not exercise this authority. The 1987 WQA dealt with this and created a framework for the permitting of municipal and industrial stormwater discharges. In doing so, however, it explicitly excluded agricultural stormwater discharges (Ruhl, 2000, pp. 295-296).

In summary, the 1972 CWA was the first comprehensive effort by the federal government to control water pollution in the United States. It set the ambitious goal of eliminating, by 1985, the discharge of all water pollutants. But, in implementing that goal, a rather narrow approach was adopted which focused on controlling discharges from primarily nonagricultural point sources. There were some reasons for this as a first approach. Regulating some 60,000 point sources was a much more manageable task than trying to regulate other forms of pollution that

[1] Nonpoint pollution also includes sediment from erosion, runoff, and siltation from mining activities, saltwater intrusion, and changes in the flow of water caused by the construction of dams and flow diversion facilities (hydromodification).

were pervasive, diffuse, and highly site specific in nature. The control of nonpoint source pollution was seen as challenging and not susceptible to a conventional engineering solution.[2] In many cases it could be addressed only through land use controls, traditionally the prerogative of the states. Agriculture, a major source of nonpoint source pollution, was politically powerful (Craig and Roberts, 2015, p. 2; Zaring, 1996, pp. 523-524). As Andreen (2004, p. 562) expressed the problem, "What was the EPA supposed to do, tell farmers how to farm?"

With respect to the pollution sources that were targeted, the CWA was quite successful. Two decades later, twice as many people as in 1972 were being served by municipal wastewater treatment plants providing secondary treatment or better. Less than 1 percent of municipal wastewater was being discharged with no treatment (EPA, 1990). This upgrading of sewage treatment plants resulted in a 46 percent reduction in the discharge of oxygen-consuming pollutants. Had the improvements not occurred, those discharges would have increased by 191 percent due to population growth (ASIWPCA, 1985, as cited in Baker, 1992).

But, although there was considerable progress, the goals of eliminating the discharge of pollutants or rendering the waters of the United States fishable and swimmable have not been attained. An assessment of streams and small rivers reported that 67 percent of U.S. stream miles are in poor or fair condition (EPA, 2006). Overall, about 45 percent of river and stream miles assessed by states and 47 percent of assessed lake acres do not meet applicable water quality standards and are impaired for one or more desired uses (EPA, 2007).

While the regulation of point sources has not been without flaws, continuing unregulated nonpoint source pollution is the dominant factor in the failure to attain the CWA goals (Andreen and Jones, 2008, pp. 15-17). Nonpoint sources are the primary cause of impairment for three-quarters of all impaired waters. Among nonpoint sources, agriculture is the leading source of impairment for rivers, streams, and lakes nationwide (EPA, 2011).

As early as 1977, a U.S. Government Accountability Office (GAO) report had warned that, if left uncontrolled, "nonpoint source pollution will prevent attainment of national water quality goals and will continue to grow in significance as point sources of pollution such as factories and municipal waste treatment plants are brought under control" (GAO, 1977, p. i). One might ask: What should have been done? Should EPA have indeed tried to tell farmers how to farm? Wouldn't that be like telling MHDV users how to operate their vehicles?

In both cases, what needs to be done is to find a way to measure emissions, or to measure behaviors tightly linked to emissions, and then set a performance standard, leaving it up to the regulated parties to determine how best to meet the standard. The lack of measurement is the crucial flaw: What is not measured cannot be regulated. Restricting regulation to what can readily be measured—and failing to improve measurement over time—will distort choices by the regulated parties and lead to unintended, but counterproductive, consequences. Moreover, the longer a source of emissions remains unmeasured and unregulated, the more likely it is that a strong constituency will emerge in support of prolonging the limited regulatory approach.

13.3 UNINTENDED CONSEQUENCES

Any action, including any regulatory action, can always have some unintended adverse (or beneficial) consequences. The challenge is to design the action so as to minimize the chance of unintended adverse consequences. There are at least three possible pathways by which unintended adverse consequences could arise from fuel efficiency and emission regulation of MHDVs: narrow (selective) regulation, regulatory transitions, and the rebound effect.

13.3.1 Narrow (Selective) Regulation

An important consideration here is the breadth of activities covered by the regulation relative to the full range of activities that could impact attainment of the regulatory objective. Regulating point but not nonpoint source pollution is one example of an unduly narrow approach to regulation. Another is the differentially less stringent regulation of the fuel efficiency of light trucks versus passenger cars by NHTSA under Corporate Average Fuel

---

[2] A view strongly challenged by Houck (2002, p. 87), who argues that nonpoint source pollution is no more varied, site specific, or technically difficult to control than most point sources.

Economy (CAFE) standards between 1975 and 2005. The narrower the regulatory focus, the greater the chance of some unintended consequence. This can occur because regulated parties deliberately switch away from regulated choices to unregulated (or less regulated) alternatives, a possibility emphasized by the National Research Council (NRC, 2010) in its Phase One Report, *Technologies and Approaches to Reducing the Fuel Consumption of Medium- and Heavy-Duty Vehicles,* in the context of vehicle class shifting by consumers. The report warned against adopting fuel efficiency regulations that significantly increase the cost, or decrease the performance, for one class of trucks compared to another (NRC, 2010, p. 152) so as not to invite the possibility that differential regulation across vehicle classes would distort consumer choices. A buyer may decide to purchase a different class of vehicle to offset the cost increase or the performance loss with the consequence that a less efficient vehicle is placed into use—exactly the opposite effect of what an efficiency standard aims to achieve.

Even if there were no distortion of consumer choice such as induced class shifting, new opportunities or technologies may emerge spontaneously that, with a narrow regulatory approach, have a freer rein in the unregulated sector and become established there. These are not mutually exclusive. Arguably, both factors—regulation-induced shifts in consumer choice and also exogenous innovations that flourished in the less regulated sector—were at play with both nonpoint source pollution and light trucks.[3] At any rate, while regulation-induced class shifting can occur, as yet there appears to be no empirical evidence from which to assess its magnitude.

Because of its concern with class shifting, the NRC Phase One Report concluded: "While it may seem expedient to focus initially on those classes of vehicles with the largest fuel consumption (i.e., Class 8, Class 6, and Class 2b, which together account for approximately 90 per cent of fuel consumption of MHDVs), the committee believes that selectively regulating only certain vehicle classes would lead to very serious unintended consequences and would compromise the intent of the regulation" (NRC, 2010, p. 189, Finding 8-1). In truth, however, there may be a trade-off. Some sectors may be more responsive to regulation than others. The potential regulatory burden may be lower for some sectors than others.[4] Or it may be more practical to formulate a performance standard for some sectors than others. Those could be valid reasons for an approach that regulates some sectors differently than others. Otherwise, with the same approach applied across all sectors, the regulation would be determined by the lowest common denominator. There is a trade-off between a partial but more efficacious regulation covering some sectors and the risk of class switching with some unintended consequences.

If there were to be differentiated regulation of some MHDV sectors, the important lesson of the CWA is the necessity to press forward with improving the measurement of outcomes in the less regulated and unregulated sectors so that the unintended consequences can be monitored and the regulatory framework extended over time.

13.3.2 Regulatory Transitions

Another form of unintended consequences is associated with what might be called regulatory transitions—when regulations are introduced for the first time or are being modified to a significant degree. If the new regulation, or the modified regulation, significantly raises the cost of a new vehicle, this could have counterproductive consequences in either of two ways. In anticipation of the change, some vehicle owners may decide to accelerate their purchase schedule, obtaining a new vehicle before the new standard goes into effect ("pre-buy"). Or, instead of purchasing a new vehicle, some owners may choose to maintain their existing vehicle and prolong its life ("delayed retirement"). Both behaviors can lead to the phenomenon of a reduction in vehicle purchases immediately following the new standard ("low-buy"). In either case, a less efficient vehicle is substituted for the use of a new, more efficient vehicle and the overall effect of the regulation is dampened, at least in the short run.

Delayed retirement following the adoption of new emission standards for CO and NO_x was identified as a factor for passenger cars by Gruenspecht (1982) and by Thorpe and Harrison (2008) with respect to NO_x and particulate matter (PM) standards for MHDVs in 2007. An econometric analysis by Rittenhouse and Zaragoza-

[3] A key development with the latter was the emergence in the mid-1980s of two new vehicle types, the minivan and the sport-utility vehicle, which were car-like in their performance, albeit regulated as light trucks. Greene (1998, p. 601) notes the uncertainty as to whether the rise of these new vehicle types was due to the stricter fuel economy standards imposed on passenger cars or exogenous social and demographic trends.

[4] Thus, with the CAFE standards, the differential treatment of light trucks versus passenger cars was in part intentional, "reflecting the belief that light trucks function more as utility vehicles and face more demanding load-carrying and towing requirements" (NRC, 2002, p. 21).

Watkins (2017) finds robust evidence of pre-buy and low-buy for trucks following implementation of the 2007 standards, but not following the 1998, 2002, or 2010 pollutant standards, which were less costly to comply with. The 2007 standards caused several thousand more trucks to be sold in each of the months prior to introduction of the standards and approximately the same number of fewer trucks in each of the months after the standards. The pre-buy and low-buy impacts were short-lived and small in volume relative to the estimates of Thorpe and Harrison (2008), and they roughly canceled out, leaving an insignificant net impact on sales (Rittenhouse and Zaragoza-Watkins, 2017, pp. 4, 33).

In general, delayed retirement and pre-buy are likely to be relatively short-lived phenomena. They are unavoidable if regulations are to be updated to stay abreast of changing economic and technological circumstances. Even a price signal such as a tax may require adjustment over time, opening it up to the consequences of a regulatory transition.

13.3.3 The Rebound Effect

This refers to the phenomenon whereby improved fuel efficiency lowers the cost of operating the vehicle sufficiently that it induces the operator to use the vehicle more (travel more miles), thereby causing an increase in the amount of fuel consumed.[5] The increase in fuel consumption from the extra usage of the vehicle diminishes the saving from its increased fuel efficiency, although it may not eliminate it. This is an inevitable consequence of employing what is an *indirect* approach to the regulation of fuel use or emissions via the pre-sale regulation of engine fuel efficiency. The direct approach would be to regulate fuel use or emissions in actual use, whether through a price instrument (a gas tax or a price on emissions) or a quantity instrument (e.g., a cap on emissions). An indirect approach may be justified, for example, because it is difficult to measure directly what is intended to be regulated (fuel use or emissions from individual trucks). There is a trade-off: the convenience of indirect regulation versus the opportunity for an unintended consequence through the rebound effect.

The size of the rebound effect, and therefore the magnitude of the unintended consequence, is an empirical question. It depends on (i) the extent to which the overall cost of using the vehicle is reduced, given other costs of operation, and (ii) the price elasticity of the vehicle usage decision.[6] By way of illustration, suppose that a new standard raises fuel efficiency by 20 percent; this reduces the amount of fuel used per mile traveled (and the fuel component of operating cost per mile) by about 17 percent (= 0.2/1.2). To incorporate the rebound effect, suppose that fuel accounts for about 34 percent of the total operating cost per mile for a truck (ATRI, 2015, Table 10). Then, the total operating cost per mile drops by about 6 percent (= 0.17×0.34).[7] If the price elasticity of miles traveled is 0.5 in absolute value, the reduction in the operating cost per mile triggers a 3 percent increase in miles traveled (= 0.5×0.17×0.34), and the net effect is a reduction in fuel consumption of only 14 percent (= 17 percent – 3 percent). With a smaller absolute value of the price elasticity, the rebound adjustment is smaller and conversely with a larger value of the price elasticity.

Compared to the literature for light-duty vehicles (LDVs), the literature on the price elasticity of MHDV activity is very sparse. There are a few studies of the price elasticity of ton-miles shipped but these results vary widely depending on the type of product being shipped, the geography of the shipments, trip length, and other factors. There are two recent sets of studies estimating the price elasticity of truck VMT for the United States. One pair of studies by Winebrake et al. (2015a,b) uses annual data for the period 1970-2012 on aggregate national VMT and fuel consumption by combination trucks (tractor-trailers) and also by vocational (single-unit) truck operations.

[5] In addition to a change in vehicle miles traveled (VMT), the rebound effect could also take the form of an increase in the loaded weight at which vehicles are operated, or a change in road and traffic conditions encountered as routes or schedules are altered in response to increased fuel efficiency.

[6] In the case of trucking, the usage decision probably lies not with the truck owner but with the shipper who chooses trucking as the mode of transportation. The impact on VMT, therefore, depends on how much of the cost reduction is passed on to the customer. The 2010 report suggests that, because trucking is a very competitive industry, any savings in fuel costs will likely be passed on to shippers.

[7] In the rebound analysis performed for the Phase II Regulatory Impact Analysis (RIA), the vehicle purchase cost was allocated on a per-mile basis over the expected lifetime number of vehicle miles and included in the cost per mile, thereby reducing the percentage cost saving when calculating the rebound effect due to improved fuel efficiency (EPA and NHTSA, 2016, p. 8-17).

They find that, for both types of trucks, the price elasticity of VMT is now close to zero, having changed after 1980 when trucking was deregulated. The other study by Leard et al. (2015) uses a large panel of truck-level microdata for the United States from the Vehicle Inventory and Use Survey (VIUS) in 1982, 1987, 1992, 1997, and 2002, and estimates price elasticities of VMT of 0.18 for combination trucks and 0.12 for vocational trucks. In the RIA for the Phase II rule, NHTSA and EPA chose to place more weight on the studies by Winebrake and colleagues; they assumed a price elasticity of 5 percent for combination trucks, and an elasticity of 15 percent for vocational vehicles. The reasons given were that Leard et al. (2015) was not yet published in a peer-reviewed journal and it uses older data from the VIUS surveys which terminated in 2002.

In fact, however, for estimating the price elasticity of VMT, the VIUS truck-level data are likely to be far superior to the coarse national-level aggregate data used by Winebrake and colleagues. The fact that the VIUS effort was discontinued after 2002 is particularly unfortunate, given the need for a regulatory baseline and the importance of tracking the impact of developments in the trucking industry since 2002. In its first report, *Reducing the Fuel Consumption and Greenhouse Gas Emissions of Medium- and Heavy-Duty Vehicles, Phase Two: First Report*, the committee highlighted the failure to restart VIUS as a significant weakness in the current regulation of truck fuel efficiency and safety (NRC, 2014, p. 11).

Finding: A finding of the committee's 2014 report was that there are no existing data that meet the needs of a baseline for the MHDV fleet. Unless a reliable, repeatable data collection process is established (as recommended in that report), there will be no way to obtain a future measure of whether there is meaningful progress in meeting the regulatory goals of the Phase I and Phase II rules.

Recommendation 13-2: NHTSA should establish a reliable, repeatable survey for truck operators. NHTSA and EPA should also reinstitute the VIUS, as the committee recommended in its first report, to provide an accurate baseline for forecasting vehicle number and vehicle technology penetration by class. The major manufacturers should be contacted for accurate numbers today and for their forecasted penetration rates in the next decade.

13.3.4 Other Unintended Consequences

A commonality linking the rebound effect and the selective regulation discussed above is the mismatch between the formulation of the regulation and the regulatory objective animating it: If the regulation does not target what the regulator aims to control, unintended consequences can arise that impede attainment of the underlying regulatory objective. Another situation in which unintended consequences can occur is if the regulation does not adequately account for the heterogeneity of the behaviors being regulated—in the case of MHDVs, the heterogeneity of the utilization and duty requirements for different trucks. An example was recently noted by Knittel and Sandler (2018) in the case of LDVs. They demonstrate that, if one is using a uniform gas tax as an indirect way to control vehicle emissions, a substantial economic welfare loss can arise when per-gallon emissions differ widely across passenger cars. What is needed to avoid that outcome, they show, is either a direct tax on vehicle emissions or a gas tax differentiated based on vehicle type.

The formulation of a regulatory policy often presents a trade-off: There may be physical, economic, or political obstacles which prevent direct regulation of what the regulator aims to control. The result is a regulation that may be only partially effective or excessively costly, including some costs associated with unintended consequences.

13.4 USING A PRICE SIGNAL

There are several ways a price signal could be designed so as to incentivize a reduction in MHDV fuel use or emissions. The signal could take the form of an increase in the price of fuel (raising the gas tax) or it could be implemented by setting a price on emissions, directly or implicitly. Because fuel consumptions and emissions are linked, a price on one would have some effect on the other, although not necessarily as strong and, for the reason mentioned above, not necessarily efficient in economic terms.

13.4.1 Raising the Fuel Price

Most economists would say that raising the price of fuel, whether through a higher fuel tax or by some other means, is the best way to lower fuel consumption and reduce GHG emissions. Whether this is so depends on the criterion by which the policy intervention is being evaluated.

In economics, the criterion commonly applied is that the policy chosen should maximize aggregate national net benefit (i.e., aggregate benefits minus aggregate costs, regardless of to whom they accrue). An alternative, less stringent, criterion would be that aggregate benefits exceed aggregate costs (net benefit being positive, rather than being maximized). But in the real world, governments sometimes apply a different criterion, and policies are sometimes adopted regardless of whether they pass a net benefit test. For example, a government may set a target of reducing emissions by 30 percent by 2030 (as was done by the European Union for CO_2 emissions) regardless of whether this is known to pass a net benefit test. It may be done because the available information is insufficient to determine whether a net benefit test would be passed. It may be done because meeting that target is considered a moral imperative, or is desirable in other way. This may or may not be regrettable, but it happens sometimes.

If an MHDV policy is seen as motivated by the objective of meeting a fuel use or emissions reduction target, the question to be addressed is this: If the price of fuel were raised, how much of a reduction in fuel use would that generate (this might be called the *efficacy* of a price instrument)?[8] If an MHDV policy is seen as requiring a net benefit test, the question to be addressed is this: What is the economic cost and consequent reduction in social welfare when the price of fuel is used to reduce fuel consumption (i.e., the *economic efficiency* of a price instrument)? Of course, both considerations may underlie the MHDV policy in question. But policy makers should make clear what objectives or combination of objectives is being pursued, so that the trade-offs between efficiency and efficacy can be evaluated.

13.4.1.1 Impact of Price on Fuel Consumption

The aggregate MHDV fuel consumption and emissions depend on decisions made by many actors, no one of whom alone exercises a decisive influence. They include the vehicle manufacturers, equipment manufacturers, vehicle buyers, vehicle owners, vehicle operators, shipping managers, and shipping customers. Reducing fuel consumption and emissions at an aggregate scale is a coordination problem: The multiple actors have to be in sync. For example, it does not help if vehicle owners want a particular type of fuel efficient truck but the truck manufacturers do not produce that type of truck—or, conversely, if the vehicle manufacturers produce a particular type of truck but fleet owners do not buy it. Similarly, if drivers do not operate the vehicle well, the fuel savings designed into it may not materialize. In this context, it is clear that the price of fuel has the potential to influence the broadest possible set of actors. In addition to influencing fuel use, a higher fuel price would tend to lower miles driven, thereby reducing emissions. Compared to a technology mandate, the price of fuel influences not only vehicle technology but also vehicle operation and it affects vehicles already on the road, not just new vehicles.

Just how responsive various actors are to a price signal is an empirical question. One would expect this to vary with the actor and, perhaps, with the circumstances. However, compared to light-duty vehicles, one would expect commercial trucking to be more responsive to a change in the price of fuel because the cost of fuel is a large component of total operating expenses and, therefore, more likely to be salient to decision makers in the trucking industry.[9] That said, there are still some factors that may mute the impact of price signals with MHDVs. These factors occur with varying degrees of frequency in any situation where people are making choices, whether they are consumers or firms. Nevertheless, they have received particular attention in the context of energy efficiency investments which people fail to make even though these would appear to save money—a phenomenon referred to as the *energy efficiency gap* or the *energy efficiency paradox*.[10] One explanation is that those investments really

[8] As explained below, there may be ways to influence consumption other than by raising the price; for example, it may be possible to alter nonprice attributes that influence consumption decisions.

[9] To the extent that fuel surcharges are imposed by trucking companies, this passes the burden of price spikes to shippers and reduces the trucking companies' sensitivity to fuel prices.

[10] For a survey, see Gillingham and Palmer (2014).

would not save money; another is that people are being irrational. Both are possible explanations, but they by no means exhaust the possibilities. There are some other explanations that are consistent with richer models of decision making and rationality. In no special order, these include the split incentives, capital constraints, uncertainty, and the use of heuristics and simplified decision procedures.

13.4.1.2 Split Incentives

It can happen that there is a mismatch between who pays for an investment and who receives the benefit from it, or who determines the magnitude of the benefit. Such a mismatch lies at the heart of the split-incentives problem (also known as the principal-agent problem). It can cause the investment not to be made even though it is—or it can become—profitable in the aggregate, just not for the actor who would have to pay for it. In the case of trucking, three examples have been identified involving (i) the role of the driver, (ii) leased trailers, and (ii) the resale of tractors after a few years of use by the initial purchaser.

Driver behavior, including nonoptimal route choice, idling, and driving techniques such as aggressive acceleration driving and suboptimal vehicle speed or choice of gear shift points, can significantly affect overall fuel consumption. If drivers do not pay their fuel costs, they may lack the incentive to minimize their fuel consumption, thereby creating a principal-agent problem. Vernon and Meier (2012) estimate that 95 percent of all tractors are operated by drivers who do not pay fuel costs; they account for 91 percent of the miles traveled by the trucking industry. One solution is for employers to offer drivers a bonus for reducing their consumption. In 1997, about 23 percent of drivers were offered such bonuses, especially by larger trucking firms. That leaves about 68 percent of the miles traveled by the trucking industry, which are driven by drivers with little or no motivation to reduce their fuel consumption.

Whether this is still an important problem in the trucking industry today is less clear. The survey of stakeholders by Roeth et al. (2013, p. 51) found conflicting perspectives, with some stakeholders seeing split incentives as still being a problem and others believing that it is no longer a problem.Industry participants interviewed by Klemick et al. (2015, p. 161) were aware of the issue and many had attempted to address it by offering drivers rewards for better fuel efficiency, or through driver training, or by adopting technologies such as cruise control and speed limiters that control vehicle performance and narrow the variation in fuel economy across drivers.

Another example is the use of leased or rented trailers. Vernon and Meier (2012, p. 272) estimate that about 23 percent of trailers fall into this category. The firm leasing the trailer chooses the type of trailer to buy but does not itself incur the fuel charges associated with use of the trailer, thereby giving it an incentive to ignore trailer characteristics that reduce fuel consumption.[11]

By way of a third example, Klemick et al. (2015, p. 162) suggest that split incentives may be present if the resale market does not value fuel economy features on vehicles, thereby truncating the return on investment for the first owner.[12]

13.4.1.3 Up-Front Capital Cost

Split incentives provide one explanation why what appears to be a profitable investment in fuel efficiency is not undertaken. Another possibility is that, even though the party that pays for the investment would receive a benefit greater than its cost, the item is expensive and requires an up-front investment that is too large to be readily financed. For example, a new long-haul tractor with a sleeper cab costs about $125,000 to $145,000; driver features added to the truck may add another $10,000 (Roeth et al., 2013, p. 29). Financing this large a cost may be a challenge for a small trucking company with limited access to credit.

Also, companies often self-ration their access to credit by imposing an internal requirement that an investment satisfy a payback period test (i.e., the value of the fuel savings from the investment must make up for the additional

---

[11] This could change if carriers priced differentially according to the fuel efficiency of the leased trailer, but that does not appear to happen.

[12] Klemick et al. (2015) note that this could arise because secondary owners often put tractors to different uses than those for which they were originally purchased. Whether or not the resale pricing is inefficient depends on whether or not there is continuing social value to the fuel economy of the used vehicle.

capital cost and other costs of that technology within a set period of time). This is used in many industries as a crude but simple criterion for selecting capital projects and rationing limited capital.

In trucking, according to Roeth et al. (2013, pp. 24-26), large and medium for-hire carriers and lease/rental fleets typically require a payback period of between 12 and 24 months; private fleets require a payback period of 24 to 36 months. This heuristic may be a way to deal with uncertainty regarding the effectiveness of technologies in actual use as well as uncertainty about future fuel prices. It may also reflect the fact, noted above, that fuel saving is often less valued in secondary markets, a state of affairs which makes it important that the investment pays off during the ownership of the first buyer. In trucking, there appears to be a norm to make the payback period roughly one-half of the vehicle ownership cycle. Thus, large for-hire carriers tend to keep their trucks for 4 to 6 years.[13] The trucks average about 120,000 miles per year, and the companies tend to replace them when the maintenance costs reach an unacceptable level at about 500,000 to 700,000 miles.[14] The relatively short payback period is a formidable obstacle to investment in improved fuel efficiency.

13.4.1.4 Consideration of Other Attributes

So far we have focused on the financial cost and benefit of an investment in fuel efficiency as though that were the only consideration in a firm's decision making. But firms are often concerned with other, nonmonetary aspects of an investment decision as well. In the case of trucking, such considerations may include safety, reliability, ease and convenience of maintenance, driver comfort and acceptance, the increased weight of the vehicle,[15] supplier relationship (a preference to purchase from a familiar supplier or manufacturer), and, in the case of alternative fuels, the length of refueling time or the availability of a fueling infrastructure.[16] The unfamiliarity of a new technology may itself be seen as a negative attribute. Whatever the nonmonetary attributes may be, they are weighed alongside the financial net benefit of a technology. The weight they receive is a matter of the decision maker's interests and preferences, not a matter of rationality. In any event, they can trigger a decision against adopting a fuel technology that reduces costs.

13.4.1.5 Uncertainty or Lack of Credible Information

Roeth et al. (2013) identify uncertainty and unfamiliarity as major factors reducing the willingness to invest in new technologies in the trucking industry. There are a variety of duty cycles being performed under very different operating conditions in different parts of the country, and reliability is a crucial consideration for the industry. Past experience is the surest way to know that a given piece of equipment will function reliably and provide the required service. Any new equipment or technology is necessarily suspect.[17] Fleet managers value their own testing experience or that of other firms above that of information provided by manufacturers.[18] Smaller fleets usually do not have the resources to investigate and test new technologies themselves. And, while not all innovators are large fleets, most fleets of all types choose to be followers rather than early adopters of new technologies.

---

[13] According to Roeth et al. (2013, p. 24), the ownership cycle used to be about 4 years, was extended during the Great Recession to 5, 6, and 7 years, and is now being lowered back to about 4 years.

[14] The old truck is then sold into the used truck market, where it is purchased by smaller fleets, owner/operators, or agricultural operators, and put to lower-mileage uses.

[15] Klemick et al. (2015, p. 164) note that, given federal restrictions on the gross maximum weight of heavy-duty trucks, fuel technologies that add weight to the truck (e.g., automatic transmission, compressed natural gas, auxiliary power units, and some fairings) have to pass a higher bar for adoption than those that lighten the truck.

[16] To be sure, these various attributes can have a monetary dimension. But the firm may place a weight on them that goes beyond their monetary cost; for nonmonetary reasons, it may have a willingness to pay for attributes that exceeds their out-of-pocket cost.

[17] "Another factor contributing to the payback gap is uncertainty how technologies will perform in practice. Participants expressed concerns that '*a lot of things can go wrong*'"(Klemick et al., 2015, p. 159).

[18] Roeth et al. (2013, p. 33) quote one interviewee as stating: "Fleets have a general mistrust of standardized drive cycles and the associated test results, as they will not reflect the [actual] performance of the technologies in their operations." Another respondent, a vice president in a large fleet corporation, said, "We only buy what we test—[we] only trust ourselves." The cost of performing in-house testing is an additional financial cost associated with the adoption of a new technology.

Roeth et al. (2013, p. 38) summarize the obstacles to the deployment of new technology caused by the users' lack of familiarity and uncertainty about its performance: "[V]erifying performance in the early stages of technology deployment is challenging, duplicated testing efforts by multiple fleets and manufacturers [are] commonplace, and data is not often shared [with] or trusted by outside parties." The uncertainty may impede adoption through two separate economic mechanisms: risk aversion (i.e., unwillingness to take certain gambles that nevertheless have a positive expected net payoff) and loss aversion (i.e., the potential monetary loss if the investment does not work out is given more weight than the potential upside gain if it does, even if those outcomes are seen as equally likely).

13.4.1.6 Simplified Choice Procedures

When economists think of rational decision making, they tend to think in terms of a decision maker performing a global optimization: The decision maker systematically considers all alternatives, and all the possible attributes associated with each alternative. In order to give due consideration to his budget constraint, he considers every other possible use of the money that would have to be spent on the item in question. He makes a thorough assessment of these things, and chooses the best course of action.

In reality, choices—by firms as well as by households—often fall short of this idealized conceptualization. This was first pointed out in economics by Herbert Simon (1955), who argued that the conventional characterization of decision making is unrealistic and lacks psychological veracity. To perform the implied analysis would be computationally demanding, if not intractable; would require information that might not always be available; and would demand an implausibly large amount of time and attention for processing information and calculating the optimum. Instead, Simon proposed that people display "bounded rationality" and, when making decisions, they "satisfice."[19] These terms embody two notions. First, people often find ways to simplify their decision making. They may limit the set of options considered; they reduce the attributes to a subset that is salient and easy to grasp; and they employ heuristics or rules of thumb to evaluate the options and their attributes. Second, instead of seeking the optimal choice, they look for a choice that is good enough. Commenting on decision making in business organizations, Simon wrote that decision makers can satisfice "either by finding the optimal solutions for a simplified world, or by finding satisfactory solutions for a more realistic world. Neither approach, in general, dominates the other, therefore both have continued to co-exist in the world of management science" (Simon, 1979, p. 498). In the untidy process of decision making under bounded rationality, the financially optimal choice of fuel technology may not get chosen; in fact, it may not even be considered.[20]

13.4.1.7 The Economic Efficiency of a Fuel Price Signal

The factors listed above provide reasons why decision makers, including firms, may make choices regarding technology adoption that appear to work against their own financial interest. Put simply, their own financial self-interest is in fact more complicated than a simple analysis suggests (e.g., split incentives, up-front capital cost); they are motivated by other factors in addition to financial considerations; they face uncertainty or lack information; or they find ways of simplifying their decision process. These phenomena are likely to play out differently with different decision makers. But, to the extent that they occur, they shroud the visibility of price as a signal for guiding behavior.

The notion that, in making a choice, people are inattentive to some types of incentives has received significant attention in economics and has been demonstrated empirically in many contexts. Most of the empirical literature has focused on decision making by consumers, but there are also applications to decision making in organizations.[21] Firms have more resources available for processing information than individual consumers and are less likely to be inattentive to business-relevant considerations, but they are not necessarily immune to these phenomena.

---

[19] *Satisfice* was coined as a mixture of *satisfy* and *suffice*.

[20] Marketers use the term "consideration set" to refer to the (typically small) subset of alternatives actually considered by a consumer when making a purchase decision.

[21] See, for example, DellaVigna (2009) and Jones (2003).

Nonprice regulation provides a way of circumventing some of those phenomena. It shortcuts the decision process by increasing the saliency of certain attributes, removing certain items from the set of alternatives considered by the decision maker, forcing other items into that consideration set, modifying the default outcome if no alternative is chosen explicitly, or otherwise changing what is referred to as the architecture of choice, a term popularized by Thaler and Sunstein (2009) in their book *Nudge*.

To be sure, there is a wide variety of potential regulatory interventions that differ in the extent to which they override individual freedom of choice.[22] All are potentially subject, in principle, to criticism on ethical and economic grounds. The ethical objection is along the lines that a regulatory intervention leads actors to make a choice they otherwise would not have made, thereby reducing their well-being.[23] We focus here on the economic objection.

The economic argument for using a price signal to change behavior rather than any other form of intervention is that, other things being equal, it ensures economic efficiency and minimizes the aggregate cost of whatever reduction in fuel consumption or emissions comes about. Even if another type of intervention could achieve the same overall reduction in fuel consumption or emissions, it would be liable to do so at a higher cost than with a price signal. This holds true if every actor is actively pursuing the financial objective of maximizing profit or minimizing cost. That objective should lead each actor to choose a level of fuel consumption such that the marginal profit/marginal cost from the next gallon of fuel used just equals the price of fuel. If all actors face the same price signal, the marginal profit/marginal cost from the next gallon of fuel used will be equalized across all fuel consumers. Technically, that is the condition to maximize economic efficiency. This is the basis for the argument that the use of a price signal brings about economic efficiency. Another form of regulatory intervention would not necessarily ensure equalization of the marginal profit/marginal cost from the next gallon of fuel used across all fuel consumers, and therefore fails the test of economic efficiency.

The argument rests on the assumptions that all actors are global optimizers, in the manner discussed (and questioned) above, and that they are all motivated solely by consideration of financial costs. If some actors care about nonfinancial attributes—for example, driver comfort and acceptance or relationship with a preferred supplier—the economically optimal public policy does not necessarily require conformity with the principle of equalizing the marginal profit/marginal cost from the next gallon of fuel across all fuel consumers. It is certainly true that some nonprice interventions can be economically wasteful and far less efficient that using a price signal, but the simple presumption that a price signal is always superior fails. The advantages and disadvantages of alternative regulatory approaches depend on the specific contexts in which they operate.

13.5 CAP AND TRADE

Some of the desirable features of a price signal can also be captured through what is known as a cap and trade system. This combines features of regulation via a quantity instrument with those of regulation via a price instrument. Around the time of the 2010 report, legislation employing this type of approach to regulate CO_2 emissions nationally was being discussed in the U.S. Senate, having been passed by the House of Representatives in 2009. The legislation was patterned after the system introduced to control SO_2 emissions in the 1990 Amendments to the Clean Air Act and subsequently adapted by the EU to regulate CO_2 emissions in 2005. H.R. 2454 set limits on CO_2 emissions that would have covered about 85 percent of total U.S. emissions, including emissions from domestic oil refining. Under the proposed legislation, the government would have set a limit (a cap) on the amount of CO_2 that could be emitted annually by each regulated firm, in a manner similar to the limits on emissions of criteria air pollutants by power plants under the Clean Air Act or the limits on the discharge of water pollutants by point

[22] Thaler and Sunstein argue that their policy recommendations meet a criterion of libertarian paternalism, seeking "to expand or maintain freedom of choice as far as possible."

[23] For example, see White (2013). It could be argued that whether or not the intervention reduces well-being depends on the particular factor causing the behavior that is being corrected. If it is due to ignorance or inertia, for example, overriding an actor's free choice might not actually reduce his well-being (Bovens, 2013). There is also the possible question of preference change. After the behavior change induced by the intervention, the actor comes to place a high value on an attribute that previously he valued little. Hence, while his well-being was reduced ex ante, it is increased ex post.

sources under the Clean Water Act. Unlike those limits (except for SO_2), however, the limits in a cap and trade system would not necessarily be binding: If a firm could obtain unused emission permits from another regulated firm, it could exceed its own cap up to the amount of the number of permits thus obtained. This does three things. It places a binding limit on the *total* quantity of emissions across all regulated firms (like a quantity instrument). However, it offers flexibility to an individual regulated firm, permitting it to exceed its limit if it obtains (purchases) unused permits from other firms. And, it creates an economic incentive (like a price instrument) for each regulated firm: the firm can make money by reducing emissions and selling unused permits; it thus has a financial incentive to reduce its emissions.[24]

In the end, the U.S. Senate did not enact its version of H.R. 2454 and the idea of using a cap and trade system to regulate CO_2 emissions has languished in the United States.[25] However, the cap and trade system has some merit and it is something that conceivably could be employed in the future to regulate fuel consumption or CO_2 emissions in the heavy-duty trucking sector. This would become possible only with in-use measurement of fuel use in that sector via some sort of telematics system. Telematic systems provide real-time monitoring of the location, movements, engine performance, and behavior of a vehicle or fleet of vehicles. This is achieved through a network of electronic monitoring devices and cameras on the vehicle combined with a Global Positioning System (GPS) receiver and an electronic Global System for Mobile communication device. These systems are obtainable in new trucks and also can readily be retrofitted for existing trucks. They are being adopted not only because they monitor and save fuel use but also because they can reduce accidents, improve driver performance, streamline maintenance, and provide better operational efficiency and improved customer service.

It has been estimated that 12.6 percent of all commercial vehicles in the United States now have a telematics unit on board.[26] Adoption of these systems is growing rapidly.[27] Adoption will be further stimulated when federal regulations for the use of electronic logging devices to track compliance with hours-of-service regulations in the commercial truck and bus industries take effect in 2019. Consequently, telematics might well be in common use in several MHDV sectors, especially Class 8 vehicles, by the time of the Phase III regulations. Where, and when, this type of in-use monitoring is adopted, it might then be possible to develop some form of a performance standard for reduced fuel consumption and CO_2 emissions.[28]

A cap and trade program would be one form of performance standard. It could be implemented in various ways. The cap could apply to fuel use, to vehicle emissions, or to both. The cap could be differentiated by type of vehicle and duty cycle. The cap could be set annually or for some other length of time. Small fleets could be exempted. Unused permits could be allowed to be carried over for use in a future year. The permits could be allotted for free or some portion could be auctioned (this could be set to change over time). To control price fluctuations

---

[24] The revenue from selling permits goes to the firm, not to the government as under a tax on emissions. Under H.R. 2454, 85 percent of the emissions allowances would have been allotted by the government for free, but the remainder would have been auctioned.

[25] There are exceptions at the state level. In 2005, Connecticut, Delaware, Maine, New Hampshire, New Jersey, New York, and Vermont agreed to establish the Regional Greenhouse Gas Initiative (RGGI), a cap and trade program covering CO_2 emissions from electric power generation. Subsequently, New Jersey left RGGI, while Maryland, Massachusetts, and Rhode Island joined it. The cap and trade program went into operation in January 2009 and continues to the present. In 2006, California enacted legislation that requires greenhouse gas emissions from all sources in California to be reduced to the 1990 level by 2020. This was to be accomplished through a cap and trade program together with a variety of regulatory restrictions. The cap and trade program in California holds refiners responsible for the GHG emissions from the transportation fuels they sell. The cap and trade program started in January 2013 and continues to the present.

[26] Fleetmatics. 2014. "Fleetbeat Report: Economic Impact of Telematics Adoption by Commercial Fleets," Waltham, MA: Fleetmatics. April.

[27] For example, Frost & Sullivan projects that 35 million light-, medium-, and heavy-duty trucks globally will have some sort of telematics connectivity by 2020 (Lau, 2015). "Although telematics technologies and services were first adopted by the long-haul trucking industry in the United States more than 20 years ago, adoption in the non-trucking sector is expected to equal the trucking industry in terms of numbers of connected vehicles by the end of 2014 and will dominate the commercial fleet telematics industry by the end of 2019, accounting for almost 50 million vehicles globally" (ABI, 2014a). "The number of subscribers to OnBoard Diagnostics aftermarket telematics solutions is expected to increase from 9.5 million in 2014 to 117.8 million in 2019" (ABI, 2014b). "Telematics will be an indispensable element of 2020's truck, impacting everything about the vehicle and its usage" (Rishi et al., 2009, p. 9).

[28] A telematics system is not necessarily a panacea since there may be issues with the precision of measurement and also with which variable is being measured. If the goal is to regulate fuel use per ton-mile, for example, it is necessary to measure simultaneously fuel use, vehicle weight, and distance traveled.

in the market for permits there could be a price ceiling and/or a price floor. In short, there would be considerable flexibility in determining the details of a cap and trade system for trucking.

By the time of the Phase III regulations, it is unlikely that this type of system could be applied to *all* MHDV classes, from Class 2b through Class 8. It would have to be limited to the subset of MHDVs with a high penetration of telematics systems. This would most likely be Class 8b vehicles, which account for about two-thirds of fuel consumption by MHDVs.

Finding: In dealing with MHDV fuel consumption and emissions, the question arises of whether to adopt the same regulatory strategy for all classes of MHDV vehicles or different strategies for different classes. If one adopts different strategies, there is a risk of artificially distorting owner choices. Because of differential regulatory impacts on vehicle cost or performance, a buyer may shift to a less regulated class, causing a less efficient vehicle to end up in use, thereby frustrating the purpose of the regulation. On the other hand, if there is significant heterogeneity across classes, applying the same regulatory strategy to all classes may lead to an outcome determined by the lowest common denominator. In particular, there is reason to anticipate that telematics systems which permit in-use measurement of vehicle fuel use will be widely adopted in Class 8b vehicles well ahead other classes, making it more practical to formulate a performance standard for those sectors.

Recommendation 13-3: The 2010 report concluded that selectively regulating only certain vehicle classes should be avoided because of the risk of class shifting. This is a trade-off that needs to be revisited when the Phase III rules are formulated. In particular, consideration should be given to implementing a performance standard first in Class 8 over-the-road vehicles, for which the telematics capability exists, before it is widely adopted in other vehicle classes.

Finding: A cap and trade system is a performance standard that combines some of the desirable features of a price signal (e.g., fuel tax) with those of a quantity instrument directly regulating fuel consumption and/or emissions. This could become possible only if fuel consumption or emissions were measurable in use (e.g., via some form of telematics). Cap and trade involves greater complexity than, say, a gas tax but it could possibly be more efficacious than a gas tax. It also raises design issues that would have to be resolved.

13.6 COMPLEMENTARY REGULATIONS

One could conceive of existing complementary regulations that, like the pre-sale certification currently in effect for MHDVs, would have the objective of reducing fuel consumption and emissions related to fuels produced and used in vehicles. Some such regulations lie potentially within the authority of federal agencies; others lie elsewhere. Already in the United States, there is a renewable fuel standard, and California and Oregon have a low-carbon fuel standard, which has also been considered, but not adopted, by the U.S. Congress.

13.6.1 Renewable Fuel Standard/Low Carbon Fuel Standard

The Renewable Fuel Standard (RFS) program was first authorized by the Energy Policy Act of 2005 and subsequently amended and expanded by the Energy Independence and Security Act of 2007 (the expanded program is referred to as RFS2). These programs require that a certain volume of renewable fuels be used annually and blended into the fuels being used by vehicles in the United States. The specific amount required to be used each year is determined by EPA. The initial RFS mandated that a minimum of 4 billion gallons be used in 2006, with this minimum usage to rise to 7.5 billion gallons by 2012. Under RFS2, the total amount mandated to be used in 2008 was raised to 9 billion gallons, and this will rise to 36 billion gallons in 2022. The renewable fuel amount proposed by EPA for 2015 was 16.3 billion gallons. The RFS2 requirement is broken down into individual mandates for four separate categories: total renewable fuels, advanced biofuels, biomass-based diesel, and cellulosic biofuel. Biofuels qualifying under each category must achieve certain minimum thresholds of life-cycle greenhouse gas emissions reduction (i.e., reduction in the emissions associated with the production, transportation, and use of that fuel): 20 percent reduction for conventional biofuels (primarily ethanol from corn starch), 50 percent reduction for advanced biofuels, and 60 percent for cellulosic biofuel. Under RFS2, all renewable fuel must be made from feedstocks that meet a new definition of renewable biomass, including certain land use restrictions. Issues subsequently raised regarding the operation of the RFS/REFS2 programs include whether the rapid expansion of U.S. corn ethanol production had an undesirable effect on food prices in the United States and elsewhere, whether there were environmentally harmful impacts in the United States or elsewhere due to the conversion of grassland and other noncropland for growing corn (indirect land use impacts), whether there will be sufficient distribution infrastructure to deliver the expanding biofuels mandate to users in the United States, whether the goals set for the supply of noncorn biofuels are feasible, and whether the assumed reductions in life-cycle GHG emissions are correct and, therefore, just how much emissions reduction will actually be accomplished.[29] To implement the system, fuel refiners and importers are the parties regulated; they are required to turn in (retire) Renewable Identification Numbers (RINs) when they sell gasoline or diesel into the domestic surface transportation market. RINs are generated upon production or impost of a qualifying renewable fuel and are typically detached from that fuel when it is blended or sold into the fuel supply system. Each of the four categories of renewable fuels has its own RINs. RINs are tradable and, subject to certain restrictions, durable. The trading of RINs introduces an economic component into what is a quantity mandate for renewable fuels through the potential influence of RIN prices on the market prices of gasoline or diesel.

A Low-Carbon Fuel Standard (LCFS) is a regulation to reduce the carbon intensity of transportation fuels. The regulation of fuel chemistry is not unprecedented.[30] Versions of an LCFS exist in the UK (adopted in 2008), in the EU (adopted in 2009), in British Columbia (adopted in 2010), in Oregon (adopted in 2012), and, since 2007, in California (effective in January 2011). In California, the LCFS regulates transport fuel providers, including refiners, blenders, producers, and importers. It requires them incrementally to reduce the carbon intensity of the fuel mix used in California, starting with a 0.25 percent reduction in 2011 and rising to a 10 percent reduction by 2020. There is an emission factor for each type of fuel, whether a type of fossil fuel or a renewable fuel, accounting for its life-cycle greenhouse gas emissions and including an allowance for the indirect effects on greenhouse gas emissions from changes in land use for some biofuels. Each fuel provider receives a certain number of credits per gallon of fuel, reflecting the carbon reduction target; at the end of the year, the producer must either meet the reduction target or make up the shortfall by obtaining credits from other fuel producers. As with RINs, there is an active market for these credits, which thus has the potential to generate a differential in the market prices of gasoline or diesel fuels based on their carbon intensity. Operationally, a difference between the RFS2 and the LCFS is that the former is closer to a technology standard while the latter is a performance standard. The California LCFS discriminates between fossil fuels of differing carbon intensity (e.g., oil from Canadian tar sands versus oil from Saudi Arabian oil fields):[31] the incentive to reduce emissions applies not only between renewable fuels and fossil fuels but also within the broad category of fossil fuels.

13.6.2 Other Complementary Regulations

Besides the Phase II standards, administered by NHTSA and EPA, there are regulatory authorities at the federal, state, and local levels that affect how, when, and where MHDVs are used, thereby influencing their fuel consumption and their emissions. There is a relationship between these other regulations and the regulation of MHDV fuel consumption and GHG emissions by NHTSA and EPA; in the case of regulation of tailpipe emissions, the relationship is particularly strong and can make compliance with the Phase II standard more difficult. Other regulations may focus not on vehicle technologies but rather on vehicle operation and driver behavior: The innovative technologies for intelligent transportation systems discussed in Chapter 9 might require regulatory

[29] See NRC (2011) for a discussion of some of these issues. Also see, for example, Stock (2015) and U.S. Congressional Research Service (2010).

[30] For example, the 2010 report noted that EPA currently regulates the sulfur content of on-road diesel fuel. EPA also required the use of certain oxygenates—one of which, methyl tertiary-butyl ether, was later banned—pursuant to the 1977 Clean Air Act.

[31] The carbon intensity of each fuel is calculated using a modification of the GREET (Greenhouse gases, Regulated Emissions, and Energy use in Transportation) model developed at the Argonne National Laboratory.

action by agencies other than NHTSA and EPA in order to promote their adoption. To take another example, recent experiences in several countries have demonstrated that congestion pricing could significantly improve traffic flow in cities and thereby reduce fuel consumption and emissions by vehicles in Classes 2b through 8a.[32] Authorizing congestion pricing within cities would fall squarely within the jurisdiction of local and state government.[33] The various regulatory agencies—federal, state, and local—*co-determine* the actual fuel consumption and emissions of MHDVs in use.

In this context, in-use measurement of fuel consumption post-sale becomes of great importance. If such measurement is in place and is being used in connection with the Phase III regulations, it would also provide a tool for assessing the effectiveness of regulations imposed by other agencies, whether federal, state, or local. And without such measurement, it will be difficult or impossible to assess whether the Phase III regulations, or regulations by other agencies, are having a meaningful impact on fuel consumption and emissions by MHDVs. If such measurement happens to come into being, it would benefit all efforts to improve fuel efficiency.

13.7 COSTS OF DELAY

The fuel consumption and GHG emissions in MHDVs are presently controlled with an incremental strategy organized around successive phases of regulation. Thus, measures not adopted in one phase can be reconsidered for adoption in a later phase. Nevertheless, several features of these vehicles and their GHG emissions militate against being too cautious or myopic at each phase. First, the current regulatory approach impacts only new vehicles from a given model year onward; the existing stock of vehicles on the road is unaffected. Since MHDVs typically have working lives of at least 10 years, and in some cases much longer, the overall impact of a new phase of regulations on aggregate fleet emissions and fuel consumption is highly circumscribed. Second, unlike criteria pollutants which are typically flow pollutants (i.e., the harm is associated mainly with the instantaneous flow of emissions), GHG emissions are stock pollutants: They accumulate and remain in the atmosphere (or the ocean) and cause harm for long periods of time, often hundreds of years. Third, as described in Section 12.4.4, the cumulative experience in producing a technology tends to generate improvements and cost reductions over time. The learning process is set back when the introduction of the technology is postponed. In short, delaying the adoption of a more forceful regulation can have some long-lasting consequences.

13.8 REFERENCES

ABI (Allied Business Intelligence). 2014a. Almost 50 Million Non-trucking Commercial Fleet Vehicles Equipped with Telematics by the End of 2019. March 24. Oyster Bay. https://www.abiresearch.com/press/almost-50-million-non-trucking-commercial-fleet-ve/ (accessed December 20, 2019).

ABI. 2014b. OBD Aftermarket Telematics Subscribers to Exceed 117 Million Globally by 2019. March 24. Oyster Bay. https://www.abiresearch.com/press/obd-aftermarket-telematics-subscribers-to-exceed-1/ (accessed December 20, 2019).

Andreen, W.L. 2004. Water quality today – has the Clean Water Act been a success? *Alabama Law Review* 55:537-593.

Andreen, W.L., and S.C. Jones. 2008. The Clean Water Act: Blueprint for Reform. Center for Progressive Reform White Paper 802. July.

ASIWPCA (Association of State and Interstate Water Pollution Control Administrators). 1985. America's Clean Water: The States' Evaluation of Progress, 1972-1982. ASIWPCA, Washington, DC.

ATRI (American Transportation Research Institute). 2015. An Analysis of the Operational Costs of Trucking: 2015 Update. http://atri-online.org/wp-content/uploads/2015/09/ATRI-Operational-Costs-of-Trucking-2015-FINAL-09-2015.pdf (accessed December 20, 2019).

Baker, L.A. 1992. Introduction to nonpoint source pollution in the United States and prospects for wetland use. *Ecological Engineering* 1(2):1-26.

Bovens, L. 2013. Why couldn't I be nudged to dislike a Big Mac? *Journal of Medical Ethics* 39(8):495-496. ISSN 0306-6800.

Congressional Budget Office (CBO). 2009. Using Pricing to Reduce Traffic Congestion. Publication Number 3133.

[32] See, for example, Small and Verhoef (2007, Chap. 4).

[33] Federal agencies could take steps to promote the use of congestion pricing on Interstate and other highways that receive federal funds, for example, by modifying conditions attached to federal highway funding programs (Congressional Budget Office, 2009; GAO, 2012).

Craig, R.K., and A.M. Roberts. 2015. When will governments regulate nonpoint source pollution? A comparative perspective. *Boston College Environmental Affairs Law Review* 42(1):1-64.

Dellavigna, S. 2009. Psychology and Economics: Evidence from the Field. *Journal of Economic Literature* 47(2): 315-372.

DOE (Department of Energy) and EPA (Environmental Protection Agency). No date. How Vehicles are Tested. Available at https://www.fueleconomy.gov/feg/how_tested.shtml (accessed August 9, 2016).

EPA (Environmental Protection Agency). 1990. National Water Quality Inventory: 1988 Report to Congress. EPA/440/4-90/003. EPA, Washington, DC.

EPA. 2006. Wadeable Streams Assessment: A Collaborative Survey of the Nation's Streams. EPA 841-B-05-002. Washington, DC.

EPA. 2007. The National Water Quality Inventory: Report to Congress for the 2002 Reporting Cycle—A Profile. EPA 841-F-07-003. Washington, DC.

EPA. 2011. A National Evaluation of the Clean Water Act Section 319 Program. https://www.epa.gov/sites/production/files/2015-09/documents/319evaluation.pdf (accessed December 20, 2019).

EPA and NHTSA (National Highway Traffic Safety Administration). 2011. Greenhouse Gas Emissions Standards and Fuel Efficiency Standards for Medium- and Heavy-Duty Engines and Vehicles; Final Rule. *Federal Register* 76:57106-57513. September 15.

EPA and NHTSA. 2016. Greenhouse Gas Emissions and Fuel Efficiency Standards for Medium- and Heavy-Duty Engines and Vehicles—Phase 2. *Federal Register* 81:73478-74274. October 25.

GAO (Government Accountability Office). 1997. National Water Quality Goals Cannot be Attained Without More Attention to Pollution from Diffused or "Nonpoint" Sources. CED-78-6.

GAO. 2012. Traffic Congestion Road Pricing Can Help Reduce Congestion, but Equity Concerns May Grow. Report to the Subcommittee on Transportation, Housing, and Urban Development and Related Agencies, Committee on Appropriations, House of Representatives. GAO-12-119.

Gillingham, K., and K. Palmer. 2014. Bridging the energy efficiency gap: Policy insights from economic theory and empirical evidence. *Review of Environmental Economics and Policy* 8(1):18-38. doi: 10.1093/reep/ret021.

Gould, G.A. 1990. Agriculture, nonpoint source pollution, and federal law. *University of California, Davis Law Review* 23(3):461-498.

Greene, D.L. 1998. Why CAFE worked. *Energy Policy* 26(8):595-613.

Gruenspecht, H.K. 1982. Differentiated regulation: The case of auto emissions standards. *American Economic Review*, Papers and Proceedings of the Ninety-Fourth Annual Meeting of the American Economic Association 72(2):328-331.

Houck, O. 2002. The Clean Water Act TMDL Program: Law, Policy, and Implementation. Washington, DC: Environmental Law Institute.

Jones, B. 2003. Bounded Rationality and Political Science: Lessons from Public Administration and Public Policy. *Journal of Public Administration Research and Theory: J-PART* 13(4): 395-412. Retrieved from http://www.jstor.org/stable/3525655.

Klemick, H., E. Kopits, A. Wolverton, and K. Sargent. 2015. Heavy-duty trucking and the energy efficiency paradox: Evidence from focus groups and interviews. *Transportation Research Part A: Policy and Practice* 77:154-166.

Knittel, C.R., R. Sandler. 2018. "The Welfare Impact of Second-Best Uniform-Pigouvian Taxation: Evidence from Transportation." *American Economic Journal: Economic Policy* 10(4): 211-42. DOI: 10.1257/pol.20160508.

Lau, W. 2015. "Strategic Outlook of Global Commercial Vehicle Industry: Trends and Impact on ATA Members." Presentation at American Trucking Association Management Conference and Exhibitions, Philadephia, PA, October 17-20, 2015, Lau presented for Frost & Sullivan.

Leard, B., J. Linn, V. McConnell, and W. Raich. 2015. Fuel Costs, Economic Activity, and the Rebound Effect for Heavy-Duty Trucks. Resources for the Future. RFF DP 15-43.

NRC (National Research Council). 2002. *Effectiveness and Impact of Corporate Average Fuel Economy (CAFE) Standards.* Washington, DC: The National Academies Press. https://doi.org/10.17226/10172.

NRC. 2010. *Technologies and Approaches to Reducing the Fuel Consumption of Medium- and Heavy-Duty Vehicles.* Washington, DC: The National Academies Press.

NRC. 2011. *Renewable Fuel Standard: Potential Economic and Environmental Effects of U.S. Biofuel Policy.* Washington DC: The National Academies Press.

NRC. 2014. *Reducing the Fuel Consumption and Greenhouse Gas Emissions of Medium- and Heavy-Duty Vehicles, Phase Two: First Report.* Washington, DC: The National Academies Press.

Rishi, S., K. Gyimesi, C. Burek, and M. Monday. 2009. Truck 2020 Transcending Turbulence. IBM Global Business Services Executive Report. Somers, NY: IBM Institute for Business Value.

Rittenhouse, K., and M. Zaragoza-Watkins. 2017. Anticipation and Environmental Regulation. MIT CEEPR Working Paper 2017-004.

Roeth, M., D. Kircher, J. Smith, and R. Swim. 2013. Barriers to the Increased Adoption of Fuel Efficiency Technologies in the North American On-Road Freight Sector. Report for the International Council for Clean Transportation. http://www.theicct.org/sites/default/files/publications/ICCT-NACFE-CSS_Barriers_Report_Final_20130722.pdf (accessed December 20, 2019).

Ruhl, J.B. 2000. Farms, their environmental harms, and environmental law. *Ecology Law Quarterly* 27:263-349.

Simon, H. 1955. A behavioral model of rational choice. *The Quarterly Journal of Economics* 69(1):99-118.

Simon, H. 1979. Rational decision making in business organizations. *American Economic Review* 69(4):493-513.

Small, K., and E. Verhoef. 2007. *The Economics of Urban Transportation*. Oxfordshire: Routledge.

Stock, J.H. 2015. The Renewable Fuel Standard: A Path Forward. Center for Global Energy Policy, SIPA, Columbia University.

Thaler, R.H., and C.R. Sunstein. 2009. *Nudge: Improving Decisions about Health, Wealth, and Happiness*. New York: Penguin Books.

Thorpe, A., and R.M. Harrison. 2008. "Sources and properties of non-exhaust particulate matter from road traffic: A review". *Science of the Total Environment* 400: 270-282.

U.S. Congressional Research Service. 2010. Renewable Fuel Standard (RFS): Overview and Issues (R40155, July 14), by Randy Schnepf and Brent Yacobucci. Text from: Congressional Research Digital Collection.

Vernon, D., and A. Meier. 2012. Identification and quantification of principal–agent problems affecting energy efficiency investments and use decisions in the trucking industry. *Energy Policy* 49:266-273.

White, M.D. 2013. *Manipulation of Choice, Ethics and Libertarian Paternalism*. New York: Palgrave Macmillan. ISBN 978-113728776-2.

Winebrake, J.J., E.H. Green, B. Comer, C. Li, S. Froman, and M. Shelby. 2015a. Fuel price elasticities for single-unit truck operations in the United States. *Transportation Research Part D: Transport and Environment* 38:166-177. doi: 10.1016/j.trd.2015.04.006.

Winebrake, J.J., E.H. Green, B. Comer, C. Li, S. Froman, and M. Shelby. 2015b. Fuel price elasticities for single-unit truck operations in the United States. *Transportation Research Part D: Transport and Environment* 38:178-187. doi: 10.1016/j.trd.2015.05.003.

Zaring, D. 1996. Agriculture, nonpoint source pollution, and regulatory control: The Clean Water Act's bleak present and future. *Harvard Environmental Law Review* 20:515-545.

A

Committee Biographies

Andrew Brown, Jr. (Chair) (NAE), is currently the chief technology officer of his own consulting practice, Diamond Consulting. He provides leadership, guidance, and solutions for technology and innovation for major corporations, organizations, and nongovernmental organizations. Dr. Brown has served as a member of the Board of Directors for the Center for Automotive Research since 2015, an Automotive News PACE Judge since 2016, and a CES Technology & Innovation Judge since 2016.

Dr. Brown retired as the vice president and chief technologist for Delphi Automotive, where he provided leadership on corporate innovation and technology issues to help achieve profitable competitive advantage. Prior to this assignment, Dr. Brown had responsibility for common policies, practices, processes, and performances across Delphi's 19,000-member technical community globally and its budget of $1.8 billion, including establishing Delphi's global engineering footprint with new centers in Poland, India, China, and Mexico, among other locations. In December 2014, he was inducted into the Delphi Innovation Hall of Fame.

Dr. Brown came to Delphi from the GM Research and Development Center in Warren, Michigan, where he was director- strategic futures. At Saturn, he was on the Site Selection Team and responsible for the conceptual design and engineering of this innovative manufacturing facility. Dr. Brown began his GM career as a project engineer at Manufacturing Development in 1973 and then progressed in the engineering field as a senior project engineer, staff development engineer, and manager of research and development (R&D) for the manufacturing staff. During this period, he worked on manufacturing processes and systems with an emphasis on energy systems, productivity improvement, and environmental efficiency.

He earned a B.S. degree in chemical engineering from Wayne State University in 1971, an MBA in finance and marketing from Wayne State University in 1975, and an M.S. degree in mechanical engineering focus on energy and environmental engineering from the University of Detroit-Mercy in 1978. As a registered Professional Engineer, Dr. Brown earned a doctorate of engineering in September 1992.

Members

Inês Azevedo is associate professor in energy resources engineering at Stanford University. Prior to that, she was professor of engineering and public policy as well as the co-director for the Climate and Energy Decision Making Center at Carnegie Mellon University. Dr. Azevedo's research interests lie at the intersection of environmental, technical, and economic issues, such as how to address the challenge of climate change and to move toward a more sustainable energy system. In particular, she has been looking at how energy systems are likely to be shaped in the future, which requires comprehensive knowledge not only of the technologies that can address future energy needs but also of the decision-making process followed by different agents in the economy. She received her B.S. in environmental engineering and M.S. in engineering policy and management of technology from IST University in Portugal, and her Ph.D. from Carnegie Mellon University in engineering and public policy.

Rodica Baranescu (NAE) retired as a professor from the College of Engineering, Department of Mechanical and Industrial Engineering, at the University of Illinois at Chicago. Before that, she was manager of the fuels, lubricants, and engine group of the International Truck and Engine Corporation in Melrose Park, Illinois. She is an international public speaker on technical issues related to mobility technology, environmental control, fuels, and energy. Dr. Baranescu has extensive expertise in diesel engine technology and was elected to the National Academy of Engineering (NAE) in 2001 for research leading to effective and environmentally sensitive diesel and alternative-fuel engines and leadership in automotive engineering. She is a fellow of SAE International after being the association's president in 2000. In 2003 she received the Internal Combustion Engine Award of the American Society of Mechanical Engineering (ASME). Dr. Baranescu received her M.S. and Ph.D. degrees in mechanical engineering in 1961 and 1970, respectively, from the Politehnica University in Bucharest, Romania, where she served as an assistant professor (1964-1968), lecturer (1970-1974), and associate professor (1974-1978).

Thomas Cackette retired at the end of 2012 after serving as the chief deputy executive officer of the California Air Resources Board (CARB) for over 20 years. With the Board since 1982, he managed the Board's motor vehicle emission control program, which develops regulations and other programs to reduce vehicle emissions. He also managed the Board's Monitoring and Laboratory Division, which performs ambient air quality monitoring and develops test methods. Overall, 400 professional and support staff are dedicated to these programs, which are contributing to a steady decline in air pollution in California's major urban areas. Mr. Cackette served as a legislative lobbyist for the CARB for several years, and worked 8 years for the U.S. Environmental Protection Agency Motor Vehicle Emission Laboratory in a variety of technical, management, and policy positions. Prior to that he was involved in rocket engine production, testing, and flight performance analysis at Rocketdyne in Los Angeles, where he gained first-hand knowledge of living in the smoggiest city in the United States. He holds an M.S. in engineering and a B.S. in aeronautics and astronautics. He has published papers for the Society of Automotive Engineers and Air and Waste Management Association, and is a frequent speaker on air quality issues.

Nigel N. Clark is the George B. Berry Chair of Engineering and Emeritus Professor for Mechanical and Aerospace Engineering in the Statler College of Engineering & Mineral Resources at West Virginia University (WVU) and associate vice president for academic strategic planning at WVU. He holds a Ph.D. in chemical engineering from the University of Natal, South Africa, and previously held assistant and associate professor positions in the Department of Mechanical & Aerospace Engineering at WVU. Dr. Clark's areas of interest include vehicle design, advanced vehicle concepts, alternative fuels, and the measurement and reduction of vehicle emissions. He has also published extensively in the areas of particle science and multiphase systems. He has conducted research for government and industry in the areas of fuel economy and emissions from heavy-duty vehicles, including buses and heavy hybrid drive vehicles, and works with the International Council for Clean Transportation on technology and efficiency review. Dr. Clark has contributed to understanding the influence of vehicle activity and test cycles on fuel use, and to relating engine and vehicle dynamometer data. He commenced his career with a National Science Foundation Presidential Young Investigator Award, was recognized as a Benedum Scholar by his institution, and is a Fellow of the Society of Automotive Engineers.

Ronald Graves is on staff at Oak Ridge National Laboratory (ORNL) as a technical advisor for transportation R&D and is also a part-time subject matter expert for Energetics, Inc. He retired from his full-time role at ORNL in 2016 as director of the Sustainable Transportation Program, which covered the laboratory's research in vehicle efficiency technologies, fuels, and intelligent transportation systems. He joined ORNL in 1976 after receiving his Ph.D. in mechanical engineering from the University of Tennessee. He has been a member and leader of many projects in transportation fuels and engines since the early 1980s, including work on pathways to higher engine efficiency, alcohol fuels, and the effects of fuel sulfur and fuel composition on combustion and emissions. He currently serves on the technical teams for the U.S. DRIVE Partnership and the 21st Century Truck Partnership. Over 25 years ago, he led the establishment of engine and emissions research at ORNL that continues today. He has participated in Working Groups of the Coordinating Research Council and is a Fellow of the Society of Automotive Engineers. Dr. Graves has a record of over 60 publications and reports that encompass subjects in internal combustion engines, fuels, power systems, and materials. He shares four patents with co-workers. He recently served on the National Academy of Sciences Committee on Fuel Economy of Medium- and Heavy-Duty Vehicles, he is a member of the judges panel for the Automotive News PACE award, and also serves on the selection committees for awards in SAE and ASME.

Dan Hancock (NAE) retired from General Motors effective September 1, 2011, where he was the vice president for Global Strategic Product Alliances. In this newly created position, Dan was charged with building strong product alliance relationships and speeding development and implementation of joint ventures for winning vehicles and technologies. Hancock's previous appointments included GM Powertrain vice president, global engineering and chief executive officer, Fiat-GM Powertrain, based in Turin, Italy. After joining General Motors in 1968 he held various engineering positions within Allison Transmission Division, which later became Detroit Diesel Allison Division in Indianapolis, Indiana. In 1983 he became chief engineer for Detroit Diesel in Redford, Michigan. He became technical director for advanced powertrain at the Chevrolet-Pontiac-GM Canada Group in 1987. In 1992, he was appointed chief engineer of the Small Block V8 engine, and in 1994 was appointed director of transmission engineering at GM Powertrain. In 1997, he returned to Indianapolis where he was named president at Allison Transmission Division. Born in Indiana, Hancock received an M.S. in mechanical engineering from Massachusetts Institute of Technology in 1973 and a B.S. in mechanical engineering from the General Motors Institute in Michigan in 1974. Hancock served as Chairman of the Society of Automotive Engineers Foundation Board of Trustees from 1998 to 2008. He served as president of FISITA, the International Federation of Automotive Engineering Societies, from 2004 to 2006. He was also inducted into the National Academy of Engineering in 2011. He is a recipient of the SAE Medal of Honor, the Great Golden Medal for Service to the Republic of Austria, and the Sagamore of the Wabash recognition from the state of Indiana.

W. Michael Hanemann (NAS) joined the Arizona State University (ASU) Department of Economics and the Center for Environmental Economics and Sustainability Policy in 2011 where he is a Wrigley Chair in Sustainability. He came to ASU from the University of California, Berkeley, where he was a Chancellor's Professor in the Department of Agricultural and Resources Economics and the Goldman School of Public Policy. His research interests include non-market valuation, the economics of water and of climate change, environmental policy, adaptive management, and demand modeling for market research. Dr. Hanemann has served on many NRC committees and was elected to the National Academy of Sciences in 2011. He is currently a lead author and a contributing lead author for Working Group III of the IPCC Fifth Assessment Report on Climate Change. Dr. Hanemann received his B.A. degree from Oxford University in philosophy, politics, and economics, M.S. from the London School of Economics in development economics, and M.A. and Ph.D. from Harvard University in public finance and decision theory and economics. He received an honorary Ph.D. from the Swedish University of Agricultural Sciences and the Lifetime Award for Outstanding Achievement from the European Association of Environmental & Resource Economists. He is an inaugural Fellow of the Association of Environmental and Resource Economists and a Fellow of the American Association of Agricultural Economics.

Winston Harrington is Senior Fellow at Resources for the Future, where his research interests include urban transportation, motor vehicles and air quality, and problems of estimating the costs of environmental policy. He has worked extensively on the economics of enforcing environmental regulations, the health benefits derived from improved air quality, the costs of waterborne disease outbreaks, endangered species policy, federal rulemaking procedures, and the economics of outdoor recreation. Harrington has written or coauthored five books and numerous book chapters. In October 2000, he won the Vernon Award of the Association of Public Policy Analysis and

Management for a paper he coauthored, "On the Accuracy of Regulatory Cost Estimates." Harrington has served as a consultant to U.S. state and federal governments, the World Bank, and the Harvard Institute for International Development and has worked in Lithuania, Mexico, and Poland.

Gary Marchant is a Regents' Professor of Law and Faculty Director of the Center for Law, Science, & Innovation in the College of Law at Arizona State University. He is also a senior sustainability scientist at ASU's Global Institute of Sustainability. Dr. Marchant teaches environmental law, science and technology, genetics and the law, and environmental justice. Prior to joining the ASU faculty, he was a partner at the Washington, D.C. office of the law firm of Kirkland & Ellis, where his practice focused on environmental and administrative law. He received his B.S. and Ph.D. in genetics from the University of British Columbia, his M.P.P. from the Kennedy School of Government of Harvard University, and his J.D. from Harvard Law School.

Paul Menig is the chief executive officer of Business Accelerants™ powered by Tech-I.M., a consulting company. Previously he was employed by Freightliner, where he was responsible for daily production problems, field problems, custom work orders, and advanced engineering for electrical and electronic items such as engines, transmissions, brakes, and safety devices. Mr. Menig joined Daimler Trucks North America in July of 1994 and initially led the development of electronics for the new Freightliner Century Class truck product line. Prior to joining Freightliner, Mr. Menig spent 7 years with Eaton Truck Components, leading a team as large as 65 people in the development of electronic products for automated mechanical transmissions, brakes, and tire pressure control. These activities included some worldwide responsibility and coordination with engineering in Europe and joint venture development with Japanese companies. Mr. Menig has worked for the industrial automation part of Eaton known as Cutler-Hammer, where for 8 years he has led teams working on sensors, factory communications, programmable and motion controllers, and vision inspection equipment. Mr. Menig has also worked 5 years for General Electric in the areas of medical equipment for hospitals, remotely guided military vehicles (smart bombs), and charge coupled device imagers and signal processors. Mr. Menig graduated from MIT in 1976 with a B.S. in electrical engineering. He participated in the ABC program of General Electric, completing the A and B portions. M.S. work in electrical engineering was completed with the exception of a thesis at Marquette University.

Amelia Regan is a professor of computer science and civil (transportation systems) engineering at the University of California, Irvine (UCI). Her research interests include dynamic and stochastic network optimization, parallel and distributed computing, optimal contracting, port operations, logistics systems analysis, freight industry analysis, shipper behavior modeling, freight transportation planning, combinatorial and on-line auction mechanism and algorithm design, transportation economics, data mining, vehicle-to-vehicle and vehicle-to-roadside communication systems (VANets), network design under uncertain demand, humanitarian cyber physical systems, and cyber physical transportation systems. Dr. Regan's research has been published in more than 120 refereed journal articles and conference proceedings papers in *Transportation Research* (A, B, C and E), *Transportation Science*, *Operations Research*, *INFOR*, the *Transportation Research Record*, the *Transportation Journal*, *Transportation*, *Computers and Industrial Engineering*, *IEEE Network* and *IIE Transactions*, and others. She has been at UCI since 1997, where she has had primary faculty appointments in the Departments of Computer Science and Civil and Environmental Engineering and a courtesy appointment in the Paul Merage School of Business (formerly the Graduate School of Management). She was the associate dean for student affairs for the Bren School of Information and Computer Sciences from 2005 to 2009. Previously, she earned M.S. and Ph.D. degrees in transportation systems engineering from the University of Texas, Austin, an M.S. in applied mathematics from the Johns Hopkins University, and a B.S. in systems engineering from the University of Pennsylvania. She also worked as a research engineer, software engineer, and operations research analyst for the Association of American Railroads and United Parcel Service prior to joining the Ph.D. program at the University of Texas.

Mike Roeth is the executive director of the North American Council for Freight Efficiency and leads the Trucking Efficiency Operations for the Carbon War Room. He has worked in the commercial vehicle industry for more than 30 years, with a specialty in brokering green truck collaborative technologies into the real world at scale.

He has a B.S. in engineering from the Ohio State University and a master's in organizational leadership from the Indiana Institute of Technology. Mr. Roeth served as the Chairman of the Board for the Truck Manufacturers Association and as a Board member of the Automotive Industry Action Group. He currently serves on the second National Academy of Sciences Committee on Technologies and Approaches for Reducing the Fuel Consumption of Medium- and Heavy-Duty Vehicles and is a Department of Energy Merit Reviewer for the SuperTruck programs. He understands the customers, operations, and intricacies of the commercial vehicle industry, having held various positions in product development, engineering, reliability, quality, sales, materials, and plant management with Navistar and Behr/Cummins.

Gary W. Rogers is vice president, Advanced Technology, of Roush Industries, Inc. His previous positions included director of research, School of Engineering and Computer Science, Oakland University; president, chief executive officer, and sole director of FEV, Inc.; director of Power Plant Engineering Services Division, senior analytical engineer for Failure Analysis Associates, Inc.; design development engineer for Garrett Turbine Engine Company; and exploration geophysicist for Shell Oil Company. He has extensive experience in research, design, and development of advanced engine and powertrain systems, including homogeneous and direct-injection gasoline engines, high-speed direction-injection passenger car diesel engines, heavy-duty diesel engines, hybrid vehicle systems, gas turbines, pumps, and compressors. He provides corporate leadership for a multinational research, design, and development organization specializing in engines and energy systems. He is a Fellow of the SAE, an advisor to the U.S. Army Tank Automotive Research Development and Engineering Center (TARDEC) on heavy-fuel engines, and previously sat on the President's Advisory Board of Clemson University and on the advisory board to the College of Engineering and Computer Science, Oakland University. He has served as a member of the NRC Committee on Review of DOE's Office of Heavy Vehicle Technologies Program, Committee on the Effectiveness and Impact of Corporate Average Fuel Economy (CAFE) Standards, Panel on Benefits of DOE's Light-Duty Hybrid Vehicle R&D Program, Committee for the Assessment of Technologies for Improving Light Duty Vehicle Fuel Economy, and Committee to Review the 21st Century Truck Partnership. He holds a B.S. in mechanical engineering from Northern Arizona University and an M.S. in mechanical engineering from the University of Colorado.

Charles K. Salter is retired after working 39 years with Mack Trucks, Inc. and Volvo PowerTrain North America for 3.5 years. His experience covers a wide range of heavy-duty diesel engine engineering and development. His most recent positions included executive director at Engine Development, where he was responsible for all engine/system functions (design and analysis; emissions control and fuel economy optimization; electronics system development, performance, and durability test, manufacturing, supplier, sales, and service liaison). This responsibility included the design and production introduction of the world's first fully electronically controlled diesel unit injector (for 12-liter, 6-cylinder) engine in 1990. He was also the executive director at Advanced Engine Engineering, and collaborated with three sites (Volvo, Goteborg; Renault, Lyon; Volvo/Mack, Hagerstown) in advanced heavy-duty diesel engine research projects, setting objectives and managing the budget. He jointly initiated with Detroit Diesel and developed with the Environmental Protection Agency (EPA) and various industry participants, a urea infrastructure for targeted 2007 calendar year engine production. He participated in industry collaborative research through the Department of Energy (DOE) Diesel Crosscut Committee, which was part of the 21st Century Truck Partnership. He was a consultant to Volvo PowerTrain NA from 2005 to June 2007 on advanced large truck diesel exhaust-gas recirculation cooler vibration study/amelioration, as well as on heavy-duty truck hybrid powertrain duty cycle test procedure development for comparative fuel consumption (EPA/industry/High Efficiency Truck Users Forum), and recently, a study of regulatory boundaries for the EPA heavy-duty truck and engine nonconformance penalty rule, particularly concerning criteria for declaring technological laggard. He holds a B.S. in mechanical engineering from Penn State University and an M.S. in mechanical engineering and solid mechanics from the University of Maryland.

Christine Vujovich retired in 2009 from Cummins Inc. as its vice president, Marketing and Environmental Policy. During much of her 31 years at Cummins, she collaborated with other industry representatives to develop industry positions on legislative and regulatory matters affecting engine emissions and the practicality of those laws and regulations given developing technologies. She oversaw all engine certification and regulatory compliance at Cummins and developed a broad-reaching worldwide platform for CO_2 emission reductions across the company. Her work at Cummins also included overseeing the development and marketing of a new light-duty diesel engine for the light pickup and work truck markets. Since retiring from Cummins, Mrs. Vujovich has co-chaired the Health Effects Special Committee on Emerging Technology. She has also served as an adjunct faculty member at the Indiana University School of Public and Environmental Affairs. Mrs. Vujovich has an undergraduate degree in the teaching of earth sciences and an M.S. degree in environmental engineering in civil engineering, both from the University of Illinois. She attended the Yale Executive Management Program.

John Woodrooffe retired in 2016 from the University of Michigan Transportation Institute (UMTRI), where he headed Transportation Safety Analytics and is director of the Commercial Vehicle Research and Policy Program. He was further responsible for the Center for National Truck and Bus Statistics, which conducts nationwide surveys of Trucks Involved in Fatal Accidents and Buses Involved in Fatal Accidents, and the Statistical Analysis Group, which performs analytical modeling and conducts research to advance statistical methods for road and vehicle safety analysis. He is an international expert on policy and safety evaluation of large vehicles including stability and control, accident reconstruction, vehicle productivity, fuel use, and environmental impact. He has participated in many large international technical projects and has been a member of vehicle-related Organisation for Economic Co-operation and Development technical expert working groups. Prior to joining UMTRI, Mr. Woodrooffe founded the Road Vehicle Research Program at the National Research Council of Canada and developed it into a successful, internationally active heavy-truck research laboratory. He was a consultant to Australia's National Road Transport Commission for a unique 3-year performance-based standards development project that produced a new performance-based regulatory system for large vehicle combinations. Mr. Woodrooffe holds both a B.S. and M.S. in mechanical engineering from the University of Ottawa.

Martin Zimmerman is the Ford Motor Company Clinical Professor of Business Administration Emeritus, at the Ross School of Business of the University of Michigan. His career has spanned academia, government, and business. He served as a chief economist as well as group vice president at Ford Motor Company, where he was responsible for corporate economics, governmental affairs, environmental and safety engineering, and corporate social responsibility. Before joining Ford, he taught at the Sloan School of Management at MIT and at the business school at the University of Michigan. He has served on the National Commission on Energy Policy and also served as a Senior Staff Economist on the President's Council of Economic Advisors and as a member of the Panel of Economic Advisors to the Congressional Budget Office. He is presently a member and past Chair of the Board of the National Bureau of Economic Research. His research is concerned with energy policy, government regulation of business, and economic developments in the automotive industry. Dr. Zimmerman earned a Ph.D. in economics from MIT and an A.B. degree from Dartmouth College.

B

Disclosure of Conflicts of Interest

The conflict-of-interest policy of the National Academies of Sciences, Engineering, and Medicine (www.nationalacademies.org/coi) prohibits the appointment of an individual to a committee like the one that authored this Consensus Study Report if the individual has a conflict of interest that is relevant to the task to be performed. An exception to this prohibition is permitted only if the National Academies determine that the conflict is unavoidable and the conflict is promptly and publicly disclosed.

When the committee that authored this report was established a determination of whether there was a conflict of interest was made for each committee member given the individual's circumstances and the task being undertaken by the committee. A determination that an individual has a conflict of interest is not an assessment of that individual's actual behavior or character or ability to act objectively despite the conflicting interest.

Mr. Gary Rogers was determined to have a conflict of interest because the firm of which he was CEO, FEV, has consulting relationships with makers of medium- and heavy-duty truck engines. Beginning in 2013, Mr. Rogers became vice president of Roush, which provides engineering services to manufacturers of medium- and heavy-duty engines, electric powertrains, and medium- and heavy-duty commercial vehicles.

Mr. Dan Hancock was determined to have a conflict of interest because he owns stock in General Motors Corporation, which makes pickup trucks and medium-duty truck engines. Beginning in 2017, Mr. Hancock was elected to the board of Westport Fuel Systems, which develops and supplies components and systems for natural gas fueling of light-, medium-, and heavy-duty vehicles.

Ms. Christine Vujovich was determined to have a conflict of interest because she owns stock in Cummins Inc., a maker of medium- and heavy-duty truck engines.

The National Academies determined that the experience and expertise of the above individuals was needed for the committee to accomplish the task for which it was established. The National Academies could not find other available individuals with the equivalent experience and expertise who did not have a conflict of interest. Therefore, the National Academies concluded that the above conflicts were unavoidable and publicly disclosed them through the National Academies Projects & Activities Repository (NAPAR) (http://webapp.nationalacademies.org/napar/).

C

Committee Activities

FIRST COMMITTEE MEETING
March 20-21, 2013

NHTSA's Presentation on MHDV Phase II
Jim Tamm, National Highway Traffic Safety Administration

Medium & Heavy Duty Fuel Efficiency and GHG Emission Standards, Phase I
Matthew Spears, U.S. Environmental Protection Agency

21st Century Truck Partnership and SuperTruck Initiative
Ken Howden, U.S. Department of Energy

Reflections on GHG Phase I and Considerations for GHG Phase II
John Wall, Cummins

Navistar Fuel Economy and Emissions
Greg Fadler, Navistar

Lessons Learned from FE/GHG Phase I Regulations and Ways to Incorporate the Most Likely Future Technologies into FE/GHG Phase II Regulations
David Kayes, Daimler Trucks North America

Overview of ICCT Heavy-Duty Vehicles Research Activities
Nic Lutsey, The International Council on Clean Transportation

Roadmap Findings and Technology Availability: Implications for GHG/Fuel Economy Regulations
Bill van Amburg, CALSTART

NAS Phase II Study of MHDV Truck Fuel Consumption
Timothy Blubaugh, Truck & Engine Manufacturers Association

Strategic Outlook on North America MHDV Commercial Truck Powertrain Market
Sandeep Kar, Frost & Sullivan

SECOND COMMITTEE MEETING
June 20-21, 2013

SwRI Work in Support of NHTSA and Its Relevance to NAS Study
Tom Reinhart, Southwest Research Institute

Heavy-Duty Greenhouse Gas from a Full-Line Manufacturer's Perspective
Ken McAlinden, Ford

An Integrated, Heavy-Duty Vehicle Approach to GHG Regulation
Sam McLaughlin, Volvo North America

Fleet Operations and Fuel Consumption
Mike Roeth and Paul Menig, Committee Members

OEM Experience with GHG Phase I and Recommendations for Phase II
Michael Christianson, Daimler Trucks

Use of Vehicle Modeling in Engine Development at Cummins
Wayne Eckerle, Cummins

Engine Models and Maps for Truck Certification
Nigel Clark, West Virginia University

THIRD COMMITTEE MEETING
July 31-August 1, 2013

CalHEAT Technology Roadmap
Fred Silver and Tom Brotherton, CalHEAT

CA Efforts to Reduce GHG Emissions
Dan Sperling, California Air Resources Board

Natural Gas Fueled Engines
Tim Frazier, Cummins Westport

GM Heavy-Duty Presentation
Mark Allen and Barbara Kiss, General Motors

Natural Gas Station Infrastructure
Spencer Richley, Clean Energy

ZEV Transformation
Matt Miyasoto, South Coast Air Quality Management District

Alternative and Renewable Vehicle Fuel Program
Janea Scott, California Energy Commission

Transmission Perspectives for the Phase II GHG Rule
Mihai Dorobontu, Eaton

Regulatory Options for Trailers: A Manufacturer's Perspective
Jeff Bennett, Utility Trailer

Trailer Technologies for Increased Heavy-Duty Vehicle Efficiency: Technical, Market, and Policy Considerations
Ben Sharpe, International Council on Clean Transportation

FOURTH COMMITTEE MEETING
September 25, 2013

No open session presentations were held during this meeting.

FIFTH COMMITTEE MEETING
November 21-22, 2013

Test Methods for Truck Tire Rolling Resistance & Reducing Fuel Consumption in MHDVs
Stan Lew, Michelin North America

Demonstrated Efficiency of achatesPower OP2S Engine
David M. Johnson, achatesPower

Transmission Technology and Fuel Consumption
Michael Howenstein and Deborah Gordon, Allison Transmissions

Continental Tire Presentation
Jay Spears and Curtis Decker, Continental

SIXTH COMMITTEE MEETING
January 30-31, 2014

No open session presentations were held during this meeting.

SEVENTH COMMITTEE MEETING
April 28-29, 2014

All of the Above—a Multipronged Approach to Fuel Efficiency
Randy Mullet, Con-way Inc.

WalMart Transportation Efficiency
Elizabeth Fretheim, WalMart

FedEx Fuel Economy and Sustainability Improvements
Tom Griffin, FedEx Express

Frito Lay Fleet Supply Chain
Mike O'Connell, Pepsico

The Difference This Time: Last Cycle's Inflation Leads to This Cycle's Productivity
Kenny Vieth, ACT Research Co., LLC

NHTSA Fuel Efficiency Project Update
Thomas Reinhart, Southwest Research Institute

EIGHTH COMMITTEE MEETING
June 4-5, 2014

Fuel Economy Regulation Impacts on Vocational Vehicles
Chris Yakes, Oshkosh Corporation

NAS Presentation
Eric Woods, Waste Management Inc.

Daimler Trucks North America's Feedback on GHG Phase I and Suggestions for Phases II and III
David Kayes, Daimler Trucks North America LLC

Volvo Phase II Regulation Structure and Technology
Tony Greszler, Volvo Group Truck Product Planning

Meritor WABCO Medium/Heavy Vehicle Efficiency and Safety Technologies
Alan Korn, Meritor WABCO

Advanced Technology and Fuel M/HD Vehicles—To and Beyond 2020
Bill Van Amburg, CalSTART

NINTH COMMITTEE MEETING
July 29-30, 2014

The Trucks of Tomorrow
Darren Gosbee, Navistar International Corporation

TARDEC Fuel Economy Studies
David Gorsich, U.S. Army Tank-Automotive Research, Development and Engineering Center

MSBE Requirements to Support MD & HD Regulations
Aymeric Rousseau, Argonne National Laboratory

Greenhouse Gas Emission Model: A Compliance Vehicle Model for Certification
Houshon Zhang, U.S. Environmental Protection Agency

Future Vehicle Technology
John Maddox, University of Michigan Transportation Research Institute

Preventing Accidents & Saving Fuel by Connecting Trucks
Stephen Boyd, Peloton Technology

Automated Vehicles
Myra Blanco, Virginia Tech Transportation Institute

Costs and Cost Reductions of Medium- and Heavy-Duty Vehicle Fuel Efficiency and Emissions Reduction Technologies
Michael Chan, Tetra Tech
Thomas Reinhart, Southwest Research Institute

An Integrated Approach to Heavy-Duty GHG and NO_x Control
Todd Sax, California Air Resources Board

Impact of NO_x Control on Heavy Duty Vehicle Efficiency
Anthony Greszler, Volvo Trucks North America

Tradeoffs in Improving Fuel Consumption and Emissions
Gary Salemme, Cummins, Inc.

Review of Heavy-Duty Truck Emissions and Efficiency Trends
Timothy Johnson, Corning

Challenges Associated with Improving Fuel Economy and NO_x Emissions for Heavy-Duty and Medium-Duty Applications
James Parks, Oak Ridge National Laboratory

Mobility Transformation Center Overview and Discussion of Connected Technology Enabling Fuel Consumption Improvements
John Maddox, University of Michigan Transportation Research Institute

TENTH COMMITTEE MEETING
December 3-4, 2014

CFD for Tractor-Trailer Aerodynamics: Current Trends and Future Outlook
Ramesh Pankajakshan, UT-Chattanooga

A Comprehensive Approach to Reducing Fuel Consumption (and Everything Else You Don't Like About Trucks)
Jose Holguin-Veras, Rensselaer Polytechnic Institute

National Research Council Presentation
Scott Greene, Federal Railroad Administration

Energy Efficiency Investment Decisions: Lessons from the Heavy-Duty Trucking Sector
Heather Klemick and Elizabeth Kopits, U.S. Environmental Protection Agency National Center for Environmental Economics

Perspectives on Commercial Duty Vocational Vehicle Hybrids
Michael Howenstein, Allison Transmission

Eaton Commercial Vehicle Hybrids Experience
Mihai Dorobantu, Eaton Vehicle Group

Electric Hybrid Powertrains: Past, Present, and Future
Bart Mancini, BAE Systems

Heavy-Duty Vehicle Hybrid Transmission Testing at ORNL
Paul Chambon, Oak Ridge National Laboratory

Assessment of Technologies and Approaches to Reduce Fuel Consumption: DTNA SuperTruck Findings on Hybrids
David Kayes, Daimler Trucks North America

Update on Phase II of the Heavy-Duty Greenhouse Gas and Fuel Efficiency Standards
Ken Katz, National Highway Traffic Safety Administration

The Future of Truck-Rail Intermodalism: Challenges and Opportunities
Dan Murray, American Transportation Research Institute

ELEVENTH COMMITTEE MEETING
March 17-18, 2015

No open session presentations were held during this meeting.

TWELFTH COMMITTEE MEETING
June 16-17, 2015

Safety and Regulatory Impacts Analysis of MD/HDV Fuel Efficiency (FE) Technologies and Alt-Fuels
Alex Epstein, Volpe, The National Transportation Systems Center, U.S. Department of Transportation

THIRTEENTH COMMITTEE MEETING
September 29-30, 2015

No open session presentations were held during this meeting.

FOURTEENTH COMMITTEE MEETING
December 1-2, 2015

No open session presentations were held during this meeting.

FIFTEENTH COMMITTEE MEETING
March 16-17, 2016

Closed sessions involving only the committee.

SIXTEENTH COMMITTEE MEETING
June 20, 2016

Closed sessions involving only the committee.

SEVENTEENTH COMMITTEE MEETING
Dec 7-9, 2016

Closed sessions involving only the committee.

D

Summary of Analysis of Engine and Vehicle Combinations

In 2015, Southwest Research Institute completed a project for the National Highway Traffic Safety Administration (NHTSA) that explored a range of vehicle technologies that could be used to comply with the proposed Phase II truck fuel economy and greenhouse gas emissions regulations. A National Research Council committee is looking at technologies that could become available beyond the Phase II time frame, so a project was created to use the models developed in the NHTSA project to explore technology potential farther into the future.

Four vehicle types were evaluated to cover the range of Class 2b through Class 8. The models include a Dodge Ram 2500/3500 pickup truck (Class 2b/3), a Ford F-650 tow truck (Class 5), a Kenworth T270 box delivery truck (Class 6), and a Kenworth T700 long-haul tractor-trailer truck (Class 8). The project was limited to looking at future engine technologies, so the baseline vehicle configurations were retained throughout the project. The following engine technologies were evaluated in the Class 2b through Class 6 vehicles:

1. 2019 baseline 6.7 diesel engine, compliant with Phase I regulations
2. A stoichiometric, spark-ignited (SI) natural gas engine with efficiency comparable to current production engines of this type
3. A diesel engine with 48 percent brake thermal efficiency (BTE) at the best point
4. A gasoline engine with 42 percent BTE at the best point

For the Class 8 tractor-trailer, the following engine technologies were evaluated:

1. 2019 baseline 15 liter diesel engine, compliant with Phase I regulations
2. A stoichiometric, SI natural gas engine with efficiency comparable to current production engines of this type
3. A high-pressure direct-injection (HPDI) lean-burn, diesel pilot-ignited natural gas engine, based on the 2010 Westport 15 liter engine
4. A diesel engine with waste-heat recovery and 55 percent BTE at the best point
5. A gas turbine engine

The project evaluated vehicle fuel consumption with the engine alternatives, in terms of both gallons consumed and fuel mass consumed. Drive cycle average BTE has been determined for all engine alternatives, along with total energy consumption and CO_2 emissions.

RESULTS:

The stoichiometric SI natural gas engines have by far the lowest thermal efficiency of all the engine options, but they still provide a 10 to 20 percent reduction in CO_2 emissions compared to the 2019 diesel baseline. This CO_2 reduction needs to be treated carefully, because the greenhouse gas advantage from reduced CO_2 can easily be overwhelmed by tailpipe methane emissions and any leakage of natural gas fuel. The diesel pilot-ignited, lean-burn HPDI natural gas engine is almost as efficient as the 2019 diesel baseline. Given additional technical development, it should be possible to make this engine type competitive with the efficiency of future diesel technology. At the current (2010) level of development, the HPDI engine provides approximately a 22 percent CO_2 reduction compared to the 2019 diesel baseline in the long-haul tractor-trailer truck. HPDI is substantially more complex and expensive than a stoichiometric, spark-ignited natural gas engine. In fact, HPDI technology was withdrawn from the market about the time oil prices started to drop, although development of a second generation of HPDI technology continues. The HPDI engine requires an expensive fuel system and the full diesel exhaust aftertreatment system (DOC + DPF + SCR),[1] while the SI natural gas engine needs only a three-way catalyst for aftertreatment. The same caveats regarding methane leakage apply to the HPDI engine.

In the Class 2b through 6 trucks, the 48 percent BTE diesel engine provides fuel consumption and CO_2 reductions of 10 to 12 percent across all vehicles and drive cycles, compared to the 2019 baseline. Compared to a Phase II compliant engine, this represents an additional 6 to 8 percent fuel and CO_2 savings. The performance of the 42 percent BTE gasoline engine in Class 2b through 6 trucks is more complex. In terms of energy consumption and CO_2 emissions, the 42 percent BTE gasoline engine provides benefits ranging from 3 to 29 percent compared to the 2019 diesel baseline, depending on the payload and drive cycle. The gasoline engine performs best with zero payload on transient cycles with lower average power demand, but not so well on higher speed and load cycles. Because a gallon of gasoline has lower energy content than a gallon of diesel, the gasoline engine does not perform so well in terms of gallons consumed. The fuel consumption of the 42 percent BTE gasoline engine ranges from an 11 percent increase over the 2019 baseline diesel to an 18 percent reduction.

There are a couple of factors that influence the performance of the gasoline engine, beyond the thermal efficiency of the engine itself. The weight of the gasoline powertrain was estimated to be 500 pounds less than a diesel or natural gas powertrain. This weight difference reduces vehicle power demand by reducing rolling resistance and the power required to accelerate the vehicle. An additional, unexpected factor helping the gasoline engine is that, over most drive cycles, the driveline (transmission and axle) is more efficient for the gasoline engine. This effect is especially pronounced on lightly loaded, low-speed cycles. A careful review of the results showed that this effect is real, but the magnitude of the effect is highly dependent on details like torque converter match and transmission shift schedule.

In the long-haul Kenworth T700, the 55 percent BTE diesel provides 15 to 18 percent fuel savings and CO_2 reduction, compared to the 2019 baseline diesel (see Figure D-1). These values will be about 4 percent smaller compared to a Phase II–compliant engine. This very large fuel savings shows the large potential benefit of the complex, high-risk technologies that will be required to achieve 55 percent BTE. The results for the SI and HPDI natural gas engines in the T700 have been described above.

[1]DOC, diesel oxidation catalyst; DPF, diesel particulate filter; SCR, selective catalytic reduction.

(a)

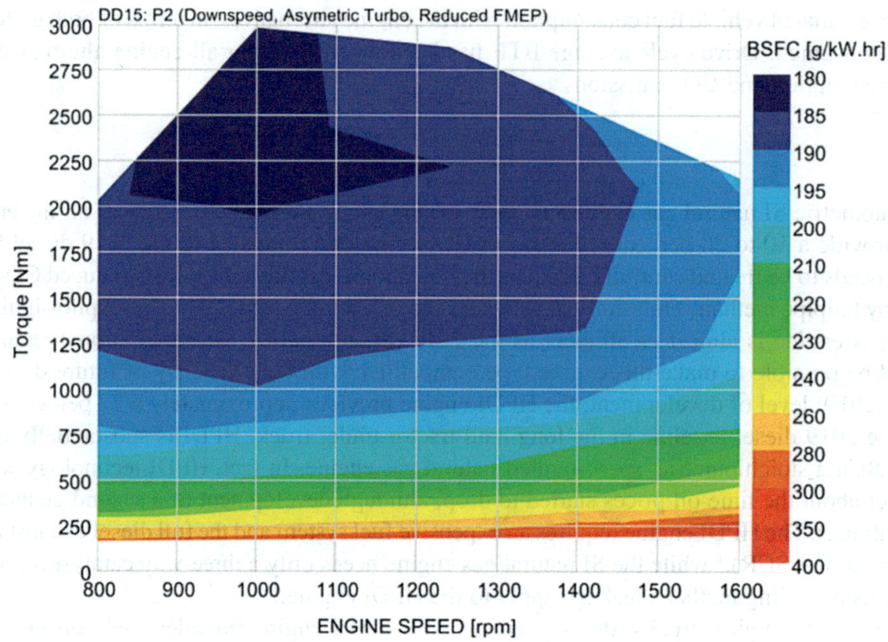

(b)

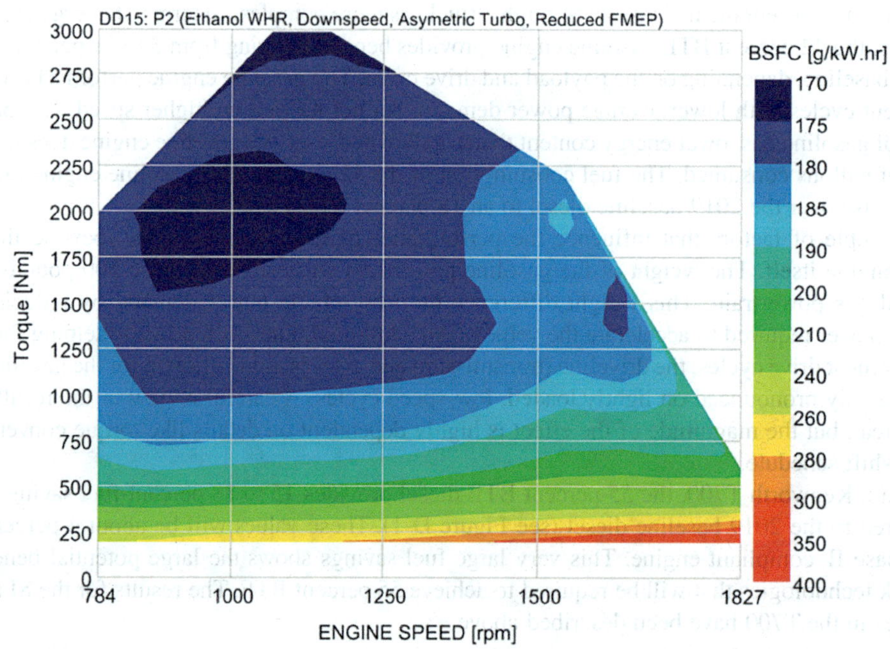

FIGURE D-1 (a) Baseline 2019 diesel plus down-speed and reduced friction mean effective pressure engine, from NHTSA project; (b) engine from (a) with waste-heat recovery added; and (c) engine map achieving 55 percent brake thermal efficiency, derived from (b).
SOURCE: Reinhart (2015).

(c)

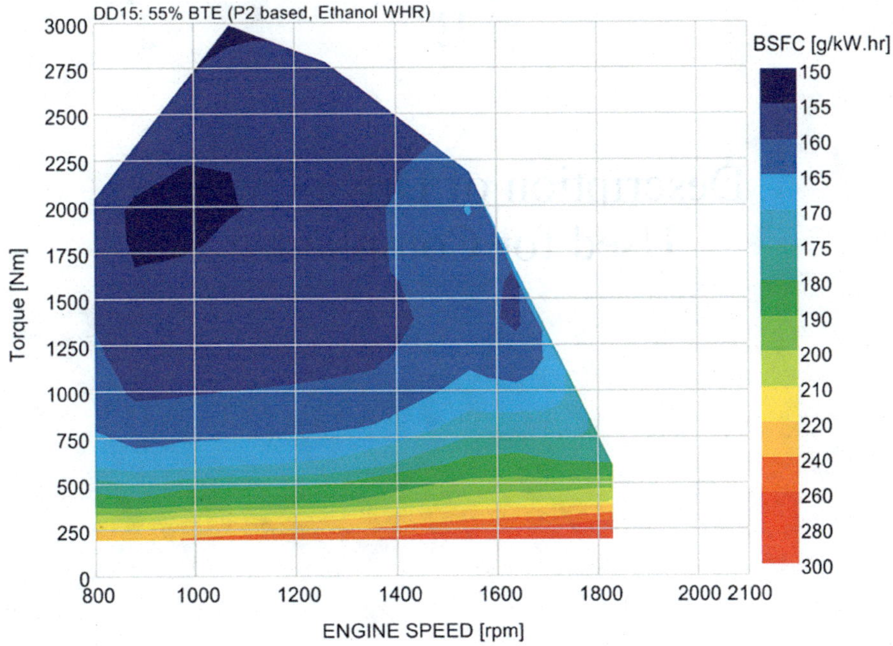

FIGURE D-1 Continued

E

Description of Drive Cycles Used for Compliance

A total of 11 vehicle drive cycles were used in the study. Each cycle is described in this section. The Northeast States Center for a Clean Air Future (NESCCAF) cycle was developed to represent a typical line-haul type of operation. There are brief urban/suburban sections at the beginning and end of the cycle, to represent getting out to the highway. The main portion of the cycle consists of five cruise segments at speeds of 65 to 70 mph. The second cruise segment has an alternating +/-1 percent grade, and the third segment has an alternating +/-3 percent grade. The NESCCAF cycle is the only cycle that includes grades. Because tractor-trailer trucks cannot normally maintain cruise speed on a 3 percent grade, the cycle was given a feature that extended the drive time to force the vehicle to complete the required distance. Figure E-1 shows the speed versus time trace, and Figure E-2 shows the grade versus time trace. This cycle was only used for tractor-trailer trucks.

Figure E-3 shows the World Harmonized Vehicle Cycle (WHVC). This cycle is intended for medium- and heavy-duty trucks and includes urban, suburban, and highway segments. The cycle assumes that a road speed governor is used to limit speed to about 53 mph, which is required in Europe. The WHVC was used for all vehicles, to provide a way to compare results across different vehicle types.

The Federal Test Procedure (FTP)-Highway cycle shown in Figure E-4 has been used for many years to evaluate light-duty vehicle fuel economy. This cycle was developed in the time of the 55 mph speed limit, so speeds are low compared to typical modern highway driving, and accelerations are gentle. This cycle was only used for the pickup truck.

The FTP-City cycle shown in Figure E-5 has been used for many years to evaluate light-duty vehicle fuel economy. The cycle uses relatively gentle accelerations compared to what is found in typical city driving, and idle time is less than would be expected in typical city driving. This cycle was only used for the pickup truck.

The California Air Resources Board (CARB) urban cycle shown in Figure E-6 is used in the Greenhouse Gas Emissions Model (GEM) for medium- and heavy-duty vehicles. This cycle simulates urban driving for trucks. The CARB cycle was used for all vehicles simulated, except the pickup truck.

The US06 cycle shown in Figure E-7 is a more aggressive drive cycle for light-duty vehicles. It was introduced to help compensate for the overly optimistic fuel economy values generated by the older FTP city and highway cycles. This cycle was only used for the pickup truck.

The SC03 cycle shown in Figure E-8 was developed to evaluate the effect of air conditioner use in hot conditions on fuel economy. In this study, the air conditioner was run on all drive cycles, which is not the standard approach. This cycle was used only with the pickup truck.

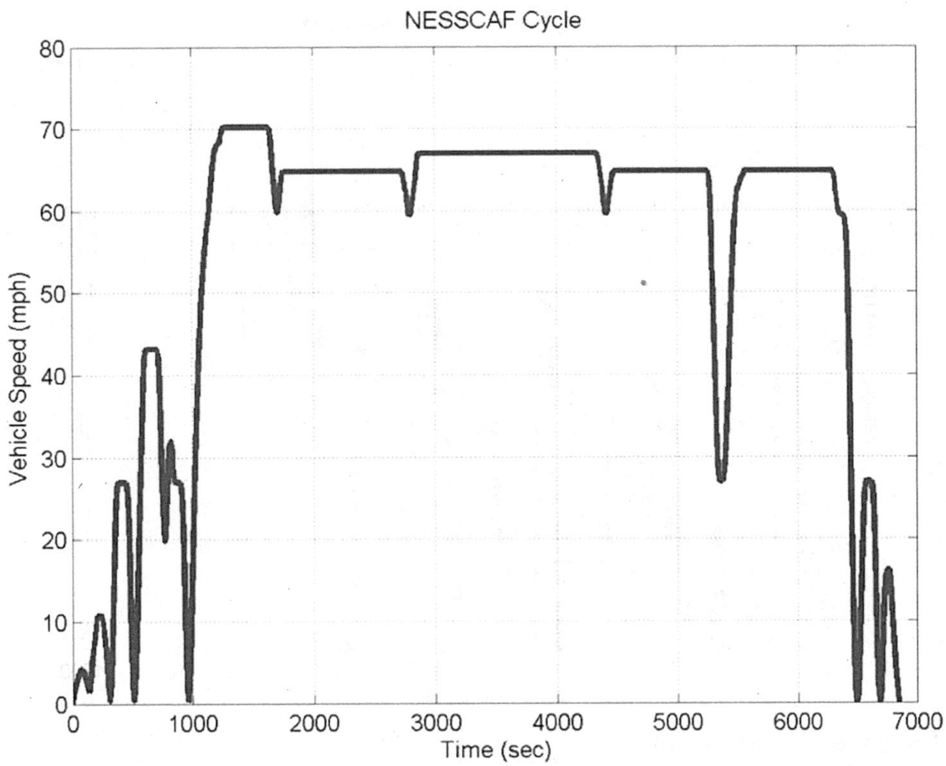

FIGURE E-1 Speed versus time trace for the NESCCAF cycle.
SOURCE: Reinhart (2015).

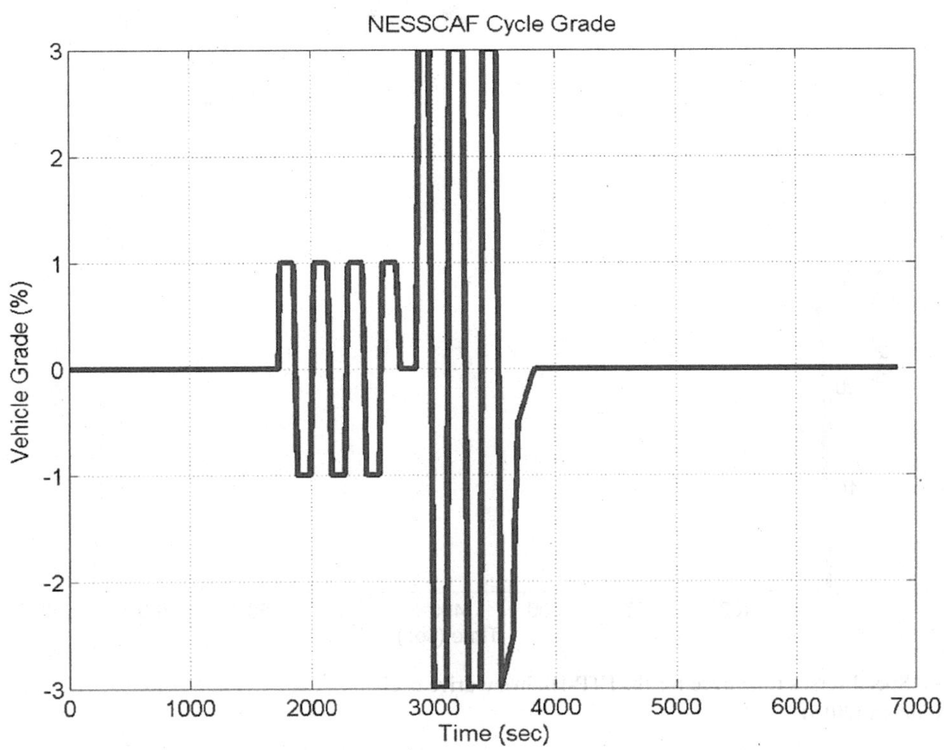

FIGURE E-2 Grade versus time trace for the NESCCAF cycle.
SOURCE: Reinhart (2015).

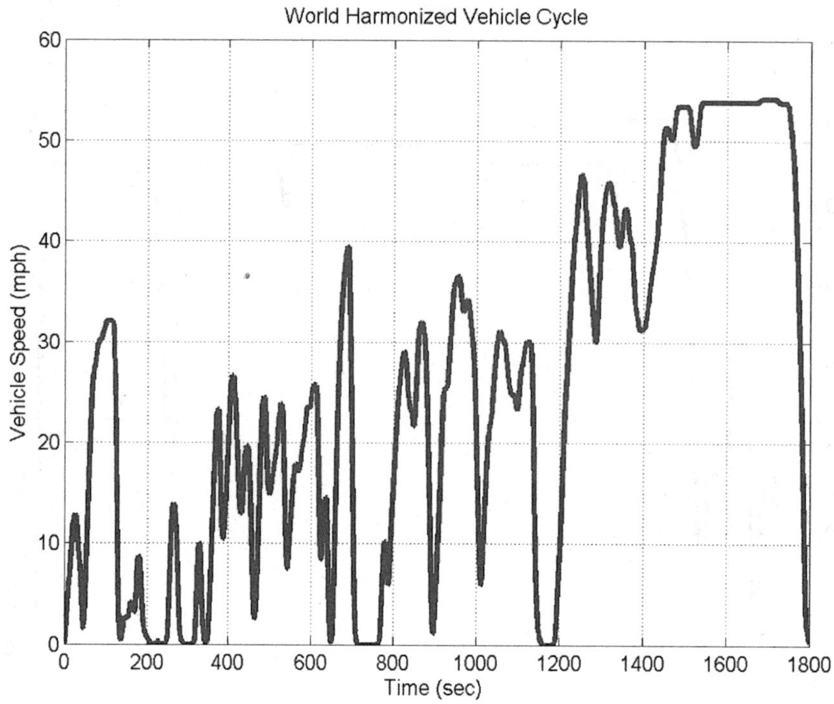

FIGURE E-3 Speed versus time trace for WHVC.
SOURCE: Reinhart (2015).

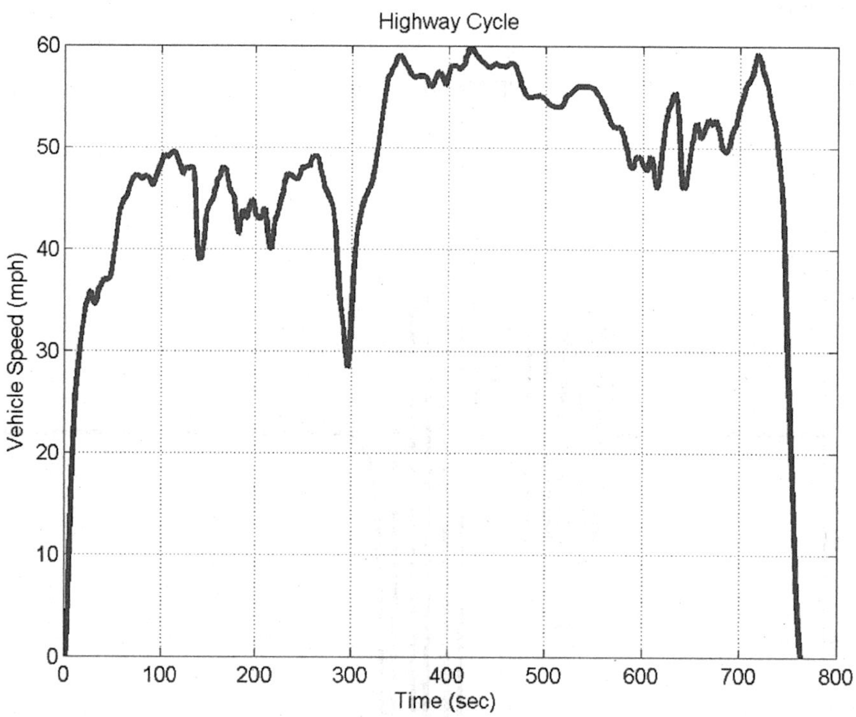

FIGURE E-4 Speed versus time trace for the FTP-Highway drive cycle.
SOURCE: Reinhart (2015).

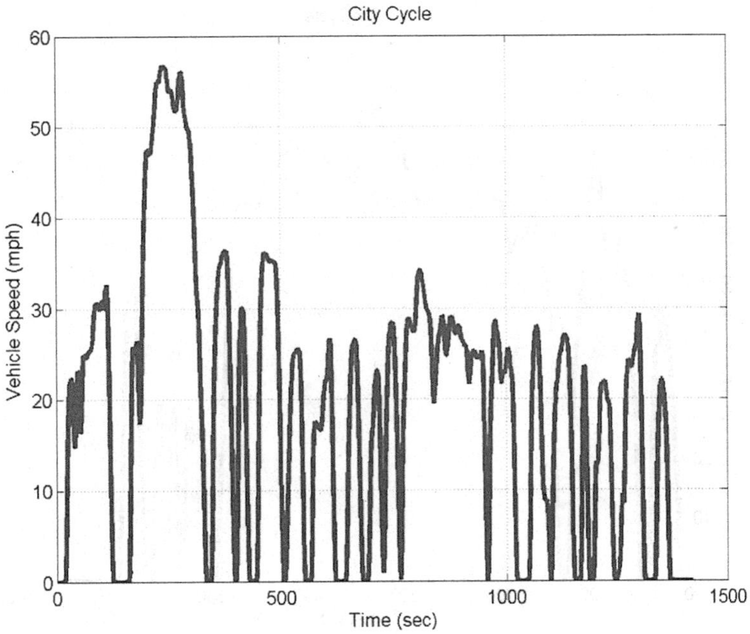

FIGURE E-5 Speed versus time trace for the FTP-City drive cycle.
SOURCE: Reinhart (2015).

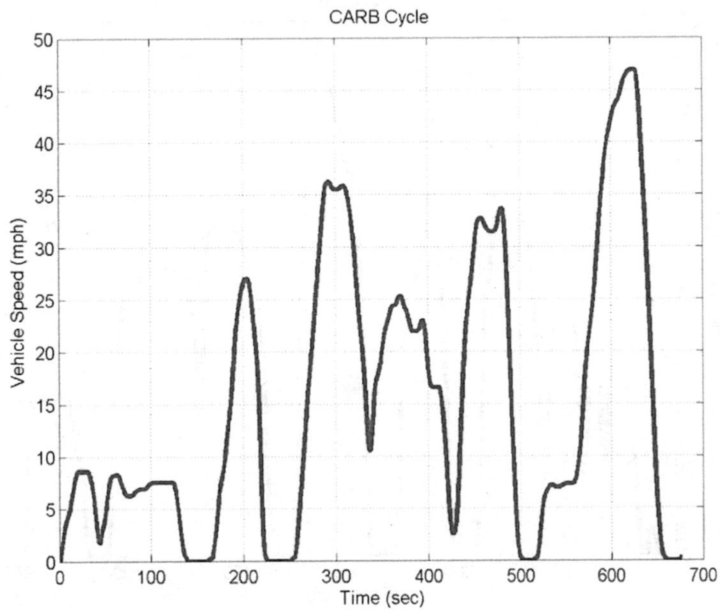

FIGURE E-6 Speed versus time trace for the CARB urban truck driving cycle.
SOURCE: Reinhart (2015).

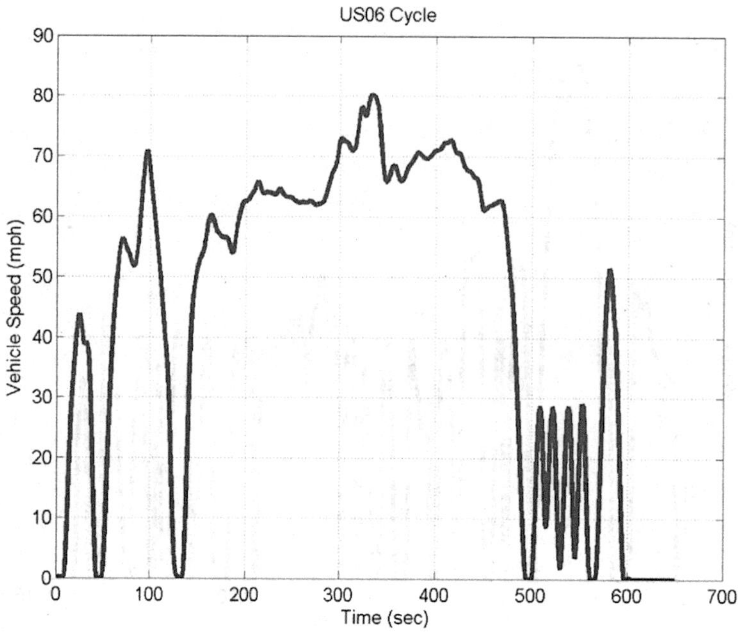

FIGURE E-7 Speed versus time trace for the US06 aggressive light-duty drive cycle.
SOURCE: Reinhart (2015).

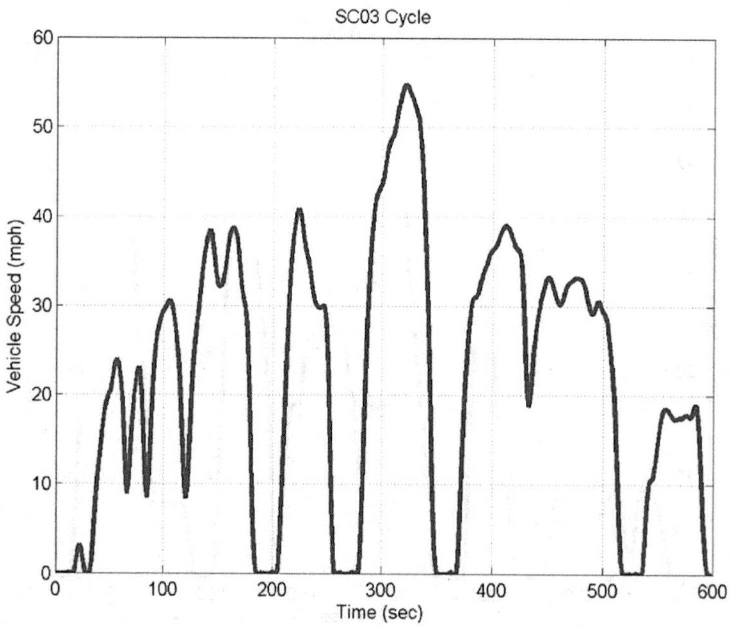

FIGURE E-8 Speed versus time trace for the SC03 air conditioner drive cycle.
SOURCE: Reinhart (2015).

The Parcel cycle shown in Figure E-9 was developed to model the drive cycle of a parcel delivery truck. About 50 percent of the drive cycle time is at idle. This causes fuel consumption at idle to be significant, especially with an automatic, where the torque converter load is significant. The Parcel cycle was only used for the T270 and F-650 medium trucks.

The Combined International Local and Commuter Cycle (CILCC) shown in Figure E-10 was developed to simulate urban driving. The general approach, with steady, very gentle accelerations, steady speed operation, and then gradual deceleration, is similar to the New European Driving Cycle. This cycle results in very light loads on the engine, and the cycle appears to be designed to favor hybrid systems. The braking sequences are gentle enough that almost all kinetic energy can be recovered by dynamic braking. The CILCC was only used for the T270 and F-650 medium trucks.

There are two remaining drive cycles that were used: the 55 mph and 65 mph cruise cycles. These cycles are strictly steady state, with no grades or other changes in load, so they effectively only operate the engine at one speed/load point. These two cycles are part of the GEM vehicle certification model. The 65 mph cruise was used for all vehicles in the study, while the 55 mph cycle was used for all vehicles except the pickup truck.

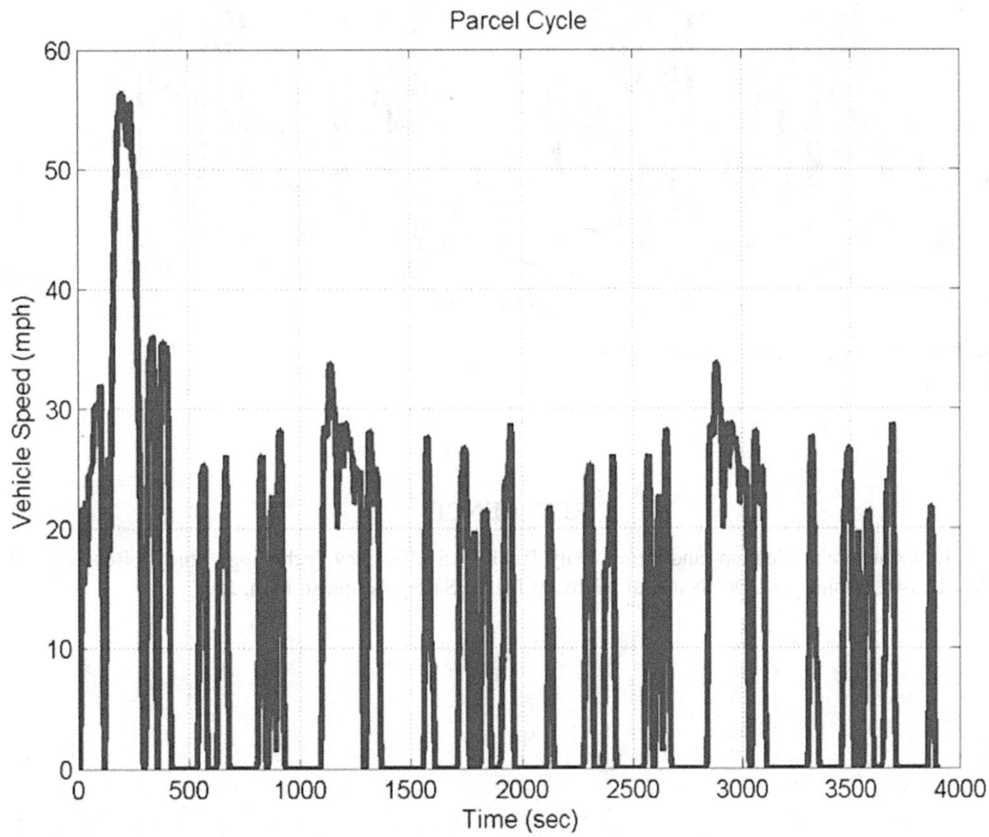

FIGURE E-9 Speed versus time trace for the parcel delivery drive cycle.
SOURCE: Reinhart (2015).

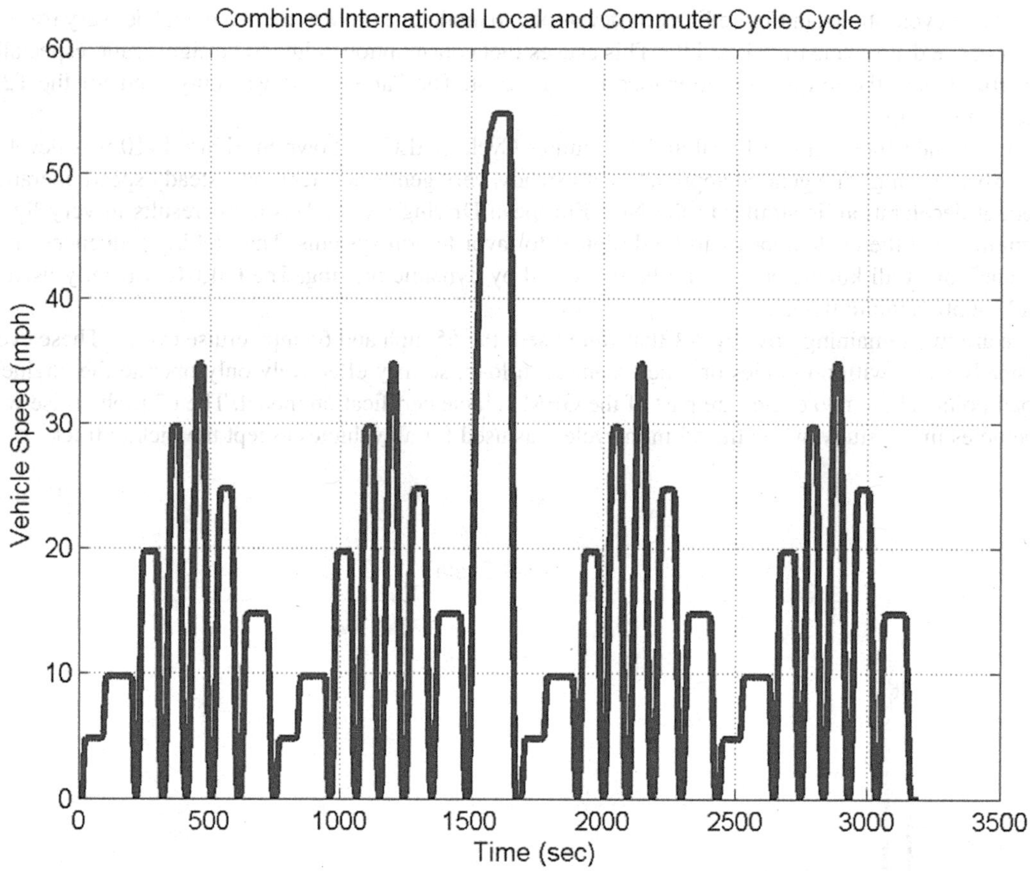

FIGURE E-10 Speed versus time trace for the CILCC.
SOURCE: Reinhart (2015).

REFERENCE

Reinhart, T. 2015. Commercial Medium- and Heavy-Duty Truck Fuel Efficiency Technology Study—Report #1. Report No. DOT HS 812 146. Washington, DC: National Highway Traffic Safety Administration, June.

F

Summary of Committee's First Report: Reducing the Fuel Consumption and Greenhouse Gas Emissions of Medium- and Heavy-Duty Vehicles, Phase Two: First Report

BACKGROUND

Medium- and heavy-duty trucks, motor coaches, and transit buses—collectively, medium- and heavy-duty vehicles (MHDVs)[1]—are used in every sector of the economy. The purpose of these vehicles ranges from carrying passengers to moving goods. The fuel consumption and greenhouse gas (GHG) emissions of MHDVs have become a focus of legislative and regulatory action in the past few years. This report is a follow-on to the National Research Council's (NRC's) *Technologies and Approaches to Reducing the Fuel Consumption of Medium- and Heavy-Duty Vehicles* (NRC, 2010; henceforth called the NRC Phase One Report) issued March 31, 2010. That report provided a series of findings and recommendations on the development of regulations for reducing fuel consumption of MHDVs.

On September 15, 2011, the National Highway Traffic Safety Administration (NHTSA) and the U.S. Environmental Protection Agency (EPA) jointly published a *Federal Register* notice (76 Fed. Reg. 57105) finalizing rules to establish a comprehensive Heavy-Duty National Program to reduce GHG emissions and fuel consumption for on-road MHDVs (the "Phase I rule").

Subsequently, NHTSA entered into a cooperative agreement with the NRC to issue a report by 2016 on technologies and approaches to reducing the fuel consumption of MHDVs with a view toward beginning work on a revision to the Phase I rule. The committee has since learned that NHTSA and EPA have commenced work on a second round (Phase II) of fuel consumption and GHG emission standards for MHDVs. This first report by the committee provides guidance for the Phase II rule, which is directed at the post-2018 time frame.[2] The committee's final report to be issued in 2016 will cover a broader range of technologies and issues and will address the 2025-2030 time frame.

COMPARISON OF PHASE I RULE WITH THE NRC PHASE ONE REPORT

In 2010, the NRC prepared a report in preparation for the Phase I regulation, proposing several recommendations. On the whole, NHTSA was quite responsive to the Phase I report, in particular on the matter of taking the NRC

[1] More precisely, these vehicles include those classified as Class 2b through Class 8, which as a group range in gross (combined) vehicle weight from 8,500 to 80,000 pounds.

[2] This Summary contains the key recommendations from the committee's report.

Phase One Report's advice on basing the standards on load-specific fuel consumption (LSFC)[3] and using modeling to determine compliance. Some recommendations were not adopted by NHTSA, and the committee encourages NHTSA and EPA to do so. In particular, the following aspects remain valid: the importance of data acquisition on the baseline MHDV stock, sales, and performance, without which conclusions on whether the goals of the regulation are met will necessarily be very uncertain; the formation of an expert group to review vehicle simulations, in particular for the Greenhouse Gas Emissions Model (GEM); implementation of driver training programs that may enable reduced fuel consumption; and incremental fuel efficiency gains that could be obtained by dieselization of Class 2b through 7 vehicles. There were other recommendations that NHTSA did not act on, but the above seem to the committee to be the most compelling.

CERTIFICATION USING MODELING AND SIMULATION

From both technical and operational perspectives there is an objective to improve, and even optimize, efficiency metrics for trucks in each class. The existing NHTSA and EPA rules have already addressed the metrics to be employed for engine efficiency, vehicle efficiency, and the associated carbon dioxide (CO_2) emissions levels, both for gasoline and diesel. From a practical standpoint at the time of writing, the rules also drive change to provide economic benefit to the truck user. However, as truck efficiency regulation advances, there are trade-offs that must be addressed. Metrics of interest are fuel efficiency, GHGs, cost, criteria pollutants, and energy security. The primary trade-off is GHGs versus fuel efficiency, when several fuels and their associated technologies are considered.

Recommendation S-1: NHTSA, in consultation with EPA, should consider carefully the impact on related metrics when attempting to optimize for a single metric, or should otherwise establish a clearly articulated objective that weights, or places limits on, relevant metrics. (Recommendation 3.1)

GEM is a relatively simple model focused on aerodynamics, rolling resistance, speed, weight, and idle control, and as such it is not capable of acknowledging efficiency and GHG emissions changes associated with engine and transmission design, the integration of advanced powertrains, alternative fuels, hybrid and electric vehicles, and optimal component management. The weight reduction input in GEM is limited to a fixed set of technologies and parts. GEM upgrades are required to provide more realistic prediction of fuel use and GHG emissions, particularly as detailed in the next three recommendations.

Models should be capable of simulating real-world component behavior accurately and should not be over-simplified. GEM specifically does not allow for synergy between components, the operation or control of components in a most efficient way, or the engendering of efficiency through operation of a smaller component at higher relative load. GEM specifies the performance maps for major components such as the engine and transmission and does not credit the vehicle manufacturer with benefits of using a potentially superior engine or transmission.

Recommendation S-2: NHTSA should investigate allowing the original equipment manufacturer (OEM) to substitute OEM-specific models or code for the fixed models in the current GEM, including substituting a power pack (the engine, aftertreatment, and transmission). These models, whether provided by OEMs or fixed in the code, should be configured to reflect real-world operation accurately. (Recommendation 3.7)

Calibration of equipment traceable to national standards for passenger car tires is in place, and self-calibration of equipment is in place for the grading of tires in Europe. Calibration of tire characterization equipment for U.S. regulations should be extended to cover all measurements of coefficients of rolling resistance used in GEM. For tire rolling resistance, there needs to be high confidence in input values to GEM.

[3] The precise metric for measuring fuel consumption, the LSFC is measured in gallons of fuel per payload ton per 100 miles. The lower the fuel consumption of the vehicle and the higher the payload the vehicle carries, the lower will be the LSFC.

Recommendation S-3: A mechanism needs to be implemented for obtaining accurate tire rolling resistance factors, including equipment calibration, and maintaining that information in a public database. This might be managed in the same way that tread wear, temperature, and traction data are displayed through the federal Uniform Tire Quality Grading system. (Recommendation 3.4)

GEM employs a limited set of cycles to challenge the simulated truck. These cycles do not include real-world road grade and neglect varying operating weights and aerodynamic yaw angles. Being speed/time based, these cycles also do not allow for the faster acceleration of more powerful trucks, or the longer time that might be taken by less powerful trucks to complete some real-world distance-based routes. This deficiency is not evident in the current model, where a hard-coded engine map, rather than a real OEM engine model, is used.

Recommendation S-4: The choice of test cycles/routes or schedules used in GEM needs to be readdressed thoroughly to avoid creating designs that are optimized for the test rather than achieving real-world performance in the design process. (Recommendation 3.6)

FLEET DATA

A further issue of importance is the need to collect data on vehicles such as would permit regulators to evaluate the regulatory efficacy and improve both the accuracy of the Phase I rule and any subsequent phases. NHTSA did not include a pilot phase when it promulgated the Phase I rule in 2011. Thus, there were no baseline data from even a few representative national fleets prior to the rulemaking, such as would have enabled comparison with post-rulemaking fuel efficiency. This would have also started to facilitate the comparison of real-world test data with compliance data. The committee nonetheless recognizes that NHTSA has begun the process of designing surveys and seeking the necessary Office of Management and Budget approvals to allow it to assemble a picture of the fleet characteristics, including the collection of R.L. Polk registration data and forecasts, which is appropriate for Class 2b through Class 8 vehicles.

Recommendation S-5: NHTSA should establish a repeatable, reliable collection process as soon as possible. In addition to continuing data procurement with Southwest Research Institute and R.L Polk, NHTSA should investigate outside sources (such as FTR, ACT Research, SmartWay, the North American Council for Freight Efficiency, and the American Transportation Research Institute) to obtain a repeatable, reliable baseline as well as future data. These sources could use the R.L. Polk data and conduct clarifying, deeper interviews with truck and trailer builders, manufacturers, and fleets for specific ongoing data on technologies procured, and fuel consumption. (Recommendation 4.2)

NATURAL GAS

Natural gas accounts for about 25 percent of all U.S. energy use, yet only 0.1 percent is used in transportation, equivalent to about 0.5 billion gallons per year of petroleum fuel. However, in the short time since the release of the NRC Phase One Report (NRC, 2010), natural gas has emerged as an economically attractive option for commercial vehicles. This has been driven by the rapid development of low-cost production of unconventional natural gas.

In order for medium and heavy trucks to use natural gas fuel rather than diesel, the most significant changes from diesel-fueled vehicles are the on-board fuel storage method and the means of introducing and igniting the fuel in the engine. On-vehicle storage is by high pressure, effected by either compressed natural gas cylinders (3,600 pounds per square inch is typical) or cryogenic containers filled with liquefied natural gas. For using natural gas in place of gasoline, the spark-ignition engine carries over with modest changes, but fuel storage is still by one of the above two methods.

Natural gas engines are well developed although improvements can be pursued in engine efficiency, maintenance costs, and on-board vehicle storage costs. Natural gas's inherent GHG benefit by its low carbon content (approximately 28 percent) is partially negated by lower efficiency in currently available engines and the higher GHG impact of methane emissions. In addition, a natural gas leakage correction to GHG impact could negate the inherent tailpipe CO_2 advantage.

Recommendation S-6: NHTSA and EPA should develop a separate standard for natural gas vehicles as is presently the case for diesel- and gasoline-fueled vehicles. Factors NHTSA and EPA should consider in setting the standard include the maximum feasible ability of natural gas engines to achieve reductions in GHG emissions and fuel consumption, the uncertainties involved with the various alternatives, the impact of duty cycles on the ability to comply with the vehicle standards, the cost of the technology, and rapid growth of the market for natural gas engines and vehicles. This may require additional focused studies. (Recommendation 5.2)

Recommendation S-7: More studies and data are needed to determine the well-to-tank GHG emissions of natural gas vehicles, since current estimates vary significantly regarding quantification of emissions leakage of methane. EPA and NHTSA should assemble a best estimate of well-to-tank GHG emissions to be used as a context for developing future rulemakings. (Recommendation 5.1)

Due to the economics-driven rapid adoption of natural gas, there is urgency to develop an optimum solution in Phase II rule standards for both GHG emissions and fuel consumption (as well as criteria emissions) that will accommodate this fuel without artificially disrupting prevailing commercial transportation business models. As a specific example, the GEM certification tools need to include natural gas engine maps to more accurately quantify the emissions and fuel economy of natural gas vehicles.

Recommendation S-8: To benefit fully from the GHG and petroleum displacement potential of natural gas, government and the private sector should support further technical improvements in engine efficiency and operating costs, reduction of storage costs, and emission controls (as is done for diesel engines). NHTSA and EPA should also evaluate the need for and benefits and costs of an in-use natural gas fuel specification for motor vehicle use. (Recommendation 5.4)

REGULATING TRAILERS

Aerodynamic Devices for Trailers

There are four regions of the tractor-van trailer combination truck that are amenable to aerodynamic design improvements including the various tractor details, the tractor-trailer gap, the trailer underbody, and the trailer tail. Side skirts constitute 90 percent of devices sold to improve aerodynamics. Most side skirts provide a 5 percent fuel saving at 65 mph (3 percent at 55 mph) on EPA testing as part of SmartWay.[4] A recent study identified barriers to use of trailers that are more aerodynamically efficient (NACFE, 2013), primarily related to understanding of cost-benefit data and lack of robust application information. Yet, early adopting carriers indicate there is a good return on investment from their use. Nonetheless, manufacturers of trailer side skirts report sales doubled between 2012 and 2013. The manufacturers also report installed prices of side skirts have declined by half.

A California regulation requires operators of van trailers to use aerodynamic devices to reduce the energy required to pull them. Observations made in California and Arizona showed a greater proportion of trailers with aerodynamic devices than did those observations made in Oregon, Texas, Michigan, Pennsylvania, and Maryland. Side skirts were overwhelmingly the predominant aerodynamic device strategy. Other strategies (underbody fairings and rear fairings) were observed just in few instances.

Recommendation S-9: NHTSA, in coordination with EPA, should adopt a regulation requiring that all new, 53 foot and longer dry van and refrigerated van trailers meet performance standards that will reduce their fuel consumption

[4] SmartWay is a voluntary program administered by EPA that was formed in 2004 with the objective of improving efficiency and reducing fuel consumption and pollution from movement of freight across the supply chain. Currently it focuses mainly on over-the-road trucking.

and CO_2 emissions. The lead time to implement this regulation should be evaluated independently from lead-time requirements applicable to the next set of standards for new engines and tractors, because less time is needed to perform compliance testing and install aerodynamic devices on new trailers. NHTSA and EPA should also collect real-world data on fleet use of aerodynamic trailers to help inform the regulation. (Recommendation 6.1)

The current SmartWay program and California Air Resources Board regulation only address the most commonly used 53+ foot van trailer, which accounts for about 60 percent of the trailers that could potentially benefit from use of aerodynamic devices. Use of aerodynamic devices on other types of trailers, such as container/chassis and shorter vans including dual trailers ("pups") could provide additional fuel savings of 4 to 9 percent per tractor-trailer, according to industry estimates. Fuel savings from use of side skirts have also been demonstrated on flatbed trailers. The cost effectiveness of using aerodynamic devices on these additional categories of trailers depends on their annual mileage accumulation and average speed, among other considerations such as access to the trailer underbody, and needs further assessment and quantification.

Recommendation S-10: NHTSA, in coordination with EPA, should determine whether it would be practical and cost effective to include with the regulation of van trailers the regulation of other types of trailers such as pups, flatbeds, and container carriers, as doing so could substantially increase overall fuel savings. (Recommendation 6.2)

Various complete truck test procedures, such as the Society of Automotive Engineers (SAE) J1321 and coast-down procedure SAE J1263, used for determining the effectiveness of aerodynamic devices have less precision than necessary to detect small incremental changes. Alternatives such as wind-tunnel testing and computational fluid dynamics simulation should be evaluated because they provide fidelity and better precision. These methods can reduce the cost of development validation, may avoid building of multiple full-size prototypes, and can accelerate development time for the final product. Better test precision and repeatability may induce otherwise skeptical end users to adopt such technologies.

Recommendation S-11: NHTSA should evaluate the relative fidelities of the coast-down procedure and candidate powered procedures to define an optimum prescribed full-vehicle test procedure and process, and should validate the improved procedure against real-world vehicle testing. Furthermore, NHTSA and EPA should assess if adding yaw loads to the validation process provides significantly increased value to the C_d result. In addition, NHTSA and EPA should disseminate to end users updated test data and fuel savings of efficient trailers, aerodynamic devices, and tires, especially to those not participating in the SmartWay program. This should increase end-user confidence in fuel savings and device reliability. (Recommendation 6.3)

Tires

Many new tractors and most new trailers are equipped with low-rolling-resistance tires that meet the SmartWay performance standard, and this is likely to increase due to regulatory requirements. However, 70 percent of new tires sold in 2012 for use on tractors and trailers were for replacement of existing tires, and only 42 percent of these are SmartWay verified. There is no assurance in the future that replacement tires will be as energy efficient as the original equipment tires they replace. Manufacturers have also introduced wide-base single tires (WBSTs), many of which feature lower rolling resistance than most dual-tire sets. WBSTs make up less than 10 percent of the commercial truck tire market, but their use is increasing as fleets strive to reduce fuel consumption and GHG emissions.

Recommendation S-12: NHTSA, in coordination with EPA, should further evaluate and quantify the rolling resistance of new tires, especially those sold as replacements. If additional, cost-effective fuel savings can be achieved, NHTSA should adopt a regulation establishing a low-rolling-resistance performance standard for all new tires designed for tractor and trailer use. (Recommendation 6.5)

Precision in tire rolling resistance coefficients (C_{rr}) measurement is mandatory. Furthermore, while the ISO28580 test procedure is given good grades by most in the industry, a robust machine cross-correlation does not exist for commercial vehicle tires in the United States. Carriers cannot depend on the comparability of C_{rr} for measurements from approximately 60 tire suppliers verified by SmartWay.

Recommendation S-13: NHTSA, supported by EPA, should expeditiously establish and validate the equipment and process of a tire industry machine alignment laboratory and mandate the use of that laboratory by each tire manufacturer seeking C_{rr} validation for any tires being offered as candidates in the GEM computation process, just as the C_{rr}s of light-duty vehicle tires were validated. (Recommendation 6.6)

OTHER APPROACHES TO REDUCING FUEL CONSUMPTION

The Phase I rule had the effect of forcing the adoption of technologies for reducing fuel consumption. Such reductions can be achieved by technological improvements to the vehicle as well as by improvements in operations, changes in behavior of drivers, and so forth. The NRC Phase One Report considered other approaches (referred to, perhaps imprecisely, as nontechnical approaches), such as intelligent transportation systems, construction of lanes exclusively for trucks, congestion pricing, driver training, and intermodal operations (NRC, 2010, pp. 159 et seq.). Also considered were market-based instruments such as fuel taxes. Another viable approach would entail adjusting size and weight restrictions on trucks. For example, this might include greater use of vehicles that have favorable LSFC such as longer combination vehicles, which have greater freight capacity than the notional tractor-trailer, which can have a combined gross vehicle weight of 80,000 pounds.[5] While some nonvehicle alternative approaches for reducing fuel consumption and greenhouse gas emissions may be beyond NHTSA's delegated authority, the Agency can work with other agencies with appropriate authority as well as encourage private actors to consider such strategies to complement and support NHTSA's standards.

Recommendation S-14: NHTSA should consider additional strategies to encourage the adoption of measures that reduce fuel consumption and attempt to quantify the impacts of nontechnological factors on the costs and feasibility of future efforts to improve fuel consumption. (Recommendations 1.9 and 1.11)

REGULATORY PROCESSES

The committee also has made several observations about the regulatory process. Currently NHTSA and EPA standards consider fuel efficiency of the vehicle (in gallons per ton-mile) and tailpipe CO_2 emissions (in grams of CO_2 per ton-mile) that need to be achieved, on average, by the mix of vehicles sold each year by each manufacturer. Manufacturers are likely to achieve these vehicle standards using a variety of different energy fuels and technologies. Failure to consider the well-to-wheel emissions of each combination of fuel and vehicle technology may lead to regulations that do not achieve the anticipated energy and GHG emissions savings. Furthermore, there is the possibility that regulations could produce incentives and behaviors that may result in unintended consequences that could be beneficial or detrimental, such as the water contamination that resulted from the addition of methyl tertiary-butyl ether to gasoline as an oxygenate intended to reduce ground-level ozone.

Recommendation S-15: NHTSA, in coordination with EPA, should begin to consider the well-to-wheel life-cycle energy consumption and greenhouse gas emissions associated with different vehicle and energy technologies to ensure future rulemakings best accomplish their overall goals. (Recommendation 1.10)

Recommendation S-16: NHTSA should conduct an analysis, including methods such as expert surveys and scenario analysis or red teaming, as appropriate, to anticipate and analyze potential unintended consequences of its

[5] This follows from the observation that "weighing out" gives advantage for fuel consumption versus "cubing out"—the latter referring to filling up the cargo area before reaching the combined gross vehicle weight limit.

regulations and to determine whether additional actions are warranted to try to minimize such impacts. NHTSA should undertake this analysis concurrently with its next revision to its regulation. (Recommendation 1.4)

Regarding the potential for technological change in the model year 2019-2022 time frame, the committee, in its investigations to date, has not identified any combustion or other engine technologies beyond those identified in the NRC Phase One Report (2010) that would provide significant further fuel consumption reduction during the Phase II rule time frame. However, those technologies identified in the NRC Phase One Report should be updated with current projections for fuel consumption reduction and adjusted for system interactions when used in combination with each other.

A further issue relates to timing and rate at which technology enters the marketplace. In establishing the stringency of a Phase II rule, careful evaluation of technology penetrations is necessary. While the NHTSA-stated intent of the Phase I rule was to base targets on off-the-shelf technologies, the new baseline for setting a Phase II rule will be drawn from more current vehicles and will consider the different degrees of penetration attained by both off-the-shelf and future and advanced technologies by 2018.

Recommendation S-17: NHTSA's Phase II rule should take the current and projected incremental fuel consumption reductions and penetration rates of the various technologies into careful consideration: These incremental reductions and penetration rates should be updated from those that were projected in the Phase I rulemaking. Furthermore, system interactions should be evaluated for the effect on the projected incremental reductions whenever combinations of technologies are considered. (Recommendation 2.2)

REFERENCES

NACFE (North American Council for Freight Efficiency) and Cascade Sierra Solutions. 2013. Barriers to the Increased Adoption of Fuel Efficiency Technologies in the North American On-Road Freight Sector. July.

NRC (National Research Council). 2010. *Technologies and Approaches to Reducing the Fuel Consumption of Medium- and Heavy-Duty Vehicles*. Washington, DC: The National Academies Press.

G

Acronyms and Abbreviations

21CTP	21st Century Truck Partnership
A	ampere
AAR	Association of American Railroads
ABI	Allied Business Intelligence Research
ABS	anti-lock braking system
AC	alternating current
ACC	American Control Conference
ACEA	European Automobile Manufacturers Association
ACEEE	American Council for an Energy-Efficient Economy
ACT	America's Commercial Transportation Research Company
AEO	Annual Energy Outlook
AFDC	Alternative Fuels Data Center
Ah	amp-hours
ANL	Argonne National Laboratory
ARPA-e	Advanced Research Projects Agency–Energy
ASIWPCA	Association of State and Interstate Water Pollution Control Administrators
ASME	American Society of Mechanical Engineering
AT	automatic transmission
ATA	American Trucking Associations
ATI	automatic tire inflation
ATIS	automatic tire inflation systems
ATRI	American Transportation Research Institute
BAU	business-as-usual
BEV	battery electric vehicle
BMP	best management plan
BSG	belt-starter-generator
BTE	brake thermal efficiency

BTL	biomass-to-liquid
CAA	Clean Air Act
CACC	cooperative adaptive cruise control
CAD	computer-aided design
CAFE	corporate average fuel economy
CalHEAT	California Hybrid, Efficient and Advanced Truck Research Center
CARB	California Air Resources Board
C_{dwa}	wind-averaged drag coefficient
CFD	computational fluid dynamics
CGVW	combined gross vehicle weight
CH_4	methane
CI	compression-ignition
CILCC	Combined International Local and Commuter Cycle
C-ITS	cooperative intelligent transportation systems
CLEERS	crosscut lean exhaust emissions reduction simulations
CMU	Carnegie Mellon University
CN	cetane number
CNG	compressed natural gas
CO	carbon monoxide
CO_2	carbon dioxide
CRC	Coordinating Research Council
C_{rr}	rolling resistance coefficient
CVT	continuously variable transmission
CWA	Clean Water Act
DBE	double carbon bond equivalent
DC	direct current
DDFS	direct dual-fuel stratification
DDPP	Deep Decarbonization Pathways Project
DEF	diesel exhaust fuel
DICE	Dynamic Integrated Climate-Economy Model
DME	dimethyl ether
DOC	diesel oxidation catalyst
DOE	Department of Energy
DOT	Department of Transportation
DPF	diesel particulate filter
EAA	European Aluminum Association
EC	European Commission
EDF	Environmental Defense Fund
EDI	Efficient Powertrains, Inc.
EDT	electric drive technologies
EERE	Office of Energy Efficiency and Renewable Energy
EESTT	Electrochemical Energy Storage Technical Team
EGR	exhaust-gas recirculation
EIA	Energy Information Administration
EISA	Energy Independence and Security Act
EPA	Environmental Protection Agency
EPCA	Energy Policy and Conservation Act

EPS	electronic power steering
ET	Economic Times Auto
EU	European Union
FACE	Fuels for Advanced Combustion Engines
FAME	fatty acid methyl esters
FC	fuel cell
FCEV	fuel cell electric vehicle
FE	fuel efficiency
FHEV	full hybrid electric vehicle
FHWA	Federal Highway Administration
FMCSA	Federal Motor Carrier Safety Administration
FMEP	friction mean effective pressure
FTP	Federal Test Procedure
FUDS	Federal Urban Driving Schedule
GAO	Government Accountability Office
GCM	Global Climate Model
GDCI	gasoline direct-injection compression ignition
GDI	gasoline direct fuel injection
GDP	gross domestic product
GEM	Greenhouse Gas Emissions Model
GHG	greenhouse gas
GPS	Global Positioning Systems
GVW	gross vehicle weight
GVWR	gross vehicle weight rating
GWP	global warming potential
H_2O	water
HC	hydrocarbon
HCCI	homogenous charge compression ignition
HDRD	hydrogenation-derived renewable diesel
HEV	hybrid-electric vehicle
HHD	heavy heavy duty
HHV	hydraulic hybrid powertrain vehicle
hp	horsepower
HPDI	high-pressure direct injection
HVAC	heating, ventilation, and air conditioning
IAM	Integrated Assessment Model
IC	indirect cost
ICCT	International Council on Clean Transportation
ICE	internal combustion engine
ICM	indirect cost multiplier
IEA	International Energy Agency
IMA	integrated motor assist
i-MMD	intelligent multi-mode drive
IPCC	Intergovernmental Panel on Climate Change
ISG	integrated starter/generator
ISO	International Organization for Standardization

ITS	intelligent transportation system
IWG	Interagency Working Group
kg	kilogram
kg/t	kilograms per ton
kW	kilowatt
kWh	kilowatt-hour
L	liter
LCA	life-cycle assessment
LCD	liquid-crystal display
LCFS	low-carbon fuel standard
LCO	lithium cobalt oxide
LCV	longer combination vehicles
LDV	light-duty vehicle
LFP	lithium iron phosphate
LHD	light heavy duty
Li-ion	lithium-ion
$LiNiMnCoO_2$	lithium-nickel-manganese-cobalt-oxide
LMO	lithium manganese oxide
LNG	liquefied natural gas
LRR	low rolling resistance
LSFC	load-specific fuel consumption
LTC	low-temperature control
LTO	lithium titanate oxide
m^2	meters squared
m^2/g	meters squared per gram
mAh/g	charge per gram
MBRC	miles between road calls
MECA	Manufacturers of Emission Controls Association
MH	mild hybrid
MHD	medium heavy duty
MHDV	medium- and heavy-duty vehicle
mi	miles
MJ	megajoule
$MMTCO_2e$	metric tons of carbon dioxide equivalent
mpg	miles per gallon
mph	miles per hour
MRD	multi-ring domed
MY	model year
NACFE	North American Council for Freight Efficiency
NAE	National Academy of Engineering
NAFTA	North American Free Trade Agreement
NAMII	National Additive Manufacturing Innovation Institute
NAS	National Academy of Sciences
NASA	National Aeronautics and Space Administration
NASEM	National Academies of Sciences, Engineering, and Medicine
NCA	nickel-cobalt-aluminum-oxide

NCM	nickel-cobalt-manganese
NEMS	National Energy Modeling System
NESCCAF	Northeast States Center for a Clean Air Future
NG	natural gas
NGV	natural gas vehicle
NGWBS	next-generation wide-based single
NHTSA	National Highway Traffic Safety Administration
Ni-Cd	nickel-cadmium
NiMH	nickel-metal hydride
Nm	Newton-meter
NNI	National Nanotechnology Initiative
NNMI	National Network for Manufacturing Innovation
NO	nitric oxide
NO_2	nitrogen dioxide
NO_x	nitrous oxides
NPC	National Petroleum Council
NPRM	notice of proposed rulemaking
NRC	National Research Council
NREL	National Renewable Energy Laboratory
NTEA	Association for the Work Truck Industry
OBD	on-board diagnostics
OECD	Organisation for Economic Co-operation and Development
OEM	original equipment manufacturer
OPEC	Organization of the Petroleum Exporting Countries
ORC	organic Rankine cycle
ORNL	Oak Ridge National Laboratory
OTAQ	Office of Transportation and Air Quality
P	power
PCCI	premixed charge compression-ignition
PEM	polymer exchange membrane
P_{eng}	engine power demand
PFI	port fuel injection
PGM	platinum group metal
PHEV	plug-in hybrid electric vehicle
PM	particulate matter
$PM_{2.5}$	fine particulate matter
PMI	particulate matter index
PPC	partially premixed combustion
PS	power split
psi	pounds per square inch
PSI	Power Solutions International
PTO	power-take-off
R&D	research and development
RCCI	reactivity controlled compression ignition
RFS	Renewable Fuel Standard
RFS2	Renewable Fuel Standard Program (2)
RGGI	Regional Greenhouse Gas Initiative

RIA	Regulatory Impact Analysis
RIN	Renewable Identification Number
ROI	return-on-investment
RON	research octane number
RPE	retail price equivalent
RPM	revolutions per minute
RSW	resistance spot welding
SAE	Society of Automotive Engineers
SCAQMD	South Coast Air Quality Management District
SCC	social cost of carbon
SCR	selective catalytic reduction
SES	Stored Energy Solutions
SI	spark-ignition
SLM	selective laser melting
SOC	state of charge
SOT	statement of task
SO_x	sulfur oxides
SS	stop-start
ST	SuperTruck
STAA	Surface Transportation Assistance Act
SwRI	Southwest Research Institute
TARDEC	Tank Automotive Research Development and Engineering Center
TMC	Technology and Maintenance Council
TP	tractive power
TPMS	tire pressure monitoring system
TTW	tank to wheel
TWC	three-way catalyst
UPS	United Parcel Service of America
USABC	United States Automotive Battery Consortium
U.S. DRIVE	United States Driving Research and Innovation for Vehicle Efficiency and Energy Sustainability
V	volt
V2P	vehicle-to-pedestrian
V2V	vehicle-to-vehicle
VANET	Vehicular Ad-Hoc Network
VIUS	Vehicle Inventory and Use Survey
VMT	vehicle miles traveled
VOC	volatile organic compound
VPD	vehicle power demand
VVA	variable valve actuation
VVT	variable valve timing
W/kg	watts per kilogram
WBG	wide band gap
WBST	wide-base single tire
Wh/kg	watt-hours per kilogram
Wh/l	watt-hours per liter

WHR	waste-heat recovery
WQA	Water Quality Act
WTT	well-to-tank
WTW	well-to-wheels
WVU	West Virginia University